TEXTBOOKS in MATHEMATICS

Series Editor: Al Boggess

PUBLISHED TITLES

ABSTRACT ALGEBRA: AN INQUIRY-BASED APPROACH
Jonathan K. Hodge, Steven Schlicker, and Ted Sundstrom

ABSTRACT ALGEBRA: AN INTERACTIVE APPROACH
William Paulsen

ADVANCED CALCULUS: THEORY AND PRACTICE
John Srdjan Petrovic

COLLEGE GEOMETRY: A UNIFIED DEVELOPMENT
David C. Kay

COMPLEX VARIABLES: A PHYSICAL APPROACH WITH APPLICATIONS AND MATLAB®
Steven G. Krantz

ESSENTIALS OF TOPOLOGY WITH APPLICATIONS
Steven G. Krantz

INTRODUCTION TO ABSTRACT ALGEBRA
Jonathan D. H. Smith

INTRODUCTION TO MATHEMATICAL PROOFS: A TRANSITION
Charles E. Roberts, Jr.

INTRODUCTION TO PROBABILITY WITH MATHEMATICA®, SECOND EDITION
Kevin J. Hastings

LINEAR ALBEBRA: A FIRST COURSE WITH APPLICATIONS
Larry E. Knop

LINEAR AND NONLINEAR PROGRAMMING WITH MAPLE™: AN INTERACTIVE, APPLICATIONS-BASED APPROACH
Paul E. Fishback

MATHEMATICAL AND EXPERIMENTAL MODELING OF PHYSICAL AND BIOLOGICAL PROCESSES
H. T. Banks and H. T. Tran

ORDINARY DIFFERENTIAL EQUATIONS: APPLICATIONS, MODELS, AND COMPUTING
Charles E. Roberts, Jr.

REAL ANALYSIS AND FOUNDATIONS, THIRD EDITION
Steven G. Krantz

TEXTBOOKS in MATHEMATICS

ABSTRACT ALGEBRA

AN INQUIRY-BASED APPROACH

JONATHAN K. HODGE
STEVEN SCHLICKER
TED SUNDSTROM

Grand Valley State University
Allendale, Michigan, USA

CRC Press
Taylor & Francis Group
Boca Raton London New York

CRC Press is an imprint of the
Taylor & Francis Group an **informa** business
A CHAPMAN & HALL BOOK

CRC Press
Taylor & Francis Group
6000 Broken Sound Parkway NW, Suite 300
Boca Raton, FL 33487-2742

International Standard Book Number-13: 978-1-4665-6706-1 (Hardback)

Library of Congress Cataloging-in-Publication Data

Hodge, Jonathan K., 1980- author.
Abstract algebra : an inquiry-based approach / Jonathan K. Hodge, Steven Schlicker, Ted Sundstrom.
pages cm. -- (Textbooks in mathematics)
Includes bibliographical references and index.
ISBN 978-1-4665-6706-1 (hardback)
1. Algebra, Abstract--Textbooks. I. Schlicker, Steven, 1958- author. II. Sundstrom, Ted, author. III. Title.

QA162.H63 2014
512'.02--dc23 2013039515

Visit the Taylor & Francis Web site at
http://www.taylorandfrancis.com

and the CRC Press Web site at
http://www.crcpress.com

Contents

Note to Students

This book may be unlike other mathematics textbooks you have read or used in previous courses. The investigations contained in it are designed to facilitate your learning by inviting you to be an active participant in the learning process. This is a book that is not meant to be simply *read*, but rather *engaged*. It includes numerous activities within the text that are intended to motivate new material, illustrate definitions and theorems, and help you develop both the intuition and rigor that is necessary to understand and apply ideas from abstract algebra.

As professors of mathematics, we have found (and research confirms) that mathematics is not a spectator sport. To learn and understand mathematics, one must engage in the process of *doing* mathematics. This kind of engagement can be challenging and even frustrating at times. But if you are up to the challenge and willing to take responsibility for your own learning, you will indeed learn a great deal.

Obviously, this is a book about abstract algebra, and you will learn more about what that means as we begin our investigations. Our goal, however, is that you will not only learn about abstract algebra, but that you will also develop a deeper understanding of what mathematics is, how mathematics is done, and how mathematicians think. We hope that you will see that the way mathematics is developed is often different than how it is presented; that definitions, theorems, and proofs do not simply appear fully formed in the minds of mathematicians; that mathematical ideas are highly interconnected; and that even in a field like abstract algebra, there is a considerable amount of intuition to be found.

Thank you for joining us on this journey. We hope you enjoy both the challenges and the rewards that await you in these pages.

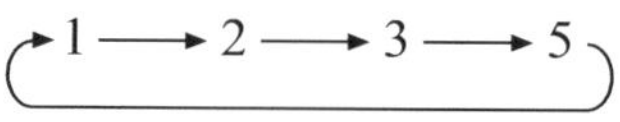

G_1 :

	a	b	c	d	e	f	g	h
a	a	b	c	d	e	f	g	h
b	b	c	d	a	h	e	f	g
c	c	d	a	b	g	h	e	f
d	d	a	b	c	f	g	h	e
e	e	f	g	h	a	b	c	d
f	f	g	h	e	d	a	b	c
g	g	h	e	f	c	d	a	b
h	h	e	f	g	b	c	d	a

G_2 :

	α	β	γ	δ	ϵ	ζ	η	θ
α	α	β	γ	δ	ϵ	ζ	η	θ
β	β	γ	δ	ϵ	ζ	η	θ	α
γ	γ	δ	ϵ	ζ	η	θ	α	β
δ	δ	ϵ	ζ	η	θ	α	β	γ
ϵ	ϵ	ζ	η	θ	α	β	γ	δ
ζ	ζ	η	θ	α	β	γ	δ	ϵ
η	η	θ	α	β	γ	δ	ϵ	ζ
θ	θ	α	β	γ	δ	ϵ	ζ	η

G_3 :

	s	t	u	v	w	x	y	z
s	s	t	u	v	w	x	y	z
t	t	u	v	s	x	y	z	w
u	u	v	s	t	y	z	w	x
v	v	s	t	u	z	w	x	y
w	w	x	y	z	t	u	v	s
x	x	y	z	w	u	v	s	t
y	y	z	w	x	v	s	t	u
z	z	w	x	y	s	t	u	v

G_4 :

	$\spadesuit$	$\heartsuit$	$\diamondsuit$	$\clubsuit$	$\maltese$	$\sharp$	$\dagger$	$\mho$
$\spadesuit$	$\spadesuit$	$\heartsuit$	$\diamondsuit$	$\clubsuit$	$\maltese$	$\sharp$	$\dagger$	$\mho$
$\heartsuit$	$\heartsuit$	$\diamondsuit$	$\clubsuit$	$\spadesuit$	$\sharp$	$\dagger$	$\mho$	$\maltese$
$\diamondsuit$	$\diamondsuit$	$\clubsuit$	$\spadesuit$	$\heartsuit$	$\dagger$	$\mho$	$\maltese$	$\sharp$
$\clubsuit$	$\clubsuit$	$\spadesuit$	$\heartsuit$	$\diamondsuit$	$\mho$	$\maltese$	$\sharp$	$\dagger$
$\maltese$	$\maltese$	$\sharp$	$\dagger$	$\mho$	$\spadesuit$	$\heartsuit$	$\diamondsuit$	$\clubsuit$
$\sharp$	$\sharp$	$\dagger$	$\mho$	$\maltese$	$\heartsuit$	$\diamondsuit$	$\clubsuit$	$\spadesuit$
$\dagger$	$\dagger$	$\mho$	$\maltese$	$\sharp$	$\diamondsuit$	$\clubsuit$	$\spadesuit$	$\heartsuit$
$\mho$	$\mho$	$\maltese$	$\sharp$	$\dagger$	$\clubsuit$	$\spadesuit$	$\heartsuit$	$\diamondsuit$

$\langle a\rangle \cap H \neq \{e\}$. Let z be a non-identity element in $\langle a\rangle \cap H$. Then there exist a nonnegative integer j, an element k_1 in K, and an integer q so that

$$z = a^j = k_1 d^q. \tag{31.5}$$

Recall that $d^p \in K$ by Claim 2. So if $p \mid q$, then for some integer t, we have $z = k_1 d^q = k_1(d^p)^t \in K$. Thus, $z \in \langle a\rangle \cap K$. Since z is not the identity element, this cannot happen. So p cannot divide q. Therefore, $\gcd(p, q) = 1$, and Corollary 3.11 (see page 29) implies that there exist integers u, v so that

$$pu + qv = 1.$$

Substituting from (31.2) and (31.4) gives us

$$c = c^{pu+qv} = (c^p)^u (c^q)^v = (a^m k)^u ((da^w)^q)^v = (a^m k)^u (d^q a^{wq})^v.$$

Substituting from (31.5) now yields

$$c = (a^m k)^u (d^q a^{wq})^v = (a^m k)^u (a^j k_1^{-1} a^{wq})^v = (a^{mu} a^{jv} a^{wqv})(k^u k_1^{-v}).$$

Now $c = (a^{mu} a^{jv} a^{wqv})(k^u k_1^{-v}) \in \langle a\rangle \times K$, which contradicts (31.1). Therefore, we can conclude that every element in G is in $\langle a\rangle \times K$. Since every element in $\langle a\rangle \times K$ is also in G, it follows that $G = \langle a\rangle \times K$, as desired. ■

There are two key steps in applying Theorem 31.3:

- First, we must find an element $a \in G$ of maximal order.
- Next, we must find a maximal subgroup K of G satisfying $\langle a\rangle \cap K = \{e\}$. Then, for this K, we will have
$$G = \langle a\rangle K = \{a^m k : m \in \mathbb{Z}, k \in K\}.$$

Activity 31.4. We will illustrate the proof of Theorem 31.3 using $G = U_{60}$ as an example. Note that

$$U_{60} = \{[1], [7], [11], [13], [17], [19], [23], [29], [31], [37], [41], [43], [47], [49], [53], [59]\},$$

and so $|U_{60}| = 16 = 2^4$. The orders of the elements of U_{60} are given in Table 31.2.

$\lvert[1]\rvert = 1$	$\lvert[17]\rvert = 4$	$\lvert[31]\rvert = 2$	$\lvert[47]\rvert = 4$
$\lvert[7]\rvert = 4$	$\lvert[19]\rvert = 2$	$\lvert[37]\rvert = 4$	$\lvert[49]\rvert = 2$
$\lvert[11]\rvert = 2$	$\lvert[23]\rvert = 4$	$\lvert[41]\rvert = 2$	$\lvert[53]\rvert = 4$
$\lvert[13]\rvert = 4$	$\lvert[29]\rvert = 2$	$\lvert[43]\rvert = 4$	$\lvert[59]\rvert = 2$

Table 31.2
Orders of elements in U_{60}.

(a) Find an element a in U_{60} of maximal order. Compute $\langle a\rangle$.

(b) Find a maximal subgroup K of G so that $\langle a\rangle \cap K = \{[1]\}$. (Hint: Recall that we want to have $\langle a\rangle \times K = G$, so what must the order of K be? Also note that $[7]^2 = [13]^2 = [17]^2 = [23]^2 = [37]^2 = [43]^2 = [47]^2 = [53]^2 = [49] \in U_{60}$.)

Preface

The impetus for this book lies in our approach to teaching abstract algebra. We place an emphasis on active learning and on developing students' intuition through their investigation of examples. For us, active learning involves students—they are ***doing*** something instead of just being passive learners. What students are doing when they are actively learning might include discovering, processing, discussing, applying information, writing intensive assignments, and engaging in common intellectual in-class experiences or collaborative assignments and projects. We support all of these activities with peer review and substantial faculty mentoring. According to Meyers and Jones [2], active learning derives from the assumptions that learning is an active endeavor by nature and that different people learn in different ways. A number of reports and studies show that active learning has a positive impact on students. For example, active learning is described as a high-impact learning activity in the latest report from the Association of American Colleges and Universities' Liberal Education and America's Promise (LEAP) initiative [1]. Results of a study [3] testing the active learning findings in liberal arts education show, in part, that students who experience the type of instruction we describe as active learning show larger "value-added" gains on a variety of outcomes than their peers. Although it is difficult to capture the essence of active learning in a textbook, this book is our attempt to do just that.

Our goals for these materials are several:

- To carefully introduce the ideas behind definitions and theorems in order to help students develop intuition and understand the logic behind them.

- To help students understand that mathematics is not done as it is often presented. We expect students to experiment through examples, make conjectures, and then refine or prove their conjectures. We believe it is important for students to learn that definitions and theorems don't pop up completely formed in the minds of mathematicians, but are the result of much thought and work.

- To help students develop their communication skills in mathematics. We expect our students to read and complete activities before class and come prepared with questions. In-class group work, student presentations, and peer-evaluation are a regular part of our courses. Of course, students also individually write solutions (mostly proofs) to exercises and receive significant feedback. Communication skills are essential in any discipline, and we place a heavy emphasis on developing students' abilities to effectively communicate mathematical ideas and arguments.

- To have students actively involved in realizing each of these goals through in-class and out-of-class activities, common in-class intellectual experiences (which, for us, include student presentations and collaborative group work), and challenging problem sets.

Layout

This text is formatted into investigations, each of which contains preview activities, in-class activities, concluding activities, exercises, and connections. The various types of activities serve different purposes.

- Preview activities are designed for students to complete before class to motivate the upcoming topic and prepare them with the background and information they need for the class activities and discussion.
- We generally use the regular activities to engage students in common in-class intellectual experiences. These activities provide motivation for new material, opportunities for students to prove substantial results on their own, and examples to help reinforce the meanings of definitions, theorems, and proofs. The ultimate goal is to help students build their intuition and develop a deep understanding of abstract algebra concepts. In our own practice, students often complete these activities—either during or before each class meeting—and then present their results to the entire class.
- Concluding activities are used to summarize, extend, or enhance the topics in a particular investigation. Concluding activities sometimes serve to foreshadow ideas that will be explored in more detail in subsequent investigations.

Each investigation contains a collection of exercises. The exercises occur at a variety of levels of difficulty, and most force students to extend their knowledge in different ways. While there are some standard, classic problems that are included in the exercises, many problems are open-ended and expect students to develop and then verify conjectures. Exercises that are highlighted with an asterisk (*) are referred to in the investigations and should be given special attention when assigning problems. Complete solutions to all activities and exercises are available to instructors at the authors' web site. In addition, the web site contains applets that can be used with the preview activities on ring and group isomorphisms (in Investigations 10 and 29) and the preview activity on normal subgroups and quotient groups (in Investigation 27).

Most investigations conclude with a short discussion of the connections between the topics in that investigation and the corresponding topics in ring theory or group theory. These discussions are intended to help students see the relationships between the two main types of algebraic objects studied throughout the text.

Organization

At Grand Valley State University, the first course we teach in modern algebra is focused on rings rather than the more simple structure of groups. Most of our majors intend to become elementary or secondary mathematics teachers, and the structure of the integers (and rings in general) is familiar to these students and therefore provides a comfortable entry point into the study of abstract algebra. Of course, a good argument can be made that groups, with their simpler structure, offer students an easier entrance to the subject. Both points are valid, and so we have designed this book so that, after completing some necessary background material, it is possible to begin with either rings or groups.

One of the consequences of this flexibility is that investigations that treat similar topics for rings and groups have very similar formats. We feel that this is an asset in that students should naturally recognize the similarities and make connections between these topics in rings and groups.

A foundations course in reading and writing mathematical proofs is a prerequisite for modern algebra for all of our students, so these materials have been formatted with that in mind. Even with this background, we aim to help students learn the new algebra content by gradually building both their intuition and their ability to write coherent proofs in context. Early investigations include many situations where students are prompted to comment on or provide missing details in proofs to help them develop their proof-writing skills, while the activities help them develop their intuition. As the investigations proceed, it is expected that students will be able to better read and write proofs without this prompting, and so it is no longer provided.

As previously mentioned, this text is organized in such a way that it is possible to begin with either rings or groups.

Rings First: For a course that begins with ring theory, the organizational structure is linear. Investigations 1 – 6 provide background, specific examples, and motivation for ring theory. Investigations 7 – 10 contain the basics of the subject, from the definitions of rings, integral domains, and fields to subrings, field extensions and direct sums, concluding with isomorphisms of rings. The majority of our mathematics majors are aspiring elementary or secondary school teachers (for whom this class is required), and for them the study of polynomial rings develops a deeper understanding of an important subject that they will themselves teach. Investigations 11 – 14 deal in depth with polynomial rings and comprise an important and relevant conclusion to our first semester course. Investigations 15 and 16 introduce the concepts of ideals, ring homomorphisms, and quotient rings for those who wish to have their students explore these topics. The ring theory portion of the text concludes with two additional investigations that require only some of the material preceding them.

- Investigation 17 treats divisibility and factorization in integral domains, proving in two different ways that every Euclidean domain is a unique factorization domain. The first approach relies primarily on the material from Investigations 1 – 7, with a few references to results about polynomials from Investigations 12 and 13. The second requires a more advanced understanding of ring theory, including results about ideals and principal ideal domains (from Investigation 16).
- Investigation 18 begins with the Peano axioms and then proceeds through the construction of $\mathbb{Q}$, $\mathbb{R}$, and $\mathbb{C}$. This investigation concludes with the characterization of the integers as the only ordered integral domain with a well-ordered set of positive elements. It requires an understanding of the material in Investigations 1 – 10.

Groups First: To begin a course with group theory, the background material needed is contained in Investigations 1 – 5. This material includes the Division Algorithm (Investigation 2); primes and prime factorizations (Investigation 4); equivalence relations, congruence, and $\mathbb{Z}_n$ (Investigation 5); and units and zero divisors in $\mathbb{Z}_n$ (Investigation 5). The instructor can choose from these investigations the material required for his/her students. We introduce groups with symmetries of planar objects (Investigation 19), and then the basic topics—groups, subgroups, cyclic groups, dihedral and symmetric groups, Lagrange's Theorem, normal subgroups and quotient groups, group isomorphisms and homomorphisms, the Fundamental Theorem of Finite Abelian Groups, and the Sylow theorems—follow (Investigations 20 – 33). This is an ambitious collection of investigations to complete in one semester.

The book concludes with several supplemental investigations in the Special Topics section. These investigations present applications of abstract algebra or investigations into additional topics in abstract algebra. They require knowledge of material from ring theory, group theory, or both.

- **Investigation 34: RSA Encryption.** This investigation describes the RSA algorithm and assumes familiarity with modular congruence and prime numbers from Investigations 1 – 4.

- **Investigation 35: Check Digits.** This investigation introduces the idea of check digits in several contexts and assumes familiarity with modular congruence (Investigation 2) and the dihedral groups (Investigation 24).

- **Investigation 36: Games: NIM and the 15 Puzzle.** This investigation applies group theory to develop a winning strategy in the game of NIM and to determine which 15 Puzzles are solvable. It assumes knowledge of groups (Investigation 20) and subgroups (Investigation 22), along with the symmetric groups (Investigation 25).

- **Investigation 37: Finite Fields, the Group of Units in $\mathbb{Z}_n$, and Splitting Fields.** In this investigation, we characterize finite fields and show how to decompose the group of units in $\mathbb{Z}_n$ as a direct product of cyclic groups. This investigation requires familiarity with rings and fields (Investigation 7), polynomials and polynomial rings (Investigation 11), field extensions (Investigation 9), ring isomorphisms (Investigation 10), roots of polynomials (Investigation 13), irreducible polynomials (Investigations 13 and 14), quotients of polynomial rings (Investigation 15), ideals (Investigation 16), groups (Investigation 20), cyclic groups (Investigation 22), and direct products of groups (Investigation 28).

- **Investigation 38: Groups of Order 8 and 12: Semidirect Products of Groups.** In this investigation, we classify all groups of order 8, introduce semidirect products of groups, and then classify all groups of order 12. We assume familiarity with the earlier classification of groups of various orders (Investigation 29) and with products of groups (Investigation 28).

Acknowledgments

We wish to thank the Academy of Inquiry Based Learning and the Educational Advancement Foundation for their generous financial support of this project. We also wish to thank Grand Valley State University for providing the necessary time and resources to complete this project. Finally, we thank the many colleagues and students within the GVSU Mathematics Department who have inspired us to be better teachers and who have given us valuable feedback on preliminary drafts of this book.

References

[1] George D. Kuh. *High-impact educational practices: What they are, who has access to them, and why they matter*. Association of American Colleges and Universities, 2008.

[2] C. Meyers and T. Jones. *Promoting active learning: Strategies for the college classroom.* Jossey-Bass, 1993.

[3] Ernest T. Pascarella, Gregory C. Wolniak, Tricia A. D. Seifert, Ty M. Cruce, and Charles F. Blaich. Liberal arts colleges and liberal arts education: New evidence on impacts. *ASHE Higher Education Report* **31**(3), 2005.

Part I

The Integers

Investigation 1

The Integers: An Introduction

Focus Questions

By the end of this investigation, you should be able to give precise and thorough answers to the questions listed below. You may want to keep these questions in mind to focus your thoughts as you complete the investigation.

- What are the integers, and what properties of addition and multiplication hold within the integers?
- What is the difference between an axiom and a theorem? In practice, how are axioms treated differently than theorems?
- How is subtraction defined within the integers, and how can the axioms of addition and multiplication be used to prove results involving subtraction?
- What ordering axioms hold within the integers, and what do these axioms imply?

Preview Activity 1.1. When doing arithmetic, we often use certain properties of addition, subtraction, and multiplication to make our calculations easier or more efficient. We don't usually state these properties explicitly, but in order to learn more about the integers, it will be helpful for us to do so now. As a first exercise, find the value of each of the following expressions, without using a calculator. As you do so, explicitly identify any shortcuts you take, and state the properties that make these shortcuts possible. Don't worry if you don't know or can't remember the formal names of the properties you use; simply describe them as precisely as you can. The first part is completed for you as an example.

(a) $(2^4 - 4^2)(5^7 - 7^5)$

Solution: *Since* $2^4 - 4^2 = 0$*, it follows that* $(2^4 - 4^2)(5^7 - 7^5) = 0$*. This is because any integer times zero is equal to zero; in other words,*

$$0 \cdot x = 0$$

for every integer x*.*

(b) $(67 - 11 + 925 - 81) + (81 + 11 - 925 - 67)$

(c) $(125 - 982) + (982 - 43) + (43 - 620) + (620 - 79) + (79 - 125)$

(d) $75(147 - 229) + 229(75) - 147(75)$

Introduction

Every journey has a beginning, and ours will begin with the integers. For likely as long as you can remember, you have been using the integers. When you first learned to count, your concept of number included only *natural numbers*, or what we might now refer to as *positive integers*. The notions of zero and negative numbers came later on, just as they did throughout the historical development of the integers. In fact, while the integers may seem elementary to us now, it actually took mathematicians thousands of years to formally develop and understand them. This historical development was rife with controversy, and it led to serious philosophical and even theological debates.

The daunting task of formally defining the integers played a key role in the development of much of modern mathematics, and in particular the field of set theory. It might surprise you to learn the most common modern construction of the integers is based entirely on sets and set operations. Such a rigorous development of the integers is not necessary for our investigations, but we should at least define the terminology and notation that we will be using.*

Definition 1.2.

- The set of **natural numbers**, denoted $\mathbb{N}$, contains the counting numbers (1, 2, 3, and so on); that is,
$$\mathbb{N} = \{1, 2, 3, \ldots\}.$$
- The set of **whole numbers**, denoted $\mathbb{W}$, contains the counting numbers and zero; that is,
$$\mathbb{W} = \{0, 1, 2, 3, \ldots\}.$$
- The set of **integers**, denoted $\mathbb{Z}$, contains the whole numbers and their opposites (or negatives); that is,
$$\mathbb{Z} = \{\ldots, -3, -2, -1, 0, 1, 2, 3, \ldots\}.$$

In addition to these basic definitions, we will also assume that addition and multiplication are defined on the set of integers in the usual way. Likewise, we will assume certain familiar facts about the way arithmetic works in the integers. The next section identifies some of these facts and explores their consequences.

*The symbols $\mathbb{N}$ for the natural numbers and $\mathbb{W}$ for the whole numbers probably seem reasonable. The symbol $\mathbb{Z}$ for the integers is from the German word *Zahlen* for number. This symbol appeared in Bourbaki's *Algébre*, Chapter 1. (Nicolas Bourbaki was a name adopted by a group of mostly French mathematicians who wrote a series of books intended to thoroughly unify mathematics through set theory.)

Integer Arithmetic

In Preview Activity 1.1, you probably used at least some of the following properties, or *axioms*, which we will assume to be true from this point forward.

Axioms of Integer Arithmetic

- **The integers are closed under addition and multiplication**, meaning that for all integers a and b, both $a + b$ and ab are also integers.
- **Addition and multiplication are commutative**, meaning that for all integers a and b, $a + b = b + a$ and $ab = ba$.
- **Addition and multiplication are associative**, meaning that for all integers a, b, and c, $(a + b) + c = a + (b + c)$ and $(ab)c = a(bc)$.
- **Multiplication distributes over addition**, meaning that $a(b + c) = ab + ac$ for all integers a, b, and c.
- **The integer 0 is an additive identity**, meaning that $a + 0 = a$ for every integer a.
- **The integer 1 is a multiplicative identity**, meaning that $1a = a$ for every integer a.
- **Every integer a has an additive inverse**, typically denoted $-a$; in particular, $a + (-a) = 0$ for every integer a.

One thing you may notice in looking at this list of axioms is that it says nothing about subtraction or division. This is actually an important observation, and one worth exploring in more detail. We will consider division of integers extensively in the next few investigations, but for now, let's focus on subtraction.

Typically, subtraction is defined in terms of addition as follows:

$$a - b = a + (-b)$$

This definition of subtraction is probably quite familiar to you, or at the very least not terribly surprising. Using it, along with the axioms of addition and multiplication, we can prove many useful facts about subtraction. For instance, let's consider the following result, which formalizes a property we didn't state above—namely, that multiplication distributes over subtraction.

Theorem 1.3. *Let a, b, and c be integers. Then*

$$a(b - c) = ab - ac.$$

Note that we stated this result as a *theorem*, which suggests that we can prove it from the axioms we have already assumed. In fact, the main difference between an axiom and a theorem is that an axiom is assumed to be true without proof, whereas a theorem must be proved from axioms and other previously established results. It's worth noting that none of the axioms we assumed can be proved from the others. In other words, our axioms are *independent* of each other. This is a desirable feature, and it suggests that we are beginning our investigations with a minimal set of assumptions, one that is robust but not redundant.

So how might we go about proving Theorem 1.3? At first glance, it may seem hard to know where to start, and you might not even be convinced that a proof is necessary. This would be a legitimate objection, especially since we *assumed* a very similar property for addition. A difference here, however, is that Theorem 1.3 *can* be proved from our other axioms. Thus, assuming it would not only be unnecessary, but would also add an undesirable redundancy to our axiom system. Thus, a proof is in order.

The only thing we really know about subtraction right now is the definition, and so it makes sense that we should start there. Perhaps we could begin by rewriting $a(b-c)$ as $a(b+(-c))$. Doing so would allow us to use the fact that multiplication distributes over addition, which would then yield

$$a(b-c) = a(b+(-c)) = a(b) + a(-c). \tag{1.1}$$

Take a close look at what we've proved so far. Are we done yet?

As it turns out, we are not. What we would like to be able to do is substitute $a(-c) = -(ac)$ into (1.1). If we could do this, then we would just have to apply the definition of subtraction once more to complete the proof. Unfortunately, none of our axioms about the integers tell us that such a substitution is valid. Of course, we suspect from past experience that it is, but how could we prove this?

What we want to show is that for all integers a and c, $a(-c) = -(ac)$. If we were to read this statement without using the words "minus" or "negative," we might say that *a times the additive inverse of c is the additive inverse of the quantity a times c.* Or, stated in a slightly different way, *the additive inverse of ac is $a(-c)$.* This wording suggests that what we need in order to proceed is a good working definition of additive inverse.

Activity 1.4. Discuss the pros and cons of each of the following potential definitions of the additive inverse of an integer x.

(a) The additive inverse of x is $-x$.

(b) The additive inverse of x is $0 - x$.

(c) The additive inverse of x is an integer y such that $x + y = 0$.

(d) The additive inverse of x is $(-1)x$.

There are advantages and disadvantages to each of the definitions in Activity 1.4. However, the definition in part (c) is the one that turns out to be the most useful for proving results involving additive inverses. We can state this definition formally as follows:

Definition 1.5. Let x be an integer. Then an **additive inverse** of x is an integer y such that $x+y = 0$.

You may notice that we used the article *an* instead of *the* when defining additive inverse. This is because Definition 1.5 alone does not imply that additive inverses must be unique. In other words, the definition is not enough to rule out the possibility that an integer might have two distinct additive inverses. Fortunately, we will be able to dispose of this potential absurdity fairly easily, as Exercise 1 suggests one way to prove that, at least in the integers, additive inverses must be unique. This is why we can use the notation $-x$ for the unique additive inverse of x.

The notion of uniqueness will arise naturally throughout our investigations of the integers and related number systems. Consequently, we will have a chance to study and prove uniqueness properties in a variety of contexts. For now, however, let's return to Theorem 1.3. Parts (a) through (c) of the next activity suggest a strategy for completing the proof we started earlier, and parts (d) through (g) ask you to prove several related results.

Activity 1.6. Let a, b, and c be integers.

(a) Prove that $a \cdot 0 = 0$. (Hint: This is not completely obvious. Start with the fact that $0+0=0$, and multiply both sides by a.)

(b) Use Definition 1.5 along with part (a) to prove that $-(ac) = a(-c)$. (Hint: You want to show that $x + y = 0$ for an appropriate choice of x and y.)

(c) Use part (b) to complete the proof of Theorem 1.3.

(d) Prove that $-(ac) = (-a)c$.

(e) Prove that $-(a+b) = -a - b$.

(f) How can $-(-a)$ be simplified? Prove your answer.

(g) How can $(-a)(-c)$ be simplified? Prove your answer.

Ordering Axioms

So far, we have stated axioms that specify how addition and multiplication work within the integers. These arithmetic axioms, however, are not the only types of axioms that we will have reason to call upon. For instance, the integers also satisfy each of the *ordering axioms* shown on the next page. Of course, there are other properties pertaining to the ordering of the integers that we have not included in our list of axioms. This is because these properties can be proved from the four axioms we have stated. As a simple example, consider the following theorem (and its simple proof), which uses the ordering axioms to establish a useful fact about additive inverses:

Theorem 1.7. *Let a be an integer. If $0 < a$, then $-a < 0$.*

Proof. Let a be an integer, and assume $0 < a$. Then, by the translation invariance axiom,

$$0 + (-a) < a + (-a).$$

Since 0 is the additive identity in the integers, and since $-a$ is the additive inverse of a, we can simplify both sides of this inequality to obtain $-a < 0$, as desired. ∎

We stated and proved Theorem 1.7 using only the "less than" ($<$) relation, writing, for example, $0 < a$ instead of $a > 0$. We made this choice to be consistent with the way our ordering axioms are stated above. Of course, analogous axioms also hold for the "greater than" ($>$) relation, and we will use both versions throughout future investigations. We will also use the $\leq$ and $\geq$ symbols as they are normally used, interpreting $a \leq b$ to mean "$a < b$ or $a = b$," with $a \geq b$ interpreted similarly. Finally, it's worth noting that the arithmetic and ordering axioms we've stated here apply to other number systems as well, such as the rational numbers and the real numbers. We'll revisit this observation in later investigations.

Ordering Axioms of the Integers

The "less than" relation on the integers, denoted $<$, satisfies all of the following properties:

- **Trichotomy**: For all integers a and b, exactly one of the following is true: $a < b$, $b < a$, or $a = b$.
- **Transitivity:** For all integers a, b, and c, if $a < b$ and $b < c$, then $a < c$.
- **Translation Invariance:** For all integers a, b, and c, if $a < b$, then $a + c < b + c$.
- **Scaling:** For all integers a, b, and c, if $a < b$ and $c > 0$, then $ac < bc$.

What's Next

In this investigation, we have identified some important axioms of the integers, and we have used these axioms to prove a few simple results. The results we proved were not terribly significant or profound, but the approach we took illustrated the difference between axioms and theorems, and it demonstrated the importance of starting with good assumptions and definitions.

Throughout the next few investigations, we will consider several other important aspects of the integers, such as division and prime factorization. Although we may not always explicitly reference the axioms we have stated here, our work will rely heavily upon them.

In the course of these investigations, we will learn not only *how* the integers work, but also *why* they work the way they do. Even more importantly, we may begin to wonder why the integers are so important in the first place. Our questions will naturally lead us to explore other number systems, some that are very similar to the integers, and some that are very different. Eventually, we will become less interested in the specifics of each particular number system, and more interested in the properties that they all seem to satisfy. We will see that the integers, and many of our other favorite number systems, all share a certain common structure, and that this common structure is in fact essential to making the integers behave in the way we expect them to. We will also see that additions to or deviations from this structure produce different behaviors that we might not expect.

Concluding Activities

Activity 1.8. We took as one of our axioms of the integers that multiplication distributes over addition.

(a) What would it mean for addition to distribute over multiplication? Write a precise definition.

(b) In the integers, does addition distribute over multiplication? Give a proof or counterexample to justify your answer.

(c) In the integers, does addition distribute over addition? Give a proof or counterexample to justify your answers.

Activity 1.9. Consider the following theorem:

> **Theorem 1.10.** *There do not exist nonzero integers a and b such that $ab = 0$.*

(a) Explain why Theorem 1.10 is equivalent to each of the following:

- For all integers a and b, if $ab = 0$, then $a = 0$ or $b = 0$.
- For all integers a and b, if $ab = 0$ and $a \neq 0$, then $b = 0$

(b) Use the ordering axioms of the integers to prove Theorem 1.10 or one of its equivalent forms. (Hint: Use the trichotomy axiom to set up cases.)

(c) Use Theorem 1.10 to prove the following result, which establishes the validity of multiplicative cancellation of a nonzero integer:

> **Theorem 1.11.** *For all integers a, b, and c, if $ac = bc$ and $c \neq 0$, then $a = b$.*

Note that "dividing by c" is not an option, as we have not yet defined division in the integers.

Activity 1.12. One of the properties of integer arithmetic is that the set of integers contains an additive inverse for each of its elements. The existence of additive inverses allows us to define an operation of subtraction on the set of integers. Although we have no operation of division on the integers, we can still ask if there are any integers that have a *multiplicative* inverse within the integers. We will call such integers *units*.

(a) State a formal definition of what it would mean for an integer a to have a multiplicative inverse within $\mathbb{Z}$.

(b) Determine all of the units in $\mathbb{Z}$. (Hint: There is more than one.) Use your definition from part (a) to verify your answer.

(c) Use the ordering axioms of the integers to prove that the units you found are the only integer units. (Warning: We have no operation of division in the integers, so you cannot "divide" in your proof.)

Activity 1.13. Write a short summary that describes the important concepts, and the relationships between these concepts, that were introduced in this investigation. Explain how the material in this investigation is related to your prior understanding of the integers.

Exercises

* (1) **Uniqueness of additive inverses.** Suppose that some integer a has two additive inverses, say b and c. Without using the symbol $-a$, prove that $b = c$.

⋆ (2) **Additive cancellation.** Let a, b, and c be integers such that $a + b = a + c$. Using only the axioms and theorems from this investigation, prove that $b = c$.

(3) **Addition and multiplication.** Let a and n be integers, with $n > 0$. Prove that the sum of n copies of a is equal to na. That is, prove that

$$\underbrace{a + a + \cdots + a}_{n \text{ terms}} = na.$$

(4) Let a, b, and c be integers. Is it always the case that $(a + b)c = ac + bc$? Prove your answer using only the axioms stated in this investigation.

(5) Find all of the integer solutions to the equation

$$x^3 + 3x^2 - 4x = 12.$$

Justify each step in your solution with one or more of the axioms or theorems from this investigation (possibly including Theorem 1.10).

(6) **Antisymmetry of the $\leq$ relation.** Prove that the $\leq$ relation is antisymmetric; that is, prove that for all integers a and b, if $a \leq b$ and $b \leq a$, then $a = b$.

(7) Let a, b, and c be integers. Prove that if $a < b$ and $c < 0$, then $ac > bc$. Deduce that if $a < b$, then $-b < -a$.

(8) Let a, b, and c be integers. Prove that if $ac > bc$ and $c > 0$, then $a > b$.

(9) Let a, b, c and d be integers.

(a) Prove that if $a < b$ and $c < d$, then $a + c < b + d$.

(b) Prove that the result from part (a) still holds if $a < b$ and $c \leq d$.

Investigation 2

Divisibility of Integers

Focus Questions

By the end of this investigation, you should be able to give precise and thorough answers to the questions listed below. You may want to keep these questions in mind to focus your thoughts as you complete the investigation.

- What does it mean for one integer to divide another? What are some important properties of divisibility within the integers, and why do these properties hold?
- What is the Division Algorithm? What does the Division Algorithm say about division of integers, and how can it be proved?
- What is the Well-Ordering Principle, and how can it be used to prove the Division Algorithm?
- What does it mean for two integers to be congruent modulo n? What are some important properties of congruence in the integers, and why do these properties hold?

Preview Activity 2.1. In Investigation 1, we learned about addition, subtraction, and multiplication of integers. The questions below will help us to shift our focus and begin to think about how division works within the integers. Try to answer these questions using only your informal or intuitive understanding of terms like "factor" and "divisor." We will give precise definitions of these terms later on.

(a) Which integers divide 360? List all such divisors.

(b) Which integers are divisors of 1? Which integers are divisible by 1?

(c) Which integers are divisors of 0? Which integers are divisible by 0?

(d) Let a, b, and d be integers. If ab is a multiple of d, does it follow that either a or b is a multiple of d?

(e) Let a and b be integers. Suppose that a is a factor of b and b is a factor of a. What can we conclude about the relationship between a and b?

(f) If a student in elementary school was asked to divide 43 by 5, what solution do you think the student would obtain, and what would his or her reasoning be?

Preview Activity 2.2. Now that we have started to think about how division works in the integers, we will begin to make our thinking a bit more precise by using algebraic equations to represent certain division problems. The questions below are related to both the *existence* and *uniqueness* of quotients and remainders, and they foreshadow an important theorem called the *Division Algorithm*.

(a) Let a and b be nonzero integers. Suppose that for some integers q_1 and q_2, $b = aq_1$ and $b = aq_2$. What can you conclude about the relationship between q_1 and q_2?

(b) Which integers q satisfy the equation $0 = 0q$?

(c) Find several pairs of integers q and r that satisfy the equation $43 = 5q + r$.

(d) How many pairs of integers q and r satisfy the equation $43 = 5q + r$ and the inequality $0 \leq r < 5$?

(e) How are your answers to parts (c) and (d) above related to part (f) of Preview Activity 2.1?

Introduction

In Preview Activity 2.1, we considered several questions related to the operation of division within the integers. Your answers to these questions likely relied on your past experience with division and your intuitive ideas of what it means for one integer to divide another. However, there are situations for which these intuitive ideas may not provide satisfactory answers. Thus, we will begin our formal investigations of integer division with a more precise definition. Just as we used addition to define subtraction in Investigation 1, here we will use multiplication to define division within the integers. Our precise definition is as follows:

Definition 2.3. An integer a **divides** an integer b, denoted $a \mid b$, if there is an integer q such that $b = aq$.

Note that when a divides b, we may also say that a is a *divisor* or *factor* of b, or that b is a *multiple* of a. Note also that the notation $a \mid b$ does not represent the rational number $\frac{b}{a}$. Rather, it expresses in shorthand a *relationship* between the integers a and b—namely, that a divides b.

Preview Activity 2.2 illustrates this important distinction. When we divide one integer by another, the quotient that we obtain is exactly the q from Definition 2.3. So, for instance, we know that $7 \mid 84$ since $84 = 7 \cdot 12$. In this case, we would write $84 \div 7 = 12$, or $\frac{84}{7} = 12$. This makes sense because 12 is the only integer q for which $84 = 7q$. In other words, the division yields a *unique* quotient.

In contrast, consider the problem of dividing 0 by 0. Notice that *every* integer q satisfies the equation $0 = 0q$. Thus, it is certainly the case that $0 \mid 0$. However, because there is more than one possibility for q, we do not obtain a unique quotient. If we said that $0 \div 0 = 1$, or $0 \div 0 = 0$, then we would have to say by the same logic that $0 \div 0 = 17$ and $0 \div 0 = -94$, and so on. This kind of reasoning quickly leads to nonsense, and so it makes sense for us to say that although $0 \mid 0$, the quantity $\frac{0}{0}$ is undefined.

To summarize, note that in order to say that $a \mid b$, there must *exist* an integer q for which $b = aq$. But to say that $b \div a = q$, or to use the notation $\frac{b}{a}$ to represent q, it must be the case that q is *unique*. It is also important to remember that **the integers are not closed under division.** In particular, there are many integers a and b for which $b \div a$ is **not** an integer. Thus, one must always heed the following warning:

WARNING!!! Within the integers, the notation $\frac{b}{a}$ only makes sense when both a and b are integers, and a divides b with a unique quotient.

Quotients and Remainders

In the previous section, we discussed division as it pertains to one integer *evenly* dividing another, i.e., with no remainder. But even in elementary school, children learn that division of integers often yields both a quotient and a nonzero remainder.

Consider again the problem of dividing 43 by 5 (from Preview Activity 2.1). A child in elementary school may view this problem as one of dividing 43 items (say apples) among 5 friends. She might begin by removing groups of 5 apples at a time, and then seeing how many are left. For instance, taking 5 apples away from 43 yields 1 group of 5, with 38 apples left over. So should we say that $43 \div 5$ is 1 with a remainder of 38?

Well, probably not. Even a child first learning about division would probably say that there are still more than 5 apples left, so we can take away another group of 5. This would leave 33 apples, and we could continue taking away groups of 5 apples until there were no longer 5 apples left to take away. Doing so would yield 8 groups of 5 apples, with 3 apples left over. Thus, we would say that 43 divided by 5 is 8 with a remainder of 3. Note that we could use an equation to express this relationship by writing $43 = 5 \cdot 8 + 3$.

Let's now consider how we might generalize this intuitive process. Suppose we have positive integers a and b with $b \geq a$, and we want to find out what quotient and remainder would result from dividing b by a. We could begin by subtracting a from b, just as the child in our previous example took away a group of 5 apples from the 43 she started with. If what is left $(b - a)$ is still greater than or equal to a, then we will subtract a again, and we will continue subtracting a until we obtain a number that is less than a (but still greater than or equal to 0). The result of this final subtraction will be our remainder.

Putting this in slightly more formal terms, we will calculate $b - am$ for increasingly large integer values of m, stopping when we find a value of m for which $0 \leq b - am < a$. This special value of m will be called q, or the *quotient*. Likewise, the corresponding quantity $b - aq$ will be called r, or the *remainder*, so that $r = b - aq$, or equivalently, $b = aq + r$. Using the latter equation, we can see that our problem of dividing b by a is really a problem of finding integers q and r for which $b = aq + r$ and $0 \leq r < a$. Of course, it would not make sense to have two different quotients and remainders for the same division problem, and so we also want q and r to be unique. The *Division Algorithm*, stated formally below, guarantees this.

The Division Algorithm. *Let a and b be integers, with $a > 0$. Then there exist unique integers q and r such that*

$$b = aq + r \text{ and } 0 \leq r < a.$$

There are a few things worth noting about the Division Algorithm before we discuss why it is true. The first is that it asserts both the existence and uniqueness of a quotient q and a remainder r, but provides no actual mechanism for finding q and r. In this sense, the Division Algorithm is not an algorithm at all, and perhaps would be better called a theorem. Of course, there are many algorithms for actually carrying out the operation of division. Long division is one that you have undoubtedly used many times in the past; it simply formalizes and makes more efficient the repeated subtraction technique that we discussed earlier.

The second fact to note about the Division Algorithm is that it requires a positive divisor ($a > 0$). This condition is actually a bit stronger than it needs to be, and it could be weakened by simply requiring $a \neq 0$. Doing so, however, necessitates changing the subsequent inequality to $0 \leq r < |a|$.

Finally, recalling our discussion from Investigation 1, it is probably worth asking whether we should simply assume the Division Algorithm (as an axiom), or try to prove it. At first glance, the conclusion of the Division Algorithm may seem obvious, or even self-evident. On the other hand, this conclusion is stated in terms of addition and multiplication, and so we may be inclined to at least try to prove it using the axioms and other results we considered in Investigation 1. In order to do so, we will also need another important axiom known as the *Well-Ordering Principle*.

The Well-Ordering Principle

Preview Activity 2.4. As we will see shortly, the Well-Ordering Principle allows us to conclude that certain sets of numbers must contain a smallest, or least, element. The questions below will help us to begin thinking about which types of sets do contain least elements, and which do not.

(a) Which of the following sets contain a least element? Which contain a greatest element?

- $A = \{1, 2, 3, 4\}$
- $B = \{x \in \mathbb{Z} : x > 4\}$
- $C = \{x \in \mathbb{Z} : x < 4\}$
- $D = \{x \in \mathbb{W} : x > 4\}$
- $E = \{x \in \mathbb{W} : x < 4\}$

(b) Does every nonempty subset of $\mathbb{Z}$ contain a least element? If not, give a counterexample.

(c) Does every nonempty subset of $\mathbb{W}$ contain a least element? If not, give a counterexample.

(d) Let $\mathbb{R}^*$ denote the set of all nonnegative real numbers. That is,

$$\mathbb{R}^* = \{x \in \mathbb{R} : x \geq 0\}.$$

Does $\mathbb{R}^*$ contain a least element? Why or why not?

(e) Again define $\mathbb{R}^*$ as in part (d). Does every nonempty subset of $\mathbb{R}^*$ contain a least element? If so, explain why. If not, give a counterexample.

Preview Activity 2.5. Now that we are at least somewhat familiar with the idea of a least element, let's see how least elements are related to the Division Algorithm. To begin, let a and b be integers, with $a > 0$, and define the set S as follows:

$$S = \{x \in \mathbb{Z} : x \geq 0 \text{ and } x = b - am \text{ for some } m \in \mathbb{Z}\}.$$

(a) For $a = 5$ and $b = 43$, list at least 5 different elements of S. Which integer appears to be the least element of S?

(b) How is your answer to part (a) related to our earlier discussion of how an elementary school student might divide 43 by 5?

(c) Repeat part (a), but this time assume that $a = 10$ and $b = -58$.

(d) Prove that if $b \geq 0$, then $b \in S$.

(e) Suppose $b < 0$. For what values of m will $b - am$ be an element of S? Prove your answer.

(f) What do your answers to parts (d) and (e) allow you to conclude about S, and how might this conclusion be related to S having a least element?

In Preview Activity 2.4, we were asked to consider whether certain sets, and their subsets, had least elements. Furthermore, Preview Activity 2.5 suggests why this task is particularly important to our goal of proving the Division Algorithm.

In our earlier discussion of division, we observed that when dividing an integer b by a positive integer a, the remainder can be obtained by repeatedly subtracting a from b until we reach the point where further subtractions would yield a negative result. In other words, the remainder is exactly the least element of the set S defined in Preview Activity 2.5. But how do we know that this set always has a least element? The answer to this question comes from the following principle, which we will take as an axiom:

The Well-Ordering Principle. *Every nonempty subset of the whole numbers contains a least element.*

The Well-Ordering Principle is actually equivalent to the Principle of Mathematical Induction, and a proof of this equivalence is provided in Appendix B. In the next section, we will use the Well-Ordering Principle as a tool to prove the Division Algorithm.

Proving the Division Algorithm

Our first step toward the goal of proving the Division Algorithm is to consider the set S defined in Preview Activity 2.5:

$$S = \{x \in \mathbb{Z} : x \geq 0 \text{ and } x = b - am \text{ for some } m \in \mathbb{Z}\}.$$

By its very definition, S is a subset of the whole numbers. (The condition that $x \geq 0$ guarantees this.) Thus, in order to apply the Well-Ordering Principle to S, we must show that S is nonempty. Parts (d) – (f) of Preview Activity 2.5 suggest one way to do so. In particular, if $b \geq 0$, then $b \in S$ since $b = b - a \cdot 0$. On the other hand, if $b < 0$, then we can simply choose any negative integer m for which $am \leq b$ and let $x = b - am$. Choosing $m = b$ is particularly convenient, since

$$b - ab = b(1 - a) \geq 0.$$

Thus, $x = b - ab \in S$. (Note that this argument holds because $a > 0$, and so $1 - a \leq 0$.)

In either case, whether $b \geq 0$ or $b < 0$, we have shown that S contains at least one element.

The set S is therefore a nonempty subset of the whole numbers, and so the Well-Ordering Principle allows us to conclude that S has a least element. Knowing that we want this least element to be our remainder, we will call it r. Furthermore, since $r \in S$, we can find an integer, say q, for which $r = b - aq$. This establishes one part of the Division Algorithm—namely, that there exist integers q and r such that $b = aq + r$.

Two assertions now remain to be shown: first, that $0 \leq r < a$; and second, that q and r are the only integers that satisfy the two aforementioned conditions. For the former, observe that, by the definition of the set S, it must be the case that $0 \leq r$. Thus, we must show that $r < a$. The next activity suggests one method for doing so.

Activity 2.6. Suppose, to the contrary, that $r \geq a$.

(a) Beginning with the fact that $r = b - aq$, show that $r - a \in S$. (Hint: Two things must be shown here—that $r - a \geq 0$ and that $r - a$ can be written in the form $b - am$ for some integer m.)

(b) Why would your answer to part (a) be a contradiction? (Hint: How was r defined?)

(c) Try to explain the reasoning from parts (a) and (b) in the context of an elementary school division problem. What does the set S represent? How is r chosen from S, and what would happen intuitively if r was not less than a?

Now we must show that the q and r we have found are unique. In particular, we want to show that if there are integers q' and r' for which

$$b = aq' + r' \text{ and } 0 \leq r' < a,$$

then it must be the case that $q' = q$ and $r' = r$. Incidentally, this technique is fairly standard for proving the uniqueness of a mathematical object: we simply assume that there are two objects (in this case, two pairs of integers) that satisfy the desired conditions, and then try to show that these objects are actually the same. The next activity suggests how the details of this method might work for our proof of the Division Algorithm. When followed, the steps suggested there complete the uniqueness argument, and thus the entire proof.

Activity 2.7. We found integers q and r for which $b = aq + r$ and $0 \leq r < a$. Suppose that for some integers q' and r', it is also the case that $b = aq' + r'$ and $0 \leq r' < a$.

(a) Use algebra to show that $a(q - q') = r' - r$.

(b) By adding the corresponding sides of two inequalities, show that $-a < r' - r < a$. (Hint: First argue that $-a < -r \leq 0$.)

(c) Use parts (a) and (b) to argue that $r' - r$ is both an integer multiple of a and strictly between $-a$ and a.

(d) What does your answer to part (c) allow you to conclude about r and r'?

(e) What do your answers to parts (a) and (d) allow you to conclude about q and q'? (Hint: You may need to use a result from Activity 1.9 on page 9.)

Putting It All Together

We are now ready to use what we have learned so far to write a complete and coherent proof of the Division Algorithm. In the proof outlined below, we have left several blanks for you to fill in as you read the proof. We have also written this proof fairly concisely, leaving some of the more minor justifications to you, the reader.

In this proof (and many others throughout the text), we will use the (?) symbol to denote places where more elaboration or justification may be desirable. When you encounter a (?), you may want to pause and ask yourself, "Wait—why is that true?" If you can convince a classmate or peer that the statement or suggested technique is valid, then you are probably ready to continue reading. On the other hand, if you cannot provide a convincing explanation, then you may not fully understand the concepts behind the proof.

Proof of the Division Algorithm. Let a and b be integers, with $a > 0$. For the e__________ portion of the proof, define the set S as follows:

$$S = \{\_\_\_\_\_\_ : \_\_\_\_\_\_ \text{ and } \_\_\_\_\_\_\_\_\_ \text{ for some } \_\_\_\_\_\_\}$$

We will use the _______________ Principle to show that S has a least element. Since S is clearly a subset of the whole numbers, we need only to show that S is nonempty.(?) If $b \geq 0$, then $b \in S$.(?) Furthermore, if $b < 0$, then _________ $\in S$.(?) In either case, S has a least element, which we will call r. It follows that $r = b - aq$ for some $q \in \mathbb{Z}$.(?) Thus, we have found integers q and r such that $b = aq + r$.

To show that $0 \leq r < a$, we will assume, to the contrary, that ________.(?) (It must be the case that $0 \leq r$, since ________.) This implies, however, that $r - a \in S$, since $r - a \geq 0$(?) and

$$r - a = (b - aq) - a = b - a(q + 1).$$

But it is also the case that $r - a < r$,(?) and so we have arrived at a contradiction.(?) It follows that $0 \leq r < a$.

To prove u___________, assume that there exist integers q' and r' such that

$$\_\_\_\_\_\_\_\_\_\_\_ \text{ and } \_\_\_\_\_\_\_\_\_\_\_.$$

It follows that

$$a(q - q') = r' - r. \text{(?)}$$

But since $0 \leq r' < a$ and $-a < -r \leq 0$,(?) it is also the case that

$$-a < r' - r < a. \text{(?)}$$

Thus $r' - r$ is both an integer multiple of a and strictly between $-a$ and a. As such, the only possibility is that $r' - r =$ ____, which implies that $q - q' =$ ____ as well.(?) Thus, the integers q and r determined by the Division Algorithm are unique, which completes the proof. ∎

Congruence

We'll conclude this section by using what we have learned about division to investigate congruence within the integers—a concept that we will use regularly in later investigations. The following preview activity will get us started.

Preview Activity 2.8. In life, whether we realize it or not, we often use congruence relationships and modular arithmetic. The questions below give an example of this and also foreshadow some of the theory that we will study shortly. To begin, suppose that it is currently Friday.

(a) What day will it be 4 days from now?

(b) What day will it be 11 days from now?

(c) What day will it be 18 days from now?

(d) Find 5 other natural numbers x for which the answer to the question, "What day will it be x days from now?" is the same as your answers to parts (a) – (c).

(e) Repeat part (d), but this time find negative values of x. In this context, what would be a more natural way of phrasing the question quoted in part (d)?

(f) Combine the numbers you found in parts (d) and (e) to create a list of 10 integers. Then find the remainder when each of these integers is divided by 7. What do you notice?

(g) Pick any two numbers on your list from part (f) and subtract them. Repeat this several times, keeping track of your results.

(h) What do all of the differences you found in part (g) have in common?

The idea of congruence is used by mathematicians to describe cyclic phenomena in the world of the integers. For instance, time is a cyclic phenomenon in that the time of day repeats every 12 or 24 hours, depending on the clock we are using. As we saw in Preview Activity 2.8, the days of the week also cycle in this same fashion, with the same day occurring every 7 days. We can use this latter observation to determine what day of the week it will be any number of days from now. For instance, if today were a Tuesday, then it would be Friday in another 3 days, and then again in another 10 days, 17 days, 24 days, and so on. We also know that it was Friday 4 days ago (or -4 days from now), 11 days ago (or -11 days from now), and so on. In other words, for every value of x in the list below, it will be Friday x days from now (or it was Friday $|x|$ days ago in the case of negative numbers):

$$\ldots, -18, -11, -4, 3, 10, 17, 24, \ldots$$

In Preview Activity 2.8, you may have noticed that since the days of the week follow a 7-day cycle, the difference between any two numbers on this list is divisible by 7. You may have also noticed that all of the numbers on the list have the same remainder (as specified by the Division Algorithm) when divided by 7. These two observations are important and useful; the first forms the basis of our definition of congruence, and the second is a consequence of this definition.

Definition 2.9. Let n be a natural number, and let a and b be integers. Then a **is congruent to** b **modulo** n, denoted $a \equiv b \pmod{n}$, provided that n divides $a - b$.

Applying Definition 2.9 to our list above, we could say that all of the numbers on the list are congruent modulo 7. The fact that all have the same remainder when divided by 7 is made formal by the next theorem.

Theorem 2.10. *Let n be a natural number, and let a and b be integers. Then $a \equiv b \pmod{n}$ if and only if a and b yield the same remainder when divided by n.*

Activity 2.11 below suggests one way to prove Theorem 2.10.

Activity 2.11. Let n be a natural number, and let a and b be integers.

(a) Use the Division Algorithm to write equations (together with the appropriate inequalities) that represent the result of dividing each of a and b by n. For convenience, use q_1, q_2, r_1, r_2 to denote the resulting quotients and remainders.

(b) If you haven't already done so, write your equations from part (a) so that they are in the form $a = \ldots$ and $b = \ldots$. Then use subtraction to obtain a new equation of the form $a - b = \ldots$.

(c) Now assume that $n \mid (a - b)$. Use your equation from part (b) to argue that $n \mid (r_1 - r_2)$ as well.

(d) Use the result you proved in part (c) to deduce that $r_1 = r_2$. (Hint: Both r_1 and r_2 satisfy a certain inequality. Use these inequalities to argue that $r_1 - r_2$ is a multiple of n and is strictly between $-n$ and n.)

(e) Which direction of the biconditional statement from Theorem 2.10 did you prove in parts (c) and (d)? What remains to be shown?

(f) Use your equation from part (b) to prove that if $r_1 = r_2$, then $n \mid (a - b)$. Explain how this argument finishes the proof of Theorem 2.10.

Theorem 2.10 is one of many results about congruence that we could prove using only what we have learned so far about divisibility. We will study congruence in much more detail later in the text, but for now, let's conclude this investigation by exploring some properties that will allow us to treat congruence much like we treat equality, at least for the purposes of doing arithmetic. Each of the results stated in Activity 2.12 can be proved by first translating the given statement into one that involves divisibility. The first part is completed for you as an example.

Activity 2.12. Let n be a natural number, and let a, b, c, and d be integers. Prove each of the following results.

(a) If $a \equiv b \pmod{n}$ and $c \equiv d \pmod{n}$, then $(a + c) \equiv (b + d) \pmod{n}$.

Solution: *Using the definition of congruence, the given result is equivalent to the following:*

If $n \mid (a - b)$ and $n \mid (c - d)$, then $n \mid [(a + c) - (b + d)]$.

Thus, assume that $n \mid (a - b)$ and $n \mid (c - d)$. Then there exist integers j and k such that $a - b = nj$ and $c - d = nk$. Simple algebra (in particular, the associative and distributive axioms) then implies that

$$\begin{aligned}(a + c) - (b + d) &= (a - b) + (c - d)\\ &= nj + nk\\ &= n(j + k).\end{aligned}$$

Thus, $n \mid [(a + c) - (b + d)]$, as desired.

(b) If $a \equiv b \pmod{n}$ and $c \equiv d \pmod{n}$, then $ac \equiv bd \pmod{n}$.

(c) If $a \equiv b \pmod{n}$ and $m \in \mathbb{N}$, then $a^m \equiv b^m \pmod{n}$.

(d) For every integer a, $a \equiv a \pmod{n}$. (This property is called the **reflexive** property of congruence.)

(e) If $a \equiv b \pmod{n}$, then $b \equiv a \pmod{n}$. (This property is called the **symmetric** property of congruence.)

(f) If $a \equiv b \pmod{n}$ and $b \equiv c \pmod{n}$, then $a \equiv c \pmod{n}$. (This property is called the **transitive** property of congruence.)

Concluding Activities

Activity 2.13. Let a and b be integers with $a > 0$, and let r be the remainder when b is divided by a. Prove that if an integer d divides both a and b, then d also divides r.

Activity 2.14. In a popular high-school mathematics textbook, students are told that one of the first theorems in number theory is the following:

> If a, b, and c are integers where a is a factor of b and a is a factor of c, then a is a factor of $b + c$

Students are then asked about generalizing the theorem to:

> If a, b, and c are integers where a is a factor of b and a is a factor of c, then a is a factor of $bm + cn$ for all integers m and n.

Is this second statement true? Verify your answer.

Activity 2.15. Write a short summary that describes the important concepts, and the relationships between these concepts, that were introduced in this investigation. Explain how the material in this investigation is connected to the material in Investigation 1.

Exercises

(1) In a popular seventh-grade mathematics textbook, students are asked to investigate the following conjecture:

> The sum of any three consecutive whole numbers will always be divisible by 3.

(a) Is the conjecture true or false? Provide a proof or a counterexample to justify your answer.

(b) If the conjecture is true, can it be generalized in any way? If it is false, are there any special cases for which it does hold? Prove your answer.

(2) Let a, b, and c be integers. What conclusions, if any, can be drawn in each of the following situations? Prove your answers.

(a) $a \mid c$ and $b \mid c$

(b) $a \mid b$ and $b \mid c$

(c) $a \mid b$ and $a \mid c$

(3) Let a and b be integers. Prove that if $a \mid b$ and $b \mid a$, then $|a| = |b|$.

(4) Let a and b be positive integers, and suppose that $a \mid b$. Prove that $(a+1) \mid (b + \frac{b}{a})$.

(5) Let a, $b \in \mathbb{N}$. Use the arithmetic and ordering axioms of the integers to prove that if $a \mid b$, then $a \leq b$.

(6) A nonempty subset S of $\mathbb{R}$ is said to be **well-ordered** if every nonempty subset of S contains a least element.

(a) Use this definition to concisely restate the Well-Ordering Principle. (Hint: You should be able to do so in no more than six words.)

(b) Is $\mathbb{R}$ well-ordered? Why or why not?

(c) Is the set $\mathbb{R}^* = \{x \in \mathbb{R} : x \geq 0\}$ well-ordered? Why or why not?

(d) Is $\{-9, -7, -5, \ldots\}$ well-ordered? Why or why not?

(e) Prove or disprove: If a set S is well-ordered, then S contains a least element.

(f) Prove or disprove: If a set S contains a least element, then S is well-ordered.

(7) Re-read the proof of the Division Algorithm, identifying each instance in which the proof relied on an axiom from Investigation 1. Specifically cite which axioms were used and where they were used.

(8) Prove or disprove: For every integer a, if $a \not\equiv 0 \pmod{3}$, then $a^2 \equiv 1 \pmod{3}$. (Hint: Consider two cases.)

(9) (a) Is the following theorem true or false?

For every integer n, if n is odd, then $8 \mid (n^2 - 1)$.

Give a proof or a counterexample to justify your answer.

(b) Translate the statement from part (a) into a corresponding statement dealing with congruence modulo 8.

(10) Prove or disprove: Let a, $b \in \mathbb{Z}$. If 3 divides $(a^2 + b^2)$, then 3 divides a and 3 divides b.

Investigation 3

Greatest Common Divisors

Focus Questions

By the end of this investigation, you should be able to give precise and thorough answers to the questions listed below. You may want to keep these questions in mind to focus your thoughts as you complete the investigation.

- What is the greatest common divisor of two integers, and how can "greatest" be defined using division?
- What is the Euclidean Algorithm, and how is it related to the Division Algorithm?
- What is the relationship between greatest common divisors and linear combinations? Why does this relationship hold?

Preview Activity 3.1. To begin our studies of greatest common divisors, let's attempt to calculate the greatest common divisor of several different pairs of integers. For the purposes of this task, we will define the greatest common divisor of two integers a and b, denoted $\gcd(a, b)$, to be the largest integer that divides both a and b. Later on, we will make this definition more precise by describing in detail exactly what we mean by the word *largest*. For now, however, use your intuitive understanding of the definition to find the greatest common divisor of each of the following pairs of integers. You may use any technique that is mathematically correct.

(a) 60 and 95

(b) 540 and 765

(c) 462 and 1080

(d) 765 and -540

(e) 1080 and 0

Preview Activity 3.2. As you may have noticed in Preview Activity 3.1, calculating greatest common divisors can be tedious and time-consuming. The steps outlined below suggest one way that the Division Algorithm can be used to make the process of finding greatest common divisors more systematic and efficient.

(a) Find the quotient and remainder that results from dividing 765 by 540.

(b) Let r denote the remainder you found in part (a). Find $\gcd(540, r)$.

(c) How does your answer to part (b) compare to your answer to part (a) of Preview Activity 3.1?

(d) If the relationship you observed in part (c) were to hold in general, how could it be used to more efficiently calculate greatest common divisors?

Introduction

Greatest common divisors are a standard part of most elementary school mathematics curricula, but their applications go far beyond this. In fact, most secure communications over the internet use an encryption method that relies heavily on the theory behind greatest common divisors. (See Investigation 34 for more details.) In this investigation, we will explore both the theoretical and practical aspects of greatest common divisors. These explorations will eventually lead us to the *Euclidean Algorithm*, which will allow us to calculate greatest common divisors easily and efficiently.

Let's begin with a definition:

Definition 3.3. Let a and b be integers, not both zero. A **common divisor** of a and b is any integer that divides both a and b. The largest integer that divides both a and b is called the **greatest common divisor** of a and b, denoted $\gcd(a, b)$.

Note that if a, b, and d are integers, and a and b are not both zero, then $d = \gcd(a, b)$ if and only if both of the following properties hold:

- $d \mid a$ and $d \mid b$. That is, d is a common divisor of a and b.
- If k is any other integer such that $k \mid a$ and $k \mid b$, then $k \leq d$. In other words, any other common divisor of a and b must be less than or equal to d.

These two conditions clarify exactly what is meant by the terms *common divisor* and *greatest*. Note if either of these conditions are violated, then $d \neq \gcd(a, b)$. In particular, $d \neq \gcd(a, b)$ if

- $d \nmid a$ or $d \nmid b$; or
- there exists some integer $k > d$ such that $k \mid a$ and $k \mid b$.

While Definition 3.3, and its negation, may seem quite natural, the inequalities involved in them can be somewhat troublesome, especially when it comes to writing proofs. Later on in this investigation, we will see how these inequalities can be replaced with conditions that involve divisibility instead. This will allow us to rephrase Definition 3.3 in a way that is both easier to work with in proofs and easier to generalize to other number systems.

It's also important to note that the wording of Definition 3.3 suggests not only that $\gcd(a, b)$ exists whenever a and b are not both zero, but also that $\gcd(a, b)$ is unique—that is, there is exactly one largest integer that divides both a and b. If we wanted to proceed in a more formal way, we could prove that for each pair of integers a and b (not both zero), there is exactly one integer d that satisfies the two properties listed immediately after Definition 3.3. The uniqueness of greatest common divisors also follows from a theorem (Theorem 3.10) that we will prove later in the investigation.

Calculating Greatest Common Divisors

In Preview Activity 3.1, we were asked to calculate the greatest common divisors of several pairs of numbers. For numbers that are small enough, we can do so by simply listing the positive factors of each number and picking the largest one that appears in both lists. For instance, to find $\gcd(28, 42)$, we might first note that the positive factors of 28 are

$$1, 2, 4, 7, 14, \text{and } 28,$$

and the positive factors of 42 are

$$1, 2, 3, 6, 7, 14, 21, \text{and } 42.$$

Thus, the largest common factor is 14, and $\gcd(28, 42) = 14$. This method works just fine for small numbers, but for larger numbers, it can be extremely time-consuming, if not practically impossible.

Another method that we might try involves factoring each number into primes and then multiplying the prime factors that are common to both numbers. Figure 3.1 illustrates how *factor trees* can be used in conjunction with this method to find $\gcd(396, 780)$. Notice that $396 = 2^2 \cdot 3^2 \cdot 11$ and $780 = 2^2 \cdot 3 \cdot 5 \cdot 13$. Since the only prime factors common to both numbers are 2 (squared in each case) and 3, it follows that $\gcd(396, 780) = 2^2 \cdot 3 = 12$.

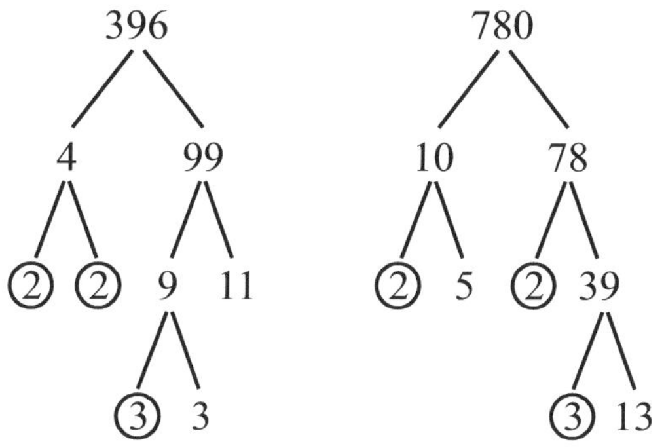

Figure 3.1
Factor trees for 396 and 780.

Of course, applying this process to larger numbers would again be considerably more difficult. For instance, if we tried to use the factor tree method to find $\gcd(17947, 17161)$, we would likely get stuck on the first step, or we would at least spend a very long time trying to find the prime factors of each number. What we need is a more efficient method for calculating greatest common divisors, and that is exactly where our work in Preview Activity 3.2 comes into play.

In Preview Activity 3.2, we began by dividing 765 by 540. If we were to revisit our previous example (of 780 and 396) in the same manner, we would obtain a quotient of 1 and a remainder of 384. In other words,

$$780 = 396 \cdot 1 + 384.$$

Of course, we know that $\gcd(780, 396) = 12$. But since $384 = 2^7 \cdot 3$ (and, as we saw before, $396 = 2^2 \cdot 3^2 \cdot 11$), it follows that $\gcd(396, 384) = 12$ also.

Is this a coincidence? Actually, it is not, and the following theorem states the general result illustrated by our example.

Theorem 3.4. *Let a and b be integers, not both zero, and suppose that*

$$b = aq + r$$

for some integers q and r. Then $\gcd(b, a) = \gcd(a, r)$.

To prove Theorem 3.4, let's begin by choosing a convenient abbreviation for $\gcd(b, a)$; we'll say that $\gcd(b, a) = d$. What we need to show is that $\gcd(a, r) = d$ as well. To do so involves two steps: we need to show (1) that d divides both a and r, and (2) that d is the largest such common divisor. The former is fairly straightforward and was proved in Activity 2.13. (See page 20.) For the latter, it is helpful to consider what would happen if d was not the greatest common divisor of a and r.

Activity 3.5. Let a, b, q and r be as in Theorem 3.4. Recall that we let $d = \gcd(b, a)$.

(a) Suppose that for some integer $k > d$, $k \mid a$ and $k \mid r$. Show that $k \mid b$ also. Deduce that k is a common divisor of b and a.

(b) Explain how part (a) contradicts the assumption that $d = \gcd(b, a)$.

The Euclidean Algorithm

Armed with Theorem 3.4, let's now return to the problem we mentioned earlier of finding $\gcd(17947, 17161)$. It is easy to verify that

$$17947 = 17161 \cdot 1 + 786.$$

But then Theorem 3.4 tells us that $\gcd(17947, 17161) = \gcd(17161, 786)$. This fact effectively reduces our hard problem into one that is a bit easier, or at least one that involves smaller numbers. It may still be difficult to find $\gcd(17161, 786)$, but we can apply the Division Algorithm again to obtain

$$17161 = 786 \cdot 21 + 655.$$

Doing so simplifies our problem once more, reducing it to that of finding $\gcd(786, 655)$. Continuing a few more steps, we find that

$$\gcd(786, 655) = \gcd(655, 131) = \gcd(131, 0) = 131.$$

Putting all of these steps together, we obtain the following:

$$\begin{aligned}
\gcd(17947, 17161) &= \gcd(17161, 786) && \text{since} \quad 17947 = 17161 \cdot 1 + 786\\
&= \gcd(786, 655) && \text{since} \quad 17161 = 786 \cdot 21 + 655\\
&= \gcd(655, 131) && \text{since} \quad 786 = 655 \cdot 1 + 131\\
&= \gcd(131, 0) && \text{since} \quad 655 = 131 \cdot 5 + 0\\
&= 131.
\end{aligned}$$

The process outlined above is known as the *Euclidean Algorithm*; it works by repeatedly applying the Division Algorithm, with each application reducing the original problem into one that involves smaller numbers. This description leads to two natural questions. First, is the Euclidean Algorithm guaranteed to terminate—in other words, does it always eventually produce an answer? Second, how can we know how many iterations of the algorithm must be carried out in order to obtain the answer? Is there a rule that tells us when to stop or lets us know that we have obtained the solution we seek?

To answer the first question, notice that each iteration of the Euclidean Algorithm produces a quotient and a remainder. For instance, to find $\gcd(b, a)$, we begin by dividing b by a to obtain two integers q_0 and r_0 such that

$$b = aq_0 + r_0 \text{ and } 0 \leq r_0 < a.$$

In the next step, however, we divide a by r_0 to obtain a new quotient and remainder, say q_1 and r_1. The Division Algorithm allows us to choose q_1 and r_1 so that $0 \leq r_1 < r_0$. Furthermore, as we continue to apply the Division Algorithm, we obtain a sequence $r_0, r_1, r_2, \ldots$ of remainders for which

$$0 \leq \cdots < r_2 < r_1 < r_0 < a.$$

Note that each of these remainders is a nonnegative integer, and each one is strictly less than the remainder obtained in the previous iteration of the algorithm. The only way for this to happen is if $r_n = 0$ for some n. Such a result forces the algorithm to terminate, since any subsequent iterations would involve dividing by zero. Thus, Theorem 3.4 implies that

$$\gcd(b, a) = \gcd(a, r_0) = \gcd(r_0, r_1) = \cdots = \gcd(r_{n-1}, 0) = r_{n-1},$$

where r_{n-1} is the last nonzero remainder obtained.

The next activity addresses several issues related to the Euclidean Algorithm and its application. The following section then explores the relationship between greatest common divisors and linear combinations.

Activity 3.6.

(a) Explain why $\gcd(r_{n-1}, 0) = r_{n-1}$ for all $r_{n-1} > 0$.

(b) Is $\gcd(x, 0) = x$ for every integer x? Why or why not? Can you make a general statement about the value of $\gcd(x, 0)$?

(c) Let a and b be any integers, not both zero. What do you think is the relationship between $\gcd(a, b)$, $\gcd(-a, b)$, $\gcd(a, -b)$, and $\gcd(-a, -b)$?

(d) Our development of the Euclidean Algorithm relied on the assumption that the initial divisor (we called it a) was positive. Use part (c) to explain how the Euclidean Algorithm can be used to find greatest common divisors when negative integers are used.

GCDs and Linear Combinations

Preview Activity 3.7. In this section, we will use what we have learned about the Euclidean Algorithm to discover a useful result that relates greatest common divisors and linear combinations. To begin, consider the following definition:

Definition 3.8. Let a and b be integers. A **linear combination** of a and b is an integer that can be written as $ax + by$ for some integers x and y.

Using this definition, find at least 10 different linear combinations of 60 and 95. How are your results related to the value of $\gcd(60, 95)$ that you found in Preview Activity 3.1?

Preview Activity 3.7 suggests that there is in fact a relationship between greatest common divisors and linear combinations. To explore this relationship in more detail, let's revisit the process we used to find $\gcd(17947, 17161)$ in the previous section.

Recall that we found $\gcd(17947, 17161) = 131$, and in the second to last step of the Euclidean Algorithm, we were able to write

$$786 = 655 \cdot 1 + 131.$$

Rearranging this equation, we obtain

$$131 = 786 \cdot 1 + 655 \cdot (-1). \tag{3.1}$$

In other words, we can write 131 as a linear combination of 786 and 655.

Notice also that in the previous step of the Euclidean Algorithm, we found that

$$17161 = 786 \cdot 21 + 655.$$

Rearranging this equation, we obtain

$$655 = 17161 \cdot 1 + 786 \cdot (-21). \tag{3.2}$$

Combining equations (3.1) and (3.2) yields

$$\begin{aligned} 131 &= 786 \cdot 1 + [17161 \cdot 1 + 786 \cdot (-21)] \cdot (-1) \\ &= 17161 \cdot (-1) + 786 \cdot 22 \end{aligned}$$

Continuing in this fashion, we obtain

$$\begin{aligned} 131 &= 17161 \cdot (-1) + [17947 \cdot 1 + 17161 \cdot (-1)] \cdot 22 \\ &= 17947 \cdot 22 + 17161 \cdot (-23) \end{aligned}$$

Thus, by solving for the remainders obtained in each step of the Euclidean Algorithm, we were able to find a way to write $\gcd(17947, 17161)$ as a linear combination of 17947 and 17161. This process can be applied whenever the Euclidean Algorithm is used to find the greatest common divisor of two integers (not both zero). Thus, we have the following theorem, which is sometimes called Bezout's Identity:

Theorem 3.9 (Bezout's Identity). *Let a and b be integers, not both zero. Then $\gcd(a, b)$ can be written as a linear combination of a and b. That is, there exist integers x and y such that*

$$\gcd(a, b) = ax + by.$$

We will consider a formal proof of Theorem 3.9 in the next section. Before we do so, however, it is useful to note two important observations that suggest an even stronger version of the theorem. The first is that $\gcd(a, b)$ not only can be written as a linear combination of a and b, but also *divides* every linear combination of a and b. This result is easy to show using the definition of the divides relation, and you were asked to do so in Activity 2.14. (See page 20.)

The second observation is that the greatest common divisor of two integers is always positive. This is true because 1 divides every integer, and so the greatest common divisor of any two integers must be greater than or equal to 1.

To summarize, we have argued the following:

- $\gcd(a, b)$ is a linear combination of a and b.
- $\gcd(a, b)$ divides every linear combination of a and b.
- $\gcd(a, b)$ is positive.

These three facts together imply that $\gcd(a, b)$ is not only a positive linear combination of a and b, but in fact the *smallest* positive linear combination of a and b. Theorem 3.10 formalizes this result, and Corollary 3.11 states a useful consequence.

Theorem 3.10. *Let a and b be integers, not both zero. Then $\gcd(a, b)$ is equal to the smallest positive linear combination of a and b.*

Corollary 3.11. *Let a and b be integers, not both zero. Then $\gcd(a, b) = 1$ if and only if there exist integers x and y such that*

$$ax + by = 1.$$

Activity 3.12. Explain why Corollary 3.11 is true. (Hint: The forward implication follows immediately from Theorem 3.9. The reverse implication requires the stronger Theorem 3.10.)

Corollary 3.11 provides a mechanism for showing that two integers a and b share no common positive divisors other than 1. Such pairs of integers are said to be *relatively prime*, defined formally below.

Definition 3.13. Let a and b be integers, not both zero. Then a and b are said to be **relatively prime** if and only if $\gcd(a, b) = 1$.

The following theorem illustrates how Corollary 3.11 can be used to establish results involving relative primality. Please note that in this theorem, we use fractional notation to represent the unique quotient that results from dividing two integers. We would be wise to again heed the warning from page 13: this notation only makes sense when the denominator divides the numerator with a unique quotient (as is the case below).

Theorem 3.14. *Let a and b be integers, not both zero, and let $d = \gcd(a, b)$. Then $\frac{a}{d}$ and $\frac{b}{d}$ are relatively prime integers.*

Proof. Since $d = \gcd(a, b)$, it follows that both $\frac{a}{d}$ and $\frac{b}{d}$ are integers.ⓘ Furthermore, there exist integers x and y such that

$$d = ax + by.ⓘ$$

From this it follows that

$$1 = \frac{a}{d} \cdot x + \frac{b}{d} \cdot y,ⓘ$$

which implies that $\gcd\left(\frac{a}{d}, \frac{b}{d}\right) = 1$, ⓘ as desired. ■

Well-Ordering, GCDs, and Linear Combinations

In the previous section, we argued that the greatest common divisor of any two integers is the smallest positive linear combination of those two integers. Our argument was fairly intuitive; it relied on solving the equations produced by the Euclidean Algorithm and performing a sequence of substitutions in order to explicitly construct the desired linear combination.

In this section, we will use the Well-Ordering Principle to present a formal proof of Theorem 3.10. Throughout this proof, we have again used the (?) symbol to indicate places where you may want to pause and provide for yourself some additional information or explanation.

Proof of Theorem 3.10. Let a and b be integers, not both zero, and define the set L as follows:

$$L = \{ax + by \in \mathbb{N} : x, y \in \mathbb{Z}\}.$$

Since $a^2 + b^2 \in L$,(?) it follows that L is nonempty. Furthermore, since L contains only positive integers, the Well-Ordering Principle implies that L has a least element, say d. We will show that $d = \gcd(a, b)$.

First we will show that $d \mid a$ and $d \mid b$. Since $d \in L$, there exist integers m and n such that

$$d = am + bn.$$

Furthermore, the Division Algorithm implies that there exist integers q and r such that

$$a = dq + r \text{ and } 0 \leq r < d.$$

Thus,

$$\begin{aligned} r &= a - dq \\ &= a - (am + bn)q \\ &= a(1 - mq) + b(nq), \end{aligned}$$

which implies that either $r \in L$ or $r = 0$.(?) Since $r < d$, it cannot be the case that $r \in L$.(?) Thus, it must be that $r = 0$, which implies that $d \mid a$. A similar argument establishes that $d \mid b$ as well.

Now suppose that for some integer k, $k \mid a$ and $k \mid b$. Then there exist integers h and j such that

$$\begin{aligned} d &= am + bn \\ &= (kh)m + (kj)n \text{ (?)} \\ &= k(hm + jn), \end{aligned}$$

and so $k \mid d$. But since $d > 0$,(?) this implies that $k \leq d$.(?) Therefore, $d = \gcd(a, b)$.(?) ■

In the proof above, we showed that d was the greatest common divisor of a and b by first showing that d was a common divisor of a and b, and then arguing that any other common divisor k would also have to be a divisor of d. This allowed us to conclude that d was the largest of all of the common divisors of a and b. The next theorem formalizes this reasoning by stating an equivalent, and in some ways preferable, form of the definition of greatest common divisor. We alluded to this alternate form earlier in the investigation when we noted that the inequalities in the original Definition 3.3 could be both difficult to work with and difficult to generalize to number systems that do not have a natural ordering relation like the integers do. Theorem 3.15 avoids inequalities altogether, producing a definition that is both easier to use and more robust. Activity 3.16 then suggests a strategy for proving Theorem 3.15.

Theorem 3.15. *Let a, b, and d be integers, with a and b not both zero. Then $d = \gcd(a, b)$ if and only if all of the following conditions hold:*

(i) *$d \mid a$ and $d \mid b$.*

(ii) *If k is an integer such that $k \mid a$ and $k \mid b$, then $k \mid d$ also.*

(iii) *d is positive.*

Activity 3.16. Let a, b, and d be integers, with a and b not both zero.

(a) Suppose $d = \gcd(a, b)$. Explain why conditions (i) and (iii) from Theorem 3.15 are automatically satisfied. Then use Bezout's Identity to prove condition (ii).

(b) Now suppose d is an integer that satisfies all three of the conditions from Theorem 3.15. Explain why there cannot exist an integer $k > d$ such that $k \mid a$ and $k \mid b$.

Concluding Activities

Activity 3.17. Throughout this investigation, the definitions and theorems involving $\gcd(a, b)$ have all assumed that a and b are not both zero. Why is this assumption important? Is it possible to find $\gcd(0, 0)$? Why or why not?

Activity 3.18. Decide whether each of the following statements is true or false. For those that are true, explain why. For those that are false, give a counterexample and then change **one word or symbol** in the statement to make it true. For each statement, assume that a, b, and d are positive integers.

(a) If $ax + by = 1$ for some integers x and y, then $\gcd(a, b) = 1$.

(b) If $ax + by \neq 1$ for some integers x and y, then $\gcd(a, b) \neq 1$.

(c) If $ax + by = d$ for some integers x and y, then $\gcd(a, b) = d$.

Activity 3.19. Write a short summary that describes the important concepts, and the relationships between these concepts, that were introduced in this investigation. Explain how the material in this investigation is connected to the material in Investigations 1 and 2.

Exercises

(1) Let a be an integer. After looking at several examples, make a general conjecture about the value of $\gcd(a - 1, a + 1)$. Then prove your conjecture.

(2) Fill in the blank, and prove your answer: For every integer a,

$$\gcd(a, a+1) = \_\_\_\_.$$

(3) For each of the following values of a and b, use the Euclidean algorithm to determine $\gcd(a,b)$. Then find integers x and y such that $ax + by = \gcd(a,b)$.

(a) $a = 525$, $b = 252$

(b) $a = 54321$, $b = 12345$

(c) $a = 27182$, $b = -3141$

(d) $a = -61880$, $b = -60678$

(e) $a = 12906$, $b = 42905$

(4) Determine all values of n for which the following statement is true:

There exist integers x and y such that $63x + 147y = n$.

Give a convincing argument to justify your answer.

(5) (a) Prove or disprove: For all nonzero integers a, b, and c, $\gcd(a, bc) = 1$ if and only if $\gcd(a,b) = 1$ and $\gcd(a,c) = 1$.

(b) Now take this a step farther. Let n be a positive integer and let $a, b_1, b_2, \ldots, b_n$ be nonzero integers. Prove or disprove: $\gcd(a, b_1 b_2 \cdots b_n) = 1$ if and only if $\gcd(a, b_i) = 1$ for all $1 \le i \le n$.

(6) Let a and b be integers, not both zero. Prove that if $\gcd(a,b) = 1$, then $\gcd(a^2, b^2) = 1$. Is the converse true? Verify your answer.

(7) Let a be any integer. What is $\gcd(a, a+2)$? Prove your answer.

(8) Let a and x be integers, with $x > 0$. Prove that $\gcd(a, a+x) = \gcd(a, x)$.

(9) Let a and b be integers, not both zero. Prove that if $\gcd(a,b) = 1$, then $\gcd(a+b, ab) = 1$. Is the converse true? Verify your answer.

(10) Are the integers guaranteed by Theorem 3.9 unique? Explain.

Investigation 4

Prime Factorization

Focus Questions

By the end of this investigation, you should be able to give precise and thorough answers to the questions listed below. You may want to keep these questions in mind to focus your thoughts as you complete the investigation.

- What does it mean for a natural number to be prime?
- What does the Fundamental Theorem of Arithmetic say about prime factorizations of integers, and how can it be proved?
- What are some ways of generalizing the definition of prime to apply to number systems other than the integers? What are some ways that prime factorizations can behave differently in other number systems?

Preview Activity 4.1. In Investigation 3, we used our knowledge of division in the integers to prove several important results about greatest common divisors. We are now ready to use these results to investigate another important topic: prime factorization. As a first step, let's begin to think about what it means for a number to be prime. The questions below will get us started.

(a) Based on your past knowledge and experience, what do you think it means for an integer p to be prime? Write a precise definition.

(b) Using your answer to part (a), write down the first 10 prime numbers.

(c) Should 1 be considered prime? Why or why not?

(d) Find a way to write the integer 420 as a product of primes, or explain why it is impossible to do so.

(e) Find *another* way (different from your answer to part (d)) to write the integer 420 as a product of primes, or explain why it is impossible to do so.

Introduction

In many ways, prime numbers are the building blocks of the integers. In elementary school, students learn that every natural number can be written as a product of primes. In fact, they often become very proficient at making factor trees like the one we used in Investigation 3. (See Figure 3.1 on page 25.) But why is it always possible to break a number down into its prime factors, and is there ever more than one way to complete such a factorization? On a more basic level, which numbers should and shouldn't be considered prime? Does our definition of prime depend on properties that are specific to the integers, or could it be generalized to apply to other number systems?

We will explore all of these questions in this investigation, and our work will allow us to understand and prove the *Fundamental Theorem of Arithmetic*, which characterizes prime factorizations in the integers. We will also see an example of a number system for which one of the conclusions of the Fundamental Theorem of Arithmetic does not hold.

Defining Prime

It is fairly common (for instance, in elementary school) to define a prime number to be a number whose only factors are 1 and itself. You might have given a definition similar to this in Preview Activity 4.1. This definition, however, raises several questions, such as:

- What do we mean by *number*? A real number? An integer? A positive integer?
- What do we mean by *factors*? Are we considering only positive factors, or are negative factors included as well? (If the latter is the case, then numbers like 3 would not be considered prime, since -3, -1, 1, and 3 are all factors of 3.)
- When we say "1 and itself," are we specifying that a prime number must have exactly two factors? Could a prime number have only one factor, if that single factor was 1? (Note that the answer to this question affects whether we consider 1 to be prime or not.)

To answer these questions, we will adopt a slightly more precise definition:

Definition 4.2. A **prime number** is an integer $p > 1$ whose only positive divisors are 1 and p. A positive integer that is greater than 1 and not prime is said to be **composite**.

Definition 4.2 answers all three of the questions raised by its less formal predecessor. In particular, it rules out the possibility of 1 being considered prime. This exclusion has a number of important theoretical implications, but it also makes sense intuitively. Although it may not be immediately apparent, 1 is different in several regards from the other numbers that meet our definition of prime. For instance, 1 has exactly *one* positive divisor (itself), whereas every other prime has exactly *two* positive divisors (1 and itself). In addition, $1 \cdot 1 = 1$, a property that none of the other prime numbers satisfy.*

The Fundamental Theorem of Arithmetic

Having defined what a prime number is, we will now turn our attention to the problem of prime factorization. In Preview Activity 4.1, we considered the problem of writing 420 as a product of

*This property implies that 1 has a *multiplicative inverse* (itself) in the integers, which cannot be said of any of the other primes. We will study multiplicative inverses more extensively in later investigations.

primes. Using either a factor tree or simple trial and error, it is easy to see that

$$420 = 2^2 \cdot 3 \cdot 5 \cdot 7. \tag{4.1}$$

Of course, we could also write

$$420 = 7 \cdot 5 \cdot 2 \cdot 3 \cdot 2,$$

or

$$420 = 3 \cdot 7 \cdot 2^2 \cdot 5,$$

or a number of other options that involve rearranging the factors in equation (4.1). Since multiplication of integers is commutative, all of these different factorizations are effectively equivalent. But are there factorizations of 420 that involve prime factors other than 2, 3, 5, and 7? In other words, setting aside the order of the factors, is the prime factorization we found for 420 unique? And can every number be written as a product of primes in the first place? The Fundamental Theorem of Arithmetic, stated formally below, answers these questions in the affirmative for all integers greater than 1.

The Fundamental Theorem of Arithmetic. *Every integer greater than* 1 *is either prime or a product of primes. Furthermore, this factorization is unique up to the order of the factors.*

Note that, like the Division Algorithm, the Fundamental Theorem of Arithmetic makes two separate assertions: one pertaining to the *existence* of prime factorizations, and one pertaining to their *uniqueness*. In the next two sections, we will consider techniques for proving each of these two aspects of the Fundamental Theorem.

Proving Existence

Preview Activity 4.3. In order to prove the existence portion of the Fundamental Theorem of Arithmetic, we will need to be able to precisely negate the definition of prime, and work with this negation in the context of an induction proof. The questions below foreshadow some of the ideas and techniques that we will use within our argument.

(a) What does it mean for a positive integer n to *not* be prime? Negate Definition 4.2 to give a precise answer.

(b) Is 6360 prime? Use part (a) to justify your answer.

(c) Find positive integers x and y such that $6360 = xy$.

(d) Are the integers you found in part (c) prime? Are they the only integers whose product is 6360?

(e) Do your answers to either of the questions in part (d) contradict the Fundamental Theorem of Arithmetic? Why or why not?

(f) Suppose that you were able to verify that every integer greater than 1 and less than 6360 was either prime or a product of primes. What would this fact, along with your answer to part (c), allow you to conclude about 6360?

Preview Activity 4.3 suggests a strategy for establishing the existence portion of the Fundamental Theorem of Arithmetic. This strategy relies on the strong form of the Principle of Mathematical Induction. In particular, if we let $P(n)$ denote the predicate

$$P(n): n \text{ is either prime or a product of primes,}$$

then the Fundamental Theorem of Arithmetic states simply that $P(n)$ is true for every integer $n > 1$. As with all induction proofs, our argument will involve both a *base case* and an *inductive step*, as described below:

- **Base case:** Show that $P(2)$ is true.
- **Inductive step:** Show that if, for some integer $n \geq 2$, $P(2)$, $P(3)$, $\ldots$, $P(n)$ are all true, then $P(n+1)$ is also true.

Our base case is simple: since 2 is a prime number, $P(2)$ is true. For the inductive step, our reasoning, as outlined in Activity 4.4, will mirror that of Preview Activity 4.3.

Activity 4.4. Suppose that, for some integer $n \geq 2$, $P(2)$, $P(3)$, $\ldots$, $P(n)$ are all true. That is, suppose that for every integer k with $2 \leq k \leq n$, k is either prime or a product of primes. (This assumption is called the *induction hypothesis*.)

(a) Suppose $n+1$ is prime. Explain why $P(n+1)$ is true in this case.

(b) Suppose $n+1$ is not prime. Explain why it is possible to find integers x and y such that $n+1 = xy$ and $2 \leq x, y \leq n$.

(c) What does the induction hypothesis allow you to conclude about the integers x and y found in part (b)?

(d) Explain how your answers to parts (b) and (c) prove that $n+1$ is a product of primes. Deduce that $P(n+1)$ is true.

Proving Uniqueness

In order to prove the uniqueness portion of the Fundamental Theorem of Arithmetic, we will first need to establish an important lemma related to part (d) of Preview Activity 2.1. (See page 11.) This lemma is attributed to Euclid; although his famous work, the *Elements*, is most commonly known for its contributions to geometry, it also contains many basic results in number theory. One of these results is the following:

Euclid's Lemma. *Let a and b be integers, and let p be prime. If $p \mid ab$, then $p \mid a$ or $p \mid b$.*

To prove Euclid's Lemma, we will first prove the following more general result:

Theorem 4.5. *Let a, b, and c be integers. If $c \mid ab$ and $\gcd(c, a) = 1$, then $c \mid b$.*

The next activity outlines one way to prove Theorem 4.5 and then suggests a strategy for using Theorem 4.5 to prove Euclid's Lemma.

Activity 4.6. Let a, b, and c be integers, and suppose that $c \mid ab$ and $\gcd(c, a) = 1$.

(a) Use Bezout's Identity to translate the assumption that $\gcd(c, a) = 1$ into an equation involving a linear combination.

(b) Multiply both sides of your equation from part (a) by an appropriate quantity in order to obtain an equation of the form
$$b = \underline{\quad} + \underline{\quad}.$$

(c) Explain why each of the terms on the right-hand side of the equation from part (b) are divisible by c. Deduce that $c \mid b$. (Note that this conclusion establishes Theorem 4.5.)

(d) Now suppose that a, b, and p are integers with p prime. Assume further that $p \mid ab$ and $p \nmid a$. Explain why $\gcd(p, a) = 1$.

(e) Explain how Theorem 4.5 and the result from part (d) establish Euclid's Lemma.

Euclid's Lemma can also be generalized as follows:

Euclid's Lemma (Strong Form). *Let $a_1, a_2, \ldots, a_n$ be integers, and let p be prime. If $p \mid a_1 a_2 \cdots a_n$, then $p \mid a_i$ for some i with $1 \le i \le n$.*

The proof of this generalization is left as an exercise. (See Exercise 1 at the end of this investigation.)

With Euclid's Lemma in hand, we are now ready to prove the uniqueness portion of the Fundamental Theorem of Arithmetic. As with the existence portion, we will use strong induction to complete the proof. However, before proceeding with this formal method, it may be helpful to consider the ideas behind it from a more intuitive vantage point.

Activity 4.7. Let $n > 1$ be an integer, and suppose that n has two prime factorizations. That is, suppose that for some primes $p_1, p_2, \ldots, p_j$ and $q_1, q_2, \ldots, q_k$,
$$p_1 p_2 \ldots p_j = n = q_1 q_2 \ldots q_k. \tag{4.2}$$

(a) Explain why Euclid's Lemma implies that $p_1 \mid q_i$ for some i. Deduce that $p_1 = q_i$ for some i.

(b) Explain how your answer to part (a) would allow you to cancel a factor from each side of equation (4.2). (Hint: Consider Theorem 1.11.)

Activity 4.7 suggests that, given two potentially different prime factorizations of a number n, Euclid's Lemma can be applied to find a factor that appears in both factorizations. Canceling this common factor will yield two factorizations of a new number (one that is smaller than n), and we could conceivably apply Euclid's Lemma again to find another common factor. If we continued this process, we would end up showing that each of the p_i's is equal to exactly one of the q_i's, and vice versa. It may not be that $p_1 = q_1, p_2 = q_2$, and so on, but that doesn't matter because all we want to show is that the factorization of n into primes is unique *up to the order of the factors*. This simply means that the *only* difference between the two factorizations is possibly the order in which the factors are listed. Because order doesn't matter, in our formal proof of uniqueness we will allow ourselves to re-order and/or re-number the factors in equation (4.2) before we cancel the first factor. Doing so will allow us to assume, *without loss of generality*,† that p_1 and q_1 are in fact equal, which will simplify our argument. We will then use induction to formalize the intuitive idea of repeatedly applying Euclid's Lemma.

†The phrase *without loss of generality* can be very helpful, but it must be used with extreme caution, and only when we are certain that the simplifying assumption does not in any way affect the generality of our argument.

Putting It All Together

We will now use what we have learned to write a complete proof of the Fundamental Theorem of Arithmetic. Throughout the proof, we have again used the (?) symbol to denote locations where additional information or expanded explanations may be helpful. We have also left several blanks for you to fill in as you read the proof. You should be able to supply the omitted details based on your understanding of the ideas in the previous two sections.

Proof of the Fundamental Theorem of Arithmetic. We must show both existence and uniqueness; for each, we will proceed by induction. To establish existence, first note that since __________, the base case is trivial. Now assume that, for some $n \geq 2$, every integer between 2 and n, inclusive, is either _______ or _________________. If $n + 1$ is _______, then the result holds trivially. Thus, assume that $n + 1$ is not prime. Then there exist integers x and y such that $n + 1 = xy$ and ___ $\leq x, y \leq$ ___.(?) By the induction hypothesis, both x and y are either _______ or _________________. Thus, $n + 1 = xy$ is a product of primes,(?) which completes the induction step.

For uniqueness, first note that 2 is prime and therefore cannot be factored in any non-trivial way. Thus, 2 (like any prime) has a unique—and trivial—prime factorization.(?) Now assume that, for some $n \geq 2$, every integer between 2 and n, inclusive, has a factorization into primes that is unique up to the order of the factors. Suppose also that for some primes $p_1, p_2, \ldots p_j$, and $q_1, q_2, \ldots, q_k$,

$$\underline{\hspace{3cm}} = n + 1 = \underline{\hspace{3cm}}.$$

By Euclid's Lemma, $p_1 \mid q_i$ for some i with $1 \leq i \leq k$.(?) Without loss of generality, assume that $p_1 \mid q_1$.(?) Then $p_1 = q_1$,(?) and so

$$p_2 p_3 \cdots p_j = q_2 q_3 \cdots q_k \leq n.\text{(?)} \tag{4.3}$$

The induction hypothesis then implies that $j = k$,(?) and the factors on each side of equation (4.3) can be re-ordered and/or re-numbered so that $p_i = q_i$ for all i with $2 \leq i \leq j = k$.(?) Thus, the factorization of $n + 1$ into primes is unique up to the order of the factors, as desired. ■

Primes and Irreducibles in Other Number Systems

Now that we have proved the Fundamental Theorem of Arithmetic, it is natural to consider whether a similar result would hold in other number systems. But what other number systems are there? You are probably quite familiar with the rational numbers ($\mathbb{Q}$), the real numbers ($\mathbb{R}$), and perhaps even the complex numbers ($\mathbb{C}$). In spite of this familiarity, you may have never considered what it would mean for a number to be prime, or whether the notion of primality even makes sense, within these other systems. And since there are other number systems in addition to these familiar ones, many of which we have not yet investigated, it makes sense for us to consider whether our ideas about prime numbers can be generalized to broader contexts.

To begin, let's consider a number system that is very closely related to the integers: the even integers, which we will denote $\mathbb{E}$. Note that $\mathbb{E}$ satisfies many of the same axioms and properties that we have attributed to $\mathbb{Z}$ in past investigations. There are, however, several notable exceptions. For instance, the even integers do not contain 1, or any multiplicative identity for that matter. This important distinction is implied by the fact that no nonzero element of $\mathbb{E}$ can be a divisor of itself. In other words, if n is a nonzero even integer, then it is impossible to write $n = nm$ for some other even integer m.

The fact that 1 is not an element of the even integers also renders our previous definition of prime (Definition 4.2) inapplicable to $\mathbb{E}$. Fortunately, there are a number of other reasonable alternatives. One possibility would be to say that, as in $\mathbb{Z}$, a positive element n of $\mathbb{E}$ is prime provided that n has exactly two positive divisors in $\mathbb{E}$. By this definition, 8 would be prime (since 2 and 4 are the only divisors of 8 in $\mathbb{E}$), but 2, 4, and 6 would not be (since 2 and 6 have *no* divisors[‡] in $\mathbb{E}$, and 4 has only one divisor, 2, in $\mathbb{E}$). Does this seem consistent with the way prime numbers work in $\mathbb{Z}$? Probably not, and the difference again has to do with the fact that $\mathbb{E}$ does not contain a multiplicative identity. Thus, it is possible for an element $n \in \mathbb{E}$ to have exactly two positive divisors in $\mathbb{E}$ without those divisors simply being 1 and n (as would be the case in $\mathbb{Z}$).

To remedy this inconsistency, we might say that an element of $\mathbb{E}$ is prime if it has *no* divisors at all in $\mathbb{E}$. In this case, 2 and 6 would be considered prime, but 8 would not be. This definition seems to capture the notion of *irreducibility* that we intuitively associate with prime numbers in $\mathbb{Z}$—that is, prime numbers are those that can't be written as a product of other numbers, except perhaps in some trivial way such as $3 = 3 \cdot 1$. Note that such trivial factorizations are impossible in $\mathbb{E}$, since, as mentioned earlier, no element of $\mathbb{E}$ is a divisor of itself.[§]

All of these observations lead us to the following definition:

Definition 4.8. A **prime number in** $\mathbb{E}$ (the even integers) is a positive even integer p that cannot be written as a product of two other even integers. That is, $p \in \mathbb{E}$ is prime provided that there do not exist even integers x and y such that $p = xy$.

Having defined a notion of primality for $\mathbb{E}$, we are now ready to explore an example that illustrates a fundamental difference between prime factorizations in $\mathbb{Z}$ and $\mathbb{E}$. To see this difference, consider Activity 4.9 below. Although the ideas explored therein are very similar to those in Preview Activity 4.1, the conclusions that can be drawn are much different.

Activity 4.9.

(a) Using the definition of prime discussed in the preceding paragraphs, list the first 10 prime numbers in $\mathbb{E}$.

(b) Find a way to write 60 as a product of primes in $\mathbb{E}$, or explain why it is impossible to do so.

(c) Find another way (different from your answer to part (b)) to write 60 as a product of primes in $\mathbb{E}$, or explain why it is impossible to do so.

Activity 4.9 suggests that while every element of $\mathbb{E}$ is either prime or a product of primes, these prime factorizations may not be unique. Since $\mathbb{E}$ and $\mathbb{Z}$ are similar in so many regards, this observation should be somewhat surprising. At the very least, it gives us reason to think twice before making assumptions about the way prime factorization works in the number systems that we will consider in future investigations.

[‡]Note that, for example, there do not exist even integers x and y such that $6 = xy$.

[§]The notion of irreducibility can actually be defined in a way that encompasses both our original definition of prime in $\mathbb{Z}$, and the modified version for $\mathbb{E}$ that we discussed in the previous paragraph. We will study this more general notion of irreducibility in Investigation 17.

Concluding Activities

Activity 4.10. Explain how allowing 1 to be considered prime would affect the conclusions of the Fundamental Theorem of Arithmetic. Give specific examples to support your answer.

Activity 4.11. Would our original definition of prime (Definition 4.2) need to be modified at all to apply to $\mathbb{Q}$, $\mathbb{R}$, or $\mathbb{C}$? Why or why not? Which numbers should be considered prime in these number systems, and why?

Activity 4.12. Write a short summary that describes the important concepts, and the relationships between these concepts, that were introduced in this investigation. Explain how the material in this investigation is connected to the material in Investigation 3.

Exercises

$\star$ (1) Use induction to prove the strong form of Euclid's Lemma.

(2) Recall that an irrational number is one that cannot be written as a ratio $\frac{a}{b}$, where a and b are integers and $b \neq 0$. Use Euclid's Lemma to prove that for all positive integers n and q, if $\sqrt[q]{n}$ is not an integer, then $\sqrt[q]{n}$ is irrational.

(3) Prove or disprove: For all integers a and b, $a \mid b$ if and only if $a^2 \mid b^2$. (Hint: Exercise 2 may be helpful.)

(4) Let p_i denote the ith prime integer (so that $p_1 = 2$, $p_2 = 3$, $p_3 = 5$, and so on). Prove or disprove: For all $n \in \mathbb{Z}^+$, $p_1p_2p_3 \cdots p_n + 1$ is prime.

(5) **The Infinitude of the Primes.** Prove that there are infinitely many prime integers. (Hint: Suppose there are only a finite number of primes, say $p_1, p_2, \ldots p_n$. Show that $p_1p_2p_3 \cdots p_n + 1$ can be neither prime nor a product of primes.)

(6) (a) Let $n \in \mathbb{N}$. Prove that for each integer k with $2 \leq k \leq n+1$, k divides $[(n+1)! + k]$.

(b) Deduce from part (a) that for each positive integer n, there exist at least n consecutive composite numbers.

(7) Prove or disprove: If n is an odd integer and 3 does not divide n, then $24 \mid (n^2 - 1)$.

(8) Let $y \in \mathbb{N}$. Use the Fundamental Theorem of Arithmetic to prove that there exists an odd natural number x and a nonnegative integer k such that $y = 2^k x$.

(9) Find an element of $\mathbb{E}$ other than 60 that has more than one prime factorization.

(10) For any integer $n > 1$, define $n\mathbb{Z}$ to be the set of all integer multiples of n. That is,

$$n\mathbb{Z} = \{nx : x \in \mathbb{Z}\}.$$

Using the same notion of prime that we applied to $\mathbb{E}$, find (and prove) a necessary and sufficient condition for an integer p to be prime in $n\mathbb{Z}$.

(11) **Goldbach's Conjecture.** *Goldbach's Conjecture*, which was made by Christian Goldbach in a letter to Leonhard Euler in 1742, states the following:

> Every even integer greater than 2 can be expressed as the sum of two (not necessarily distinct) prime numbers.

As of this printing, it is not known whether Goldbach's Conjecture is true or false, although most mathematicians believe it to be true.

(a) Write each of 78, 90, and 138 as a sum of two primes.

(b) Is there an even integer that can be written as a sum of two primes in more than one way? If so, find the smallest such integer.

(c) Prove that Goldbach's Conjecture implies that every integer greater than 5 can be written as a sum of three primes.

(d) Prove that Goldbach's Conjecture implies that every odd integer greater than 7 can be written as a sum of three odd primes.

(12) **The Twin Prime Conjecture.** A quick look at the first dozen or so prime numbers reveals several cases in which consecutive prime numbers differ by 2 (for instance, 3 and 5, 11 and 13, etc.) Such pairs of primes are called *twin primes*, and the *Twin Prime Conjecture*, which as of this printing has been neither proved nor disproved, states that there are infinitely many twin primes. Answer each of the following questions related to the Twin Prime Conjecture:

(a) Find the first 10 pairs of twin primes.

(b) How many pairs of primes p and q satisfy $q - p = 3$? (Note that twin primes satisfy $q - p = 2$.)

(c) How many triplets of primes of the form p, $p + 2$, $p + 4$ are there? That is, how many triplets of primes are there where each prime is 2 more than the preceding prime? Prove your answer. (Hint: Set up cases using congruence modulo 3.)

Part II

Other Number Systems

Investigation 5

Equivalence Relations and $\mathbb{Z}_n$

Focus Questions

By the end of this investigation, you should be able to give precise and thorough answers to the questions listed below. You may want to keep these questions in mind to focus your thoughts as you complete the investigation.

- What is a congruence class, and what are some properties of congruence classes?
- What is an equivalence relation, and what are some strategies for proving the properties that characterize equivalence relations?
- What are equivalence classes, and how do the equivalence classes corresponding to an equivalence relation divide the underlying set into subsets?
- What is $\mathbb{Z}_n$, and what arithmetic axioms hold within $\mathbb{Z}_n$?
- What is a binary operation, and what does it mean for a binary operation to be well-defined?
- What are zero divisors and units in $\mathbb{Z}_n$? How are zero divisors and units related to solving linear equations?

Preview Activity 5.1. When working with large sets of objects, it is often useful to group these objects according to some common attribute or property. For instance, in a cooler containing 100 cans of soft drinks, there may be 30 cans of Coke, 30 cans of Pepsi, and 40 cans of 7 Up. If someone wanted to drink a can of Coke, they probably would not care exactly which can of Coke they pulled out of the cooler. In other words, they would probably consider all of the different cans of Coke to be indistinguishable, or equivalent. This same kind of grouping can be applied to a set of mathematical objects by defining an *equivalence relation*. In this preview activity, we will investigate how congruence can be used to define such a relation on the integers.

(a) For every integer a, let $[a]_3$ denote the set of all integers that are congruent to a modulo 3. Using the roster method,* list the elements in $[0]_3$.

(b) Repeat part (a) for $[1]_3$, $[2]_3$, $[3]_3$, $[4]_3$, and $[5]_3$. Do you notice anything about the relationships between these sets?

(c) What is the remainder when 734 is divided by 3? Which, if any, of the sets $[0]_3$, $[1]_3$, and $[2]_3$ contain 734?

*To specify a set using the roster method, we simply list the elements of the set between braces, as in Definition 1.2 on page 4.

(d) Repeat part (c), replacing 734 with another integer of your choosing.

(e) Which elements belong to $[0]_3 \cap [1]_3$? What about $[0]_3 \cap [2]_3$, or $[1]_3 \cap [2]_3$?

(f) What familiar set is $[0]_3 \cup [1]_3 \cup [2]_3$ equal to, and why?

(g) Based on your answers to parts (c) – (f), make as many conjectures as you can about the sets $[0]_3$, $[1]_3$, and $[2]_3$.

Congruence Classes

In Preview Activity 5.1, we investigated several sets of integers called *congruence classes*, which we can define formally as follows:

Definition 5.2. Let n be a natural number, and let a be an integer. The **congruence class of** a **modulo** n, denoted $[a]_n$, is the set of all integers congruent to a modulo n. In other words,

$$[a]_n = \{x \in \mathbb{Z} : x \equiv a \pmod{n}\}.$$

Based on our work in Preview Activity 5.1, we can make several observations and conjectures about congruence classes. In particular:

- For $0 \leq a \leq n-1$, $[a]_n$ contains all integers x for which x divided by n yields a remainder of a.
- It is possible for the congruence classes of two distinct integers to be equal to each other. That is, it is possible for $[a]_n = [b]_n$ even when $a \neq b$. This occurs exactly when $a \equiv b \pmod{n}$. In other words,

$$[a]_n = [b]_n \text{ if and only if } a \equiv b \pmod{n}.$$

- If two congruence classes are not equal, then they are disjoint. That is, $[a]_n \cap [b]_n = \emptyset$ for all integers a and b such that $[a]_n \neq [b]_n$.
- There are exactly n distinct congruence classes modulo n, which can be represented by $[0]_n$, $[1]_n$, $[2]_n$, $\ldots$, $[n-1]_n$. All of the other congruence classes modulo n (such as $[n+3]_n$, $[-1]_n$, etc.) are equal to one of these n classes.
- For every natural number n, the union of the n distinct congruence classes modulo n is equal to $\mathbb{Z}$. That is,

$$[0]_n \cup [1]_n \cup [2]_n \cup \cdots \cup [n-1]_n = \mathbb{Z}.$$

- Every integer x belongs to exactly one of the n distinct congruence classes modulo n. In other words, for every natural number n and every integer x, there exists a unique integer a with $0 \leq a \leq n-1$ such that $x \in [a]_n$.

As you may have noticed, some of these properties are related to each other or even implied by each other. Most of them, in fact, are consequences of the more general theory of equivalence relations that we will study in the next section. Before we move on, however, the last three observations on our list merit some special attention. Together, they imply that congruence modulo n divides

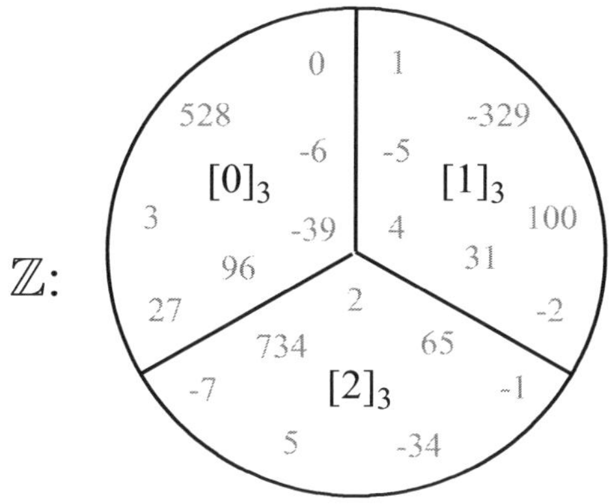

Figure 5.1
Dividing $\mathbb{Z}$ into classes based on congruence modulo 3.

the set of integers into n distinct congruence classes, and that each integer belongs to exactly one of these classes. In the context of our original example of congruence modulo 3, this allows us to picture the integers as shown in Figure 5.1.

This nice division of the integers into congruence classes is made possible by the fact that congruence satisfies three very important properties. These three properties characterize *equivalence relations*, which we will study in the next section.

Equivalence Relations

Throughout mathematics, *binary relations* (or just *relations* for short) are often used to specify certain associations or relationships between the elements of a set of mathematical objects. For instance, the equals relation ($=$), the less than or equal to relation ($\leq$), and congruence modulo n are all examples of relations on the set of integers. In geometry, similarity and congruence (of shapes—not to be confused with congruence modulo n) both define relations on the set of all triangles. In everyday life, we use the same kind of language; we talk about two people being related if they are, for lack of a better word, relatives—either by blood or by another familial relationship such as marriage.

Technically, a *binary relation* on a set S is just a set of ordered pairs, where both coordinates of each pair in the relation are elements of S. So, for example, all of the following pairs (along with many others) would belong to the congruence modulo 3 relation:

$$(0,3), (1,4), (2,-1), (3,99)$$

For our purposes, we will usually describe a relation not by listing the ordered pairs that belong to it, but rather by specifying some kind of rule, as we did when we defined congruence modulo n (i.e., $a \equiv b \pmod{n}$ provided that $n \mid (a-b)$). We will often use the $\sim$ symbol to denote a relation on a set. For instance, if we wanted to define two integers a and b to be related whenever b was the square of a, we might write something like this:

Let $\sim$ be the relation on $\mathbb{Z}$ defined by $a \sim b$ if and only if $a^2 = b$.

We could then use this notation to say that $3 \sim 9$ (since $3^2 = 9$), but $2 \not\sim 5$ (since $2^2 \neq 5$). If we had used a different symbol to define our relation, say $\propto$ instead of $\sim$, then we would simply adjust our notation accordingly, writing $3 \propto 9$ and $2 \not\propto 5$.

When we use a rule to define a relation, we are specifying exactly what it means for two elements to be related to each other. So, in the context of the preceding example, if we wanted to prove that $a \sim b$ for two integers a and b, we would need to show that $a^2 = b$. Likewise, if we were assuming that $a \sim b$, then this assumption would allow us to conclude that $a^2 = b$. Stated another way, the rules that define relations allow us to translate generic statements, such as $a \sim b$, into more specific statements, such as $a^2 = b$, and vice versa.

There are many different kinds of relations, but for now we will be mainly interested in the type of relations suggested by Preview Activity 5.1—that is, those that identify certain objects as being equivalent in some way. Recall that such relations are called *equivalence relations*. We will use equivalence relations throughout many of our later investigations. The formal definition of an equivalence relation is as follows:

Definition 5.3. Let S be a set, and let $\sim$ be a binary relation on S. Then $\sim$ is called an **equivalence relation** on S provided that $\sim$ satisfies all of the following properties:

- **Reflexivity:** For all $a \in S$, $a \sim a$.
- **Symmetry:** For all $a, b \in S$, if $a \sim b$, then $b \sim a$.
- **Transitivity:** For all $a, b, c \in S$, if $a \sim b$ and $b \sim c$, then $a \sim c$.

In other words, an equivalence relation is a binary relation that is reflexive, symmetric, and transitive.

If the properties from Definition 5.3 sound familiar, it's probably because we proved all of them for congruence modulo n back in Activity 2.12. (See page 19.) In other words, we proved that congruence modulo n is an equivalence relation. This result is very important, and we state it formally in Theorem 5.4 below. The proof that follows ties together the ideas and arguments from parts (d) – (f) of Activity 2.12. You should read this proof carefully, inserting additional details or explanations wherever you see the (?) symbol.

Theorem 5.4. *Let n be any natural number. Then congruence modulo n is an equivalence relation on $\mathbb{Z}$. In other words, the relation $\sim$ defined on $\mathbb{Z}$ by the rule*

$$a \sim b \text{ if and only if } a \equiv b \pmod{n}$$

is an equivalence relation.

Proof. We must show that the congruence modulo n relation $\sim$, as defined in the statement of the theorem, is reflexive, symmetric, and transitive.

For reflexivity, let $a \in \mathbb{Z}$. Since $a - a = 0$, it follows that $n \mid (a - a)$.(?) Thus, $a \equiv a \pmod{n}$, and congruence modulo n is reflexive.

For symmetry, suppose $a \equiv b \pmod{n}$ for some $a, b \in \mathbb{Z}$. Then $n \mid (a - b)$, which implies that $n \mid (b - a)$.(?) Thus, $b \equiv a \pmod{n}$, and congruence modulo n is symmetric.

For transitivity, suppose $a \equiv b \pmod{n}$ and $b \equiv c \pmod{n}$ for some $a, b, c \in \mathbb{Z}$. Then $n \mid (a - b)$ and $n \mid (b - c)$. Thus, $n \mid [(a - b) + (b - c)]$,(?) which implies that $n \mid (a - c)$. It follows that $a \equiv c \pmod{n}$, and so congruence modulo n is transitive.

By showing that the congruence modulo n relation is reflexive, symmetric, and transitive, we have established that congruence modulo n is an equivalence relation on $\mathbb{Z}$. ■

Equivalence Classes

A few pages ago, we showed how congruence modulo n naturally divides the integers into n congruence classes. Now we have also shown that congruence modulo n is an equivalence relation on $\mathbb{Z}$. As you might suspect, these two results are closely related. In fact, we will show that *every* equivalence relation divides the set on which it is defined into subsets called *equivalence classes*. The definition of an equivalence class, stated formally below, generalizes Definition 5.2.

Definition 5.5. Let $\sim$ be an equivalence relation on a nonempty set S, and let $a \in S$. The **equivalence class of** a (with respect to $\sim$), denoted $[a]_\sim$, is the set of all elements of S that are related to a by $\sim$. More precisely,

$$[a]_\sim = \{x \in S : x \sim a\}.$$

Note that Definition 5.5 implies that if $x \in S$, then $x \in [a]_\sim$ if and only if $x \sim a$. In other words, for all elements $x \in S$, the symbolic expressions $x \in [a]_\sim$ and $x \sim a$ are interchangeable.

When the context is clear (that is, when we have defined a particular equivalence relation and are interested only in that relation), we will often omit the subscript on the $[a]_\sim$ notation, simply writing $[a]$ instead. The same convention applies to the notation for congruence classes that we studied earlier. For instance, if we had already stated that we were working with congruence modulo 6, then statements like $[3] \neq [4]$ would make perfect sense even without the usual subscripts. On the other hand, if we were trying to compare congruence classes with respect to two different moduli, then the subscripts would be absolutely essential. To illustrate, note that the true statement $[3]_{12} \neq [3]_6$ becomes ambiguous and potentially confusing when written as $[3] \neq [3]$. Thus, while there are times when it is convenient to simplify our notation, we must be careful to do so only when the context is clear.

Now that we have defined equivalence classes, let's turn our attention to generalizing some of the observations about congruence classes that we noted earlier in the investigation. The key result that we will prove is the following:

Theorem 5.6. *Let S be a nonempty set, and let $\sim$ be an equivalence relation on S. Then S can be written as the disjoint union of the distinct equivalence classes corresponding to $\sim$. That is, the equivalence classes corresponding to $\sim$ are pairwise disjoint, and every element of S belongs to exactly one equivalence class. In particular:*

(i) *For all $a, b \in S$, if $[a] \neq [b]$, then $[a] \cap [b] = \emptyset$.*

(ii) *For all $a \in S$, $a \in [a]$.*

(iii) *For all $a \in S$, if $a \in [b]$ for some $b \in S$, then $[a] = [b]$.*

Note that, like our earlier observation about congruence classes, Theorem 5.6 acknowledges the fact that the equivalence classes of two elements of set can be equal to each other, even if the elements themselves are not. The next lemma characterizes exactly when this equality can occur; it will be particularly helpful in our proof of Theorem 5.6.

Lemma 5.7. *Let S be a nonempty set, and let $\sim$ be an equivalence relation on S. Then for all a, $b \in S$, $[a] = [b]$ if and only if $a \sim b$.*

To prove Lemma 5.7, we will need to remember exactly what it means for two sets to be equal. In particular, we will need to make use of the fact that $[a] = [b]$ if and only if $[a] \subseteq [b]$ and $[b] \subseteq [a]$. Activity 5.8 below will help us work out the rest of the details.

Activity 5.8. Suppose, as in Lemma 5.7, that $\sim$ is an equivalence relation on a nonempty set S. Let $a, b \in S$.

(a) Suppose $[a] = [b]$. Using the fact that $[a] \subseteq [b]$, argue that $a \in [b]$. Then deduce that $a \sim b$. (Hint: Use one of the three properties that define an equivalence relation.)

(b) For the converse, suppose that $a \sim b$, and let $x \in [a]$. Show that $x \in [b]$. (Hint: Use one of the properties that you didn't use in part (a).)

(c) Now suppose that $x \in [b]$. Still assuming that $a \sim b$, show that $x \in [a]$. Is there a difference between your argument here and the one you made in part (b)?

Having established Lemma 5.7, we are now ready to move on to the proof of Theorem 5.6. As you read this proof, you should try to fill in the missing details, treating the ? symbol as we have in previous proofs.

Proof of Theorem 5.6. Let $\sim$ be an equivalence relation on a nonempty set S. For part (i), we will prove the ______________. So let $a, b \in S$, and assume that $[a] \cap [b] \neq \emptyset$.? Then there exists an element $x \in [a] \cap [b]$, which implies that $x \sim a$ and $x \sim b$.? By the ____________ property, it follows that $a \sim x$. But ____________ then implies that $a \sim b$.? Since $a \sim b$, Lemma 5.7 now lets us conclude that ____________, as desired.

For part (ii), let $a \in S$, and note that by the ____________ property, $a \sim a$. Thus, by the definition of equivalence class, $a \in [a]$.

For part (iii), let $a, b \in S$, and suppose that $a \in [b]$. Then $a \sim b$,? which implies by ____________ that $[a] = [b]$. ■

Activity 5.9. Finish the proof of Theorem 5.6 by explaining why conditions (i) – (iii) imply that S can be written as the disjoint union of the distinct equivalence classes corresponding to $\sim$.

The Number System $\mathbb{Z}_n$

So far, we have shown that equivalence relations, such as the congruence modulo n relation on $\mathbb{Z}$, always divide the sets on which they are defined into pairwise disjoint subsets called equivalence classes. In this section, we will use what we have learned to investigate a family of number systems whose elements are not numbers at all, but rather congruence classes. We define this family of number systems as follows:

Definition 5.10. For every integer $n \geq 2$, the **integers modulo** n, denoted $\mathbb{Z}_n$, is the set of the n distinct congruence classes of $\mathbb{Z}$ modulo n, i.e.,

$$\mathbb{Z}_n = \{[0]_n, [1]_n, [2]_n, \ldots, [n-1]_n\}.$$

We can make $\mathbb{Z}_n$ into a number system by defining an addition and multiplication on the set. There is a seemingly natural way to do this:

$$[a] + [b] = [a+b] \text{ and } [a] \cdot [b] = [a \cdot b]$$

for all $[a]$, $[b] \in \mathbb{Z}_n$.

A few observations are worth noting. First, we haven't yet defined exactly what a *number system* is. For now, we will think of a number system as a set of mathematical objects with one or more operations, like addition and multiplication, defined on it. Later on we will add more precision and clarity to this informal definition.

Second, we must keep in mind that the elements of $\mathbb{Z}_n$ are not actually numbers, but rather *sets* of numbers, and *infinite* sets at that. It is for this reason that we must formally define how exactly addition and multiplication should work in $\mathbb{Z}_n$. Our definitions are quite natural, so much so that they may not seem like definitions at all, but rather statements of fact. Be assured that this is not the case, for while it may seem natural to write $[3] + [2] = [5]$ when working in $\mathbb{Z}_7$, what we are really defining with this notation is the set operation that specifies

$$\{\ldots, -11, -4, 3, 10, \ldots\} + \{\ldots, -12, -5, 2, 9, \ldots\} = \{\ldots, -9, -2, 5, 12, \ldots\}.$$

If we saw an expression like this outside of the context of $\mathbb{Z}_7$, we would probably be quite puzzled, since the notions of addition and multiplication of sets are not universally defined. Also note that the addition and multiplication operations we have defined for $\mathbb{Z}_n$ are quite different from some of the more familiar set operations, such as unions and intersections. In Investigation 6, we will see an example of a number system whose elements are sets and for which addition and multiplication are defined in terms of unions, intersections, and relative complements.

Finally, note that in our definition of $\mathbb{Z}_n$, we have used the same notation, namely the $+$ symbol, to represent both addition in $\mathbb{Z}_n$ and addition in $\mathbb{Z}$. The same could be said for multiplication, and we will rely on other notation (such as brackets) to make the context of our operations clear. So, for instance, if we write

$$2 + 4 = 6,$$

we will assume that the $+$ sign is indicating addition in $\mathbb{Z}$. However, if we write

$$[2]_5 + [4]_5 = [6]_5 = [1]_5,$$

then the bracket notation should indicate clearly to us that our addition is being performed within $\mathbb{Z}_5$. Note that in the latter case, we took the extra step of reducing our final answer so that the representative a chosen to denote $[a]_5$ satisfied $0 \le a < 5$. Adopting this standard reducing convention, we can construct the addition and multiplication tables for $\mathbb{Z}_n$. For example, the addition and multiplication tables for $\mathbb{Z}_3$ are as follows:

+	[0]	[1]	[2]
[0]	[0]	[1]	[2]
[1]	[1]	[2]	[0]
[2]	[2]	[0]	[1]

·	[0]	[1]	[2]
[0]	[0]	[0]	[0]
[1]	[0]	[1]	[2]
[2]	[0]	[2]	[1]

Activity 5.11.

(a) Make the addition and multiplication tables for $\mathbb{Z}_4$ and $\mathbb{Z}_5$.

(b) Do you notice any patterns or symmetries in the tables for $\mathbb{Z}_3$, $\mathbb{Z}_4$, and $\mathbb{Z}_5$? If so, what do your observations allow you to conclude about the way arithmetic works in $\mathbb{Z}_n$? State your conclusions specifically and precisely, as we did when we stated the axioms for $\mathbb{Z}$ in Investigation 1.

(c) Are there any differences between the way arithmetic works in $\mathbb{Z}$ and the way it works in either $\mathbb{Z}_3$, $\mathbb{Z}_4$, or $\mathbb{Z}_5$? If so, state these differences precisely.

(d) Do addition and multiplication behave the same in $\mathbb{Z}_3$, $\mathbb{Z}_4$, and $\mathbb{Z}_5$, or are there differences between these three number systems? Give specific examples to justify your answer.

As you may have noticed from Activity 5.11, $\mathbb{Z}_n$ and $\mathbb{Z}$ are similar in many ways, especially with regard to the axioms of addition and multiplication. In fact, for at least some of the arithmetic axioms from Investigation 1, there are analogous properties that hold in $\mathbb{Z}_n$. For instance, addition in $\mathbb{Z}_n$ is associative, just as it is in $\mathbb{Z}$. If we wanted to prove this result, our argument might look something like this:

Proof that addition in $\mathbb{Z}_n$ is associative.. Let $n \in \mathbb{N}$, and let $[a]$, $[b]$, $[c] \in \mathbb{Z}_n$. Then

$$\begin{aligned}([a]+[b])+[c] &= [a+b]+[c]^{ⓘ} \\ &= [(a+b)+c]^{?} \\ &= [a+(b+c)]^{?} \\ &= [a]+[b+c]^{?} \\ &= [a]+([b]+[c]),^{?}\end{aligned}$$

as desired. ■

As you read this proof, you should have been able to provide an explanation for each step (as indicated by the ? symbol). In particular, the first two and the last two steps were simply applications of the definition of addition in $\mathbb{Z}_n$, and the middle step used the fact that addition in $\mathbb{Z}$ is associative. In other words, the proof used associativity of addition in one number system, $\mathbb{Z}$, to prove a similar property for a related number system, $\mathbb{Z}_n$. The same strategy can be employed to prove other properties in $\mathbb{Z}_n$, and Activity 5.12 asks us to do exactly that.

Activity 5.12.

(a) For each of the arithmetic axioms listed on page 5, either prove a corresponding property for $\mathbb{Z}_n$ or give a counterexample to show that no such property holds in $\mathbb{Z}_n$.

(b) Do the ordering axioms from page 8 hold in $\mathbb{Z}_n$? Why or why not?

(c) Does an analogous version of Theorem 1.10 hold in $\mathbb{Z}_n$? (Hint: Consider the multiplication tables you made in Activity 5.11. Does your answer depend on n?)

Binary Operations

What we did in the previous section may seem very natural and may not concern you at all, but it will all be total nonsense if addition and multiplication on $\mathbb{Z}_n$ are not *well-defined*. The next activity illustrates why it is important for operations to be well-defined by showing what can happen if an operation is *not* well-defined.

Activity 5.13. Let $\diamond$ be the operation on $\mathbb{Z}_n$ defined as follows:

$$[a] \diamond [b] = \begin{cases} [1] & \text{if } a \text{ and } b \text{ have the same parity} \\ [0] & \text{if } a \text{ and } b \text{ have opposite parity} \end{cases}$$

Note that *parity* refers to whether an integer is even or odd. Thus, two integers have the same parity if they are both even or both odd. Likewise, two integers have opposite parity if one is even and the other is odd.

(a) Use the above definition to calculate each of the following quantities in $\mathbb{Z}_5$:

- $[1] \diamond [3]$
- $[1] \diamond [8]$
- $[6] \diamond [13]$
- $[11] \diamond [13]$

(b) What is the relationship between $[1]$, $[6]$, and $[11]$ in $\mathbb{Z}_5$?

(c) What is the relationship between $[3]$, $[8]$, and $[13]$ in $\mathbb{Z}_5$?

(d) In light of your answers to parts (b) and (c), does anything seem strange or unusual about your answers to part (a)? Explain.

You should have observed something unusual in Activity 5.11. In particular, the result of the operation depended on which representative we used for our input. This is a critically important observation to make in sets like $\mathbb{Z}_n$ where each of the equivalence classes can be represented in infinitely many different ways. If the result of an operation depends on which way we choose to represent one of the inputs, then two people working the same problem could, without making any mistakes in computation, arrive at different answers. This would result in chaos! What defines these equivalence classes, however, is not the way we choose to represent them, but rather the elements that they contain. Thus, if two equivalence classes contain exactly the same elements, and are therefore equal, we would expect any reasonable operation to treat them exactly the same.

This, however, is not what happened in Activity 5.11. There, in spite of the fact that $[1] = [6]$ and $[3] = [13]$, we saw that $[1] \diamond [3] = [1]$, but $[6] \diamond [13] = [0]$. In other words, the output of the $\diamond$ operation depended not only on the elements that we used (they were the same in each case), but also on the way we chose to name, or represent, those elements. Because of this, we might say that $\diamond$ is *ill-defined*. Not surprisingly, the opposite of an ill-defined operation is a *well-defined* operation.

Although we have already seen and used several different binary operations (the most familiar being standard addition and multiplication in $\mathbb{Z}$), we have not yet formally defined exactly what a binary operation is. The next definition fills in this gap. It formalizes the idea that a binary operation on a set S maps every ordered pair of elements of S to single element of S. So, for instance, the operation of addition on the integers maps the pair $(1, 4)$ to the integer 5 (since $1 + 4 = 5$). This same idea forms the basis of the definition of a binary operation on any set.

Definition 5.14. Let S be a set. The **Cartesian product** $S \times S$ is the set of all ordered pairs of elements from S; that is,

$$S \times S = \{(x, y) : x \in S \text{ and } y \in S\}.$$

A **binary operation** on S is a function $f : S \times S \to S$.

Activity 5.15. For each part below, find the image of the given pair of elements under the given binary operation. The first part is completed for you as an example.

(a) $(3, -5)$; addition in $\mathbb{Z}$

Solution: *Using function notation,* $+(3, -5) = 3 + (-5) = -2$.

(b) $([2], [5])$; addition in $\mathbb{Z}_8$

(c) $([4], [3])$; multiplication in $\mathbb{Z}_5$

You may be wondering at this point why it is important to define binary operations so formally. It is certainly not standard practice to use function notation when doing arithmetic. For instance, we would rarely if ever write

$$+(2, 3) = 5$$

in $\mathbb{Z}$. Instead, we would use the more natural notation of

$$2 + 3 = 5.$$

So, on the surface, it seems that our formal definition only makes matters more complicated. While this may be the case from a notational perspective, defining a binary operation as a function does help to make explicit a very important property that we should expect all binary operations to satisfy. This property, which we will define next, is one that you have used many times, perhaps without even realizing it. As we saw in Activity 5.11, the result of an operation must not be dependent on the particular representation or name we choose for the input, but must be independent of such superficial distinctions. That is, if we have two elements that may "look" different ($[1]$ and $[4]$ in $\mathbb{Z}_3$, for example) but are actually equal, any operation must treat them the same way. Activity 5.11 shows that this is a property of operations that cannot be taken for granted. We formalize what we mean by a well-defined operation in the next definition.

Definition 5.16. Let $\star$ denote a binary operation on a set S. Then $\star$ is said to be **well-defined** provided that whenever $a = x$ and $b = y$ in S we have $a \star b = x \star y$.

Definition 5.16 suggests that in order to prove that an operation $\star$ on S is well-defined, we should assume that $a = x$ and $b = y$ for some a, b, x, $y \in S$, and then try to show that $a \star b = x \star y$. To illustrate this technique, consider the following proof that addition in $\mathbb{Z}_n$ is well-defined:

Proof. Let $[a]$, $[b]$, $[x]$, $[y] \in \mathbb{Z}_n$, and suppose that $[a] = [x]$ and $[b] = [y]$. Then $a \equiv x \pmod{n}^{(?)}$ and $b \equiv y \pmod{n}$,$^{(?)}$ which implies that $n \mid (a - x)^{(?)}$ and $n \mid (b - y)$.$^{(?)}$ From this it follows that $n \mid [(a - x) + (b - y)]$,$^{(?)}$ or $n \mid [(a + b) - (x + y)]$. Thus,

$$[a] + [b] \overset{(?)}{=} [a + b] \overset{(?)}{=} [x + y] \overset{(?)}{=} [x] + [y],$$

as desired. ■

You will show that multiplication is well-defined on $\mathbb{Z}_n$ in Activity 5.25.

A few observations are in order before we move on. The first is that, after all of this discussion, it turns out in many cases, we don't have to worry too much about whether our operations are well-defined or not. This is because many of the number systems we have studied (and will study) do not allow for multiple representations of the same element. For such number systems, the definition of well-defined is trivially satisfied. Thus, the systems that merit more attention are those whose elements are equivalence classes, like $\mathbb{Z}_n$. In general, whenever the elements of a number system can be represented in multiple ways, we will need to verify that the operations within the number system do not depend upon these varying representations.

A second observation is that our definition of well-defined is actually redundant and even unnecessary in some sense. This is due to the fact that we defined a binary operation on S to be a *function* $f : S \times S \to S$, which implies that equal inputs must be mapped to equal outputs, regardless of their representation. Thus, for all a, b, x, $y \in S$, if $a = b$ and $x = y$, then it must be the case that $(a, b) = (x, y)$, which implies that $f(a, b) = f(x, y)$. In other words, in order for a binary operation to really be a binary operation, it must first pass the test of being well-defined. Along these same lines, because the codomain of a binary operation on S is always S itself, closure is also guaranteed by definition.

Throughout subsequent investigations, we will be referencing $\mathbb{Z}_n$ frequently. We will also be considering several other number systems, all of which satisfy at least some of the same axioms that the integers do. Our goal in these investigations will be to identify a certain common structure that is shared by a variety of number systems. Once we have done so, we will be able to prove general results about the way arithmetic works in various contexts. Before we move on, however, we will consider some important ideas related to solving linear equations in $\mathbb{Z}_n$.

Zero Divisors and Units in $\mathbb{Z}_n$

Solving equations is an important part of algebra, and the simplest type of equation is the linear equation of the form $ax = b$. These types of equations are among the first studied in beginning algebra courses, but here we will consider them from a more general perspective.

Activity 5.17. If $a \neq 0$ in the real number system, then the equation $ax = b$ always has a unique solution.

(a) Does the equation $2x = 1$ have a solution in $\mathbb{Z}$? Explain.

(b) Find all solutions to $[2]x = [4]$ in $\mathbb{Z}_6$, and then reflect on your answer. Do you notice anything interesting or surprising?

(c) Find all solutions to $[2]x = [3]$ in $\mathbb{Z}_6$, and then reflect on your answer. Do you notice anything interesting or surprising?

Activity 5.17 shows that it is possible for the equation $ax = b$ to have no solutions or more than one solution even if $a \neq 0$, depending on the number system in question. The element $[2]$ in $\mathbb{Z}_6$ has a special property that leads to the behavior in Activity 5.17.

Activity 5.18. In Investigation 1, we saw that there are no nonzero integers a and b such that $ab = 0$.

(a) Show that this is not true in general in $\mathbb{Z}_n$ by exhibiting a particular value of n and specific nonzero classes $[a]$ and $[b]$ so that $[a][b] = [0]$ in $\mathbb{Z}_n$.

(b) Is there a multiplicative cancellation law in $\mathbb{Z}_n$ as there is in $\mathbb{Z}$? That is, if $[c] \neq [0]$ and $[a][c] = [b][c]$ in $\mathbb{Z}_n$, must it be true that $[a] = [b]$? Verify your conjecture.

Elements that behave like those described in Activity 5.18 are given a special name.

Definition 5.19. A nonzero element $[a] \in \mathbb{Z}_n$ is a **zero divisor** in $\mathbb{Z}_n$ if there is a nonzero element $[b] \in \mathbb{Z}_n$ so that $[a][b] = [0]$.

So if a is a zero divisor, then the equation $ax = b$ may have no solutions or more than one solution. But when can we be sure that the equation $ax = b$ has exactly one solution?

Activity 5.20. Just as we did in $\mathbb{Z}$ (see Definition 1.5 on page 6), we can define the additive inverse of an element $[x] \in \mathbb{Z}_n$ to be an element $[y] \in \mathbb{Z}_n$ such that $[x] + [y] = [0]$.

(a) Write a similar definition of the multiplicative inverse of an element $[x] \in \mathbb{Z}_n$.

(b) Find all of the elements of $\mathbb{Z}_3$ that have a multiplicative inverse in $\mathbb{Z}_3$.

(c) Find all of the elements of $\mathbb{Z}_4$ that have a multiplicative inverse in $\mathbb{Z}_4$.

(d) In Activity 1.12 of Investigation 1 (see page 9), we saw that there are only two integers that have a multiplicative inverse in $\mathbb{Z}$, namely 1 and -1. Is the same result true in $\mathbb{Z}_n$ in general? Explain. (Hint: What element of $\mathbb{Z}_n$ is analogous to -1 in $\mathbb{Z}$?)

(e) In the set of real numbers, we know that every nonzero element has a real multiplicative inverse. Is the comparable statement true in $\mathbb{Z}_n$? Explain.

Elements that behave like those described in Activity 5.20 are also given a special name.

Definition 5.21. An element $[x] \in \mathbb{Z}_n$ is a **unit** in $\mathbb{Z}_n$ if there is an element $[y] \in \mathbb{Z}_n$ such that $[x][y] = [1]$. In this case, the element $[y]$ is called a **multiplicative inverse** of $[x]$.

It can be shown that if $[a]$ is a unit in $\mathbb{Z}_n$, then the equation $[a]x = [b]$ has exactly one solution for every $[b] \in \mathbb{Z}_n$. (See Exercise 17.) There is also a pattern to the zero divisors and units in $\mathbb{Z}_n$, as we will see in the next activity.

Activity 5.22. Construct the multiplication tables for $\mathbb{Z}_3$, $\mathbb{Z}_4$, $\mathbb{Z}_5$, $\mathbb{Z}_6$, and $\mathbb{Z}_8$.

(a) Identify the zero divisors in each set.

(b) Identify the elements in each set that have a multiplicative inverse in the set.

(c) Make a conjecture in the form of a biconditional statement, such as:

> Let $n \in \mathbb{N}$, and let $[a] \in \mathbb{Z}_n$. Then $[a] \neq [0]$ is a zero divisor in $\mathbb{Z}_n$ if and only if …

(d) Make a conjecture in the form of a biconditional statement, such as:

> Let $n \in \mathbb{N}$, and let $[a] \in \mathbb{Z}_n$. Then $[a]$ has a multiplicative inverse in $\mathbb{Z}_n$ if and only if …

Proofs of correct conjectures for parts (c) and (d) are outlined in Exercises 14 and 15.

We will now conclude our explorations of $\mathbb{Z}_n$ with a short discussion of how to solve an equation of the form $[a]x = [b]$ in $\mathbb{Z}_n$ if $[a]$ is a unit.

Activity 5.23.

(a) What tool can we use to quickly determine if 231 and 4210 are relatively prime?

(b) Bezout's Identity states that we can write

$$231x + 4210y = 1 \tag{5.1}$$

for some integers x and y. Find integers x and y that satisfy the above equation. (Hint: Think back to an algorithm we learned in a previous investigation.)

(c) Reduce both sides of equation (5.1) modulo 4210 and find a multiplicative inverse of $[231]$ in $\mathbb{Z}_{4210}$.

(d) Summarize this process to explain how to find a multiplicative inverse of $[a]$ in $\mathbb{Z}_n$. Then explain how such an inverse could be used to solve an equation of the form $[a]x = [b]$.

Concluding Activities

Activity 5.24. Consider the following proof of Lemma 5.7:

> *Proof.* Let $a, b \in S$, and suppose that $[a] = [b]$. By the reflexive property, $a \sim a$, which implies that $a \in [a]$. But since $[a] = [b]$, it then follows that $a \in [b]$, which implies that $a \sim b$.
>
> For the converse, suppose that $a \sim b$. We must show that $[a] \subseteq [b]$ and $[b] \subseteq [a]$. For the former, let $x \in [a]$. Then $x \sim a$. But since $a \sim b$ as well, transitivity implies that $x \sim b$. Thus, $x \in [b]$, and $[a] \subseteq [b]$, as desired. A similar argument establishes that $[b] \subseteq [a]$, which completes the proof. ■

(a) Is this proof correct? If so, are there any places in the proof where more detail should be provided, or where the argument could be made clearer? If not, where are the errors in the proof? Give a thorough and precise answer.

(b) As you may have noticed, the proof above made no reference to the symmetric property. Does this mean that Lemma 5.7 holds for all relations that are reflexive and transitive (instead of only those that are equivalence relations)? Why or why not?

(c) Consider the "less than or equal to" ($\leq$) relation on $\mathbb{Z}$. Is $\leq$ both reflexive and transitive? Give a proof or counterexample to justify your answer for each property.

(d) For each $a \in \mathbb{Z}$, define $[a]_{\leq}$ as in Definition 5.5. That is, let

$$[a]_{\leq} = \{x \in \mathbb{Z} : x \leq a\}.$$

Use roster notation to list the elements of $[1]_{\leq}$ and $[2]_{\leq}$. Is $1 \leq 2$? Does $[1]_{\leq} = [2]_{\leq}$? How do your answers to these questions relate to your answers to parts (a) – (c) above?

Activity 5.25. We have already shown that addition is well-defined in $\mathbb{Z}_n$. In this activity, we will show that multiplication is also well-defined.

(a) Let $n \in \mathbb{N}$, and let $[a], [b], [x], [y] \in \mathbb{Z}_n$ such that $[a] = [x]$ and $[b] = [y]$. Prove that $[a][b] = [x][y]$. (There **is** something to prove here. If the result seems trivial, then you should go back and reread the section on well-defined operations.)

(b) Why is it important for the operations on $\mathbb{Z}_n$ to be well-defined?

Activity 5.26. Write a short summary that describes the important concepts, and the relationships between these concepts, that were introduced in this investigation. Explain how the material in this investigation is connected to the material in Investigations 2 and 3.

Exercises

(1) Let $S = \{1, 2, 3, \ldots, 999, 1000\}$, and define the relation $\sim$ on S as follows:

For all $x, y \in S$, $x \sim y$ if and only if y has the same number of digits as x.

Prove that $\sim$ is an equivalence relation on S, and find all of the distinct equivalence classes corresponding to $\sim$.

(2) Let $\sim$ be the relation on $\mathbb{Z}$ defined by $a \sim b$ if and only if $3a + 4b = 7n$ for some integer n. Prove that $\sim$ is an equivalence relation, and give a precise description of the equivalence classes of $\sim$.

(3) Let $\sim$ be the relation on $\mathbb{Z}$ defined by $a \sim b$ if and only if $a + b$ is even.

(a) Is $\sim$ an equivalence relation? Verify your answer.

(b) Suppose the definition of $\sim$ is changed so that $a \sim b$ if and only if $a + b$ is odd. Would this change your answer to part (a)? Why or why not?

(4) Let $\sim$ be the relation on $\mathbb{Z}$ defined by $a \sim b$ if and only if $ab \geq 0$.

(a) Is $\sim$ an equivalence relation? Verify your answer.

(b) Suppose the definition of $\sim$ is changed so that $a \sim b$ if and only if $ab > 0$. Would this change your answer to part (a)? Why or why not?

(5) Let $S = \{(x, y) \in \mathbb{Z} \times \mathbb{Z} : y \neq 0\}$ and let $\sim$ be the relation on S defined by $(a, b) \sim (c, d)$ if and only if $ad = bc$. Prove that $\sim$ is an equivalence relation on S, and describe the equivalence classes corresponding to $\sim$. Then find and explain a connection between $\sim$ and the rational numbers.

(6) For each part below, find a binary relation on the set $S = \{1, 2, 3, 4\}$ that satisfies the given combination of properties.

(a) reflexive, symmetric, and transitive

(b) not reflexive, but symmetric and transitive

(c) not symmetric, but reflexive and transitive

(d) not transitive, but reflexive and symmetric

(e) neither reflexive nor symmetric, but transitive

(f) neither reflexive nor transitive, but symmetric

(g) neither symmetric nor transitive, but reflexive

(h) not reflexive, not symmetric, and not transitive

(7) Which of the properties of reflexive, symmetric, and transitive are satisfied by each of the following relations on the set of natural numbers? Give a proof or counterexample to justify each of your answers.

(a) The relation $\sim$ defined by $a \sim b$ if and only if $a \mid b$

(b) The relation $\sim$ defined by $a \sim b$ if and only if $a \mid b$ or $b \mid a$

(c) The relation $\sim$ defined by $a \sim b$ if and only if $a + b = 10$

(8) Which of the following sets could be the set of all equivalence classes for an equivalence relation on $\{a, b, c, d, e\}$? Cite a theorem from this investigation to justify each of your answers.

(a) $\{\{a, b, c, d, e\}\}$

(b) $\{\{a\}, \{b, c\}, \{d\}\}$

(c) $\{\{a\}, \{b, c\}, \{d, e\}\}$

(d) $\{\{a\}, \{b\}, \{c\}, \{d\}, \{e\}\}$

(e) $\{\{a, b\}, \{b, c\}, \{d, e\}\}$

(f) $\{\{a, d, e\}, \{b, c\}\}$

(9) A relation $\sim$ on a set S is said to be **circular** provided that for all a, b, $c \in S$, if $a \sim b$ and $b \sim c$, then $c \sim a$.

(a) Suppose a relation $\sim$ on a set S is both reflexive and circular. Prove that S must also be symmetric.

(b) Use part (a) to show that if a relation $\sim$ on a set S is both reflexive and circular, then S is also transitive.

(c) Parts (a) and (b) establish one direction of the biconditional statement in the following theorem:

Theorem. A relation $\sim$ on a set S is an equivalence relation if and only if $\sim$ is reflexive and circular.

Complete the proof of this theorem by proving the converse of the statement you proved in parts (a) and (b).

(10) Decide whether each of the following statements is true or false. Prove each of your answers.

(a) $572 \in [11]_{17}$

(b) $-37 \in [7]_{10}$

(c) $[5]_7 \subseteq [10]_{14}$

(d) $[3]_8 \subseteq [3]_4$

(e) $[3]_4 \subseteq [3]_8$

(11) Let a be an integer. Suppose $a \equiv 7 \pmod 9$ and $b \equiv 1 \pmod 6$. What is the remainder when $a^2 + 2b$ is divided by 3? Explain.

(12) Let $m, n \in \mathbb{N}$, and let $a \in \mathbb{Z}$. Find and prove a necessary and sufficient condition for $[a]_m \subseteq [a]_n$.

(13) Let $m, n \in \mathbb{N}$, and let $a \in \mathbb{Z}$. Show that $[a]_m \cap [a]_n$ contains infinitely many elements.

⋆ (14) **Zero divisors in $\mathbb{Z}_n$.** Let n be a positive integer.

(a) Prove that if $\gcd(a, n) > 1$ and $n \nmid a$, then $[a]$ is a zero divisor in $\mathbb{Z}_n$.

(b) Prove that if $[a]$ is a zero divisor in $\mathbb{Z}_n$, then $\gcd(a, n) > 1$.

(c) Correctly complete the biconditional statement:

Let $n \in \mathbb{N}$, and let $[a] \in \mathbb{Z}_n$. Then $[a] \neq [0]$ is a zero divisor in $\mathbb{Z}_n$ if and only if ...

⋆ (15) **Units in $\mathbb{Z}_n$.** Let n be a positive integer. Let $[a] \in \mathbb{Z}_n$.

(a) Let $[a] \in \mathbb{Z}_n$ with $\gcd(a, n) = 1$. Using Bezout's Identity we can find integers x and y so that $ax + ny = 1$. Explain how $[a]$ and $[x]$ are related in $\mathbb{Z}_n$.

(b) Prove that if $\gcd(a, n) = 1$, then $[a]$ is a unit in $\mathbb{Z}_n$.

(c) Prove that if $[a]$ is a unit in $\mathbb{Z}_n$, then $\gcd(a, n) = 1$. (Hint: Is the converse of Bezout's Identity ever true?)

(d) Correctly complete the biconditional statement:

Let $n \in \mathbb{N}$, and let $[a] \in \mathbb{Z}_n$. Then $[a]$ is a unit in $\mathbb{Z}_n$ if and only if ...

(16) For the given value of n and the given $[a] \in \mathbb{Z}_n$, show that $[a]$ is a unit in $\mathbb{Z}_n$ and find an element $[x]$ so that $[a][x] = [1]$.

(a) $n = 5$, $[a] = [2]$

(b) $n = 15$, $[a] = [7]$

(c) $n = 24672$, $[a] = [443]$

⋆ (17) Let n be a positive integer. Prove that $[a]$ is a unit in $\mathbb{Z}_n$ if and only if the equation $[a]x = [b]$ has a unique solution for each $[b] \in \mathbb{Z}_n$.

(18) **Units and zero divisors.** Let $[a] \in \mathbb{Z}_n$. Determine whether each of the following statements are true or false. Verify your answers.

(a) If $[a]$ is a zero divisor, then $[a]$ is not a unit.

(b) If $[a]$ is not a unit, then $[a]$ is a zero divisor.

(c) If $[a]$ is a unit, then $[a]$ is not a zero divisor.

(d) If $[a]$ is not a zero divisor, then $[a]$ is a unit.

(19) In this exercise, we will create a different number system from $\mathbb{Z}$ using absolute value as our relation. In other words, let $\sim$ be the relation on $\mathbb{Z}$ defined by $a \sim b$ for $a, b \in \mathbb{Z}$ if $|a| = |b|$.

(a) List all of the integers that are related to 0 under the relation $\sim$.

(b) List all of the integers that are related to 1 under the relation $\sim$.

(c) Show that $\sim$ is an equivalence relation.

(d) What is the equivalence class of -3, under the relation $\sim$?

We will now attempt to define a number system based on this equivalence relation. Let S be the set of all distinct equivalence classes under the absolute value relation $\sim$. Define addition and multiplication on S the same way that we did in $\mathbb{Z}_n$:

$$[x] + [y] = [x + y] \quad \text{and} \quad [x][y] = [xy].$$

Note here that $[x]$ denotes the set of all elements equivalent to x, according to the absolute value relation $\sim$. Now let's see if these operations are well defined.

(e) Calculate each of the following quantities in S:
   (i) $[1] + [1]$
   (ii) $[-1] + [1]$
   (iii) $[1] + [-1]$
   (iv) $[-1] + [-1]$

(f) If addition in S were well-defined, what would have to be true about all of the calculations you just performed? What can you conclude about addition in S?

(g) Is multiplication in S well-defined? Give a convincing argument or counterexample to justify your answer.

(20) Define the following operation on $\mathbb{Z}_n$:

$$[a] \star [b] = \begin{cases} [1] & \text{if } a \equiv b \pmod 5 \\ [0] & \text{if } a \not\equiv b \pmod 5 \end{cases}$$

(a) Is $\star$ a well-defined operation on $\mathbb{Z}_4$?

(b) Is $\star$ a well-defined operation on $\mathbb{Z}_5$?

Investigation 6

Algebra in Other Number Systems

Focus Questions

By the end of this investigation, you should be able to give precise and thorough answers to the questions listed below. You may want to keep these questions in mind to focus your thoughts as you complete the investigation.

- What are some familiar subsets of the real numbers, and what algebraic properties are satisfied by these various subsets?
- What are the complex numbers, and what algebraic properties do they satisfy?
- What algebraic properties are satisfied by matrix addition and multiplication? What are some common number systems that involve matrices?
- How can set operations be used to define addition and multiplication on collections of sets? What algebraic properties are satisfied by the resulting number systems?

Preview Activity 6.1. Up to this point, our study of number systems has focused mainly on systems that are related in some way to the integers. Let's now, however, shift our focus to some number systems that aren't quite as familiar. To begin, let $S = \{\gamma, \alpha, \delta, \beta\}$, and define addition and multiplication on S as shown in Table 6.1.

$+$	γ	α	δ	β
γ	α	γ	β	δ
α	γ	α	δ	β
δ	β	δ	γ	α
β	δ	β	α	γ

$\cdot$	γ	α	δ	β
γ	α	α	γ	γ
α	α	α	α	α
δ	γ	α	β	δ
β	γ	α	δ	β

Table 6.1
An unfamiliar number system.

(a) Is addition commutative in S? Is multiplication commutative in S?

(b) Does S contain an additive identity? If so, which elements of S have an additive inverse? Find the additive inverse of each such element.

(c) Does S contain a multiplicative identity? If so, which elements of S have a multiplicative inverse? Find the multiplicative inverse of each such element.

(d) Do you think that addition is associative in S? Do you think that multiplication is associative in S?

(e) Does multiplication distribute over addition in S? Does addition distribute over multiplication? (Hint: Exactly one of these potential distributive laws holds.)

(f) What makes the questions in parts (d) and (e) harder to answer than those in parts (a) – (c)?

(g) Solve each of the following equations for x. That is, find all of the elements $x \in S$ for which the equation holds. Justify each of your steps by citing one or more of the axioms from parts (a) – (e) (including associativity from part (d) and the appropriate distributive law from part (e), both of which you may assume to be true).

(i) $\alpha \cdot x = \alpha$

(ii) $\delta = \beta + x$

(iii) $\delta + x = x \cdot (\delta + \delta)$

(iv) $\delta \cdot x = \gamma \cdot (\beta + x)$

(v) $\gamma + x = \gamma + \beta$

(vi) $\gamma \cdot x = \gamma \cdot \beta$

Introduction

In Preview Activity 6.1, we investigated a number system that, at least on the surface, didn't seem to involve numbers at all. Instead of starting with the integers or even congruence classes of integers, we simply picked four letters from the Greek alphabet and defined, using tables, how addition and multiplication of these letters should work. Our definition of addition and multiplication was exhaustive in the sense that it specified exactly what the sum and product of every pair of elements of $S = \{\gamma, \alpha, \delta, \beta\}$ should be. This information allowed us to determine, or at least attempt to determine, which arithmetic axioms were satisfied by S.

Some of these axioms were easier to observe than others. For instance, the fact that S is closed under addition and multiplication can be ascertained by simply noticing that every entry in the addition and multiplication tables is an element of S. The fact that addition and multiplication are commutative in S is implied by the diagonal symmetry exhibited by each table.

Associativity, on the other hand, requires more than a simple observation. This is primarily due to the fact that the definition of associativity involves three elements instead of just two. The same could be said of the fact that multiplication distributes over addition. Each of these properties would require numerous cases to prove; in particular, every ordered triple of elements of S (for instance, (α, β, γ)) would need to be considered. This work is tedious, but not difficult. Thus, we will skip over the details and simply state the conclusion—that both addition and multiplication are associative in S, and multiplication does indeed distribute over addition.

These properties are important and very useful, just as they were in the integers. For instance, to solve the the equation

$$\delta \cdot (x + \beta) = x \cdot \gamma,$$

which is similar to one from part (g) of Preview Activity 6.1, we must take advantage of many of the axioms that we were able to use in the integers. In the solution that follows, see if you can identify the axiom or axioms being used in each of the steps marked with the ⓘ symbol:

$$\begin{aligned} \delta \cdot (x+\beta) &= x \cdot \gamma \\ \delta \cdot x + \delta \cdot \beta &= x \cdot \gamma \text{ ⓘ} \\ \delta \cdot x + \delta &= \gamma \cdot x \text{ ⓘ} \\ (\delta \cdot x + \delta) + \beta &= \gamma \cdot x + \beta \\ \delta \cdot x + (\delta + \beta) &= \gamma \cdot x + \beta \text{ ⓘ} \\ \delta \cdot x + \alpha &= \gamma \cdot x + \beta \\ \delta \cdot x &= \gamma \cdot x + \beta \text{ ⓘ} \\ \gamma \cdot x + \delta \cdot x &= \gamma \cdot x + (\gamma \cdot x + \beta) \\ (\gamma + \delta) \cdot x &= (\gamma \cdot x + \gamma \cdot x) + \beta \text{ ⓘ} \\ \beta \cdot x &= (\gamma + \gamma) \cdot x + \beta \text{ ⓘ} \\ x &= \alpha \cdot x + \beta \text{ ⓘ} \\ x &= \alpha + \beta \\ x &= \beta \end{aligned}$$

A few comments about this solution are in order. First, note that we obtained the solution by using the standard technique that is taught in almost every high school algebra course. In particular, we started with the equation we were trying to solve, and we carried out a sequence of simplifications, stopping when we had isolated x on one side of the equation. When we use this technique, we are essentially working backwards. That is, we are *assuming* that there is some solution to our equation, and then finding out what that solution would have to be. What holds this kind of argument together is the fact that each of our steps is reversible. In other words, if we had wanted to, we could have started with the fact that $x = \beta$ and shown that $\delta \cdot (x + \beta) = x \cdot \gamma$ by simply carrying out the above steps in reverse order. Of course, if we knew, or even suspected, that $x = \beta$ was a solution, then we wouldn't have needed all of those steps in the first place. Instead, we could have simply substituted $x = \beta$ into the original equation and verified that the equation was indeed satisfied.

It is almost always easier to verify a solution than it is to find that solution in the first place. This is why mathematicians have developed algebraic techniques, such as the one illustrated above, for systematically reducing equations into simpler, but logically equivalent, forms. For number systems with only a few elements, these techniques may be more trouble than they are worth. In fact, in small number systems, it may be easier to solve equations by simply checking all of the possible solutions. For instance, to solve our equation $\delta \cdot (x + \beta) = x \cdot \gamma$ in S, we could have just made a table to compare both sides of the equation for each value of $x \in S$:

x	$\delta \cdot (x+\beta)$	$x \cdot \gamma$
γ	β	α
α	δ	α
δ	α	γ
β	γ	γ

This table clearly demonstrates that $x = \beta$ is the only solution to our equation. Of course, for number systems with more elements, such a table would be difficult, if not impossible, to make. This fact underscores the limitations of "guess-and-check" methods, and it suggests that the algebraic techniques we used earlier are worth further investigation.

With that said, our main goal in this investigation will be to compare and contrast the way algebra works in a variety of different number systems, some familiar and some not. Our work will be focused on the axioms we have studied in previous investigations and the theorems that follow from them. As we will see, these axioms are the glue that holds algebra together, not only in the integers, but in many other contexts as well. When certain axioms are satisfied, algebra works exactly the way we would expect it to. But when these axioms fail to hold, even algebraic manipulations that seem trivial may not be valid.

Subsets of the Real Numbers

The first number system we will consider is the set of real numbers, denoted $\mathbb{R}$, with addition and multiplication defined in the usual way. We will refrain from giving a formal definition of $\mathbb{R}$, mainly because it is surprisingly difficult to do so without developing a more sophisticated framework that is better suited for a course on advanced calculus or real analysis.* We can, however, define certain subsets of the real numbers in a slightly more precise fashion. For instance, we have already given definitions of the natural numbers ($\mathbb{N}$), the whole numbers ($\mathbb{W}$), the integers ($\mathbb{Z}$), and the even integers ($\mathbb{E}$). The *rational numbers* can be defined as follows:

Definition 6.2. A **rational number** is a real number that can be expressed as the quotient of two integers a and b, with $b \neq 0$.

Thus, the set of all rational numbers, denoted $\mathbb{Q}$,† is defined to be

$$\mathbb{Q} = \left\{ \frac{a}{b} : a \in \mathbb{Z}, b \in \mathbb{Z}, \text{ and } b \neq 0 \right\}.$$

Two rational numbers $\frac{a}{b}$ and $\frac{c}{d}$ are considered equal if and only if

$$ad = bc.$$

Furthermore, addition and multiplication within $\mathbb{Q}$ are defined by

$$\frac{a}{b} + \frac{c}{d} = \frac{ad + bc}{bd}$$

and

$$\frac{a}{b} \cdot \frac{c}{d} = \frac{ac}{bd}$$

for all $\frac{a}{b}, \frac{c}{d} \in \mathbb{Q}$.

The rationals are a familiar number system and you are probably comfortable with the fact that a rational number can be represented in many different ways—for example, $\frac{1}{2} = \frac{2}{4} = \frac{-5}{-10}$. This

*There are two main approaches to formally defining the real numbers: one involving sequences of rational numbers called *Cauchy sequences*, and one involving sets of rational numbers called *Dedekind cuts*.

†The symbol $\mathbb{Q}$ for the rational numbers comes from the German word *Quotient*, as rational numbers are quotients of integers. This symbol appeared in Bourbaki's *Algébre*, Chapter 1.

should prompt you to ask whether the operations we have defined on $\mathbb{Q}$ are well-defined. You will show that they are in Activity 6.10.

Using the definitions of addition and multiplication in $\mathbb{Q}$, it is easy to show that the rational numbers satisfy many of the same axioms and properties as the integers. For instance, consider the following proof that the rational numbers are closed under addition:

Proof. Let $\frac{a}{b}$, $\frac{c}{d} \in \mathbb{Q}$. Then a, b, c, and d are all integers, with $b \neq 0$ and $d \neq 0$. Because the integers are closed under addition and multiplication, both $ad + bc$ and bd are integers. Furthermore, Theorem 1.10 implies that $bd \neq 0$. Thus, it follows from Definition 6.2 that

$$\frac{a}{b} + \frac{c}{d} = \frac{ad + bc}{bd} \in \mathbb{Q},$$

as desired. ■

Notice that this proof uses facts that we have already assumed or proved for $\mathbb{Z}$ in order to establish a related result for $\mathbb{Q}$. Similar arguments can be used to prove several other properties of addition and multiplication in $\mathbb{Q}$, including closure under multiplication, associativity of addition and multiplication, commutativity of addition and multiplication, and distribution of multiplication over addition.

Activity 6.3.

(a) Does $\mathbb{Q}$ contain an additive identity? If so, what is it?

(b) Does every element of $\mathbb{Q}$ have an additive inverse?

(c) Does $\mathbb{Q}$ contain a multiplicative identity? If so, what is it?

(d) Which elements of $\mathbb{Q}$ are units—that is, which elements of $\mathbb{Q}$ have a multiplicative inverse in $\mathbb{Q}$? Are there elements of $\mathbb{Q}$ (besides 0) that do not have a multiplicative inverse in $\mathbb{Q}$?

(e) Does $x \cdot 0 = 0$ for all $x \in \mathbb{Q}$? (This property is sometimes called the *zero property of multiplication*.)

(f) Does $\mathbb{Q}$ contain any *zero divisors*? That is, do there exist nonzero elements x, $y \in \mathbb{Q}$ such that $xy = 0$?

(g) Does *additive cancellation* hold in $\mathbb{Q}$? That is, if

$$x + z = y + z$$

for some x, y, $z \in \mathbb{Q}$, does it follow that $x = y$?

(h) Does *multiplicative cancellation* hold in $\mathbb{Q}$? That is, if

$$x \cdot z = y \cdot z$$

for some x, y, $z \in \mathbb{Q}$, does it follow that $x = y$? Would your answer change if z was assumed to be nonzero?

(i) Is there a natural ordering defined on the elements of $\mathbb{Q}$? If so, does this ordering satisfy the axioms listed on page 8?

In Activity 6.3, we proved a variety of important properties of the rational numbers. But what about the *irrational numbers*—that is, the real numbers that are not rational?

There is no standard symbol for denoting the set of irrational numbers, but for convenience here, we will use the symbol $\mathbb{J}$, so that $\mathbb{R} = \mathbb{Q} \cup \mathbb{J}$. If we were to go through the same list of properties that we proved for $\mathbb{Q}$ and try to prove them for $\mathbb{J}$ instead, we wouldn't get very far before we ran into trouble. In fact, the very first property we considered, closure under addition, does not hold in $\mathbb{J}$. To see this, note that if $x \in \mathbb{J}$, then $-x \in \mathbb{J}$ also. (The proof of this simple fact is left to you in Exercise 1.) But then $x + (-x) = 0 \notin \mathbb{J}$. Note that the same issue would arise if we investigated the odd integers, say $\mathbb{O}$, in more detail.

It's hard to imagine doing any kind of meaningful arithmetic or algebra in a number system that isn't even closed under its operations. For this reason, we will restrict our attention to number systems that are closed under both addition and multiplication. As we just saw, this restriction rules out both $\mathbb{J}$ and $\mathbb{O}$. Note that if we also required closure under subtraction, then we would have to eliminate both $\mathbb{N}$ and $\mathbb{W}$ from consideration as well.

The Complex Numbers

Having considered several of the most familiar subsets of the real numbers, let's now turn our attention to the complex numbers, defined formally below.

Definition 6.4. A **complex number** is any number of the form $a + bi$, where a and b are real numbers and i is an imaginary number with the property that $i^2 = -1$. For a complex number $x = a + bi$, the real number a is called the **real part** of x, and the real number b is called the **imaginary part** of x.

The set of all complex numbers is denoted $\mathbb{C}$, so that

$$\mathbb{C} = \{a + bi : a \in \mathbb{R}, b \in \mathbb{R}, \text{ and } i^2 = -1\}.$$

Two complex numbers, $a + bi$ and $c + di$, are considered equal if and only if both their real and imaginary parts are equal. In other words,

$$a + bi = c + di \text{ if and only if } a = c \text{ and } b = d.$$

Furthermore, addition and multiplication within $\mathbb{C}$ are defined by

$$(a + bi) + (c + di) = (a + c) + (b + d)i$$

and

$$(a + bi) \cdot (c + di) = (ac - bd) + (ad + bc)i$$

for all $a + bi, c + di \in \mathbb{C}$.

Note that since every real number x can be written in the form $x + 0i$, it follows that the real numbers are a subset of the complex numbers. Not surprisingly, the complex numbers satisfy many of the same properties as the real numbers, and most of these properties are fairly easy to prove. For instance, to show that addition is commutative in $\mathbb{C}$, it suffices to observe that

$$\begin{aligned}(a + bi) + (c + di) &= (a + c) + (b + d)i \ \textcircled{?} \\ &= (c + a) + (d + b)i \ \textcircled{?} \\ &= (c + di) + (a + bi) \ \textcircled{?}\end{aligned}$$

for all $a + bi$, $c + di \in \mathbb{C}$. Notice that, like some of the other arguments we have seen, this one used the fact that addition is commutative in $\mathbb{R}$ to establish an analogous result for $\mathbb{C}$. Many other properties of $\mathbb{C}$ can be proved in a similar manner, but some require more work.

Take, for example, the existence of multiplicative inverses within $\mathbb{C}$. How can we determine which elements of $\mathbb{C}$ are units? Although it is easy to show that $1 + 0i$ is a multiplicative identity for $\mathbb{C}$, it is more difficult to find the multiplicative inverse of an arbitrary nonzero element of $\mathbb{C}$. To illustrate, let $a + bi \in \mathbb{C}$. At first glance, we may be tempted to say that since

$$(a + bi) \cdot \frac{1}{a + bi} = 1 = 1 + 0i,$$

the multiplicative inverse of $a + bi$ is just $\frac{1}{a+bi}$. But is $\frac{1}{a+bi}$ even an element of $\mathbb{C}$? It certainly isn't written in the form $c + di$ for some real numbers c and d. In fact, it's not even clear what exactly $\frac{1}{a+bi}$ would represent in light of Definition 6.4.

The real question then is this: Can we find real numbers c and d such that

$$(a + bi) \cdot (c + di) = 1 + 0i?$$

In other words, can we find real numbers c and d such that

$$ac - bd = 1$$

and

$$ad + bc = 0?$$

Activity 6.5.

(a) Viewing a and b as constants, solve the following system of equations for c and d:

$$\begin{aligned} ac - bd &= 1 \\ ad + bc &= 0 \end{aligned}$$

(b) Verify that, for the values of c and d you found in part (a),

$$(a + bi) \cdot (c + di) = 1 + 0i.$$

Activity 6.5 demonstrates that every nonzero element of $\mathbb{C}$ has a multiplicative inverse. We will have a chance to consider several other important properties of $\mathbb{C}$ later on in the investigation. For now, however, let's shift gears and briefly consider two other important types of number systems.

Matrices

Given a number system, say $\mathbb{R}$, we can define a new number system, denoted $\mathcal{M}_{n\times n}(\mathbb{R})$, that consists of all $n \times n$ matrices whose entries are elements of $\mathbb{R}$, with addition and multiplication of matrices defined in the usual way. As with the complex numbers, some properties of $\mathcal{M}_{n\times n}(\mathbb{R})$ are easy to establish, and others are more difficult, mainly because the calculations involved can become fairly tedious. As an example of one of the more straightforward properties, let's try to find an additive inverse of an arbitrary element of $\mathcal{M}_{2\times 2}(\mathbb{R})$. Note that for any a, b, c, $d \in R$,

$$\begin{bmatrix} a & b \\ c & d \end{bmatrix} + \begin{bmatrix} -a & -b \\ -c & -d \end{bmatrix} = \begin{bmatrix} a + (-a) & b + (-b) \\ c + (-c) & d + (-d) \end{bmatrix} = \begin{bmatrix} 0 & 0 \\ 0 & 0 \end{bmatrix}.$$

Since $\begin{bmatrix} 0 & 0 \\ 0 & 0 \end{bmatrix}$ is the additive identity in $\mathcal{M}_{2\times 2}(\mathbb{R})$, it follows that $\begin{bmatrix} -a & -b \\ -c & -d \end{bmatrix}$ is an additive inverse of $\begin{bmatrix} a & b \\ c & d \end{bmatrix}$.

Like $\mathbb{Q}$ and $\mathbb{C}$, the set $\mathcal{M}_{n\times n}(\mathbb{R})$ has a multiplicative identity (the familiar identity matrix). Unlike $\mathbb{Q}$ and $\mathbb{C}$, however, not every element of $\mathcal{M}_{n\times n}(\mathbb{R})$ is a unit—a fact that we should recall from linear algebra. In fact, the units in $\mathcal{M}_{n\times n}(\mathbb{R})$—that is, the elements of $\mathcal{M}_{n\times n}(\mathbb{R})$ that have a multiplicative inverse within $\mathcal{M}_{n\times n}(\mathbb{R})$—are just the invertible matrices.

Collections of Sets

Most of the number systems we have looked at so far have involved operations that are at least in some way related to addition and multiplication of real numbers. The last type of number system we will consider is like $\mathbb{Z}_n$ in the sense that its elements are sets. Unlike $\mathbb{Z}_n$, however, addition and multiplication in this new number system are defined in terms of standard set operations such as unions and intersections. We define these operations, and the corresponding family of number systems, as follows:

Definition 6.6.

- Let S be a set. The **power set** of S, denoted $\mathcal{P}(S)$, is the collection of all subsets of S. That is,
$$\mathcal{P}(S) = \{T : T \subseteq S\}.$$
- For any sets A and B, the **symmetric difference** of A and B, denoted $A \triangle B$, is the set of all elements that belong to either A or B, but not both. That is,
$$A \triangle B = \{x : x \in A \cup B \text{ and } x \notin A \cap B\}.$$
- For any natural number n, the number system $\mathcal{P}_n$ is the power set of the set $\{1, 2, \ldots, n\}$, with addition defined as symmetric difference, and multiplication defined as intersection. In other words,
$$\mathcal{P}_n = \mathcal{P}(\{1, 2, \ldots, n\}),$$
with
$$A + B = A \triangle B \quad \text{and} \quad A \cdot B = A \cap B$$
for all A, $B \in \mathcal{P}_n$.

Note that we have used symmetric differences, rather than just unions, to define addition in $\mathcal{P}_n$. This is because defining addition as set union turns out to be problematic. (Exercise 7 explores why this is the case.)

To become more familiar with $\mathcal{P}_n$, consider the following addition and multiplication tables for $\mathcal{P}_2$.

$+$	$\emptyset$	$\{1\}$	$\{2\}$	$\{1,2\}$
$\emptyset$	$\emptyset$	$\{1\}$	$\{2\}$	$\{1,2\}$
$\{1\}$	$\{1\}$	$\emptyset$	$\{1,2\}$	$\{2\}$
$\{2\}$	$\{2\}$	$\{1,2\}$	$\emptyset$	$\{1\}$
$\{1,2\}$	$\{1,2\}$	$\{2\}$	$\{1\}$	$\emptyset$

$\cdot$	$\emptyset$	$\{1\}$	$\{2\}$	$\{1,2\}$
$\emptyset$	$\emptyset$	$\emptyset$	$\emptyset$	$\emptyset$
$\{1\}$	$\emptyset$	$\{1\}$	$\emptyset$	$\{1\}$
$\{2\}$	$\emptyset$	$\emptyset$	$\{2\}$	$\{2\}$
$\{1,2\}$	$\emptyset$	$\{1\}$	$\{2\}$	$\{1,2\}$

From these tables, we can observe many important properties of $\mathcal{P}_2$. For instance, it is easy to see that both addition and multiplication are commutative in $\mathcal{P}_2$. To prove a property like this in general—that is, for $\mathcal{P}_n$ and not just for $\mathcal{P}_2$—we would need to construct a more general set theoretic argument. To illustrate what such an argument might look like, let's consider a property that is somewhat more involved than commutativity. In particular, we will show that addition is associative in $\mathcal{P}_n$. Throughout the proof, we have left out some details for you to fill in as you read.

Proof. Let $A, B, C \in \mathcal{P}_n$. We will argue that $(A \triangle B) \triangle C \subseteq A \triangle (B \triangle C)$ and $A \triangle (B \triangle C) \subseteq (A \triangle B) \triangle C$.(?)

For the forward inclusion, let $x \in (A \triangle B) \triangle C$. Then either $x \in A \triangle B$ and $x \notin C$, or ________ and ____________. Consider these two cases:

Case 1: $x \in A \triangle B$ *and* $x \notin C$. In this case, $x \in A$ and $x \notin B$, or ________ and ________. If $x \in A$ and $x \notin B$, then since ________ also, it follows that $x \notin B \triangle C$. Thus, $x \in A \triangle (B \triangle C)$.(?) In the case that $x \in B$ and $x \notin A$, it must be that $x \in B \triangle C$,(?) which again implies that $x \in A \triangle (B \triangle C)$,(?) as desired.

Case 2: $x \in C$ *and* $x \notin A \triangle B$. Since $x \notin A \triangle B$, it follows that either ________ and ________, or ________ and ________. For the former, since $x \notin B$ and $x \in C$, it must be that $x \in B \triangle C$. But ________, and so it follows that $x \in A \triangle (B \triangle C)$. For the latter, note that since $x \in B$ and $x \in C$, it is again the case $x \notin B \triangle C$.(?) Thus, $x \in A \triangle (B \triangle C)$ in this case as well.

The above two cases establish that $(A \triangle B) \triangle C \subseteq A \triangle (B \triangle C)$. A similar argument establishes the reverse inclusion. Thus, $(A \triangle B) \triangle C = A \triangle (B \triangle C)$, as desired. ■

While formal proofs like the one above are necessary in order to prove conjectures involving set operations, making those conjectures in the first place often requires the use of less formal exploratory tools, such as Venn diagrams. For instance, the Venn diagrams in Figure 6.1 illustrate the associative property that we just proved. You may find Venn diagrams such as these to be helpful as you explore other properties of $\mathcal{P}_n$ in the next section.

Putting It All Together

In this investigation, we have introduced several new number systems, and we have identified some of the algebraic properties that each of these systems satisfy. Table 6.2 lists the number systems and

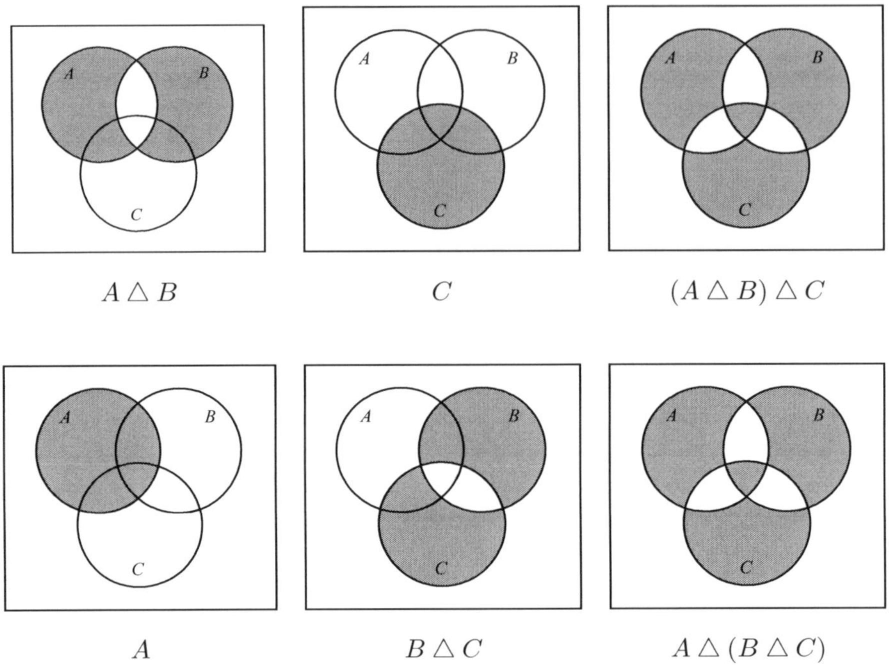

Figure 6.1
Venn diagrams illustrating $(A \triangle B) \triangle C = A \triangle (B \triangle C)$.

properties we have considered. To summarize our work thus far, fill in this table, using the following codes to indicate whether each number system satisfies the given properties:

- **Y**, to indicate that you have proved the property or are confident that it holds.
- **N**, to indicate that the property does not hold, and you have found a counterexample to illustrate its failure.
- **Y/N**, to indicate, for $\mathbb{Z}_n$, $\mathcal{M}_{n\times n}(\mathbb{R})$, or $\mathcal{P}_n$, that the property holds for some values of n but not for others. In this case, you should be able to give at least one value of n for which the property holds and at least one for which it does not. For an extra challenge, try to give a precise description of all of the values of n for which the property does hold.
- **?**, to indicate that you have doubts about whether the property holds, but you have not been able to find a counterexample to disprove it.
- **N/A**, to indicate that the property does not apply. (For instance, it wouldn't make sense to talk about multiplicative inverses in a number system that does not have a multiplicative identity.)

For the purposes of this activity, it is not necessary to prove every property that you suspect is true, but you should be able to find counterexamples for those that are not true. When the variable n is used, you may assume that $n > 1$. Furthermore, you may assume that the variable p (as in $\mathbb{Z}_p$) represents a prime number.

Property	$\mathbb{Z}$	$\mathbb{E}$	$\mathbb{Q}$	$\mathbb{R}$	$\mathbb{C}$	$\mathbb{Z}_n$	$\mathbb{Z}_p$	$\mathcal{M}_{n\times n}(\mathbb{R})$	$\mathcal{P}_n$
The system is closed under addition.									
The system is closed under multiplication.									
Addition is associative.									
Multiplication is associative.									
Addition is commutative.									
Multiplication is commutative.									
Multiplication distributes over addition.									
There is an additive identity.									
Every element has an additive inverse.									
There is a multiplicative identity.									
Every nonzero element has a multiplicative inverse.									
The zero property of multiplication holds.									
There are no zero divisors.									
Additive cancellation holds.									
Multiplicative cancellation (of nonzero elements) holds.									
The ordering axioms hold.									

Table 6.2
Properties of various number systems.

Concluding Activities

Activity 6.7. Suppose that you were asked to solve the equation

$$x^2 + [3]x = [4]$$

in $\mathbb{Z}_6$.

(a) Would the following solution be correct?

$$x^2 + [3]x = [4]$$
$$x^2 + [3]x + [2] = [0]$$
$$(x + [1])(x + [2]) = [0]$$
$$\text{Therefore, } x = [5] \text{ or } x = [4].$$

If so, explain the logic behind each step. Otherwise, explain why the solution is not correct.

(b) Evaluate $x^2 + [3]x$ for every $x \in \mathbb{Z}_6$. Are your findings consistent with your answer for part (a)?

(c) Look back to part (g) of Preview Activity 6.1. Does $\gamma \cdot x = \gamma \cdot \beta$ imply $x = \beta$? Why or why not?

Activity 6.8. When we discussed number systems whose elements were matrices, we considered only square ($n \times n$) matrices. Why is this? Would it be possible to define a meaningful number system using non-square matrices? Why or why not?

Activity 6.9. Prove that for all $n \in \mathbb{N}$ and $a \in \mathbb{Z}$, the following statements are equivalent:

(i) $\gcd(a, n) = 1$.

(ii) $[a]$ has a multiplicative inverse in $\mathbb{Z}_n$.

(iii) The function $g : \mathbb{Z}_n \to \mathbb{Z}_n$ defined by $g([x]) = [ax]$ is injective.

Activity 6.10. The rational numbers can be defined formally as equivalence classes of ordered pairs of integers. In particular, let

$$Q = \mathbb{Z} \times (\mathbb{Z} - \{0\}) = \{(a, b) \in \mathbb{Z} \times \mathbb{Z} : b \neq 0\},$$

and define the following equivalence relation on Q:

$$(a, b) \sim (x, y) \text{ if and only if } ay = bx$$

Let $\mathbb{Q}$ be the set of all equivalence classes of this relation, and define addition and multiplication on $\mathbb{Q}$ as follows:

$$[(a, b)] + [(c, d)] = [(ad + bc, bd)] \text{ and } [(a, b)][(c, d)] = [(ac, bd)]$$

Using only these definitions and the axioms of the integers, prove that addition and multiplication are well-defined binary operations on $\mathbb{Q}$. Then give two compelling mathematical reasons why we do NOT define the sum of $\frac{a}{b}$ and $\frac{c}{d}$ in $\mathbb{Q}$ to be $\frac{a+c}{b+d}$.

Activity 6.11. Write a short summary that describes the important concepts, and the relationships between these concepts, that were introduced in this investigation. Explain how the material in this investigation is connected to the material in Investigations 1 and 5.

Exercises

⋆ (1) Let $\mathbb{J}$ denote the set of all irrational numbers. Prove that if $x \in \mathbb{J}$, then $-x \in \mathbb{J}$ also.

⋆ (2) A **Gaussian integer** is a complex number whose real and imaginary parts are both integers. The set of Gaussian integers is usually denoted $\mathbb{Z}[i]$, so that

$$\mathbb{Z}[i] = \{a + bi : a \in \mathbb{Z}, b \in \mathbb{Z}, \text{ and } i^2 = -1\}.$$

Which of the properties in Table 6.2 are satisfied by $\mathbb{Z}[i]$? Use what you know about $\mathbb{C}$ to give a proof or counterexample for each property.

⋆ (3) Let $n \geq 2$ be an integer. Which properties from Table 6.2 are satisfied by each of the following number systems? Give a proof or counterexample for each property, possibly using properties that we have already stated or proved for $\mathcal{M}_{n\times n}(\mathbb{R})$.

(a) $\mathcal{M}_{n\times n}(\{0, 1\})$

(b) $\mathcal{M}_{n\times n}(\mathbb{Z})$

(c) $\mathcal{M}_{n\times n}(\mathbb{E})$

(d) $\mathcal{M}_{n\times n}(\mathbb{Z}_5)$

(4) Let $m, n \in \mathbb{N}$, and let $A, B \in \mathcal{M}_{m\times n}(\mathbb{R})$. The **Hadamard product** of A and B is the matrix $A \cdot B$ whose (i, j) entry is equal to the product of the (i, j) entries of A and B. That is,

$$(A \cdot B)_{i,j} = A_{i,j} \cdot B_{i,j}.$$

With multiplication defined by the Hadamard product, and addition defined as usual, which of the properties from Table 6.2 are satisfied by $\mathcal{M}_{m\times n}(\mathbb{R})$? Give a proof or counterexample for each property.

(5) For any subset S of $\mathbb{R}$, let S^c denote the complement of S in $\mathbb{R}$; that is,

$$S^c = \{x \in \mathbb{R} : x \notin S\}.$$

Assuming that the ordering axioms hold in $\mathbb{R}$ just as they do in $\mathbb{Z}$, answer each of the following questions:

(a) Does there exist a nonempty, proper subset S of $\mathbb{R}$ such that both S and S^c are closed under addition? Prove your answer.

(b) Does there exist a nonempty, proper subset S of $\mathbb{R}$ such that both S and S^c are closed under multiplication? Prove your answer.

(6) Find all of the units in $\mathcal{P}_n$. Verify your answer.

⋆ (7) Suppose that addition in $\mathcal{P}_n$ was defined to be set union instead of symmetric difference. Which properties from Table 6.2 would be satisfied in this case?

(8) Let n be a positive integer

(a) Is the identity matrix I_n a unit in $\mathcal{M}_{n\times n}(\mathbb{R})$? Explain.

(b) Assume A and B are units in $\mathcal{M}_{n\times n}(\mathbb{R})$. Is the product AB a unit in $\mathcal{M}_{n\times n}(\mathbb{R})$? Prove your answer. (Be careful; is matrix multiplication commutative?)

(9) Assume that $*$ is a binary operation on a set S. Suppose that a, b, and c are elements of S with $a = b$. Explain why $a * c = b * c$. (Hint: Think of $*$ as a function.)

(10) Which of the following operations are well-defined? Verify your answers.

(a) The operation $\odot$ defined on $\mathbb{Q}$ by $\frac{a}{b} \odot \frac{c}{d} = \frac{adbc}{b+d}$.

(b) The operation $*$ defined on $\mathbb{Z}_7$ by $[a] * [b] = [2a + 3b]$.

(c) The operation $\star$ defined on $\mathcal{P}_n$ by

$$\{a_1, a_2, \ldots, a_j\} \star \{b_1, b_2, \ldots, b_k\} = \{a_1, b_1\}.$$

(11) Define a **blip** to be a pair of integers, denoted $\langle a, b\rangle$, and define two blips $\langle a, b\rangle$ and $\langle x, y\rangle$ to be equal whenever $a + b = x + y$ (so that, for instance, $\langle 3, 5\rangle$ and $\langle 10, -2\rangle$ would be considered equal since $3 + 5 = 8 = 10 + (-2)$). Using these definitions, decide whether each of the following are well-defined operations on the set B of all blips. Give a proof or counterexample to justify each of your answers.

(a) The operation $*$ defined by $\langle a, b\rangle * \langle c, d\rangle = \langle a + c, b + d\rangle$.

(b) The operation $\bullet$ defined by $\langle a, b\rangle \bullet \langle c, d\rangle = \langle a^2 + c^2, b^2 + d^2\rangle$.

Part III

Rings

Investigation 7

An Introduction to Rings

Focus Questions

By the end of this investigation, you should be able to give precise and thorough answers to the questions listed below. You may want to keep these questions in mind to focus your thoughts as you complete the investigation.

- What is a ring, and what are some special types of rings?
- What uniqueness properties are satisfied by identities and inverses in rings, and why do these properties hold?
- What is a field? What is an integral domain? What is the relationship between fields and integral domains, and why does this relationship hold?

Preview Activity 7.1. In Investigation 6, we considered a variety of different number systems, some familiar and some not. While there were significant differences between these number systems, there were also common features that seemed to be shared by all of them. In this investigation, we will focus on these common features and their implications. With that in mind, look back at Table 6.2 (on page 73), and make a list of the properties that were satisfied by *all* of the number systems from Investigation 6. Include in your list the properties that you were able to prove as well as those that you suspected were true but were not able to prove.

Introduction

In Preview Activity 7.1, we identified a set of properties that seemed to be satisfied by a variety of different number systems, such as $\mathbb{Z}$, $\mathbb{E}$, $\mathbb{Q}$, $\mathbb{R}$, $\mathbb{C}$, $\mathbb{Z}_n$, $\mathcal{M}_{n\times n}(\mathbb{R})$, and $\mathcal{P}_n$. All of these number systems are examples of a special type of algebraic structure known as a *ring*.* The properties that they have in common can be used to define rings in general, as follows:

*The phrase *number ring* (from the German *Zahlring*), shortened to *ring*, is due to Hilbert in his *Zahlbericht*. Hilbert used this term to refer to certain collections of algebraic integers. Some speculate that Hilbert used the word ring because of the cyclical (ring-shaped) behavior of powers of certain algebraic integers.

Definition 7.2. A **ring** is a set R together with two binary operations, called addition ($+$) and multiplication ($\cdot$), such that all of the following axioms hold:

The Ring Axioms

- **The set R is closed under addition and multiplication**, meaning that for all x, $y \in R$, $x + y \in R$ and $x \cdot y \in R$.
- **Addition is associative**, meaning that for all x, y, $z \in R$, $(x + y) + z = x + (y + z)$.
- **Addition is commutative**, meaning that for all x, $y \in R$, $x + y = y + x$.
- **The set R contains an additive identity**, also called a **zero element**, meaning that there exists some element $0_R \in R$ such that $x + 0_R = x$ for all $x \in R$.
- **Every element of R has an additive inverse within R**, meaning that for every $x \in R$, there exists $y \in R$ such that $x + y = 0_R$.
- **Multiplication is associative**, meaning that for all x, y, $z \in R$, $(x \cdot y) \cdot z = x \cdot (y \cdot z)$.
- **Multiplication distributes over addition**, meaning that for all x, y, $z \in R$, $x \cdot (y + z) = x \cdot y + x \cdot z$ and $(x + y) \cdot z = x \cdot z + y \cdot z$.

Consistent with the usual convention, we will often omit the symbol for multiplication ($\cdot$), writing xy instead of $x \cdot y$. When the context is clear, we will also omit the subscript from our notation for the zero element, writing 0 instead of 0_R. There are some situations in which we would want to include this subscript—for instance, to differentiate between the zero elements in two different rings, or to distinguish the zero element in a ring from the integer 0. Most of the time, however, we will be able to use 0 in lieu of 0_R without any confusion or ambiguity.

Definition 7.2 raises a number of questions that will form the basis of our work in this investigation. These questions include:

- Why are the ring axioms listed the only ones included in the definition of a ring? What other properties are satisfied by all rings, and how can these properties be proved from the ring axioms?
- What are the differences between addition and multiplication in rings, and what special types of rings can be defined in light of these differences?

Before we try to answer these questions, a quick reminder is in order. Recall that in Investigation 5, we defined a binary operation on a set S to be a function from $S \times S$ to S. Because the codomain of a binary operation on a set S is always S itself, closure is guaranteed whenever we have a well-defined operation. Thus, our first ring axiom is not entirely necessary; we include it in the definition to remind us more explicitly that every ring must be closed under its operations. Indeed, if this were not the case, then the ring operations themselves would not be true binary operations in the sense of Definition 5.14. (See page 53.)

Basic Properties of Rings

The ring axioms themselves should seem quite familiar, since we have used them both implicitly and explicitly throughout previous investigations. But what about the other properties we have studied? You may have noticed that at least a few of these properties (including some that were satisfied by all of the number systems from Investigation 6) were not included in Definition 7.2. Do these properties in fact hold for all rings, and if so, why are they not included in our list of ring axioms?

To answer this question, we must think back to our discussion in Investigation 1 regarding the difference between axioms and theorems. There are certain properties, such as additive cancellation and the zero property of multiplication, that are in fact satisfied by all rings. These properties, however, can be proved from the ring axioms. Thus, to include them would cause our axiom system to be redundant. The primary benefit of our definition of a ring is that it provides a minimal set of axioms from which numerous other algebraic properties and theorems can be proved. Moreover, any property that we can prove using only the ring axioms must necessarily hold in *every* number system that satisfies these axioms. Thus, the theory of rings gives us a way to study algebra more abstractly, instead of just within the context of specific number systems. In fact, the entire field of abstract algebra revolves around the study of general algebraic structures, such as rings, and their applications.

With that said, let's now formally state and prove the two properties we just mentioned (additive cancellation and the zero property of multiplication). The integer versions of each of these properties were part of Investigation 1—the former as Exercise 2 (see page 10) and the latter as Activity 1.6. (See page 7.) As it turns out, the arguments we used there generalize easily to the context of arbitrary rings.

Theorem 7.3. *Let R be a ring. For all x, y, $z \in R$, if $x + z = y + z$, then $x = y$.*

Activity 7.4. Prove Theorem 7.3.

Theorem 7.5. *Let R be a ring. Then $0x = 0 = x0$ for all $x \in R$.*

Activity 7.6. Prove Theorem 7.5. (Hint: Start with the fact that $0_R + 0_R = 0_R$, and multiply both sides of this equation by an arbitrary element $x \in R$.)

Commutative Rings and Rings with Identity

In the definition of a ring, you may have noticed similarities between the axioms for addition and the axioms for multiplication. For instance, both addition and multiplication satisfy the closure axiom, and both are associative. With that said, there are also some significant differences between the two operations. One of these differences is that multiplication is not required to be commutative. This observation leads to the following definition:

Definition 7.7. Let R be a ring. Then R is said to be **commutative** if multiplication in R is commutative—that is, if $xy = yx$ for all x, $y \in R$.

Notice that when we call a ring *commutative*, we are always referring to the multiplication operation. This is because addition is guaranteed to be commutative in *every* ring. Thus, the important distinguishing feature with regard to commutativity is whether or not multiplication commutes.

A similar distinction can be made with regard to additive and multiplicative identities. In particular, we call an additive identity (which every ring must have) a *zero element*, and we refer to a multiplicative identity (which a ring may or may not have) as simply an identity. The next definition formalizes this language.

Definition 7.8. Let R be a ring. An **identity** for R is an element $1_R \in R$ such that $1_R \neq 0_R$ and $1_R \cdot x = x = x \cdot 1_R$ for all $x \in R$. If such an element exists, then R is said to be a **ring with identity**.

Note that, unlike the definition of a zero element, it is necessary to specify in the definition of an identity that both $1_R \cdot x$ and $x \cdot 1_R$ are equal to x. This is again due to the fact that addition is commutative in every ring, but multiplication may not be. Note also that, as with zero elements, we will often omit the subscript from our notation for an identity, writing 1 instead of 1_R whenever the context is clear.

We have seen many examples of commutative rings, including $\mathbb{Z}$, $\mathbb{E}$, $\mathbb{Q}$, $\mathbb{R}$, $\mathbb{C}$, $\mathbb{Z}_n$, and $\mathcal{P}_n$. In fact, the only family of non-commutative rings we have studied is $\mathcal{M}_{n\times n}(\mathbb{R})$. Of the commutative rings mentioned above, all but $\mathbb{E}$ have identity. The non-commutative ring $\mathcal{M}_{n\times n}(\mathbb{R})$ also has identity. Thus, the rings we have studied so far can be divided into the following categories:

- **Commutative rings with identity:** $\mathbb{Z}$, $\mathbb{Q}$, $\mathbb{R}$, $\mathbb{C}$, $\mathbb{Z}_n$, $\mathcal{P}_n$
- **Commutative rings without identity:** $\mathbb{E}$
- **Non-commutative rings with identity:** $\mathcal{M}_{n\times n}(\mathbb{R})$

Activity 7.9. Find a non-commutative ring without identity, or prove that no such ring exists.

Uniqueness of Identities and Inverses

As we suggested in Investigation 1, nothing in the definition of a zero element or a multiplicative identity requires uniqueness. That is, the definitions alone do not rule out the possibility of a ring having more than one zero element or identity. Fortunately, however, uniqueness does hold for both zero elements and multiplicative identities, as stated in the theorems below.

Theorem 7.10. *Let R be a ring, and suppose that both 0 and $0'$ are zero elements for R. Then $0 = 0'$.*

Theorem 7.11. *Let R be a ring, and suppose that both 1 and $1'$ are identities for R. Then $1 = 1'$.*

Activity 7.12. Let R be a ring, and suppose that 0 and $0'$ are zero elements for R.

(a) Let a be any element of R. What must $a + 0$ and $a + 0'$ equal, and why?

(b) Use your answer to part (a) to equate $a + 0$ and $a + 0'$.

(c) What axiom or theorem, along with your answer to part (b), allows you to conclude that $0 = 0'$?

(d) Combine your work in parts (a) – (c) to write a proof of Theorem 7.10.

(e) Explain why the strategy from parts (a) – (c) would be invalid for a proof of Theorem 7.11. (Hint: See part (c).)

(f) Prove Theorem 7.11 by evaluating $1 \cdot 1'$ in two different ways.

Because of Theorems 7.10 and 7.11, we can now refer to *the* zero element and *the* identity (when it exists) of a ring. Similar uniqueness results can be established for both additive and multiplicative inverses, the latter of which is defined formally below.

Definition 7.13. Let R be a ring with identity, and let $x \in R$. An element $y \in R$ is said to be a **multiplicative inverse** of x provided that $xy = 1 = yx$.

It is important to note that the definition of a ring does not require any particular ring element to actually have a multiplicative inverse. In fact, the very notion of a multiplicative inverse makes sense only in rings with identity, as suggested by Definition 7.13. Furthermore, we have seen several examples of rings with identity—for instance, $\mathcal{M}_{n\times n}(\mathbb{R})$, $\mathcal{P}_n$, and $\mathbb{Z}_n$ for certain values of n—in which numerous ring elements do not have a multiplicative inverse. In Investigation 5, we called elements of $\mathbb{Z}_n$ that do have a multiplicative inverse *units*. This definition generalizes to arbitrary rings as follows:

Definition 7.14. Let R be a ring with identity. An element $x \in R$ is said to be a **unit** provided that R contains a multiplicative inverse for x. In other words, $x \in R$ is a unit if and only if there exists $y \in R$ such that $xy = 1 = yx$.

The next two theorems show that both additive and multiplicative inverses (when they exist) must be unique. Their proofs use ideas similar to those in Activity 7.12. As you read these proofs, you should be able to fill in the missing details and provide additional explanation or justification wherever it is needed.

Theorem 7.15. *Let R be a ring, and let $x \in R$. Suppose that both y and y' are additive inverses for x. Then $y = y'$.*

Proof. Let R be a ring, let $x \in R$, and let y and y' be additive inverses for x. Then $x + y =$ ____ and $x + y' =$ ____. Thus, $x + y = x + y'$, and ____________________ implies that $y = y'$, as desired. ∎

Theorem 7.16. *Let R be a ring with identity, and let $x \in R$ be a unit. Suppose that both y and y' are multiplicative inverses for x. Then $y = y'$.*

Proof. Let R be a ring with identity, let $x \in R$ be a unit, and let y and y' be multiplicative inverses for x. Then $xy = 1 = xy'$, which implies that

$$y(xy) = y(xy').^{(?)}$$

But then

$$(yx)y = (yx)y',^{(?)}$$

which implies that $y = y'$,$^{(?)}$ as desired. ∎

Now that we have shown that additive and multiplicative inverses must be unique, we can use the standard notations of $-x$ (for the additive inverse of a ring element x) and x^{-1} (for the multiplicative inverse, if one exists) without any risk of ambiguity. We will use these notations regularly throughout subsequent investigations. We will also define subtraction within an arbitrary ring R just as we did in $\mathbb{Z}$—that is,

$$x - y = x + (-y)$$

for all x, $y \in R$.

Zero Divisors and Multiplicative Cancellation

In Investigation 6, you should have noticed that in some number systems, multiplicative cancellation, even of nonzero elements, does not hold. For instance, consider the following equation in $\mathbb{Z}_6$:

$$[2]x = [2]y$$

If we were to naively apply the same cancellation rules that apply in $\mathbb{Z}$ (and many other rings), we might conclude that $x = y$. This conclusion, however, would not be valid, since

$$[2][2] = [4] = [10] = [2][5],$$

but clearly $[2] \neq [5]$.

So under what circumstances does multiplicative cancellation hold? One obvious answer to this question is when the element being canceled is a unit. In this case, the proof of Theorem 7.3 generalizes easily to multiplication, yielding the following result:

Theorem 7.17. *Let R be a ring with identity, and let z be a unit in R. For all x, $y \in R$, if $xz = yz$, then $x = y$. Similarly, if $zx = zy$, then $x = y$.*

Theorem 7.17 provides a sufficient condition for multiplicative cancellation to hold, but this condition is far from being necessary. For instance, the only units in $\mathbb{Z}$ are 1 and -1, but multiplicative cancellation works in $\mathbb{Z}$ as long as the integer being canceled is nonzero. The same can be said of $\mathbb{E}$, which has no identity and therefore no units. In these rings, the validity of multiplicative cancellation rests on the fact that the elements being canceled are not *zero divisors*, defined formally as follows:

Definition 7.18. Let R be a ring. An element $x \in R$ is said to be a **zero divisor** if $x \neq 0$, and $xy = 0$ or $yx = 0$ for some nonzero $y \in R$.

We first studied zero divisors, and their relation to multiplicative cancellation, all the way back in Investigation 1. (See Activity 1.9 on page 9.) We revisited these ideas again in Investigations 5 and 6 (Activities 5.18, 5.22, and 6.7 on pages 55, 56, and 74, respectively), where we investigated zero divisors within the context of $\mathbb{Z}_n$ and another number system. The ideas from these activities can be generalized to arbitrary rings, as stated in the next two theorems. The proof of the first theorem is left as an exercise, and the proof of the second should mirror your work in part (c) of Activity 1.9.

Theorem 7.19. *Let R be a ring. The following statements are equivalent:*

- *R contains no zero divisors.*
- *For all x, $y \in R$, if $xy = 0$, then $x = 0$ or $y = 0$.*
- *For all x, $y \in R$, if $xy = 0$ and $x \neq 0$, then $y = 0$.*

- *For all $x, y \in R$, if $yx = 0$ and $x \neq 0$, then $y = 0$.*

Theorem 7.20. *Let R be a ring, and let z be a nonzero element of R that is not a zero divisor. For all $x, y \in R$, if $xz = yz$, then $x = y$. Similarly, if $zx = zy$, then $x = y$.*

Proof. Let R be a ring, let z be a nonzero element of R that is not a zero divisor, and let $x, y \in R$. Suppose that $xz = yz$. Then

$$xz + (-y)z = yz + (-y)z,$$

which implies that

$$\begin{aligned}(x + (-y))z &= (y + (-y))z^{\text{(?)}} \\ &= 0z^{\text{(?)}} \\ &= 0.^{\text{(?)}}\end{aligned}$$

Since $(x + (-y))z = 0$, and since z is neither zero nor a zero divisor, it must be the case that $x + (-y) = 0.^{\text{(?)}}$ Thus,

$$(x + (-y)) + y = 0 + y,$$

which implies that

$$x + (y + (-y)) = y.^{\text{(?)}}$$

It then follows that

$$x + 0 = y,^{\text{(?)}}$$

and so $x = y,^{\text{(?)}}$ as desired. A similar argument establishes the result when $zx = zy$. ∎

As it turns out, Theorem 7.17 is actually implied by Theorem 7.20. This is because of the following result, which states that a unit can never be a zero divisor:

Theorem 7.21. *Let R be a ring with identity, and let $x \in R$ be a unit. Then x is not a zero divisor. That is, if $xy = 0$ or $yx = 0$ for some $y \in R$, then $y = 0$.*

Proof. Let R be a ring with identity, and let $x \in R$ be a unit. Then $x \neq 0$. Thus, suppose that $xy = 0$. Then

$$x^{-1}(xy) = x^{-1}(0),$$

which implies that

$$(x^{-1}x)y = 0,^{\text{(?)}}$$

so $y = 0$. A similar argument shows that $y = 0$ in the case that $yx = 0$. Thus, x is not a zero divisor. ∎

The statement of Theorem 7.21, and the subsequent proof, relies on a particularly useful form of the negation of the definition of zero divisor. In particular, $x \in R$ is *not* a zero divisor if and only if either (i) $x = 0$; or (ii) $xy \neq 0$ and $yx \neq 0$ for all nonzero $y \in R$. Equivalently, x is not a zero divisor if and only if the following statement is true:

> If $x \neq 0$, and $xy = 0$ or $yx = 0$ for some $y \in R$, then $y = 0$.

Thus, to show that a nonzero element $x \in R$ is *not* a zero divisor, we can assume that $xy = 0$ for some $y \in R$, and show that $y = 0$. If R is non-commutative, then we must also show that $y = 0$ whenever $yx = 0$. This is exactly the strategy we used in the proof above, and the equivalence it relies upon is quite similar to the ones asserted by Theorem 7.19.

Fields and Integral Domains

Having investigated the relationships between zero divisors, units, and the property of multiplicative cancellation, we are now ready to define and give examples of two very important types of rings.

Definition 7.22. An **integral domain** is a commutative ring with identity that contains no zero divisors.

Definition 7.23. A **field** is a commutative ring with identity in which every nonzero element has a multiplicative inverse.[†]

Notice that fields and integral domains are similar in many ways. Both are commutative. Both have identity. And, by Theorems 7.17 and 7.20, both satisfy the property of multiplicative cancellation. The following result, which follows from Theorem 7.21 and some of the examples we have already considered, establishes the relationship between fields and integral domains.

Corollary 7.24. *Every field is an integral domain, but not every integral domain is a field.*

We have already investigated several examples of fields (for instance, $\mathbb{Q}$, $\mathbb{R}$, and $\mathbb{C}$) and integral domains (for instance, $\mathbb{Z}$, which is an integral domain but not a field). We will consider numerous other examples in future investigations.

Concluding Activities

Activity 7.25. As noted earlier, subtraction in arbitrary rings can be defined in the same way we defined it for $\mathbb{Z}$. Using such a definition when needed, prove the following theorem:

Theorem 7.26. *Let R be a ring, and let x, y, $z \in R$. Then:*

(i) $-(-x) = x$

(ii) $x(-y) = -(xy) = (-x)y$

(iii) $(-x)(-y) = xy$

(iv) $x(y - z) = xy - xz$

(v) $(y - z)x = yx - zx$

(vi) $-(x + y) = -x - y$

Activity 7.27. Let $\mathbb{Z}^\star$ be the number system consisting of the set of all integers, with addition ($\oplus$) and multiplication ($\otimes$) defined as follows:

$$x \oplus y = x + y - 1 \quad \text{and} \quad x \otimes y = x + y - x \cdot y$$

Note that $+$ and $\cdot$ denote the normal operations of addition and multiplication in $\mathbb{Z}$. Which of the ring axioms are satisfied by $\mathbb{Z}^\star$, and which are not? Is $\mathbb{Z}^\star$ a ring? If so, is $\mathbb{Z}^\star$ commutative? Does $\mathbb{Z}^\star$ have an identity? Prove your answers.

[†]Dedekind introduced the term *field* (from the German *Köper*, or body).

Activity 7.28.

(a) Find all values of n for which $\mathbb{Z}_n$ is a field. State (and prove) your answer in the form of an if and only if statement. (Hint: See Activity 5.20 on page 56.)

(b) How, if at all, would your answer to part (a) change if you were instead asked to find the values of n for which $\mathbb{Z}_n$ is an integral domain?

Activity 7.29. Write a short summary that describes the important concepts, and the relationships between these concepts, that were introduced in this investigation. Explain how the material in this investigation is connected to the material in Investigations 1, 5, and 6.

Exercises

(1) Let R be a ring. Suppose that, due to a printer error, the addition and multiplication tables for R were printed with several entries missing, as shown below:

$+$	a	b	c
a	a	b	c
b		c	
c		a	

$\times$	a	b	c
a			
b		a	
c			

Using only the ring axioms, complete the tables. Explain how each entry you add can be determined from the ring axioms and the entries already completed.

(2) Let R be a ring. Suppose that, due to a printer error, the addition and multiplication tables for R were printed with several entries missing, as shown below:

$+$	a	b	c
a	a	b	c
b		c	
c		a	

$\times$	a	b	c
a			
b			
c			b

Using only the ring axioms, complete the tables. Explain how each entry you add can be determined from the ring axioms and the entries already completed.

(3) Let R be the number system consisting of the set of all integers, with addition ($\oplus$) and multiplication ($\otimes$) defined as follows:

$$x \oplus y = x \cdot y \quad \text{and} \quad x \otimes y = x + y$$

Note that $+$ and $\cdot$ denote the normal operations of addition and multiplication in $\mathbb{Z}$. Which of the ring axioms are satisfied by R, and which are not? Is R a ring? If so, is R commutative? Does R have an identity? Prove your answers.

(4) Let $\mathbb{R}^+$ denote the set of all positive real numbers. For all x, $y \in \mathbb{R}^+$, define

$$x \oplus y = xy \quad \text{and} \quad x \otimes y = x^{\log y}.$$

(a) With these operations, does $\mathbb{R}^+$ have an additive identity? If so, what is it?

(b) Does $\mathbb{R}^+$ have a multiplicative identity? If so, what is it?

(c) Is $\mathbb{R}^+$ a ring with the operations defined above? Prove your answer.

⋆ (5) Let n be a nonnegative integer, and let $n\mathbb{Z} = \{nx : x \in \mathbb{Z}\}$, with addition and multiplication defined as in $\mathbb{Z}$. Is $n\mathbb{Z}$ a ring? If so, is $n\mathbb{Z}$ commutative? Does $n\mathbb{Z}$ have an identity? Does your answer depend on the value of n? Explain.

(6) Let $\mathbb{Z}(\sqrt{2}) = \{a + b\sqrt{2} : a, b \in \mathbb{Z}\}$. Is $\mathbb{Z}(\sqrt{2})$ a ring? If so, is $\mathbb{Z}(\sqrt{2})$ commutative? Does $\mathbb{Z}(\sqrt{2})$ have an identity? Verify your answers.

(7) Let n and k be natural numbers, both greater than 1.

(a) How many elements does $M_{n\times n}(\mathbb{Z}_k)$ have?

(b) Is $M_{n\times n}(\mathbb{Z}_k)$ a ring? If so, is $M_{n\times n}(\mathbb{Z}_k)$ commutative? Does $M_{n\times n}(\mathbb{Z}_k)$ have an identity? Verify your answers.

(8) For all elements $x_1, x_2, \ldots, x_n$ in a ring R, we can define $x_1 + x_2 + \cdots + x_n$ recursively as follows:

$$x_1 + x_2 + \cdots + x_n = (x_1 + x_2 + \cdots + x_{n-1}) + x_n$$

Use this definition, along with mathematical induction, to prove the following **generalized distributive laws**:

Theorem 7.30 (Generalized Distributive Laws). *Let R be a ring, and let a, b_1, b_2, $\ldots$, $b_n \in R$. Then*

(i) $a(b_1 + b_2 + \cdots + b_n) = ab_1 + ab_2 + \cdots ab_n$.

(ii) $(b_1 + b_2 + \cdots + b_n)a = b_1a + b_2a + \cdots b_na$.

(9) Let R be a ring with at least two elements. Prove that $\mathcal{M}_{2\times 2}(R)$ is always a ring (with addition and multiplication of matrices defined as usual).

(10) **Rings of functions.** Let $\mathcal{F}(\mathbb{R})$ denote the set of all functions from $\mathbb{R}$ to $\mathbb{R}$. Define addition and multiplication on $\mathcal{F}(\mathbb{R})$ as follows:

- For all $f, g \in \mathcal{F}(\mathbb{R})$, $(f + g) : \mathbb{R} \to \mathbb{R}$ is the function defined by
$$(f + g)(x) = f(x) + g(x)$$
for all $x \in \mathbb{R}$.
- For all $f, g \in \mathcal{F}(\mathbb{R})$, $(fg) : \mathbb{R} \to \mathbb{R}$ is the function defined by
$$(fg)(x) = f(x)g(x)$$
for all $x \in \mathbb{R}$.

Prove that $\mathcal{F}(\mathbb{R})$ is a commutative ring with identity.

(11) A **Boolean ring** R is one in which $x^2 = x$ for all $x \in R$.

(a) Prove that in a Boolean ring, every element is its own additive inverse. Deduce that in a Boolean ring, addition and subtraction are the same. (Hint: Square a convenient element of R.)

(b) Prove that every Boolean ring is commutative. (Hint: Square *another* convenient element of R. You may want to eventually use part (a).)

(12) Let R be a ring, and suppose there exists a positive even integer n such that $x^n = x$ for all $x \in R$. Prove that $-x = x$ for all $x \in R$.

(13) Let n and k be positive integers, both greater than 1. State and prove a necessary and sufficient condition for a matrix A to be a unit in $M_{n\times n}(\mathbb{Z}_k)$.

(14) Let R be a ring with identity, and let x and y be units in R. Prove or disprove each of the following statements:

(a) $x + y$ is a unit in R.

(b) xy is a unit in R.

(c) Let z be any element of R such that $xz = 1$. Then z is a unit in R.

(15) Let R be a ring with identity, and let $x, y \in R$. Prove or disprove: If xy is a unit in R, then both x and y are units in R.

(16) (a) For which values of n is $\mathcal{M}_{n\times n}(\mathbb{R})$ a field? An integral domain? Find all such values, and prove your answer.

(b) For which values of n is $\mathcal{P}_n$ a field? An integral domain? Find all such values, and prove your answer.

(17) **Finite integral domains and fields.** Prove that if R is a finite ring with identity, then every nonzero element of R is either a zero divisor or a unit. Deduce that every finite integral domain is a field. (Hint: Let x be a nonzero element of R that is not a zero divisor. Show that $x^n = 1$ for some $n \in \mathbb{N}$, and deduce from this that x must be a unit.)

(18) Prove that if R is a finite ring that contains at least one nonzero element that is not a zero divisor, then R has identity. Use this result to state a stronger version of the result from Exercise 17.

Connections

This investigation introduced the concept of a ring. Rings are algebraic objects that share the same basic structure as the integers that we saw in Investigation 1 and the different number systems discussed in Investigation 6. There is a great deal of power to be had in recognizing the features these number systems have in common and then creating a larger category (rings) that encapsulates all of these features. By doing so, we can learn about all of these number systems at one time by studying arbitrary rings.

If you studied group theory before ring theory, you should notice connections between the topics in this investigation and those in Investigation 20. Rings and groups are both algebraic objects—that is, sets on which an operation or operations are defined, yielding a particular algebraic structure. The main difference between a ring and a group is that a ring comes with two binary operations and a group comes with only one. In fact, every ring is a group under its operation of addition, but not every group is a ring.

From a structural standpoint, rings may be more familiar to you than groups in that many of the sets with which you have worked in your mathematical past (e.g., $\mathbb{Z}$, $\mathbb{Q}$, $\mathbb{R}$, and sets of polynomials) are all rings. For this reason, starting our exploration of modern algebra with rings is a reasonable choice. On the other hand, since there is only one operation in a group, groups are simpler objects than rings and for that reason a good argument can be made that the study of modern algebra should begin with groups. In either case, many of the concepts we will encounter in these investigations will apply to both rings and groups.

Investigation 8

Integer Multiples and Exponents

Focus Questions

By the end of this investigation, you should be able to give precise and thorough answers to the questions listed below. You may want to keep these questions in mind to focus your thoughts as you complete the investigation.

- How can integer multiplication be defined in an arbitrary ring, and what properties are satisfied by integer multiplication in rings?
- How can integer exponentiation be defined in an arbitrary ring? What properties are satisfied by integer exponentiation in rings, and what special considerations must be taken into account when using nonpositive exponents?
- What is the characteristic of a ring? What are some examples of rings with characteristic zero? What are some examples of rings with nonzero characteristic?

Preview Activity 8.1. Our study of rings began in Investigation 1, where we learned about the integers and their various axioms. Our next example of a ring was $\mathbb{Z}_n$, a set of equivalence classes of integers. As you might suspect from these two examples, the integers play an important role in the general theory of rings. In fact, even in rings whose elements are not integers, it is possible to define notions of integer multiplication and integer exponentiation. In other words, it is possible to multiply and exponentiate ring elements by integers, even though the ring elements themselves may not be integers. In fact, it turns out that integer multiplication and exponentiation work exactly the way we would expect them to. To see this, use your intuition to calculate as many of the quantities listed below as you can. For those that you are not able to calculate, explain why. Throughout your calculations, you will be applying the definitions that we will formally develop in this investigation.

In $\mathbb{Z}_6$:	In $\mathcal{M}_{2\times 2}(\mathbb{R})$:	In $\mathcal{P}_3$:
$0[4]$	$0\begin{bmatrix} 1 & \pi \\ -\pi & 3 \end{bmatrix}$	$0\{1,2\}$
$3[5]$	$3\begin{bmatrix} e^2 & e^3 \\ e^3 & e^4 \end{bmatrix}$	$3\{2\}$
$(-4)[2]$	$(-4)\begin{bmatrix} 1 & 1+\sqrt{2} \\ 1-\sqrt{2} & -1 \end{bmatrix}$	$(-4)\{2,3\}$
$[3]^4$	$\begin{bmatrix} \pi & 0 \\ 0 & -\pi \end{bmatrix}^4$	$\{3\}^4$
$[2]^0$	$\begin{bmatrix} 1 & 1+\sqrt{2} \\ 1-\sqrt{2} & -1 \end{bmatrix}^0$	$\{1,3\}^0$
$[3]^{-5}$	$\begin{bmatrix} e^2 & e^3 \\ e^3 & e^4 \end{bmatrix}^{-3}$	$\{1,2,3\}^{-3}$
$[5]^{-3}$	$\begin{bmatrix} \pi & 0 \\ 0 & -\pi \end{bmatrix}^{-5}$	$\{1\}^{-5}$

Introduction

In Preview Activity 8.1, we began to intuitively develop the notions of integer multiplication and exponentiation for arbitrary rings. You may have performed the requested calculations by simply thinking of multiplication as repeated addition, and exponentiation as repeated multiplication. For instance, in $\mathcal{P}_3$, we can calculate $3\{1,2\}$ as follows:

$$\begin{aligned} 3\{1,2\} &= \{1,2\} + \{1,2\} + \{1,2\} \\ &= (\{1,2\} + \{1,2\}) + \{1,2\} \\ &= \emptyset + \{1,2\} \\ &= \{1,2\}. \end{aligned}$$

Likewise,

$$\begin{aligned} \{1,2\}^3 &= \{1,2\}\{1,2\}\{1,2\} \\ &= (\{1,2\}\{1,2\})\{1,2\} \\ &= \{1,2\}\{1,2\} \\ &= \{1,2\}. \end{aligned}$$

This intuitive formulation of integer multiplication and exponentiation makes sense as long as we are multiplying or exponentiating by a positive integer. For nonpositive integers, however, we will need to be a bit more careful. Furthermore, in order to prove that integer multiplication and exponentiation work the way we would expect them to, we will need to make use of a more formal definition.

We will develop such a definition in the next section, and we will use this definition to prove several fundamental properties of integer multiplication and exponentiation. We will then use integer multiplication to define the *characteristic* of a ring, which, intuitively speaking, measures the extent to which the addition operation exhibits cyclic behavior, as in $\mathbb{Z}_n$.

Integer Multiplication and Exponentiation

In the previous section, we observed that multiplication and exponentiation by a positive integer could be defined in terms of repeated addition or multiplication, respectively. Note, however, that in order to actually carry out such a repeated operation, we must first parenthesize the corresponding expression so that we will only be dealing with two ring elements at a time. Addition and multiplication are, after all, *binary* operations. Thus, expressions of the form

$$x_1 + x_2 + \cdots + x_n$$

or

$$x_1 x_2 \cdots x_n$$

only make sense if we define them in terms of a sequence of binary additions or multiplications.

As an example, suppose you were asked to calculate $37+63+29$. You might begin by adding 37 and 63 to get 100, and then add 29 to 100 to yield a final answer of 129. In other words, you would perform a sequence of two binary additions, effectively carrying out the computation as follows:

$$37 + 63 + 29 = (37 + 63) + 29 = 100 + 29 = 129.$$

The next definition formalizes this type of process for calculating sums of three or more ring elements.

Definition 8.2. Let R be a ring, let $n \geq 3$ be an integer, and let $x_1, x_2, \ldots, x_n \in R$. Then we define

$$x_1 + x_2 + \cdots + x_n = (x_1 + x_2 + \cdots + x_{n-1}) + x_n$$

and

$$x_1 x_2 \cdots x_n = (x_1 x_2 \cdots x_{n-1}) x_n.$$

Note that Definition 8.2 is what mathematicians typically call a *recursive* definition. In essence, this means that the definition refers to a simpler case of itself and would typically need to be applied repeatedly in order to actually yield a final answer. For instance, to calculate

$$x_1 + x_2 + x_3 + x_4 + x_5,$$

one must first calculate

$$x_1 + x_2 + x_3 + x_4,$$

which requires

$$x_1 + x_2 + x_3,$$

and so on.

We can formally define multiplication and exponentiation by a positive integer in a similar manner:

Definition 8.3. Let R be a ring, and let $x \in R$. Then the expressions $1x$ and x^1 are both defined to be equal to x; that is,

$$1x = x \text{ and } x^1 = x.$$

Furthermore, for every integer $n > 2$, we define the expressions nx and x^n recursively as follows:

- $nx = \underbrace{x + x + \cdots + x}_{n \text{ terms}} = (\underbrace{x + x + \cdots + x}_{n-1 \text{ terms}}) + x = (n-1)x + x$
- $x^n = \underbrace{x \cdot x \cdots x}_{n \text{ factors}} = (\underbrace{x \cdot x \cdots x}_{n-1 \text{ factors}})x = x^{n-1}x$

Nonpositive Multiples and Exponents

Definition 8.3 is quite natural, but unfortunately it applies only to multiplication or exponentiation by a *positive* integer. To extend the definition to nonpositive multiples and exponents, let's begin by making a few observations, some of which you may have noted in your answers to Preview Activity 8.1.

First, we would like multiplication by an integer to possess the same properties as other notions of multiplication (such as multiplication of ring elements). For instance, if n is a positive integer and x is an element of a ring R, then we would expect

$$-(nx) = (-n)x = n(-x),$$

since we stated a similar property for multiplication of ring elements in Theorem 7.26. (See page 86.) Likewise, it would make sense for the integer 0 times x to be the same as the zero element of R (0_R) times x. In other words, it should be the case that

$$0x = 0_R x = 0_R.$$

We would also like exponentiation by an integer in an arbitrary ring to behave similarly to exponentiation in more familiar number systems. For instance, we know that $x^0 = 1$ (the multiplicative identity) for every nonzero element x of $\mathbb{Z}$, $\mathbb{Q}$, $\mathbb{R}$, or $\mathbb{C}$. It would therefore be natural to define x^0 to be equal to 1_R in an arbitrary ring R. Of course, in order for such a definition to make sense, R would need to be a ring with identity. Likewise, if $x \in R$ and n is a positive integer, it would seem natural to define

$$x^{-n} = (x^{-1})^n.$$

This definition, however, requires R to be a ring with identity *and* x to be a unit in R.

So, to summarize, there are natural ways to extend Definition 8.3 to nonpositive multiples and exponents, provided that certain conditions are met. The next definition incorporates these conditions, which are based on the observations that we noted above.

Definition 8.4. Let R be a ring, and let n be a positive integer.

- For all $x \in R$, we define $0x = 0_R$ and $(-n)x = n(-x)$.
- If R is a ring with identity, then for each nonzero $x \in R$, we define $x^0 = 1_R$. If R does not have identity, then x^0 remains undefined.

- If R is a ring with identity, then for each unit $x \in R$, we define $x^{-n} = (x^{-1})^n$, where x^{-1} denotes the multiplicative inverse of x. If R does not have identity, or if x is not a unit in R, then x^{-n} remains undefined.

Note that if n is a negative integer, then Definition 8.4 implies that $nx = (-n)(-x)$ and $x^n = (x^{-1})^{-n}$, where $-n > 0$. This alternative definition will be particularly useful in the proofs in the next section.

Properties of Integer Multiplication and Exponentiation

Now that we have precisely defined integer multiplication and exponentiation for both positive and nonpositive integers, we are ready to state and prove the familiar properties that we commonly associate with these operations. We begin with the following theorem:

Theorem 8.5. *Let R be a ring, let x, $y \in R$, and let m and n be integers. Then*

(i) $m(x + y) = mx + my$

(ii) $-(mx) = m(-x) = (-m)x$

(iii) $(m + n)x = mx + nx$

(iv) $m(nx) = (mn)x$

(v) $m(xy) = (mx)y = x(my)$

(vi) $(mx)(ny) = (mn)(xy)$.

Because the proof of Theorem 8.5 is fairly long, we will only consider parts of it here. In particular, we will explore the inductive argument behind part (i) in Activity 8.6, and we will then consider a more complete proof of part (ii). The proofs of parts (iii) – (vi) will be left as exercises for you to complete at the conclusion of the investigation.

Activity 8.6.

(a) Parts (i) and (iii) look very similar to the distributive axiom in the definition of a ring. Explain why, in spite of these similarities, these properties still must be proved. (Hint: What is the fundamental difference between these properties and the ring axioms?)

(b) We will prove part (i) of Theorem 8.5 by considering three separate cases: $m = 0$, $m > 0$, and $m < 0$. Apply Definition 8.4 to prove the first case (that is, when $m = 0$).

(c) Now let $m > 0$, and let $P(m)$ be the predicate, "$m(x + y) = mx + my$." Prove that $P(1)$ is true.

(d) Let k be a positive integer, and assume that $P(k)$ (as defined in part (c)) is true. Use this assumption, along with Definition 8.3 and one or more of the ring axioms, to show that

$$(k + 1)(x + y) = (k + 1)x + (k + 1)y.$$

(e) Deduce from parts (b) – (d) that the statement from part (i) holds for all $m \geq 0$.

(f) Now assume that m is a negative integer. Give a justification for each step in the following argument:

$$\begin{aligned} m(x+y) &= (-m)[-(x+y)]^{ⓘ} \\ &= (-m)[(-x)+(-y)]^{ⓘ} \\ &= (-m)(-x)+(-m)(-y)^{ⓘ} \\ &= mx+my.^{ⓘ} \end{aligned}$$

(g) Use your work from parts (b) – (f) to write a clear and convincing proof of part (i) of Theorem 8.5.

Now that we have proved part (i) of Theorem 8.5, we can use this result in the proofs of the remaining parts. To illustrate, try to fill in the missing details in the following proof of part (ii). In particular, make sure that you are able to identify all of the instances in which the proof relies on part (i).

Proof of Theorem 8.5, part (ii). Let m be an integer. We will first prove that $m0_R = 0_R$. By ___________, we know that

$$m0_R = m(0_R + 0_R) = m0_R + m0_R,$$

which implies that $m0_R = 0_R$.ⓘ Now let $x \in R$. Then

$$mx + m(-x) \overset{?}{=} m(x + (-x)) \overset{?}{=} m0_R \overset{?}{=} 0_R.$$

This proves that $m(-x)$ is the additive inverse of mx; that is, $-(mx) = m(-x)$.

To complete the proof, we will now show that $m(-x) = (-m)x$. Note that if $m \geq 0$, then $-m \leq 0$, and so $(-m)x = m(-x)$ by Definition _____. If $m < 0$, then

$$m(-x) \overset{?}{=} (-m)[-(-x)] \overset{?}{=} (-m)x.$$

Thus, we have shown that for every integer m, $-(mx) = m(-x) = (-m)x$. ■

Theorem 8.5 asserted several familiar properties for integer multiplication within an arbitrary ring. The next theorem deals with integer exponentiation.

Theorem 8.7. *Let R be a ring, let $x, y \in R$, and let m and n be positive integers. Then:*

(i) $x^{m+n} = x^m x^n$

(ii) $(x^m)^n = x^{mn}$

If R is a ring with identity, then the above properties hold for all nonnegative integers m and n (provided that $x \neq 0$, as 0^0 remains undefined). Furthermore, if R is a ring with identity and x is a unit in R, then the above properties hold for all integers m and n.

We will prove part (i) of Theorem 8.7 and leave part (ii) as an exercise. For convenience, we will begin with the case where both m and n are positive. Our proof will employ an induction argument, but instead of using induction on m or n, we will instead induct on the sum $m + n$.

Proof of Theorem 8.7, part (i), for m, n positive. Let m and n be positive integers, and let $x \in R$. We will proceed by induction on $m + n$. For the base case, suppose $m + n = 2$. Then $m = 1$ and $n = 1$, ⓘ and so

$$x^{m+n} = x^2 \overset{\textcircled{?}}{=} x^1 x \overset{\textcircled{?}}{=} x^1 x^1 = x^m x^n,$$

as desired. Now suppose that for some integer $k \geq 2$,

$$x^{m+n} = x^m x^n$$

whenever $m + n = k$. Let m' and n' be positive integers such that $m' + n' = k + 1$. Then

$$\begin{aligned} x^{m'+n'} &= x^{(m'+n'-1)+1} \;\textcircled{?} \\ &= x^{(m'+n'-1)} x^1 \;\textcircled{?} \\ &= (x^{m'} x^{n'-1}) x \;\textcircled{?} \\ &= x^{m'} (x^{n'-1} x) \;\textcircled{?} \\ &= x^{m'} x^{n'}. \;\textcircled{?} \end{aligned}$$

This completes the induction argument, and so the result holds for all positive integers m and n. ■

The preceding proof established part (i) of Theorem 8.7 for the case in which m and n are positive. It is easy to show that, in a ring with identity, this result also holds when m or n is zero. (See Exercise 2.) To show that it is true for negative values of m and/or n, however, we will need the following lemma:

Lemma 8.8. *Let R be a ring with identity, and let x be a unit in R. Then:*

(i) *The element x^{-1} is a unit in R, and $(x^{-1})^{-1} = x$.*

(ii) *For every integer n, $x^n = x \cdot x^{n-1}$.*

(iii) *For every integer n, x^n is a unit and $(x^n)^{-1} = (x^{-1})^n$.*

Although the statements in Lemma 8.8 may seem obvious, their proofs are not trivial (with the possible exception of (i)). The next activity suggests some strategies for proving these results.

Activity 8.9. Let R be a ring with identity, and let x be a unit in R.

(a) Apply the definition of unit to prove part (i) of Lemma 8.8.

(b) Explain why part (ii) is satisfied by definition for $n = 0$, $n = 1$, and $n = 2$.

(c) The following argument can be used, along with part (b), to establish part (ii) of the lemma for every nonnegative integer n. Explain what proof technique is being used and what assumptions would need to be made for this technique to be valid. Also provide a justification for each step in the argument.

$$x^n \overset{\textcircled{?}}{=} x^{n-1} x \overset{\textcircled{?}}{=} (x \cdot x^{n-2}) x \overset{\textcircled{?}}{=} x(x^{n-1})$$

(d) The following argument can be used to prove part (ii) of Lemma 8.8 for negative values of n.

Provide a justification for each step in the argument, being sure to identify explicitly where the assumption that $n < 0$ would be used.

$$\begin{aligned} x \cdot x^{n-1} &= x \cdot (x^{-1})^{-n+1} \text{ (?)} \\ &= x \cdot (x^{-1})^{1+(-n)} \text{ (?)} \\ &= x \cdot [(x^{-1})^{1}(x^{-1})^{-n}] \text{ (?)} \\ &= (x \cdot x^{-1})(x^{-1})^{-n} \text{ (?)} \\ &= 1_R \cdot (x^{-1})^{-n} \text{ (?)} \\ &= x^{n} \text{ (?)} \end{aligned}$$

(e) To prove part (iii) of Lemma 8.8, it suffices to show that

$$x^n \cdot (x^{-1})^n = (x^{-1})^n \cdot x^n = 1_R.$$

Explain why this equality holds for $n = 0$ and $n = 1$.

(f) Assume that the equality from part (e) holds for some integer $n \geq 1$. Complete the following inductive argument to establish that

$$x^n \cdot (x^{-1})^n = 1_R$$

for every positive integer n:

$$\begin{aligned} x^n \cdot (x^{-1})^n &= (x^{n-1} \cdot x) \cdot [x^{-1} \cdot (x^{-1})^{n-1}] \text{ (?)} \\ &= \cdots \\ &= 1_R \end{aligned}$$

(g) An argument similar to the one you used in part (f) can be used to show that $(x^{-1})^n \cdot x^n = 1_R$ for every nonnegative integer n. Thus, all that remains to show is that

$$x^n \cdot (x^{-1})^n = (x^{-1})^n \cdot x^n = 1_R$$

for $n < 0$. Use what you have proved in the previous parts of this activity to establish this equality. You may want to begin your proof by noting that

$$x^n \cdot (x^{-1})^n = (x^{-1})^{-n} \cdot [(x^{-1})^{-1}]^{-n},$$

with $-n > 0$ whenever $n < 0$.

(h) Combine your work in parts (a) – (g) to write a clear and convincing proof of Lemma 8.8.

With Lemma 8.8 in hand, we are now ready to finish our proof of part (i) of Theorem 8.7, which entails proving the result for negative values of m and/or n. Fortunately, we can use what we have proved in the case where m and n are both positive to accomplish this goal.

Proof of Theorem 8.7, part (i), for m and/or n negative.. Without loss of generality, assume that $m < 0$. We will consider three cases.

Case 1: $m + n > 0$. In this case, $-m > 0$, $n > 0$, and $-m < n$. (?) Thus,

$$\begin{aligned} x^{m+n} &= [(x^{-1})^{-m} x^{-m}] x^{m+n} \text{ (?)} \\ &= (x^{-1})^{-m}(x^{-m} x^{m+n}) \text{ (?)} \\ &= x^m x^n. \text{ (?)} \end{aligned}$$

Case 2: $m + n = 0$. In this case, $n = -m > 0$,ⓘ and so

$$\begin{aligned} x^{m+n} &= 1_R \text{ⓘ} \\ &= (x^{-1})^n x^n \text{ⓘ} \\ &= (x^{-1})^{-m} x^n \text{ⓘ} \\ &= x^m x^n. \text{ⓘ} \end{aligned}$$

Case 3: $m + n < 0$. In this case, either $n < 0$, or $-m > n \geq 0$.ⓘ For the former, note that

$$\begin{aligned} x^{m+n} &= (x^{-1})^{-(m+n)} \text{ⓘ} \\ &= (x^{-1})^{(-m)+(-n)} \text{ⓘ} \\ &= (x^{-1})^{-m}(x^{-1})^{-n} \text{ⓘ} \\ &= x^m x^n. \text{ⓘ} \end{aligned}$$

For the latter, assume that $n > 0$. (The case where $n = 0$ is treated in Exercise 2.) Then

$$\begin{aligned} x^{m+n} &= (x^{-1})^{-(m+n)} \text{ⓘ} \\ &= (x^{-1})^{-(m+n)}[(x^{-1})^n x^n] \text{ⓘ} \\ &= [(x^{-1})^{-(m+n)}(x^{-1})^n]x^n \text{ⓘ} \\ &= (x^{-1})^{-m} x^n \text{ⓘ} \\ &= x^m x^n. \text{ⓘ} \end{aligned}$$

Since we have established the desired result in all three cases, our proof is complete. ■

The Characteristic of a Ring

We will conclude this investigation by introducing a way to classify rings like $\mathbb{Z}_n$ whose addition operations are cyclic in some sense. Preview Activity 8.10 demonstrates what we mean by this and provides the motivation for the more formal treatment that follows.

Preview Activity 8.10.

(a) Suppose that a standard 12-hour clock reads 9:00. What time will it be in 6 hours?

(b) Write an equation in $\mathbb{Z}_{12}$ that represents the question from part (a) and your answer to it.

(c) Repeat parts (a) and (b), but this time add 12 hours, 18 hours, and 24 hours.

(d) Again using a 12-hour clock, describe all of the positive integers k that make the following statement true:

No matter what time it is currently, it will be the same time k hours from now.

(e) Write an equation in $\mathbb{Z}_{12}$ that represents the statement from part (d).

(f) For which positive integers k does $k[3] = [0]$ in $\mathbb{Z}_{12}$?

(g) For which positive integers k does $k[4] = [0]$ in $\mathbb{Z}_{12}$?

(h) For which positive integers k does $k[5] = [0]$ in $\mathbb{Z}_{12}$?

(i) For which positive integers k does $k[x] = [0]$ for *all* $[x] \in \mathbb{Z}_{12}$? What is the smallest of these values?

(j) What is the relationship between your answers to parts (d) and (i)? Why does this relationship hold?

(k) Is there a positive integer k such that $kx = 0$ for all $x \in \mathbb{Z}$? Why or why not?

The questions in Preview Activity 8.10 begin to formalize an important feature of rings like $\mathbb{Z}_n$—namely, that they behave cyclically when it comes to addition and integer multiplication. Just like on a clock, if we add the right element in $\mathbb{Z}_n$, or if we add the same element the right number of times, we'll always end up right back where we started. Figure 8.1 illustrates this cyclic behavior for $\mathbb{Z}_6$:

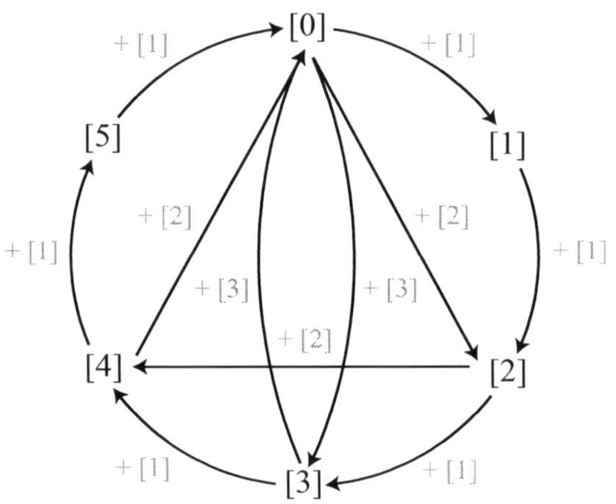

Figure 8.1
Additive cycles in $\mathbb{Z}_6$.

Notice that, for any element $[x]$ in $\mathbb{Z}_6$, if we add $[1]$ to $[x]$ exactly 6 times, the resulting sum will be equal to $[x]$. In other words,

$$[x] + [1] + [1] + [1] + [1] + [1] + [1] = [x] + 6[1] = [x].$$

This, of course, implies that $6[1] = [0]$ which should come as no surprise to us, since we know that $6[x] = [0]$ for *all* $[x] \in \mathbb{Z}_6$. In fact, the integer 6 is special in this sense. Notice that 6 is the first positive integer that yields a result of $[0]$ when multiplied by *any* element of $\mathbb{Z}_6$. Other integers may yield a similar result in some cases (for instance, $3[x] = [0]$ for $[x] = [0], [2], [4]$, but not for any other $[x] \in \mathbb{Z}_6$), but 6 is the first positive integer that does so universally. Furthermore, it is easy to show that any other integer k for which

$$k[x] = [0] \text{ for all } [x] \in \mathbb{Z}_6$$

must be an integer multiple of 6.

Putting all of these observations together, we might say that the integer 6 in some way *characterizes* the cyclic nature of the ring $\mathbb{Z}_6$. The next definition uses similar language to describe cyclic behavior in arbitrary rings.

Definition 8.11. Let R be a ring. The **characteristic** of R, denoted $\mathrm{char}(R)$, is the smallest positive integer k such that

$$kx = 0_R \text{ for all } x \in R.$$

If no such integer exists, then R is said to have **characteristic zero**.

Using the language and notation from Definition 8.11, we can now state, as we argued earlier, that $\mathrm{char}(\mathbb{Z}_6) = 6$. In fact, our prior observations can be generalized to establish that $\mathrm{char}(\mathbb{Z}_n) = n$ for every integer $n \geq 2$. We state this result and several others in the next theorem. The proof that follows uses a simple lemma that we have not yet stated or proved. This lemma plays an essential role in at least three different locations marked by the ⓘ symbol. Pay careful attention, as you will be asked to state and prove the missing lemma in Activity 8.13.

Theorem 8.12.

- *For every integer* $n \geq 2$*,* $\mathrm{char}(\mathbb{Z}_n) = n$*.*
- *The rings* $\mathbb{Z}$*,* $\mathbb{Q}$*,* $\mathbb{R}$*, and* $\mathbb{C}$ *all have characteristic zero.*

Proof. Let $n \geq 2$ be given. To show that $\mathrm{char}(\mathbb{Z}_n) = n$, we must show three things:

- that n is a positive integer;
- that $n[x] = [0]$ for all $[x] \in \mathbb{Z}_n$; and
- that there is no positive integer $k < n$ for which $k[x] = [0]$ for all $[x] \in \mathbb{Z}_n$.

The first is immediate. For the second, note that for all $[x] \in \mathbb{Z}_n$,

$$n[x] \overset{?}{=} [nx] \overset{?}{=} [n][x] \overset{?}{=} [0][x] = [0].$$

Now let k be any positive integer with $k < n$. It follows that $n \nmid k$,ⓘ and so

$$k[1] = [k] \neq [0].ⓘ$$

Therefore, $\mathrm{char}(\mathbb{Z}_n) = n$, as desired.

To show that $\mathbb{Z}$, $\mathbb{Q}$, $\mathbb{R}$, and $\mathbb{C}$ all have characteristic zero, it suffices to show that there does not exist a positive integer k such that $k \cdot 1 = 0$.ⓘ This, however, follows immediately from the ordering axioms of the integers.ⓘ ■

Concluding Activities

Activity 8.13. State and prove the missing lemma that is used in the proof of Theorem 8.12.

Activity 8.14. A classmate of yours claims that $\mathrm{char}(\mathcal{P}_n) = 2$ and offers the following argument in support of her claim:

Let $S \in \mathcal{P}_n$. Then

$$2S = S + S = S \triangle S = \emptyset = 0_{\mathcal{P}_n}.$$

Thus, $\text{char}(\mathcal{P}_n) = 2$.

Is your classmate's claim correct? Is her argument correct? Why or why not?

Activity 8.15. Consider the following alternative definition of the characteristic of a ring with identity.

> **Definition 8.16.** Let R be a ring with identity. Then $\text{char}(R)$ is the smallest positive integer k such that $k \cdot 1_R = 0_R$. If no such integer exists, then $\text{char}(R) = 0$.

Is this definition equivalent to Definition 8.11? If so, prove the equivalence. Otherwise, give an example to show that the definitions are not equivalent.

Activity 8.17. Is the following statement true for all rings, some rings, or no rings?

> Let R be a ring, and let $x, y \in R$. For every positive integer m,
>
> $$(xy)^m = x^m y^m.$$

Prove your answer, including a necessary and sufficient condition if you think that the result holds in only some rings.

Activity 8.18. Find all of the rings that have characteristic 1. Use Definition 8.11 to justify your answer.

Activity 8.19. Write a short summary that describes the important concepts, and the relationships between these concepts, that were introduced in this investigation. Explain how the material in this investigation is connected to the material in Investigations 5 and 7.

Exercises

(1) **Cancellation of integer multiples.**

  (a) For which types of rings is the following statement true?

  If $x, y \in R$ and m is a nonzero integer, then $mx = my$ implies $x = y$.

  Find a sufficient condition for the statement to be true, and prove the resulting theorem.

  (b) Is your sufficient condition from part (a) also necessary? Explain.

* (2) Prove Theorem 8.7 for the case where at least one of m or n is zero. (You may assume that R is a ring with identity.)

(3) Prove part (ii) of Theorem 8.7.

(4) Let R be a ring with the following addition table:

+	a	b	c	d
a	a	b	c	d
b	b	a	d	c
c	c	d	a	b
d	d	c	b	a

What is the characteristic of R? Prove your answer.

(5) Let m and n be integers, both greater than 1. What is the characteristic of $\mathcal{M}_{n\times n}(\mathbb{Z}_m)$? Prove your answer.

(6) A **Boolean ring** R is one in which $x^2 = x$ for all $x \in R$. Prove that every non-trivial (i.e., not $\{0\}$) Boolean ring has characteristic 2. (Hint: See Exercise 11 from Investigation 7.)

⋆ (7) **The characteristic of an integral domain.** Prove that the characteristic of an integral domain is either zero or prime. (Hint: Reason by contradiction.)

(8) **The characteristic of a finite ring.** Prove that a finite ring R cannot have characteristic zero. (Hint: Begin by showing that for each element $x \in R$, there must exist some positive integer k_x such that $k_x x = 0_R$.) Deduce from this and Exercise 7 that every finite integral domain (and consequently, every finite field) has prime characteristic.

⋆ (9) **The Binomial Theorem.** If you have ever expanded an expression of the form $(x+y)^n$, then you probably made use of the *Binomial Theorem.* The Binomial Theorem for real numbers is stated formally in Investigation 34, but here we will prove the following generalization for commutative rings.

> **Theorem 8.20** (Binomial Theorem for Commutative Rings). *Let R be a commutative ring, and let n be a positive integer. Then for all x, $y \in R$,*
>
> $$(x+y)^n = \sum_{k=0}^{n} \binom{n}{k} x^{n-k} y^k,$$
>
> *where*
>
> $$\binom{n}{k} = \frac{n!}{k!(n-k)!}.$$

(a) For several values of n and k, calculate $\binom{n}{k}$, $\binom{n}{k-1}$ and $\binom{n+1}{k}$. Use your calculations to state a conjecture about the relationship between these three quantities.

(b) Prove the conjecture you made in part (a).

(c) Use induction to prove the Binomial Theorem for commutative rings. (Hint: You will need to use your conjecture from part (a).)

(d) Where in your proof did you make use of the fact that R was commutative? Would your proof still be valid if R was not commutative?

⋆ (10) **The Freshman's Dream.** How often have you been tempted to expand $(x+y)^n$ as $x^n + y^n$? Wouldn't life be so much simpler if this were true in general? In this exercise, we will state and prove a related theorem that is sometimes amusingly called the "Freshman's Dream." The theorem requires x and y to be elements of a ring with prime characteristic. (So, if you ever mistakenly simplify $(x+y)^n$ as $x^n + y^n$, you can recover from your mistake by simply saying, "Oh, I thought we were working in a ring with prime characteristic.")

(a) Let R be a commutative ring with characteristic 3, and let $x, y \in R$. Expand and simplify the expression $(x+y)^3$. Do you notice anything interesting? (Hint: Remember that 3 times any element of R is equal to 0_R.)

(b) Let R be a commutative ring with characteristic 4, and let $x, y \in R$. Expand and simplify the expression $(x+y)^4$. Compare and contrast your results to those from part (a).

(c) Prove that if p is a prime number, then $p \mid \binom{p}{k}$ for every integer k such that $1 \leq k \leq p-1$.

(d) Explain why the assumption that p is prime is essential to the result in part (c). Use this insight to explain the difference between the simplifications in parts (a) and (b).

(e) Use the result from part (c), along with the Binomial Theorem (see Exercise 9) to prove the following general version of the Freshman's Dream:

> **Theorem 8.21** (Freshman's Dream). *Let p be a prime number, and let R be a commutative ring with characteristic p. Then for all $x, y \in R$,*
>
> $$(x+y)^p = x^p + y^p.$$

(f) Explain why the assumption that p is prime is an essential part of the statement of the Freshman's Dream. Give a specific example (that is, an example from a particular ring with non-prime characteristic) to illustrate the failure of the Freshman's Dream when p is not prime.

Connections

In this investigation, we studied integer multiples and integer powers of elements in rings. If you studied group theory before ring theory, you should notice connections between the topics in this investigation and those in Investigation 21. The major difference between rings and groups in this context is that there are two operations in a ring and only one in a group. As a result, we need to understand both integer multiples (under addition) and integer powers (under multiplication) of ring elements. With only one operation in a group, we only need one of these ideas. However, we use integer multiples when we represent a group operation as addition, and we use integer powers when we write a group operation multiplicatively. Consequently, we need to understand both notations even when working with algebraic structures—such as groups—that have only one operation.

Investigation 9

Subrings, Extensions, and Direct Sums

Focus Questions

By the end of this investigation, you should be able to give precise and thorough answers to the questions listed below. You may want to keep these questions in mind to focus your thoughts as you complete the investigation.

- What is a subring? What conditions must be verified in order to show that a subset of a ring is a subring?
- In what ways are a ring and all of its subrings guaranteed to be similar? In what ways can a ring and its subrings be different?
- What is a field extension, and how can field extensions be used to construct larger rings from smaller ones?
- What is a direct sum, and how can direct sums be used to construct larger rings from smaller ones?
- How are the properties of field extensions and direct sums related to the properties of the individual rings used to construct them?

Preview Activity 9.1. Throughout mathematics, the relationship between mathematical objects and their sub-objects is of central importance. For instance, in linear algebra, we study vector spaces and their subspaces. In discrete mathematics, many graph theory problems can be solved by finding a subgraph that is optimal in some sense. Furthermore, many other applied optimization problems involve minimizing or maximizing a certain function subject to certain constraints. These constraints define what is known as a *feasible region*, which is nothing more than a subset of the space of all possible solutions.

In light of these examples and our recent investigations of rings, it seems natural that we would be interested in defining and characterizing *subrings*. To begin thinking along these lines, we will consider a set of numbers that is larger than $\mathbb{Q}$ but smaller than $\mathbb{R}$. The set, denoted $\mathbb{Q}(\sqrt{2})$, is defined as follows:

$$\mathbb{Q}(\sqrt{2}) = \{a + b\sqrt{2} : a, b \in \mathbb{Q}\}$$

(a) Show that $\mathbb{Q} \subset \mathbb{Q}(\sqrt{2}) \subset \mathbb{R}$ (that is, $\mathbb{Q}$ is a proper subset of $\mathbb{Q}(\sqrt{2})$, which is a proper subset of $\mathbb{R}$).

(b) With addition and multiplication defined as in $\mathbb{R}$, which of the ring axioms does $\mathbb{Q}(\sqrt{2})$ satisfy? Give a brief explanation to justify your answer for each axiom.

(c) Critique the following proof that multiplication distributes over addition in $\mathbb{Q}(\sqrt{2})$. Is the proof correct? If so, could it be improved in any way? If not, what is the main error in the argument?

Proof. Let $x, y, z \in \mathbb{Q}(\sqrt{2})$. Since $\mathbb{Q}(\sqrt{2}) \subseteq \mathbb{R}$, it follows that $x, y, z \in \mathbb{R}$ as well. Thus,

$$x(y+z) = xy + xz$$

and

$$(x+y)z = xz + yz,$$

since multiplication distributes over addition in $\mathbb{R}$. This, however, proves that multiplication distributes over addition in $\mathbb{Q}(\sqrt{2})$. ■

(d) Critique the following proof that $\mathbb{Q}(\sqrt{2})$ is closed under addition. Is the proof correct? If so, could it be improved in any way? If not, what is the main error in the argument?

Proof. Let $x, y \in \mathbb{Q}(\sqrt{2})$. Since $\mathbb{Q}(\sqrt{2}) \subseteq \mathbb{R}$, it follows that $x, y \in \mathbb{R}$ as well. But $\mathbb{R}$ is closed under addition, so $x + y \in \mathbb{R}$. This shows that $\mathbb{Q}(\sqrt{2})$ is closed under addition. ■

(e) Which of the ring axioms were easiest to establish for $\mathbb{Q}(\sqrt{2})$? Which required the most work?

Introduction

In Preview Activity 9.1, we considered $\mathbb{Q}(\sqrt{2})$, a subset of the real numbers that was formed, loosely speaking, by "adding" $\sqrt{2}$ (an irrational number) to $\mathbb{Q}$. As it turns out, this special set is not only a subset of $\mathbb{R}$, but a *subring* as well. The next definition formalizes this terminology.

Definition 9.2. Let R be a ring, and let S be a subset of R. Then S is said to be a **subring** of R provided that S itself is a ring with the operations of addition and multiplication defined the same as in R.

Definition 9.2 is not terribly surprising. Nevertheless, there is one important caveat to note—namely, the condition that both R and S have the same operations. Although this condition is often satisfied trivially, it is still important to verify. For instance, Activity 7.27 (see page 86) describes an alternative way to define addition and multiplication on the integers. Let $\mathbb{Z}^\star$ denote the set of integers with these new operations. Although $\mathbb{Z}^\star$ is both a ring and a subset of $\mathbb{R}$, $\mathbb{Z}^\star$ is not a subring of $\mathbb{R}$. This is because the operations of addition and multiplication are defined differently in $\mathbb{Z}^\star$ than in $\mathbb{R}$.

Putting strange examples like the previous one aside, let's assume that we are given a ring R and a subset S of R, with the operations in S defined the same way as in R. What must we verify in order to show that S is a subring of R? The most obvious answer to this question is exactly what is suggested by Definition 9.2: we must show that S satisfies all of the ring axioms. But is there any information we can use to simplify this task? As it turns out, there is. If we are careful, we can use the fact that S is a subset of R to automatically establish more than half of the ring axioms. This leaves only a handful of axioms to prove, and these axioms form the basis of a result known as the *Subring Test*, which we will investigate in the next section.

After we have proved the Subring Test, we will turn the question around and consider ways of constructing larger rings from smaller ones. In particular, we will see how *field extensions* and *direct sums* can be used to construct rings that contain one or more prespecified subrings.

The Subring Test

As you may have observed in Preview Activity 9.1, some of the ring axioms are satisfied almost trivially by *any* subset of a ring R. These include associativity of addition and multiplication, commutativity of addition, and distribution of multiplication over addition. The proof of the latter is given in part (c) of Preview Activity 9.1, and the others follow in a similar fashion. Notice that all four of these axioms share a common form or structure. In particular, they all assert that a particular equality holds for *all* elements in a particular set. When an axiom of this form is *assumed* for a set R, it can be automatically *inferred* for every subset of R.

But what about the remaining three axioms: closure under addition and multiplication, the existence of a zero element, and the existence of additive inverses? What do they have in common, and why do they require more attention than the others?

The answer to this question lies in one small phrase: "there exists." Notice that two of the three axioms mentioned above contain this phrase, which, as you may remember from previous courses, is called an *existential quantifier*. Note also that the condition

$$x + y \in R,$$

which appears in the closure axiom, is equivalent to the following:

$$\text{There exists } z \in R \text{ such that } x + y = z.$$

A similar equivalence holds for the statement $xy \in R$. Thus, the closure axiom can be rephrased in an equivalent form as follows:

$$\text{For all } x, y \in R, \text{ there exists } z_1, z_2 \in R \text{ such that } x + y = z_1 \text{ and } xy = z_2.$$

So, to summarize, of the seven ring axioms, four are satisfied automatically by any subset of a ring R. The statements of the remaining three all contain an existential quantifier, and these existential axioms (as we might call them) are what need to be established in order to show that a subset of R is in fact a subring. The following theorem states these observations more formally:

Theorem 9.3. *Let R be a ring, and let S be a subset of R. Then S is a subring of R if*

(i) *S is closed under addition;*

(ii) *S is closed under multiplication;*

(iii) *S contains 0_R (that is, $0_R \in S$); and*

(iv) *S is closed under additive inverses (that is, for every $x \in S$, $-x \in S$).*

Theorem 9.3 significantly simplifies the task of showing that a particular set is a subring. That being said, the next theorem (commonly called the Subring Test) makes the task even easier by combining closure under addition and closure under additive inverses into a single condition: closure under subtraction.

Theorem 9.4 (Subring Test). *Let R be a ring, and let S be a subset of R. Then S is a subring of R if and only if*

(i) *S is nonempty;*

(ii) *S is closed under multiplication; and*

(iii) *S is closed under subtraction.*

The Subring Test allows us to prove that a subset S of a ring R is a subring of R by verifying only three essential conditions: that S is nonempty, that S is closed under multiplication, and that S is closed under subtraction. To illustrate, consider the set $\mathbb{Q}(\sqrt{2})$ introduced in Preview Activity 9.1. This set is clearly nonempty. Furthermore, for all $x = a + b\sqrt{2}$ and $y = c + d\sqrt{2}$ in $\mathbb{Q}(\sqrt{2})$, notice that

$$x - y = (a + b\sqrt{2}) - (c + d\sqrt{2}) = (a - c) + (b - d)\sqrt{2},$$

and

$$xy = (a + b\sqrt{2})(c + d\sqrt{2}) = (ac + 2bd) + (ad + bc)\sqrt{2}.$$

Since a, b, c, and d are all elements of $\mathbb{Q}$, and since $\mathbb{Q}$ is closed under addition, subtraction, and multiplication, it follows that both $x - y$ and xy are elements of $\mathbb{Q}(\sqrt{2})$. Thus, by the Subring Test, $\mathbb{Q}(\sqrt{2})$ is a subring of $\mathbb{R}$.

Now that we have seen how the Subring Test can be used, we are ready to investigate its proof, which we will present shortly. You may notice that, unlike previous proofs we have considered, we have not included any instances of the ? symbol. This is because you have now gained enough experience reading proofs to be able to decide for yourself where additional details or explanations are necessary. Although we will still occasionally use the ? symbol throughout the remainder of our investigations on ring theory, we will do so less frequently than in past investigations. You should still try to fill in missing details and add clarifying information to the proofs we consider, even when you are not explicitly prompted to do so. You may even want to use the ? symbol as we have in the past to remind yourself where these additional details are necessary.

Proof of the Subring Test. Let R be a ring, and let S be a subset of R. If S is a subring of R, then S clearly satisfies conditions (i) – (iii) of the theorem. For the converse, suppose that S is nonempty and closed under both multiplication and subtraction. By Theorem 9.3, it suffices to show that S contains 0_R, S is closed under additive inverses, and S is closed under addition. Since S is nonempty, S contains some element, say z. But then, since S is closed under subtraction, we know that

$$0_R = z - z \in S.$$

Now let x be any element of S. Since $0_R \in S$ and S is closed under subtraction, it follows that

$$-x = 0_R - x \in S.$$

Thus, S is closed under additive inverses. To show closure under addition, let x, $y \in S$. Since we have assumed that S is closed under subtraction and proved that S is closed under additive inverses, it follows that

$$x + y = x - (-y) \in S.$$

Thus, S is closed under addition, and the proof is complete. ■

Now that we have proved the Subring Test, we will conclude our study of subrings with a series of three activities. The first provides an opportunity to practice using the Subring Test, while the second and third explore the relationship between rings and their subrings, in particular with regard to the properties satisfied by each.

Activity 9.5. Use the Subring Test to prove or disprove each of the following statements:

(a) The set $\mathbb{E}$ of even integers is a subring of $\mathbb{Z}$.

(b) The set

$$K = \left\{ \begin{bmatrix} a & b \\ c & 0 \end{bmatrix} : a, b, c \in \mathbb{R} \right\}$$

is a subring of $\mathcal{M}_{2\times 2}(\mathbb{R})$.

(c) The set $2\mathbb{Z}_6 = \{[0], [2], [4]\}$ (with addition and multiplication defined as in $\mathbb{Z}_6$) is a subring of $\mathbb{Z}_6$.

Activity 9.6.

(a) Does $\mathbb{Z}_6$ have a multiplicative identity? If so, what is it?

(b) Does $2\mathbb{Z}_6$ (defined in part (c) of Activity 9.5) have a multiplicative identity? If so, what is it?

(c) Is $\mathbb{Z}_6$ a field? Why or why not?

(d) Is $2\mathbb{Z}_6$ a field? Why or why not?

Activity 9.7. Let R be a ring, and let S be a subring of R. Which of the following conjectures do you think are true, and which do you think are false? Whenever possible, provide brief arguments or examples to justify your answers.

(a) If R has a multiplicative identity, then S has a multiplicative identity.

(b) If S has a multiplicative identity, then R has a multiplicative identity.

(c) If both R and S have a multiplicative identity, then $1_R = 1_S$.

(d) If R is commutative, then S is commutative.

(e) If S is commutative, then R is commutative.

(f) If R and S both have identity and x is a unit in R, then x is a unit in S. (Does your answer depend on whether $1_R = 1_S$?)

(g) If R and S both have identity and x is a unit in S, then x is a unit in R. (Does your answer depend on whether $1_R = 1_S$?)

(h) If R is a field, then S is a field.

(i) If S is a field, then R is a field.

(j) If R is an integral domain, then S is an integral domain.

(k) If S is an integral domain, then R is an integral domain.

Activities 9.6 and 9.7 demonstrate that the relationships between rings and subrings are not always as clear-cut as we might like them to be. For instance, a ring and its subrings can have different multiplicative identities. A ring that is not a field can contain a subring that is a field. And the set of units in a ring may be completely disjoint from the set of units in one of its subrings. These results are somewhat surprising, and there are numerous others like them. The bottom line is that we must be very careful not to make unwarranted assumptions when dealing with rings and subrings. A ring may satisfy a property that its subrings do not, and vice versa.

Subfields and Field Extensions

Preview Activity 9.8. In Preview Activity 9.1, we considered the set

$$\mathbb{Q}(\sqrt{2}) = \{a + b\sqrt{2} : a, b \in \mathbb{Q}\},$$

with addition and multiplication defined in the usual way (i.e., as in $\mathbb{R}$). On the one hand, we can (and did) show that $\mathbb{Q}(\sqrt{2})$ is a subring of $\mathbb{R}$. On the other hand, since $\mathbb{Q}$ is a subring of $\mathbb{Q}(\sqrt{2})$, we can also think of $\mathbb{Q}(\sqrt{2})$ as being an *extension* of $\mathbb{Q}$. (We will define this term more precisely in just a bit.) In either case, since both $\mathbb{Q}$ and $\mathbb{R}$ are fields, it makes sense to ask whether $\mathbb{Q}(\sqrt{2})$ is also a field. The questions below do exactly that, and they also explore in more detail the general methods used to construct $\mathbb{Q}(\sqrt{2})$.

(a) We have already shown that $\mathbb{Q}(\sqrt{2})$ is a subring of $\mathbb{R}$. What would we still need to prove in order to establish that $\mathbb{Q}(\sqrt{2})$ is a *subfield* of $\mathbb{R}$ (that is, a subring of $\mathbb{R}$ that is also a field)?

(b) Prove each of the properties you identified in part (a). (Hint: For one of these properties, you will need to show that for all nonzero $a + b\sqrt{2} \in \mathbb{Q}(\sqrt{2})$, there exists $c + d\sqrt{2} \in \mathbb{Q}(\sqrt{2})$ such that

$$(a + b\sqrt{2})(c + d\sqrt{2}) = 1 + 0\sqrt{2}.$$

It may be helpful to look back at Activity 6.5 and the discussion that precedes it on page 69.)

(c) Is there a proper subfield of $\mathbb{Q}(\sqrt{2})$ that contains all of the rational numbers and the real number $\sqrt{2}$? That is, can we find a field that is smaller than $\mathbb{Q}(\sqrt{2})$ and yet contains both $\mathbb{Q}$ (as a subfield) and $\sqrt{2}$? Give a convincing argument to justify your answer.

(d) Suppose we defined $\mathbb{Q}(\sqrt[3]{2})$ analogously to $\mathbb{Q}(\sqrt{2})$—that is,

$$\mathbb{Q}(\sqrt[3]{2}) = \{a + b\sqrt[3]{2} : a, b \in \mathbb{Q}\}.$$

With this definition, is $\mathbb{Q}(\sqrt[3]{2})$ a field? Why or why not?

(e) How is the definition of the complex numbers, $\mathbb{C}$, similar to that of $\mathbb{Q}(\sqrt{2})$? In what ways do these definitions differ?

The topic of field extensions is of central importance to much of abstract algebra. Many of the questions pertaining to field extensions have to do with finding solutions to polynomial equations. For instance, there are numerous polynomials with rational coefficients that fail to have any roots in the rational numbers. One such polynomial is $f(x) = x^2 - 2$. This polynomial does have two roots in $\mathbb{R}$ (that is, two values of $x \in \mathbb{R}$ for which $f(x) = 0$): $\sqrt{2}$ and $-\sqrt{2}$. But there are no values of $x \in \mathbb{Q}$ for which $x^2 - 2 = 0$.

Now suppose that we wanted to "enlarge" $\mathbb{Q}$ to create a new ring that also contains both of these irrational roots. One possibility would be to just use the set $\mathbb{Q} \cup \{-\sqrt{2}, \sqrt{2}\}$. Unfortunately, this set is not even closed under addition, so we would have a hard time doing even basic algebra within it. What we really want is to construct a set that has the same algebraic properties as $\mathbb{Q}$, but that also contains $\pm\sqrt{2}$. Since $\mathbb{Q}$ is a field, this means that we are looking for a larger field that contains both $\mathbb{Q}$ (as a subset) and $\pm\sqrt{2}$. Of course, there are many fields, such as $\mathbb{R}$, that satisfy these properties. But some of these fields contain other elements that we might not be interested in. So the question becomes this:

> Given a field F and a polynomial* $p(x)$ with coefficients from F, what is the smallest field that contains every element of F and one or more solutions to the equation $p(x) = 0$?

In general, this is a surprisingly difficult question to answer. First off, it is not always immediately obvious that such a field even exists. For example, in the history of mathematics, imaginary numbers got their name for a reason. For a long time, it seemed impossible for mathematicians to conceive of any kind of number x that would satisfy the equation $x^2 + 1 = 0$. But the complex numbers are formed exactly by "adding" such a solution (we call it i) to $\mathbb{R}$, just as $\mathbb{Q}(\sqrt{2})$ is formed by "adding" a solution to $x^2 - 2 = 0$ to $\mathbb{Q}$. In each case, the resulting number system is what is known as a *field extension*, which we define formally as follows:

Definition 9.9. Let F be a field.

- A **subfield** of F is a subring of F that is also a field.
- If F is a subfield of another field E, then E is said to be a **field extension** (or simply an **extension**) of F.
- If E is an extension of F, and S is a subset of E, then the set $F(S)$, called the extension of F **generated by** S, is defined to be the smallest subfield of E that contains all of the elements of both F and S. In the case that S contains a single element $\alpha \in E$, then $F(\alpha)$ is called a **simple extension**.

You may have noticed that the notation used in Definition 9.9 to denote a simple extension is the same as the notation used in Preview Activities 9.1 and 9.8, where we defined $\mathbb{Q}(\sqrt{2})$. The next theorem explains the reason for this similarity in the case that α is the root of a quadratic polynomial. You should read its proof very carefully, adding additional details and explanations as you see fit.

Theorem 9.10. *Let F be a field, and let $p(x)$ be a quadratic polynomial with coefficients from F such that $p(x)$ has no roots in F. Suppose also that $p(x)$ does have a root α in some extension E of F. Then the simple extension of F generated by α (that is, the smallest field containing both F and α) can be described as follows:*

$$F(\alpha) = \{u + v\alpha : u, v \in F\}.$$

Proof. Let $p(x) = ax^2 + bx + c$ for some a, b, $c \in F$. Without loss of generality, we may assume that $a = 1$, so that $p(x) = x^2 + bx + c$. Since α is a root of $p(x)$, it follows that

$$\alpha^2 + b\alpha + c = 0,$$

or equivalently,

$$\alpha^2 = -b\alpha - c.$$

To prove the theorem, we must show that the set $S = \{u + v\alpha : u, v \in F\}$ is the smallest subfield of E that contains both α and all of the elements of F. We will first show that S is a subfield of E. We will do so by showing that S is a subring of E and S is a field. By definition, S is

*We have not yet formally defined what a polynomial is, but for the purposes of this investigation, your intuitive understanding should suffice. If you are interested in a formal definition, we define a polynomial with coefficients from a field F to be an expression of the form

$$p(x) = c_0 + c_1x + c_2x^2 + c_3x^3 + \cdots c_nx^n,$$

where each c_i belongs to F. We will study polynomials in much more detail in Investigations 11 – 15.

a nonempty subset of E. Thus, to establish that S is a subring of E, we must show that S is closed under subtraction and multiplication.

Let $u + v\alpha$ and $w + z\alpha$ be elements of S. Then

$$(u + v\alpha) - (w + z\alpha) = (u - w) + (v - z)\alpha \in S.$$

Thus, S is closed under subtraction. For closure under multiplication, note that

$$\begin{aligned}(u + v\alpha)(w + z\alpha) &= uw + vw\alpha + uz\alpha + vz\alpha^2 \\ &= uw + vw\alpha + uz\alpha + vz(-b\alpha - c) \\ &= (uw - vzc) + (vw + uz - vzb)\alpha,\end{aligned}$$

which is an element of S. Thus, we have shown that S is a subring of E.

Since E is a field, it follows that S is commutative. Furthermore, S clearly contains 1. Thus, to show that S is a subfield of E, we need only to show that every nonzero element of S is a unit. To this end, consider an arbitrary element $u + v\alpha \in S$, with u and v not both zero. We must show that there exists an element $w + z\alpha \in S$ such that

$$(u + v\alpha)(w + z\alpha) = 1.$$

First, we will argue that $\beta = -u^2 + buv - cv^2$ is a nonzero element of F. If $v = 0$, then $u \neq 0$, which implies that $\beta = -u^2 \neq 0$. Now suppose $v \neq 0$, and suppose also that $\beta = -u^2 + buv - cv^2 = 0$. Then

$$\left(-\tfrac{u}{v}\right)^2 + b\left(-\tfrac{u}{v}\right) + c = 0.$$

This, however, is a contradiction to our assumption that $p(x)$ has no roots in F. Thus, β is nonzero, which implies that β is a unit in F.

Now let $w = (-u + vb)\beta^{-1}$ and $z = v\beta^{-1}$. Then

$$\begin{aligned}(u + v\alpha)(w + z\alpha) &= (u + v\alpha)[(-u + bv)\beta^{-1} + v\beta^{-1}\alpha] \\ &= (u + v\alpha)(-u + bv + v\alpha)\beta^{-1} \\ &= (-u^2 + buv + uv\alpha - uv\alpha + bv^2\alpha + v^2\alpha^2)\beta^{-1} \\ &= [-u^2 + buv + bv^2\alpha + v^2(-b\alpha - c)]\beta^{-1} \\ &= (-u^2 + buv + bv^2\alpha - bv^2\alpha - cv^2)\beta^{-1} \\ &= (-u^2 + buv - cv^2)\beta^{-1} \\ &= \beta\beta^{-1} \\ &= 1\end{aligned}$$

So far, we have shown that S is a subfield of E. To finish the proof, we must argue that S is the smallest subfield of E that contains both α and all of the elements of F. To do so, it suffices to note that if a field K contains both α and F, then K must also contain $u + v\alpha$ for all $u, v \in F$. Thus, S is a subfield of any such K, which is what we needed to prove.

Since we have shown that S is the smallest subfield of E that contains both α and F, it follows by Definition 9.9 that

$$F(\alpha) = S = \{u + v\alpha : u, v \in F\}.$$

■

Activity 9.11. The proof of Theorem 9.10 was long and somewhat involved. In this activity, we will take a closer look at two of the important details within this proof.

(a) At the very beginning of the proof, we made the assumption that the leading coefficient of $p(x)$ was equal to 1. (Incidentally, such polynomials are said to be *monic*.) To see why this assumption is valid, show that if α is a root of any polynomial $p(x)$ with coefficients from F, then α is also a root of a monic polynomial $\tilde{p}(x)$ with coefficients from F.

(b) When we argued that every nonzero element $(u + v\alpha) \in S$ was a unit, it may have seemed as if we pulled our choice for $(u + v\alpha)^{-1}$ out of thin air. This was intentional, and it was done to illustrate an unfortunate feature of many of the proofs that you will read throughout your mathematical career. In particular, while the method we used was entirely correct, and we certainly proved that $u + v\alpha$ was a unit, the way we presented our argument provided virtually no insight into the way we actually found out what $(u + v\alpha)^{-1}$ should be. Perhaps we just guessed and got lucky. Or perhaps we remembered a similar technique from another proof. While either of these is a possibility, it seems more likely that we would have worked backwards, starting with the *assumption* that $(u + v\alpha)(w + z\alpha) = 1$ and then determining from this assumption what w and z would need to be.

    Try this technique; that is, assuming that $(u + v\alpha)(w + z\alpha) = 1$, set up and solve an appropriate system of equations, writing w and z in terms of u and v. How does your solution compare to that presented in our proof?

(c) Is your work from part (b) an acceptable alternative to our original argument? Does it establish that every nonzero element of S is a unit, or is the original proof still necessary? Explain.

(d) Are there any changes you would make to the proof of Theorem 9.10 to make it easier to understand or more insightful? If so, what would these changes be?

Before moving on, it is worth noting that Theorem 9.10 is actually a very special case of a much more general result on field extensions—in particular, extensions generated by roots of polynomials. The method we used to prove it is adequate for *quadratic extensions* (that is, extensions generated by the roots of quadratic polynomials), but it is not easy to generalize to broader contexts. For this, we will need to develop some more sophisticated tools, which we will do in Investigation 37.

Finally, the statement of Theorem 9.10 *assumed* the existence of an extension field E containing a root of $p(x)$. As it turns out, this assumption is always valid, thanks to the following result:

Theorem 9.12 (Kronecker's Theorem). *Let F be a field, and let $p(x)$ be a non-constant polynomial with coefficients from F. Then there exists an extension E of F and an element $\alpha \in E$ such that $p(\alpha) = 0$.*

Like the more general version of Theorem 9.10, the proof of Kronecker's Theorem requires a more thorough treatment of polynomials and field extensions than we have considered up to this point. Thus, we will omit the proof for now and return to it (in Investigation 15) after we have developed more fully the necessary theoretical foundations.

Direct Sums

Preview Activity 9.13. In the previous section, we considered a way to enlarge a given field so that the resulting extension would contain an additional element of interest, such as the root of a polynomial. We will now consider a different, but related, problem.

Suppose that we have two rings R and S, and we want to construct a new ring that contains both R and S as subrings. One way to do so is by using what is known as a *direct sum*. We will soon define direct sums formally, but before doing so, let's take a look at an example. The addition table for the direct sum of $\mathbb{Z}_2$ and $\mathbb{Z}_3$, denoted $\mathbb{Z}_2 \oplus \mathbb{Z}_3$, is shown in Table 9.1.

$+$	$([0]_2,[0]_3)$	$([0]_2,[1]_3)$	$([0]_2,[2]_3)$	$([1]_2,[0]_3)$	$([1]_2,[1]_3)$	$([1]_2,[2]_3)$
$([0]_2,[0]_3)$	$([0]_2,[0]_3)$	$([0]_2,[1]_3)$	$([0]_2,[2]_3)$	$([1]_2,[0]_3)$	$([1]_2,[1]_3)$	$([1]_2,[2]_3)$
$([0]_2,[1]_3)$	$([0]_2,[1]_3)$	$([0]_2,[2]_3)$	$([0]_2,[0]_3)$	$([1]_2,[1]_3)$	$([1]_2,[2]_3)$	$([1]_2,[0]_3)$
$([0]_2,[2]_3)$	$([0]_2,[2]_3)$	$([0]_2,[0]_3)$	$([0]_2,[1]_3)$	$([1]_2,[2]_3)$	$([1]_2,[0]_3)$	$([1]_2,[1]_3)$
$([1]_2,[0]_3)$	$([1]_2,[0]_3)$	$([1]_2,[1]_3)$	$([1]_2,[2]_3)$	$([0]_2,[0]_3)$	$([0]_2,[1]_3)$	$([0]_2,[2]_3)$
$([1]_2,[1]_3)$	$([1]_2,[1]_3)$	$([1]_2,[2]_3)$	$([1]_2,[0]_3)$	$([0]_2,[1]_3)$	$([0]_2,[2]_3)$	$([0]_2,[0]_3)$
$([1]_2,[2]_3)$	$([1]_2,[2]_3)$	$([1]_2,[0]_3)$	$([1]_2,[1]_3)$	$([0]_2,[2]_3)$	$([0]_2,[0]_3)$	$([0]_2,[1]_3)$

Table 9.1
The addition table for $\mathbb{Z}_2 \oplus \mathbb{Z}_3$.

(a) Describe precisely how the elements of $\mathbb{Z}_2 \oplus \mathbb{Z}_3$ are related to the elements of $\mathbb{Z}_2$ and $\mathbb{Z}_3$.

(b) How does addition in $\mathbb{Z}_2 \oplus \mathbb{Z}_3$ seem to be related to the addition operations in $\mathbb{Z}_2$ and $\mathbb{Z}_3$?

(c) Defining multiplication analogous to the way addition was defined in Table 9.1, make the multiplication table for $\mathbb{Z}_2 \oplus \mathbb{Z}_3$.

(d) With the given addition table and the multiplication table you made in part (c), does $\mathbb{Z}_2 \oplus \mathbb{Z}_3$ seem to be a ring? If so, is $\mathbb{Z}_2 \oplus \mathbb{Z}_3$ an integral domain and/or a field? Why or why not?

(e) Consider the set S defined by

$$S = \{([0]_2, [x]_3) \in \mathbb{Z}_2 \oplus \mathbb{Z}_3\}.$$

Is S a subring of $\mathbb{Z}_2 \oplus \mathbb{Z}_3$? Why or why not?

(f) What is the relationship between $\mathbb{Z}_3$ and the set S defined in part (e)? Explain.

Preview Activity 9.13 introduces a new type of number system called a *direct sum*, which can be defined formally as follows:

Definition 9.14. Let R and S be rings. The **Cartesian product** of R and S is the set

$$R \times S = \{(r, s) : r \in R, s \in S\}.$$

The **direct sum** of R and S, denoted $R \oplus S$, is the set $R \times S$, with addition and multiplication defined componentwise—that is,

$$(r_1, s_1) + (r_2, s_2) = (r_1 + r_2, s_1 + s_2)$$

and

$$(r_1, s_1)(r_2, s_2) = (r_1 r_2, s_1 s_2).$$

It is important to note in Definition 9.14 that the $+$ sign really denotes three distinct operations: addition in R, addition in S, and addition in the new number system $R \oplus S$. The same could be said for multiplication; although the notation does not explicitly indicate this fact, there are really three different multiplication operations being used. In either case, the context in which the notation is used should alleviate any potential ambiguities. For instance, when the $+$ symbol is used *between* two ordered pairs of elements, we know that it is referring to addition in $R \oplus S$. Likewise, when $+$ is used *within* an ordered pair of elements, we know that it is referring to addition either within R or within S, depending on the coordinate in which it is used.

As you may have observed in Preview Activity 9.13, direct sums satisfy a number of important properties. The most fundamental of these is the following:

Theorem 9.15. *Let R and S be rings. Then $R \oplus S$ is also a ring.*

Theorem 9.15 is not difficult to prove. In fact, each of the ring axioms for $R \oplus S$ can be established by simply invoking the corresponding axioms within R and S. To illustrate, consider the following argument that addition in $R \oplus S$ is commutative:

Let $x = (r_1, s_1)$, $y = (r_2, s_2) \in R \oplus S$. Then

$$\begin{aligned} x + y &= (r_1, s_1) + (r_2, s_2) \\ &= (r_1 + r_2, s_1 + s_2) \quad \text{(by Definition 9.14)} \\ &= (r_2 + r_1, s_2 + s_1) \quad \text{(by commutativity of } + \text{ in } R \text{ and } S\text{)} \\ &= (r_2, s_2) + (r_1, s_1) \quad \text{(again, by Definition 9.14)} \\ &= y + x. \end{aligned}$$

The other ring axioms for $R \oplus S$ can be established in a similar manner. Therefore, we will leave the rest of the proof of Theorem 9.15 as an exercise. (See Exercise 12.)

Finally, it is worth noting that $R \oplus S$ always contains both R and S (or, more accurately, rings that look and behave just like R and S) as subrings. The next theorem formalizes this fact. Its proof is left as an exercise. (See Exercise 13.)

Theorem 9.16. *Let R and S be rings. Then*

$$R \oplus \{0_S\} = \{(r, 0_S) : r \in R\}$$

and

$$\{0_R\} \oplus S = \{(0_R, s) : s \in S\}$$

are both subrings of $R \oplus S$.

You may have noticed that $R \oplus \{0_S\}$ contains the elements of R, each juxtaposed with the zero element for S. Thus, while $R \oplus \{0_S\}$ is not technically the same ring as R, the two are virtually identical, both in their makeup and in the way they behave with respect to addition and multiplication. We might say that R and $R \oplus \{0_S\}$ are essentially the same, a notion that we will make more precise in the next investigation when we study ring isomorphism. Likewise, S and $\{0_R\} \oplus S$ can also be considered to be essentially the same ring, which implies that $R \oplus S$ in some sense contains a copy of both R and S. Thus, direct sums provide a way to construct a larger ring that contains each of two smaller rings. This type of construction is useful in numerous examples, and it also plays an essential role in the classification of algebraic objects called *finite groups*, which we will study in Investigation 31.

Concluding Activities

Activity 9.17. Fill in the blank to make the following statement true, and prove the resulting theorem.

> **Theorem 9.18.** *Every non-trivial ring contains at least ______ subrings.*

Activity 9.19. Theorem 9.3 states that every subring S of a ring R must contain the zero element of R, 0_R. However, this statement seems to be stronger than what is required by Definition 9.2. In particular, the definition requires S to be a ring with the same operations as R, which implies that S must contain *some* zero element, say 0_S, which could conceivably be different than 0_R. You may recall from Activity 9.7 that it is possible for a ring and one or more of its subrings to have different multiplicative identities. Could this same behavior occur with respect to additive identities? In other words, does Theorem 9.3 need to be modified to account for the possibility that $0_S \neq 0_R$? Why or why not?

Activity 9.20. Under what conditions is $R \oplus S$ commutative? Under what conditions does $R \oplus S$ have an identity? Prove your answers.

Activity 9.21.

(a) Compare the addition and multiplication tables for $\mathbb{Z}_2 \oplus \mathbb{Z}_3$ and for $\mathbb{Z}_6$. How are these rings similar? How are they different?

(b) Make the addition and multiplication tables for $\mathbb{Z}_2 \oplus \mathbb{Z}_2$, and compare these tables to the tables for $\mathbb{Z}_4$. How are $\mathbb{Z}_2 \oplus \mathbb{Z}_2$ and $\mathbb{Z}_4$ similar? How are they different?

Activity 9.22. Write a short summary that describes the important concepts, and the relationships between these concepts, that were introduced in this investigation. Explain how the material in this investigation is connected to the material in Investigations 7 and 8.

Exercises

(1) Let S denote the set of all 2×2 matrices of the form

$$\begin{bmatrix} x & 0 \\ y & 0 \end{bmatrix},$$

where $x, y \in \mathbb{R}$. Is S a subring of $\mathcal{M}_{2\times 2}(\mathbb{R})$? Prove your answer.

(2) Let S denote the set of all 2×2 matrices of the form

$$\begin{bmatrix} x & y \\ -y & x \end{bmatrix},$$

where $x, y \in \mathbb{R}$. Is S a subring of $\mathcal{M}_{2\times 2}(\mathbb{R})$? Prove your answer.

(3) The set H of **quaternions** is defined as follows:

$$H = \left\{ \begin{bmatrix} a+bi & c+di \\ -c+di & a-bi \end{bmatrix} : a,b,c,d \in \mathbb{R} \right\}.$$

Prove that H is a subring of $\mathcal{M}_{2\times 2}(\mathbb{C})$. Then prove that H is a non-commutative ring with identity in which every nonzero element has a multiplicative inverse. (Such rings are called *division rings* or *skew fields*. They differ from fields only in the fact that their multiplication operation is not required to be commutative.)

(4) **Subrings and set operations.** Let S and T be subrings of a ring R. For each of the questions below, give a proof or a pair of examples (whichever is most appropriate) to justify your answer.

(a) Is $S \cup T$ always, sometimes, or never a subring of R?

(b) Is $S \cap T$ always, sometimes, or never a subring of R?

(c) Is $S \bigtriangleup T$ always, sometimes, or never a subring of R?

(5) **Subrings and commutativity.** Is every subring of a commutative ring necessarily commutative? Give a proof or counterexample to justify your answer.

(6) Let R be a ring. The **center** of R is defined to be the set of all elements $x \in R$ such that $xr = rx$ for all $r \in R$. Prove that the center of R is always a commutative subring of R.

(7) Let R be a ring, and let $r \in R$. The **centralizer** of r is defined to be the set of all $x \in R$ such that $xr = rx$.

(a) Prove that the centralizer of r is a subring of R.

(b) Is the centralizer of r necessarily commutative? Explain.

(8) Let R be a ring, and let $r \in R$. Prove that $S = \{x \in R : rx = 0\}$ is a subring of R.

(9) Let R be a ring, and let $n \in \mathbb{Z}$. Prove that $S = \{x \in R : nx = 0\}$ is a subring of R.

(10) **Subrings and units.** Let R be a ring with identity, and let S be the set of all units of R. Is S always, sometimes, or never a subring of R? Give a proof or a pair of examples (whichever is most appropriate) to justify your answer.

(11) **Subrings of $\mathbb{Z}$.** For every nonnegative integer n, let $n\mathbb{Z} = \{nx : x \in \mathbb{Z}\}$, with addition and multiplication defined as in $\mathbb{Z}$. (See Exercise 5 on page 88.)

(a) Prove that $n\mathbb{Z}$ is always a subring of $\mathbb{Z}$.

(b) Prove that every subring of $\mathbb{Z}$ is equal to $n\mathbb{Z}$ for some nonnegative integer n.

⋆ (12) Prove Theorem 9.15.

⋆ (13) Prove Theorem 9.16.

(14) **Direct sums of integral domains and fields.**

(a) Suppose that R and S are both integral domains. Is $R \oplus S$ always, sometimes, or never an integral domain? Give a proof or a pair of examples (whichever is most appropriate) to justify your answer.

(b) Would your answer to part (a) change if the words *integral domain* were replaced by *field*? Why or why not?

(15) **Direct sums and characteristic.**

(a) What is the characteristic of $\mathbb{Z}_m \oplus \mathbb{Z}_n$? Prove your answer.

(b) Let R and S be rings with characteristic m and n, respectively, where $m, n > 0$. Generalize your work in part (a) to determine (with proof) the characteristic of $R \oplus S$.

(16) **Direct sums and rings with characteristic zero.** Let R and S be rings. Decide whether each of the following statements is true or false. Verify your answers.

(a) If both R and S have nonzero characteristic, then $R \oplus S$ has nonzero characteristic.

(b) If $R \oplus S$ has nonzero characteristic, then both R and S have nonzero characteristic.

(c) If both R and S have characteristic zero, then $R \oplus S$ has characteristic zero.

(d) If $R \oplus S$ has characteristic zero, then both R and S have characteristic zero.

(17) For each natural number i, let R_i be a ring. Define the **infinite direct sum**

$$\bigoplus_{i=1}^{\infty} R_i = R_1 \oplus R_2 \oplus R_3 \oplus \cdots$$

to be the set of all sequences of the form

$$x = (x_1, x_2, x_3, \ldots),$$

where $x_i \in R_i$ for all i, and $x_i = 0_{R_i}$ for all but finitely many values of i. To illustrate, let $R_i = \mathbb{Z}_{i+1}$ for every natural number i, and define

$$R = \bigoplus_{i=1}^{\infty} R_i = \bigoplus_{i=1}^{\infty} \mathbb{Z}_{i+1} = \mathbb{Z}_2 \oplus \mathbb{Z}_3 \oplus \mathbb{Z}_4 \oplus \cdots$$

Then

$$x = ([1]_2, [0]_3, [0]_4, [4]_5, [3]_6, [0]_7, [0]_8, [0]_9, \ldots) \in R,$$

since x_i is the zero element in $\mathbb{Z}_{i+1}$ for all but 3 values of i. In contrast,

$$y = ([1]_2, [1]_3, [1]_4, [1]_5, \ldots) \notin R,$$

since y_i is nonzero for infinitely many values of i.

(a) With addition and multiplication defined componentwise (as in the definition of a finite direct sum), prove that every infinite direct sum of rings is also a ring.

(b) An **infinite direct product** is defined similarly to an infinite direct sum, but without the restriction that $x_i = 0_{R_i}$ for all but finitely many values of i. Is every infinite direct product of rings also a ring? Prove your answer.

(18) **The characteristic of an infinite direct sum.** Let R be the infinite direct sum

$$R = \bigoplus_{i=1}^{\infty} \mathbb{Z}_{i+1} = \mathbb{Z}_2 \oplus \mathbb{Z}_3 \oplus \mathbb{Z}_4 \oplus \cdots,$$

as defined in Exercise 17.

(a) Show that for every $x = (x_1, x_2, x_3, \ldots) \in R$, there exists a positive integer k such that

$$kx = 0_R = ([0]_2, [0]_3, [0]_4, \ldots).$$

(Hint: Recall that x_i is nonzero for only finitely many values of i.)

(b) Show that, in spite of your conclusion in part (a), the characteristic of R is zero.

(c) Explain the apparent contradiction between your answers to parts (a) and (b).

Connections

In this investigation, we studied subrings and direct sums of rings. If you studied group theory before ring theory, you should notice connections between the topics in this investigation and those in Investigations 22 and 28. The idea of a subgroup is the same is that of a subring; in particular, a subgroup of a group G is just a subset of G that is also a group using the operation from G. The only significant difference is that a ring comes with two operations and a group comes with only one, so it is a bit easier to determine if a subset of a group is a subgroup than if a subset of a ring is a subring.

The constructions of direct sums of rings and groups are also analogous. To create a direct sum of a pair of objects (either rings or groups) we make the Cartesian product of the sets into a ring or group using operations defined componentwise. Because of the simpler structure of groups, we can define more than one type of direct sum. In addition, we often use multiplicative notation to denote the operation in a group. For this reason, the group theoretic analog of a direct sum of rings is often called an *external direct product* to distinguish it from from other direct sums.

Investigation 10

Isomorphism and Invariants

Focus Questions

By the end of this investigation, you should be able to give precise and thorough answers to the questions listed below. You may want to keep these questions in mind to focus your thoughts as you complete the investigation.

- Intuitively, what does it mean for two rings to be "essentially the same?"
- What does it mean for two rings to be isomorphic? How does the definition of isomorphism reflect the informal definition of "essentially the same?"
- What strategies can be used to prove that two rings are isomorphic? How are these strategies motivated by the definition of isomorphism?
- What is an invariant, and how does one prove that a property is an invariant?
- How can invariants be used to prove that two rings are not isomorphic? Can invariants be used to prove that two rings are isomorphic?

Preview Activity 10.1. In Investigation 5, we saw how equivalence relations can be used to identify mathematical objects that are the same (or equivalent) in one way or another. The notion of "sameness" is very important in mathematics, for it allows us to identify when two objects should be considered indistinguishable, and thus treated identically. Identifying sameness also makes our analysis more efficient, since it allows us to consider entire classes of objects at the same time, instead of dealing with each object individually.

In this investigation, we will define precisely what it means for two rings to be the same, or *isomorphic*. Before we do so, however, let's apply our intuitive ideas about sameness to a few examples.

The addition and multiplication tables for four rings are shown below. Which of these rings would you consider to be essentially the same, and which would you consider to be different? Consider each possible pair of rings, and give a convincing argument to justify your answer for each.

R_1:

$+$	γ	α	δ	β
γ	α	γ	β	δ
α	γ	α	δ	β
δ	β	δ	γ	α
β	δ	β	α	γ

$\cdot$	γ	α	δ	β
γ	α	α	γ	γ
α	α	α	α	α
δ	γ	α	β	δ
β	γ	α	δ	β

R_2:

$+$	w	x	y	z
w	w	x	y	z
x	x	w	z	y
y	y	z	w	x
z	z	y	x	w

$\cdot$	w	x	y	z
w	w	w	w	w
x	w	x	w	x
y	w	y	w	y
z	w	z	w	z

R_3:

$+$	q	r	s	t
q	q	r	s	t
r	r	q	t	s
s	s	t	q	r
t	t	s	r	q

$\cdot$	q	r	s	t
q	q	q	q	q
r	q	r	q	r
s	q	q	s	s
t	q	r	s	t

R_4:

$+$	i	j	k	l
i	i	j	k	l
j	j	k	l	i
k	k	l	i	j
l	l	i	j	k

$\cdot$	i	j	k	l
i	i	i	i	i
j	i	j	k	l
k	i	k	i	k
l	i	l	k	j

Introduction

In Preview Activity 10.1, you were asked to decide which of the four rings shown were essentially the same, and which were not. At first glance, it would be easy to think that all of the rings are different. After all, their elements are certainly different. But is this enough to conclude that the rings themselves are different?

To answer this question, let's consider the addition and multiplication tables for $\mathbb{Z}_4$:

$\mathbb{Z}_4$:

$+$	[0]	[1]	[2]	[3]
[0]	[0]	[1]	[2]	[3]
[1]	[1]	[2]	[3]	[0]
[2]	[2]	[3]	[0]	[1]
[3]	[3]	[0]	[1]	[2]

$\cdot$	[0]	[1]	[2]	[3]
[0]	[0]	[0]	[0]	[0]
[1]	[0]	[1]	[2]	[3]
[2]	[0]	[2]	[0]	[2]
[3]	[0]	[3]	[2]	[1]

Let's suppose also that we decided to abbreviate the names of the equivalence classes in $\mathbb{Z}_4$ by assigning a variable to each one. In particular, we'll let $i = [0]$, $j = [1]$, $k = [2]$, and $l = [3]$.

Activity 10.2. Substitute i, j, k, and l for [0], [1], [2], and [3] in the tables for $\mathbb{Z}_4$. That is, each time [0] appears in the tables, replace it with i. Do the same for the other classes as well, replacing [1] with j, [2] with k, and [3] with l. What do you notice?

If you completed Activity 10.2 correctly, you probably observed that the addition and multiplication tables for $\mathbb{Z}_4$ can be made to look exactly like those of R_4 in Preview Activity 10.1, simply by renaming the elements. In other words, the only differences between $\mathbb{Z}_4$ and R_4 are the names of the elements. We might even say that $\mathbb{Z}_4$ and R_4 are essentially the same ring.

As it turns out, we can carry out a similar renaming to show that $\mathbb{Z}_4$ and R_1 are also essentially the same. Here we have to be a bit more careful, however, since the elements of R_1 seem to be arranged in a different order than those of $\mathbb{Z}_4$ and R_4. To illustrate, let's see what would happen if we replaced the names of the elements of R_1 with the names of the elements of $\mathbb{Z}_4$, keeping the elements in the same order as they are listed in the tables. We would replace γ with [0], α with [1], δ with [2], and β with [3], which would yield the following addition and multiplication tables:

+	[0]	[1]	[2]	[3]
[0]	[1]	[0]	[3]	[2]
[1]	[0]	[1]	[2]	[3]
[2]	[3]	[2]	[0]	[1]
[3]	[2]	[3]	[1]	[0]

·	[0]	[1]	[2]	[3]
[0]	[1]	[1]	[0]	[0]
[1]	[1]	[1]	[1]	[1]
[2]	[0]	[1]	[3]	[2]
[3]	[0]	[1]	[2]	[3]

A quick look at these tables reveals that they certainly do not appear to be the addition and multiplication tables for $\mathbb{Z}_4$. In this case, simply renaming the elements of R_1 did not yield a ring that looked essentially the same as $\mathbb{Z}_4$. The ring that resulted from such a renaming had the same elements as $\mathbb{Z}_4$, but its operations seem different from those of $\mathbb{Z}_4$.

So what can we conclude from this? Does our failed attempt at renaming the elements of R_1 necessarily imply that R_1 and $\mathbb{Z}_4$ are different rings? In fact, it does not. All that we know at this point is that the particular renaming that we used yields a ring that appears to behave differently than $\mathbb{Z}_4$. But what if we used a different renaming? For instance, what if we replaced γ with [2], α with [0], δ with [3], and β with [1]? Doing so would yield the following addition and multiplication tables:

+	[2]	[0]	[3]	[1]
[2]	[0]	[2]	[1]	[3]
[0]	[2]	[0]	[3]	[1]
[3]	[1]	[3]	[2]	[0]
[1]	[3]	[1]	[0]	[2]

·	[2]	[0]	[3]	[1]
[2]	[0]	[0]	[2]	[2]
[0]	[0]	[0]	[0]	[0]
[3]	[2]	[0]	[1]	[3]
[1]	[2]	[0]	[3]	[1]

Notice that these tables are nearly identical to the addition and multiplication tables for $\mathbb{Z}_4$. In fact, the only difference is the order in which the elements have been assigned to the rows and columns of the tables. This minor detail affects only the way the tables are displayed, and not the information that they contain. A simple rearrangement of the rows and columns would put the tables in their more standard form. Thus, we can see that by renaming the elements of R_1 and possibly reordering the rows and columns of the resulting tables, we are able to produce the addition and multiplication tables for $\mathbb{Z}_4$. Because of this, we might say that R_1 and $\mathbb{Z}_4$ are essentially the same ring.

Now that we have seen a few examples, we are ready to state an informal definition, which we will use to formally define the notion of isomorphism in the next section.

Informal Definition 10.3. Let R and S be finite rings. Then R is said to be **essentially the same** as S if the addition and multiplication tables for R can be transformed into the addition and multiplication tables for S by doing nothing more than renaming the elements of R and/or reordering the rows and columns of R's addition and multiplication tables.

Note that this informal definition can easily be used to show that two rings are *not* essentially the same. For example, each of the rings from Preview Activity 10.1 contains an element that appears more often than any other element in the ring's multiplication table. For R_1, that element is α, which appears 8 times in the multiplication table for R_1. For R_2, it is w (10 times). For R_3, it is q (9 times). And for R_4, it is i (8 times).

From these observation, it is clear that R_1 and R_2 are not the same ring. This is because no matter how we rename and/or reorder the elements of R_1, we will still end up with a multiplication table whose most common element appears 8, not 10, times. Thus, no matter how we rename or reorder the elements of R_1, we will never get a ring whose addition and multiplication tables look just like those of R_2. A similar argument can be made for every other pair of rings from Preview Activity 10.1, with the exception of R_1 and R_4, both of which are the same as $\mathbb{Z}_4$ (and thus the same as each other) by our previous arguments.

To summarize, note that in order to show that two rings are the same, we must find a way to rename and/or reorder the elements of one ring so that its addition and multiplication tables are identical to those of the other ring. In order to show that two rings are different, however, it often suffices to identify a property that is different between the two rings—in particular, a property that could not possibly be different if one ring had been obtained from the other by simply renaming and reordering elements.

Activity 10.4. Consider each possible pair of rings from Preview Activity 10.1, with the exception of R_1 and R_4. For each such pair, make a list of properties that are different between the two rings. Then explain why each difference you listed would contradict Informal Definition 10.3. Include in your list at least 3 properties for each pair of rings.

Isomorphisms of Rings

Informal Definition 10.3 provides a helpful and intuitive way of thinking about what it means for two rings to be the same. This informal definition, however, has some significant limitations. First, it only works for finite rings. This is because it would be impossible to actually create the addition and multiplication tables for a ring with infinitely many elements. Second, even for finite rings, the definition can be extremely cumbersome to work with, especially if the rings in question have more than a few elements. Can you imagine trying to create a multiplication table for a ring with 50 or 1000 or even 50,000 elements? The task would be daunting at best, and practically impossible at worst.

To deal with these difficulties, we will adopt a formal definition that captures the idea behind Informal Definition 10.3, but does so in a more precise manner. In order to motivate this definition, let's consider again the two main parts of Informal Definition 10.3.

Renaming Elements

When we argued that $\mathbb{Z}_4$ and R_4 were essentially the same ring, we found a way to rename the elements of $\mathbb{Z}_4$ using the same names as the elements of R_4. This renaming was really just a bijective function (that is, a function that is both one-to-one and onto)* from $\mathbb{Z}_4$ to R_4. Denoting this function by $\varphi : \mathbb{Z}_4 \to R_4$, we could write:

$$\varphi([0]) = i$$
$$\varphi([1]) = j$$
$$\varphi([2]) = k$$
$$\varphi([3]) = l$$

Note that any function that actually corresponds to a valid renaming would have to be bijective. This is because it wouldn't make sense to give two ring elements the same name, or to leave a ring element out. If two rings are truly the same, then the elements of one should be able to be matched in a one-to-one correspondence with the elements of the other. As such, our formal definition of "sameness" will begin with a bijective function.

Preserving Operations

As we saw in our earlier example, just having a bijective function from one ring to another is not enough to say that the two rings are the same. Indeed, this bijective function must also transform the addition and multiplication tables of the first ring into those of the second. To see exactly what this means, let's look at an example.

Consider two rings R and S, each having three elements. Suppose also that we have defined a bijective "renaming" function $\varphi : R \to S$. Let's consider the multiplication table for R, which we can write generically as follows (using a, b, and c to denote the elements of R):

$\cdot$	a	b	c
a	aa	ab	ac
b	ba	bb	bc
c	ca	cb	cc

If we simply replace each entry in this table with its new name (as given by φ), we obtain the following table:

$\cdot$	$\varphi(a)$	$\varphi(b)$	$\varphi(c)$
$\varphi(a)$	$\varphi(aa)$	$\varphi(ab)$	$\varphi(ac)$
$\varphi(b)$	$\varphi(ba)$	$\varphi(bb)$	$\varphi(bc)$
$\varphi(c)$	$\varphi(ca)$	$\varphi(cb)$	$\varphi(cc)$

Is this table the multiplication table for S? Its entries are certainly elements of S, since φ maps from R to S. But the actual multiplication table for S would be defined as follows:

*The remainder of this investigation assumes an understanding of injective, surjective, and bijective functions. For a review of these topics, see Appendix A.

$\cdot$	$\varphi(a)$	$\varphi(b)$	$\varphi(c)$
$\varphi(a)$	$\varphi(a)\varphi(a)$	$\varphi(a)\varphi(b)$	$\varphi(a)\varphi(c)$
$\varphi(b)$	$\varphi(b)\varphi(a)$	$\varphi(b)\varphi(b)$	$\varphi(b)\varphi(c)$
$\varphi(c)$	$\varphi(c)\varphi(a)$	$\varphi(c)\varphi(b)$	$\varphi(c)\varphi(c)$

If the notation seems confusing here, just keep in mind that $\varphi(a)$, $\varphi(b)$, and $\varphi(c)$ are the elements of S, and we have formed the multiplication table for S in the usual way. In particular, each entry in the table is the product of the corresponding row and column headers (so, for instance, the entry in the $\varphi(a)$ row and $\varphi(b)$ column is just $\varphi(a)\varphi(b)$).

So what can we conclude? Recall that we wanted the renamed R table to be equal to the S table. In order for this to happen, each entry of the renamed R table must be equal to the corresponding entry of the S table. Thus, it must be the case that $\varphi(aa) = \varphi(a)\varphi(a)$, $\varphi(ab) = \varphi(a)\varphi(b)$, $\varphi(ac) = \varphi(a)\varphi(c)$, $\varphi(ba) = \varphi(b)\varphi(a)$, and so on.

In other words, we want φ to *preserve multiplication*, which means:

For all x, $y \in R$, $\varphi(xy) = \varphi(x)\varphi(y)$.

And, of course, since we want the renamed addition table for R to be the same as the addition table for S, we would like φ to *preserve addition* as well:

For all x, $y \in R$, $\varphi(x + y) = \varphi(x) + \varphi(y)$.

These two conditions help us to state in a more precise way exactly what it means for two rings to have the same addition and multiplication tables—or, in other words, the same algebraic structure. Any bijective function φ that satisfies both of these conditions is called an *isomorphism*, defined formally below.

Definition 10.5. Let R and S be rings. An **isomorphism** is a bijective function $\varphi : R \to S$ such that for all x, $y \in R$,

$$\varphi(x + y) = \varphi(x) + \varphi(y) \quad \text{and} \quad \varphi(xy) = \varphi(x)\varphi(y).$$

If there exists an isomorphism from R to S, then R is said to be *isomorphic* to S.

The word *isomorphic* comes from two Greek words: *isos*, which means *equal* or *same*, and *morphe*, which means *form* or *structure*. Thus, when we say that two rings are isomorphic, we mean that they have the same structure. Since *structure* in abstract algebra is really defined by the way a number system's operations work, our specific definition of isomorphism requires φ to preserve both the addition and multiplication operations. Thus, we can think of an isomorphism as being an *operation-preserving bijection* or a *structure-preserving bijection*.

It is worth noting that the isomorphism relation is symmetric. (In fact, the isomorphism relation is an equivalence relation, as we will show in Activity 10.17.) In particular, if $\varphi : R \to S$ is an isomorphism, then $\varphi^{-1} : S \to R$ is also an isomorphism. For this reason, we will often simply say that two rings R and S are isomorphic, rather than saying R is isomorphic to S, or S is isomorphic to R. When R and S are isomorphic, we denote this relationship by writing $R \cong S$. So $R \cong S$ means that there is an isomorphism from R to S (or, equivalently, an isomorphism from S to R). It's also important to note that the *function* in Definition 10.5 is called an *isomorphism*, while the *rings* R and S are said to be *isomorphic*.

Activity 10.6.

(a) Use Definition 10.5 to explain why the function $\varphi : R_1 \to R_4$ (from Preview Activity 10.1) defined by

$$\varphi(\gamma) = k, \;\; \varphi(\alpha) = i, \;\; \varphi(\delta) = l, \;\; \varphi(\beta) = j$$

is an isomorphism.

(b) Use Definition 10.5 to explain why the function $\varphi : R_1 \to R_4$ defined by

$$\varphi(\gamma) = i, \;\; \varphi(\alpha) = j, \;\; \varphi(\delta) = k, \;\; \varphi(\beta) = l$$

is **not** an isomorphism.

(c) Is the function $\varphi : R_3 \to R_2$ defined by

$$\varphi(q) = w, \;\; \varphi(r) = x, \;\; \varphi(s) = y, \;\; \varphi(t) = z$$

an isomorphism? Use Definition 10.5 to justify your answer.

(d) What does your answer to part (c) allow you to conclude about whether R_2 and R_3 are isomorphic?

(e) Are R_2 and R_3 isomorphic? Use Definition 10.5 to justify your answer.

Proving Isomorphism

Now that we have precisely defined what it means for two rings to be isomorphic, let's consider how we might use this definition in the context of rings that have more than just a few elements. In particular, we will consider the ring M defined as follows:

$$M = \left\{ \begin{bmatrix} x & 0 \\ -x & 0 \end{bmatrix} : x \in \mathbb{R} \right\}.$$

Activity 10.7.

(a) Notice that M (as defined above) is a subset of $\mathcal{M}_{2\times 2}(\mathbb{R})$. Show that M is actually a *subring* of $\mathcal{M}_{2\times 2}(\mathbb{R})$.

(b) To which familiar ring do you think M is isomorphic? You don't have to prove your answer now, but you should make a reasonable conjecture with some solid reasoning to back it up.

Looking again at the definition of M, it appears that each element of M corresponds to a unique real number (and vice versa). Thus, it seems reasonable that we would try to prove that M is isomorphic to $\mathbb{R}$. Since both M and $\mathbb{R}$ have infinitely many elements, we will not be able to simply work with their addition and multiplication tables. Instead, we must use Definition 10.5, which suggests the following steps:

(1) We must define an appropriate function $\varphi : M \to \mathbb{R}$.

(2) We must show that φ is bijective; that is, we must show that φ is both *injective* and *surjective*.

(3) We must show that φ preserves addition.

(4) We must show that φ preserves multiplication.

Activity 10.8. Carefully read the following proof that M is isomorphic to $\mathbb{R}$, filling in all of the missing details and providing additional explanations where appropriate. As you read the proof, try to identify where each of the four steps outlined above is taking place.

Theorem. *Let* $M = \left\{ \begin{bmatrix} x & 0 \\ -x & 0 \end{bmatrix} : x \in \mathbb{R} \right\}$. *Then M is isomorphic to $\mathbb{R}$.*

Proof. Let $\varphi : M \to \mathbb{R}$ be defined by

$$\varphi\left(\begin{bmatrix} x & 0 \\ -x & 0 \end{bmatrix} \right) = x.$$

To see that φ is injective, suppose that

$$\varphi\left(\begin{bmatrix} x & 0 \\ -x & 0 \end{bmatrix} \right) = \varphi\left(\begin{bmatrix} y & 0 \\ -y & 0 \end{bmatrix} \right)$$

for some $\begin{bmatrix} x & 0 \\ -x & 0 \end{bmatrix}, \begin{bmatrix} y & 0 \\ -y & 0 \end{bmatrix} \in M$. Then $x = y$, which implies that

$$\begin{bmatrix} x & 0 \\ -x & 0 \end{bmatrix} = \begin{bmatrix} y & 0 \\ -y & 0 \end{bmatrix}.$$

Thus, φ is injective.

To see that φ is surjective, observe that for all $x \in \mathbb{R}$,

$$\varphi\left(\begin{bmatrix} x & 0 \\ -x & 0 \end{bmatrix} \right) = x.$$

We must now show that φ preserves both addition and multiplication. To this end, note that for all $\begin{bmatrix} x & 0 \\ -x & 0 \end{bmatrix}, \begin{bmatrix} y & 0 \\ -y & 0 \end{bmatrix} \in M$,

$$\begin{aligned} \varphi\left(\begin{bmatrix} x & 0 \\ -x & 0 \end{bmatrix} + \begin{bmatrix} y & 0 \\ -y & 0 \end{bmatrix} \right) &= \varphi\left(\begin{bmatrix} x+y & 0 \\ (-x)+(-y) & 0 \end{bmatrix} \right) \\ &= \varphi\left(\begin{bmatrix} x+y & 0 \\ -(x+y) & 0 \end{bmatrix} \right) \\ &= x+y \\ &= \varphi\left(\begin{bmatrix} x & 0 \\ -x & 0 \end{bmatrix} \right) + \varphi\left(\begin{bmatrix} y & 0 \\ -y & 0 \end{bmatrix} \right), \end{aligned}$$

and

$$\begin{aligned}
\varphi\left(\begin{bmatrix} x & 0 \\ -x & 0 \end{bmatrix}\begin{bmatrix} y & 0 \\ -y & 0 \end{bmatrix}\right) &= \varphi\left(\begin{bmatrix} xy & 0 \\ (-x)y & 0 \end{bmatrix}\right) \\
&= \varphi\left(\begin{bmatrix} xy & 0 \\ -(xy) & 0 \end{bmatrix}\right) \\
&= xy \\
&= \varphi\left(\begin{bmatrix} x & 0 \\ -x & 0 \end{bmatrix}\right)\varphi\left(\begin{bmatrix} y & 0 \\ -y & 0 \end{bmatrix}\right).
\end{aligned}$$

Since we have shown that there exists a bijective function $\varphi : M \to \mathbb{R}$ that preserves both addition and multiplication, it follows that M and $\mathbb{R}$ are isomorphic. ■

Well-Defined Functions

Preview Activity 10.9. Let f be a mapping that assigns to each element $[a]_3$ in $\mathbb{Z}_3$ the element $[a]_6$ in $\mathbb{Z}_6$—that is, $f([a]_3) = [a]_6$ for all $[a]_3 \in \mathbb{Z}_3$.

(a) Does f preserve the operations in $\mathbb{Z}_3$? Explain.

(b) Consider the following proof that f is an injection:

> Let $[a]_3$ and $[b]_3$ be in $\mathbb{Z}_3$, and assume $f([a]_3) = f([b]_3)$. Then $[a]_6 = [b]_6$, and so 6 divides $b - a$. Thus, 3 divides $b - a$, which implies that $[a]_3 = [b]_3$.

This proof might seem to imply that $\mathbb{Z}_3$ is isomorphic to the set $f(\mathbb{Z}_3) = \{f([a]_3) : [a]_3 \in \mathbb{Z}_3\} = \{[0]_6, [1]_6, [2]_6\}$. What do you think about this conclusion? Explain your answer in detail.

There is one additional important piece of information we need to consider when proving isomorphism. Activity 10.9 shows that we can define a map that preserves a ring's structure and seems to behave like an isomorphism, but if the map treats equal elements with different representations in different ways, then whatever conclusions we might draw will make little sense. We saw this same idea in Investigation 5 when we discussed well-defined operations. To emphasize this point, any time we have multiple ways to represent the elements in a set (like in $\mathbb{Z}_n$ or $\mathbb{Q}$), we need to be sure that anything that acts on the elements of that set (like an operation or a function) is well-defined. We formalize this idea for functions in the next definition.

Definition 10.10. Let S and T be sets. A mapping $f : S \to T$ is **well-defined** if $f(a) = f(b)$ whenever $a = b$ in S.

When we use the word *function*, we always mean a well-defined mapping. Well-defined mappings or functions are also called *single-valued*. In many cases we do not need to worry about a function begin well-defined. In particular, if there is only one way to represent each element in the domain, then there is nothing to show. If, however, there are multiple ways to represent elements in the domain (like in $\mathbb{Z}_n$ or $\mathbb{Q}$), then we need to know whether our mapping is well-defined before we worry about any other properties the mapping might possess.

Activity 10.11. Let f be a map from $\mathbb{Q}$ to $\mathbb{Z}$ defined by $f\left(\frac{a}{b}\right) = ab$. Is f well-defined? Verify your answer.

Disproving Isomorphism

In a previous section, we saw an example of how Definition 10.5 can be used to show that two rings are isomorphic. Although other examples may require more sophisticated arguments, the basic structure will often be the same: first define a particular function (well-defined), and then show that this function is bijective and preserves both addition and multiplication.

What should we do, however, if we want to prove that two rings are *not* isomorphic? For instance, consider the rings (fields, actually) $\mathbb{C}$ and $\mathbb{R}$. From an algebraic standpoint, these two rings seem to be quite different. After all, $\mathbb{C}$ contains an element i such that $i^2 = -1$, while $\mathbb{R}$ does not.

If we wanted to use the definition of isomorphism to prove that $\mathbb{C}$ and $\mathbb{R}$ are not isomorphic, we would have to show that there does not exist an isomorphism $\varphi : \mathbb{C} \to \mathbb{R}$. In other words, we would have to show that every function that we could possibly define from $\mathbb{C}$ to $\mathbb{R}$ would violate at least one of the conditions that define isomorphisms. To show this directly seems daunting, if not impossible.

Let us consider, therefore, a proof by contradiction. Perhaps we could begin by assuming that there does exist a function $\varphi : \mathbb{C} \to \mathbb{R}$ that is both bijective and operation-preserving. The next activity suggests how this assumption naturally leads to a contradiction.

Activity 10.12. Assume that $\varphi : \mathbb{C} \to \mathbb{R}$ is an isomorphism.

(a) Show that $\varphi(0) = 0$. (Hint: It suffices to show that $x + \varphi(0) = x$ for all $x \in \mathbb{R}$. You may need to use the fact that φ is surjective.)

(b) Using a similar argument as in part (a), show that $\varphi(1) = 1$.

(c) Use Definition 10.5 and your answers to part (a) and (b) to argue that $\varphi(i)$ is an element of $\mathbb{R}$, and that $\varphi(i)^2 = -1$. (Hint: Apply φ to both sides of the equation $i^2 + 1 = 0$.)

(d) Explain why the result you proved in part (c) is a contradiction. Deduce that $\mathbb{C}$ cannot be isomorphic to $\mathbb{R}$.

Invariants

In Activity 10.12, we identified a property (namely, the existence of element i such that $i^2 = -1$) that $\mathbb{C}$ satisfied and $\mathbb{R}$ did not. We then argued that this difference between $\mathbb{C}$ and $\mathbb{R}$ was incompatible with the definition of isomorphism, which allowed us to conclude that $\mathbb{C}$ and $\mathbb{R}$ could not be isomorphic.

The property mentioned above is known as an *invariant*, or more specifically, an invariant of ring isomorphism. Invariants are properties that must be preserved by isomorphism. Thus, if P is an invariant and a ring R satisfies P, then every ring that is isomorphic to R must also satisfy P. Consequently, if two rings differ with respect to an established invariant, then they cannot be isomorphic.

There are many different invariants, and we provide a list of the more common ones in Table

10.1. But how does one prove that a particular property is an invariant, and how can one use this fact once it has been established? To answer these questions, consider again the result we proved in part (b) of Activity 10.12. We can state a slight generalization of this result as follows:

Theorem 10.13. *Let R and S be rings, and let $\varphi : R \to S$ be an isomorphism. If R has an identity, say 1_R, then S also has an identity. Specifically, $\varphi(1_R)$ is an identity for S.*

Activity 10.14. Fill in the missing details in the following proof of Theorem 10.13:

Proof. To show that $\varphi(1_R)$ is an identity for S, we must show that $\varphi(1_R) \cdot x = x = x \cdot \varphi(1_R)$ for all $x \in S$. Let $x \in S$. Since φ is ____________, there exists $r \in R$ such that $\varphi(r) = x$. Thus,

$$\begin{aligned}\varphi(1_R) \cdot x &= \varphi(1_R) \cdot \varphi(r)^{\text{?}} \\ &= \varphi(1_R \cdot r)^{\text{?}} \\ &= \varphi(r)^{\text{?}} \\ &= x.\end{aligned}$$

A similar argument establishes that $x \cdot \varphi(1_R) = x$. Thus, $\varphi(1_R)$ is an identity for S. ■

The following corollary of Theorem 10.13 is immediate:

Corollary 10.15. *The existence of an identity is an invariant. Specifically, if R and S are isomorphic rings and R has identity, then S has identity also.*

Corollary 10.15 can be used to show, for example, that $\mathbb{Z}$ and $\mathbb{E}$ are not isomorphic, since $\mathbb{Z}$ has an identity and $\mathbb{E}$ does not. But what about two rings that are both the same with respect to the existence of an identity? For instance, $\mathbb{Z}_4$ and $\mathbb{Z}_2 \oplus \mathbb{Z}_2$ both have an identity. Are they isomorphic?

The answer to this question is a resounding "NO!" Although $\mathbb{Z}_4$ and $\mathbb{Z}_2 \oplus \mathbb{Z}_2$ have the same number of elements, are both commutative, and both have identity, they differ with regard to a number of other invariants. For instance, $\mathbb{Z}_4$ has exactly one zero divisor ($[2]$), while $\mathbb{Z}_2 \oplus \mathbb{Z}_2$ has two ($([0],[1])$ and $([1],[0])$). Similarly, $\mathbb{Z}_4$ has exactly two units ($[1]$ and $[3]$), while $\mathbb{Z}_2 \oplus \mathbb{Z}_2$ has only one ($([1],[1])$). Finally $\mathbb{Z}_4$ has characteristic 4, while $\mathbb{Z}_2 \oplus \mathbb{Z}_2$ has characteristic 2. Any one of these properties (i.e., invariants) would be sufficient to establish that $\mathbb{Z}_4$ and $\mathbb{Z}_2 \oplus \mathbb{Z}_2$ are not isomorphic. However, even if two rings agree with respect to every invariant we can think of, this does not prove that they are isomorphic. **Although invariants can be used to prove that two rings are different, they cannot be used to prove that two rings are the same. In order to prove that two rings are isomorphic, we must find an appropriate function from one ring to the other, and prove that this function is in fact an isomorphism.**

Table 10.1 lists some of the more common and useful invariants. This list, however, is far from complete, and we will add to it as needed throughout the remainder of our study of rings.

Concluding Activities

Activity 10.16. To which familiar ring is R_3 (from Preview Activity 10.1) isomorphic? Prove your answer.

A Partial List of Invariants of Ring Isomorphism

- Number of elements
- Commutativity
- The existence of an identity
- The existence or number of zero divisors
- The existence or number of units (in a ring with identity)
- Being a field
- Being an integral domain
- Characteristic

Table 10.1
Some common invariants.

Activity 10.17.

(a) Prove that the identity function is always an isomorphism, and thus every ring is isomorphic to itself.

(b) Let R and S be rings. Prove that if a function $\varphi : R \to S$ is a bijection, then $\varphi^{-1} : S \to R$ is also a bijection. (Hint: Use the definition of bijection to guide you. Remember that you need to show two things: that φ^{-1} is injective and that φ^{-1} is surjective.)

(c) Let R and S be rings. Prove that if $\varphi : R \to S$ is an isomorphism, then φ has an inverse and $\varphi^{-1} : S \to R$ is also an isomorphism. Deduce that if R is isomorphic to S, then S is isomorphic to R. (Hint: For each $s \in S$, use the fact that φ is surjective to write $s = \varphi(r)$ for some $r \in R$. Then note that $\varphi^{-1}(s) = r$. Use this kind of reasoning to prove that φ^{-1} preserves addition and multiplication, and then use part (b) to complete the proof of isomorphism.)

(d) Let R, S, and T be rings. Prove that if two functions $\varphi : R \to S$ and $\psi : S \to T$ are both bijections, then the composition $\psi \circ \varphi : R \to T$ is also a bijection.

(e) Let R, S, and T be rings. Prove that if $\varphi : R \to S$ and $\psi : S \to T$ are both isomorphisms, then $\psi \circ \varphi : R \to T$ is also an isomorphism. Deduce that if R and S are isomorphic, and S and T are isomorphic, then R and T are isomorphic. (Hint: Part (d) establishes part of the result. What else do you need to show?)

(f) What have you proved about the isomorphism relation in parts (a), (c), and (e)?

Activity 10.18. Prove that commutativity is an invariant. That is, prove that if R and S are isomorphic rings and R is commutative, then S must also be commutative.

Activity 10.19. Write a short summary that describes the important concepts, and the relationships between these concepts, that were introduced in this investigation. Explain how the material in this investigation is connected to the material in Investigations 7 and 9.

Exercises

(1) Consider the ring R_2^* defined by the addition and multiplication tables below.

R_2^* :

$+$	w	x	y	z
w	w	x	y	z
x	x	w	z	y
y	y	z	w	x
z	z	y	x	w

$\cdot$	w	x	y	z
w	w	w	w	w
x	w	x	y	z
y	w	w	w	w
z	w	x	y	z

Is R_2^* isomorphic to R_2 (from Preview Activity 10.1)? Why or why not?

(2) **Subrings of direct sums.** Let R and S be rings. Prove that $R \oplus S$ contains a subring isomorphic to R and a subring isomorphic to S.

(3) Let R and S be rings, and let $f : R \to S$ be an isomorphism. Prove that for any integer m and any $x \in R$, $f(mx) = mf(x)$.

(4) Prove that characteristic is an invariant. (Hint: Use Exercise 3.)

(5) Prove that being a field is an invariant.

(6) Prove that being an integral domain is an invariant.

(7) For any ring R, let R_0 be the subset of R containing the zero divisors of R. Show that if R and S are isomorphic rings, then there is a one-to-one correspondence between the elements of R_0 and the elements of S_0. Conclude that the number of zero divisors in a ring is an invariant.

(8) For any ring R, let R^* be the subset of R containing the units of R. Show that if R and S are isomorphic rings, then there is a one-to-one correspondence between the elements of R^* and the elements of S^*. Conclude that the number of units in a ring is an invariant.

(9) Is it possible for two fields with the same characteristic to not be isomorphic to each other? Verify your answer.

(10) Is it possible for a ring to be isomorphic to one of its proper subrings? Prove your answer. (Hint: Use Exercise 17 on page 118 of Investigation 9 with a direct sum of copies of some familiar ring.)

(11) **Rings of order 3.** Show that there are only two non-isomorphic rings of order 3 (that is, with exactly 3 elements).

(12) (a) Prove that every integral domain with characteristic zero contains a subring isomorphic to $\mathbb{Z}$. (Hint: Use multiples of the identity to define the desired subring.)

(b) One of the hypotheses in the statement from part (a) can be weakened, and the statement will still be true. State this improved result, and explain why your proof from part (a) is sufficient to establish it.

(13) **Matrix representations of complex numbers.** Find a subring of $M_{2\times 2}(\mathbb{R})$ that is isomorphic to $\mathbb{C}$. Prove your answer.

(14) **Subsets of $\mathbb{C}$.** We have already seen that $\mathbb{C}$ and $\mathbb{R}$ are not isomorphic, but let's now consider other subsets of $\mathbb{C}$.

(a) Are $\mathbb{R}$ and $\mathbb{Q}$ isomorphic? Prove your answer.

(b) Are $\mathbb{Q}$ and $\mathbb{Z}$ isomorphic? Prove your answer.

(15) Which of the following functions are well-defined? Prove each of your answers.

(a) $f : \mathbb{Q} \to \mathbb{Q}$ defined by $f\left(\frac{a}{b}\right) = \frac{a+b}{b}$

(b) $f : \mathbb{Z}_4 \to \mathbb{Z}_8$ defined by $f([a]_4) = [3a]_8$

(c) $f : \mathbb{Z}_6 \to \mathbb{Z}_7$ by $f([a]_6) = [a + k]_7$, where k is any integer.

(16) **Multiples of $\mathbb{Z}$.** Recall that for any integer $n > 1$,

$$n\mathbb{Z} = \{nx : x \in \mathbb{Z}\}.$$

Let m and n be positive integers. Find and prove a necessary and sufficient condition for $m\mathbb{Z}$ and $n\mathbb{Z}$ to be isomorphic.

(17) **Direct sums of $\mathbb{Z}_m$ and $\mathbb{Z}_n$.** Under what circumstances is $\mathbb{Z}_n \oplus \mathbb{Z}_m$ isomorphic to $\mathbb{Z}_{mn}$? State (and prove) your answer in the form of a biconditional (if and only if) statement.

(18) Recall that, for every natural number n, $\mathcal{P}_n$ denotes the power set of $\{1, 2, \ldots, n\}$, with addition defined by symmetric difference and multiplication defined by set intersection. (See Investigation 6.)

(a) Prove that $\mathcal{P}_2$ is isomorphic to $\mathbb{Z}_2 \oplus \mathbb{Z}_2$.

(b) Prove that $\mathcal{P}_3$ is isomorphic to $\mathbb{Z}_2 \oplus \mathbb{Z}_2 \oplus \mathbb{Z}_2$.

(c) Based on parts (a) and (b), make and prove a general conjecture about $\mathcal{P}_n$.

Connections

In this investigation, we studied isomorphisms of rings. If you studied group theory before ring theory, you should notice connections between isomorphisms of rings in this investigation and isomorphisms of groups in Investigation 29. The idea is the same in both contexts: isomorphic rings (or groups) are essentially the same, and an isomorphism is a bijection that preserves the structure of the algebraic set. Since there are two operations defined on a ring and only one operation in a group, the major difference between isomorphisms of rings and isomorphisms of groups is that an isomorphism of rings needs to preserve two operations but an isomorphism of groups needs to preserve only one. The process of verifying an isomorphism is the same in both contexts, but there is an extra step required for isomorphisms of rings.

Part IV

Polynomial Rings

Investigation 11

Polynomial Rings

Focus Questions

By the end of this investigation, you should be able to give precise and thorough answers to the questions listed below. You may want to keep these questions in mind to focus your thoughts as you complete the investigation.

- What is a polynomial ring over a commutative ring? How are addition and multiplication of polynomials defined?
- What are some important definitions and terminology associated with polynomials?
- When are two polynomials over a commutative ring considered to be equal?
- Under what conditions is a polynomial ring an integral domain?
- What is a polynomial function, and how are polynomial functions different than polynomials?

A good deal of time is spent studying polynomials in algebra courses that are intended to be a preparation for the study of calculus. For the time being, we can consider polynomials to be algebraic expressions such as

$$3x^2 - 5x - 7 \quad \text{and} \quad x^4 - 3x^3 + 2x^2 - 7.$$

In algebra courses, we learned how to add, subtract, and multiply polynomials. We even studied "polynomial long division" as a way to find a quotient and a remainder when one polynomial is divided by another polynomial. This is similar to the work we have done with the integers and so it may not be surprising that we will be able to develop our study of polynomials within the context of ring theory.

The polynomials studied in elementary mathematics courses were almost always restricted to having real number coefficients. In this and subsequent investigations, we will consider polynomials that have coefficients from other commutative rings and fields. We will see that most of the familiar results from elementary algebra courses about polynomials will be true when the coefficients for the polynomials are from a field.

Preview Activity 11.1. Complete each of the following using the rules for addition and multiplication of polynomials learned in precalculus algebra courses.

(a) Let $f(x) = 2x^2 + 3x + 2$ and $g(x) = 3x^2 + 4x + 2$. Determine $f(x) + g(x)$ and $f(x)g(x)$.

(b) Let $f(x) = [2]x^2 + [3]x + [2]$ and $g(x) = [3]x^2 + [4]x + [2]$, where the coefficients of the polynomials are elements of $\mathbb{Z}_5$. Determine $f(x) + g(x)$ and $f(x)g(x)$.

(c) Let $f(x) = [2]x^2 + [3]x + [2]$ and $g(x) = [3]x^2 + [4]x + [2]$, where the coefficients of the polynomials are elements of $\mathbb{Z}_6$. Determine $f(x) + g(x)$ and $f(x)g(x)$.

Polynomial Rings

We will now give a formal definition of a polynomial over a ring. Although the definitions could be written for any ring, we will focus our attention on polynomials over commutative rings.

Definition 11.2. Let R be a commutative ring. A **polynomial in *x* over *R*** is an expression of the form

$$a_nx^n + a_{n-1}x^{n-1} + \cdots a_2x^2 + a_1x^1 + a_0x^0,$$

where n is a nonnegative integer, and $a_n, a_{n-1}, \ldots, a_2, a_1, a_0$ are elements of R. The **set of all polynomials over the ring *R*** will be denoted by $R[x]$.

Conventions and Terminology

One of our first goals will be to prove that $R[x]$ is a commutative ring. Before doing that, we will need to introduce some notation and terminology. Let R be a commutative ring, and let $p(x) = a_nx^n + a_{n-1}x^{n-1} + \cdots a_2x^2 + a_1x^1 + a_0x^0 \in R[x]$ with $a_n \neq 0$.

- The symbol x is called an **indeterminate**. It is to be regarded as a formal symbol and not as an element of the ring R. In effect, the symbols x^0, x^1, x^2, ..., x^n serve as placeholders for the ring elements a_0, a_1, ..., a_n.

- The expressions a_kx^k in the polynomial $p(x)$ are called the **terms of the polynomial**. The elements $a_0, a_1, \ldots, a_n$ in the ring R are called the **coefficients** of the polynomial $p(x)$. We call a_k the coefficient of x^k in the representation of $p(x)$.

- When working with a polynomial, instead of writing x^1, we simply write x. In addition, we usually do not write x^0 and will write a_0x^0 simply as a_0. Using these conventions, we can write $p(x)$ in the form

$$p(x) = a_nx^n + a_{n-1}x^{n-1} + \cdots a_2x^2 + a_1x + a_0$$

  where n is a nonnegative integer, $a_n, a_{n-1}, \ldots, a_2, a_1, a_0$ are elements of R, and $a_n \neq 0$. (Although this substitution may seem obvious, there are actually some subtle issues that must be addressed in order for it to be valid. Activity 11.14 explores these subtleties in more detail.)

- We will usually omit any term having a zero coefficient from the representation of a polynomial, and if the ring R has an identity, we will write a term of the form 1_Rx^k simply as x^k. For example, in $\mathbb{R}[x]$, instead of writing $f(x) = 3x^3 + 1x^2 + 0x + 7$, we will write $f(x) = 3x^3 + x^2 + 7$. We will also write terms of the form $(-a_k)x^k$ as $-a_kx^k$. So instead of writing $g(x) = (-3)x^2 + (-7)$, we will write $g(x) = -3x^2 - 7$.

- The coefficient a_0 is called the **constant term** of the polynomial $p(x)$. A polynomial of the form $p(x) = a$, where $a \in R$, is called a **constant polynomial**.

- The coefficient a_n is called the **leading coefficient** of the polynomial $p(x)$. If the ring R has an identity and the leading coefficient a_n is equal to 1_R, the polynomial $p(x)$ is called a **monic polynomial.***

- The nonnegative integer n is called the **degree** of the polynomial $p(x)$, and we write $\deg(p(x)) = n$.

- We will denote by 0 the polynomial in $R[x]$ having all of its coefficients equal to 0. This polynomial is called the **zero polynomial**, and since it does not have a leading coefficient, the degree of the zero polynomial is undefined.

- Two polynomials in $R[x]$

$$p(x) = a_n x^n + a_{n-1}x^{n-1} + \cdots a_2 x^2 + a_1 x^1 + a_0 x^0, \text{ and}$$
$$q(x) = b_m x^m + b_{m-1}x^{m-1} + \cdots b_2 x^2 + b_1 x^1 + b_0 x^0,$$

are considered to be **equal polynomials** if both of them are the zero polynomial or if both have the same degree and all pairs of corresponding coefficients are equal.

- When we write a polynomial in the form

$$p(x) = a_n x^n + a_{n-1}x^{n-1} + \cdots a_2 x^2 + a_1 x + a_0,$$

we say that we have written the polynomial in **descending powers of *x***. A polynomial in the form

$$p(x) = a_0 + a_1 x + a_2 x^2 + \cdots + a_{n-1}x^{n-1} + a_n x^n$$

is said to be written in **ascending powers of *x***.

The formal definitions of addition and multiplication of polynomials are quite technical and notationally complex, but one of the purposes of Preview Activity 11.1 was to show that when working with specific polynomials, we can simply add and multiply polynomials as we did in previous mathematics courses. For example, we might "combine like terms" as follows:

$$5x^2 + 7x^2 = 12x^2.$$

Or, stated more generally:

$$a_2 x^2 + b_2 x^2 = (a_2 + b_2)\, x^2.$$

The next activity builds on this intuitive process and foreshadows the formal definitions of polynomial addition and multiplication.

Activity 11.3. Let R be a commutative ring with identity and let

$$p(x) = a_3 x^3 + a_2 x^2 + a_1 x + a_0 \quad \text{and} \quad q(x) = b_2 x^2 + b_1 x + b_0$$

be polynomials in $R[x]$. Using the standard rules for adding and multiplying polynomials from previous mathematics courses, determine the sum $p(x) + q(x)$ and the product $p(x)q(x)$. Write both results in descending powers of x.

*In familiar rings, it is easy to determine whether a given polynomial is monic by simply looking at its leading term. For instance, in $\mathbb{R}[x]$, both $p(x) = x^3 + 3x - 5$ and $q(x) = x^{99} + 1$ are monic polynomials since their leading terms are of the form x^n (for $n = 3$ in the first case, and $n = 99$ in the second). It is important to note, however, that the definition of a monic polynomial depends on 1_R, and so certain monic polynomials in certain rings may not "look" monic. For instance, $p(x) = [4]x^2$ is monic in $2\mathbb{Z}_6[x]$, since $[4]$ is the multiplicative identity in $2\mathbb{Z}_6$.

We will now formally define polynomial addition and multiplication. It might be helpful to use the work from Activity 11.3 as examples while studying these definitions.

Definition 11.4. Let R be a commutative ring and let $p(x), q(x) \in R[x]$ with

$$p(x) = a_n x^n + a_{n-1}x^{n-1} + \cdots + a_2x^2 + a_1x + a_0 \text{ with } a_n \neq 0, \text{ and}$$
$$q(x) = b_m x^m + b_{m-1}x^{m-1} + \cdots + b_2x^2 + b_1x + b_0 \text{ with } b_m \neq 0.$$

Since it must be true that $m \leq n$ or $n \leq m$, we can assume that $m \leq n$ without loss of generality. We will then use $b_{m+1} = b_{m+2} = \cdots b_n = 0$ and so we can write

$$p(x) = a_n x^n + a_{n-1}x^{n-1} + \cdots + a_2x^2 + a_1x + a_0, \text{ and}$$
$$q(x) = b_n x^n + b_{n-1}x^{n-1} + \cdots + b_2x^2 + b_1x + b_0.$$

The **sum of the polynomials** $p(x)$ and $q(x)$ is defined to be

$$p(x) + q(x) = (a_n + b_n)\, x^n + (a_{n-1} + b_{n-1})\, x^{n-1} + \cdots + (a_1 + b_1)\, x + (a_0 + b_0)\,.$$

The **product of the polynomials** $p(x)$ and $q(x)$ is defined to be

$$p(x)q(x) = c_{m+n}x^{m+n} + c_{m+n-1}x^{m+n-1} + \cdots + c_2x^2 + c_1x + c_0,$$

where for each k with $0 \leq k \leq m+n$,

$$c_k = a_kb_0 + a_{k-1}b_1 + a_{k-2}b_2 + \cdots + a_2b_{k-2} + a_1b_{k-1} + a_0b_k.$$

In the definition of multiplication, notice that in each of the terms for c_k, the sum of the two subscripts is equal to k. So we can say that in the product $p(x)q(x)$, the coefficient of the x^k term is the sum of all products of the form a_ib_j, where $i + j = k$. We can also write

$$c_k = \sum_{i+j=k} a_ib_j = \sum_{i=0}^{k} a_{k-i}b_i.$$

For example,

$$\begin{aligned} c_0 &= a_0b_0 \\ c_1 &= a_1b_0 + a_0b_1 \\ c_2 &= a_2b_0 + a_1b_1 + a_0b_2 \\ &\vdots \\ c_{m+n} &= a_nb_m \end{aligned}$$

The work in Activity 11.3 illustrated these definitions with specific polynomials. Using

$$p(x) = a_3x^3 + a_2x^2 + a_1x + a_0 \quad \text{and} \quad q(x) = b_2x^2 + b_1x + b_0,$$

we see that

$$p(x) + q(x) = a_3x^3 + (a_2 + b_2)\, x^2 + (a_1 + b_1)\, x + (a_0 + b_0)\,,$$

and

$$\begin{aligned} p(x)q(x) = a_3b_2x^5 &+ (a_3b_1 + a_2b_2)\, x^4 + (a_3b_0 + a_2b_1 + a_1b_2)\, x^3 \\ &+ (a_2b_0 + a_1b_1 + a_0b_2)\, x^2 + (a_1b_0 + a_0b_1)\, x + a_0b_0. \end{aligned}$$

If R is a commutative ring, then in order for $R[x]$ to be a ring, $R[x]$ must be closed under polynomial addition and polynomial multiplication. So if $p(x), q(x) \in R[x]$, we see by the very definitions of addition and multiplication that $p(x) + q(x) \in R[x]$ and $p(x)q(x) \in R[x]$. Hence, $R[x]$ is closed under addition and multiplication. In addition, using the definitions for addition and multiplication of polynomials, it is possible to prove that if R is a commutative ring, then $R[x]$ is a commutative ring.

Activity 11.5. In this activity, we will explore the proofs of the properties of addition in $R[x]$. Formal proofs of these properties would be required in a proof that $R[x]$ is a commutative ring when R is a commutative ring. To understand these proofs, it is sometimes wise to work with the properties in "simple" special cases. For polynomials, this often means trying to do the proofs for polynomials with small degrees, such as degree 1, 2, or 3. That is what we will do in this activity.

So let R be a commutative ring and let

$$p(x) = a_2x^2 + a_1x + a_0, \quad q(x) = b_2x^2 + b_1x + b_0, \quad \text{and} \quad r(x) = c_1x + c_0$$

be polynomials in $R[x]$. For the following activities (and the formal proofs), since the coefficients of a polynomial are elements of the ring R, we can use the ring properties of R to prove the corresponding properties for $R[x]$.

(a) Illustrate the commutative property of addition in $R[x]$ by using the polynomials $p(x)$ and $q(x)$.

(b) Illustrate the associative property of addition in $R[x]$ by using the polynomials $p(x)$, $q(x)$, and $r(x)$.

(c) Let $z(x) = 0$ be the zero polynomial in $R[x]$. Verify that $p(x) + z(x) = p(x)$. What does this illustrate about addition in $R[x]$?

(d) Determine an additive inverse for $p(x)$ in $R[x]$.

(e) Notice that

$$\begin{aligned} p(x)q(x) = a_2b_2x^4 &+ (a_2b_1 + a_1b_2)\,x^3 + (a_2b_0 + a_1b_1 + a_0b_2)\,x^2 \\ &+ (a_1b_0 + a_0b_1)\,x + a_0b_0. \end{aligned}$$

Verify that $p(x)q(x) = q(x)p(x)$.

(f) Assume R has a multiplicative identity and let $u(x) = 1_R$. Verify that $p(x)u(x) = p(x)$. What does this illustrate about multiplication in $R[x]$?

We still have to explore the associative property for multiplication and the distributive property in $R[x]$. This will be done in Activity 11.16 and Exercise 8.

Even though we have not given a formal proof, our work in Activity 11.5, Activity 11.16, and Exercise 8 should suggest that the following theorem is true.

Theorem 11.6. *If R is a commutative ring, then $R[x]$ is a commutative ring. In addition, if the ring R has an identity, then the ring $R[x]$ has an identity.*

A formal proof of Theorem 11.6 is given in the appendix to this investigation.

Polynomials over an Integral Domain

Let R be a commutative ring with identity. Since we now know that $R[x]$ is a commutative ring with identity, it is natural to consider whether $R[x]$ is also an integral domain, or perhaps even a field. The following examples might be helpful in exploring the conditions under which $R[x]$ is an integral domain. Notice that if $p(x) = 2x^2 + 4x + 2$ and $q(x) = 3x + 3$ are considered to be polynomials in $\mathbb{Z}[x]$, we can then verify that

$$p(x)q(x) = \left(2x^2 + 4x + 2\right)(3x + 3) = 6x^3 + 18x^2 + 18x + 6.$$

However, if $f(x) = [2]x^2 + [4]x + [2]$ and $g(x) = [3]x + [3]$ are polynomials in $\mathbb{Z}_6[x]$, then we see that

$$\begin{aligned} f(x)g(x) &= [6]x^3 + [18]x^2 + [18]x + [6] \\ &= [0] \end{aligned}$$

and so $f(x)$ and $g(x)$ are zero divisors in $\mathbb{Z}_6[x]$. This shows that $\mathbb{Z}_6[x]$ is not an integral domain. Basically, what we did was to use zero divisors in $\mathbb{Z}_6$ to construct zero divisors in $\mathbb{Z}_6[x]$. In Exercise 4, we will generalize this argument and prove the following theorem.

Theorem 11.7. *If R is a commutative ring that contains zero divisors, then the polynomial ring $R[x]$ also contains zero divisors.*

A question that still remains is whether or not $\mathbb{Z}[x]$ is an integral domain. The work in Activity 11.9 will help determine the answer to this question, but in a somewhat more general context. The key will be to focus on the relationship between the degrees of the two polynomials and the degree of the product of these two polynomials. In previous algebra classes, when we worked with polynomials with real number coefficients, we learned that if $\deg(p(x)) = m$ and $\deg(q(x)) = n$, then $\deg(p(x)q(x)) = m + n$. The preceding example, however, shows that this may not be the case in polynomial rings such as $\mathbb{Z}_6[x]$, as multiplying two polynomials with positive degree can yield a polynomial with undefined degree—namely, the zero polynomial. The next activity provides another example of this type of behavior.

Activity 11.8. In $\mathbb{Z}_6[x]$, let

$$f(x) = [2]x^2 + x + [5] \quad \text{and} \quad g(x) = [3]x + [2].$$

(a) Calculate (and simplify) the product $f(x)g(x)$.

(b) What is the degree of the product $f(x)g(x)$? How does this compare to $\deg(f(x)) + \deg(g(x))$?

Activity 11.8 demonstrates that polynomials over rings other than $\mathbb{R}$ may not behave exactly the way we might expect them to with regard to multiplication. The next activity explores what conditions must be placed on R in order to ensure the expected relationship between the degrees of two polynomials in $R[x]$ and the degree of their product.

Activity 11.9. Let D be an integral domain. We then know that the product of any two nonzero elements of D is also nonzero.

(a) Let a and u be nonzero elements in D and let $b, v \in D$. Write the product

$$(ax + b)(ux + v)$$

in $D[x]$ in descending powers of x. What is the degree of this product? Justify your conclusion. (Hint: Since $a \neq 0$, $u \neq 0$, and D is an integral domain, what can be concluded about the product au?)

(b) With the same notation as in part (a), let $c \in D$ and write the product

$$\left(ax^2 + bx + c\right)(ux + v)$$

in $D[x]$ in descending powers of x. What is the degree of this product? Justify your conclusion.

(c) Use the ideas from the previous two examples to help complete the proof of the following very important theorem.

Theorem 11.10. *Let m and n be nonnegative integers. If D is an integral domain, then the product of polynomials of degree m and n in $D[x]$ is a polynomial in $D[x]$ of degree $(m+n)$.*

Proof. Let m and n be nonnegative integers, and let D be an integral domain and let $p(x), q(x) \in D[x]$ with $\deg(p(x)) = n$ and $\deg(q(x)) = m$. We can then write

$$\begin{aligned} p(x) &= a_n x^n + a_{n-1}x^{n-1} + \cdots a_2x^2 + a_1x^1 + a_0x^0, \text{ and} \\ q(x) &= b_m x^m + b_{m-1}x^{m-1} + \cdots b_2x^2 + b_1x^1 + b_0x^0, \end{aligned}$$

with $a_n \neq 0$ and $b_m \neq 0$. Using the definition of the product of two polynomials in Definition 11.4, we first note that the largest possible sum of the subscripts of a product of the form a_ib_j is $m + n$ and hence, $\deg(p(x)q(x)) \leq m + n$. (?)

(Now complete the proof by focusing on the coefficient of the x^{m+n} term.) ■

(d) Explain how to use Theorem 11.10 to prove the following corollary.

Corollary 11.11. *If D is an integral domain, then $D[x]$ is an integral domain.*

(e) Determine $([2]x + [1])^2$ in $\mathbb{Z}_4[x]$. Write the product in descending powers of x. Explain why this shows that it is necessary to assume that D is an integral domain in Theorem 11.10.

(f) How would Theorem 11.10 need to be modified if D was not assumed to be an integral domain? (Hint: Look at the proof from part (c). Only part of it relies on the assumption that D is an integral domain.)

Polynomial Functions

The definitions of polynomials and addition and multiplication of polynomials were meant to formalize and generalize the way we added and multiplied polynomials in previous mathematics courses. However, there has been one big difference between our formal study of polynomials and the idea of a polynomial function used in precalculus and calculus courses. In those courses, we often think of a polynomial as function $\overline{p} : \mathbb{R} \to \mathbb{R}$ for which $\overline{p}(x) = a_n x^n + a_{n-1} x^{n-1} + \cdots a_1 x + a_0$, for some real numbers a_0, a_1, $\ldots a_{n-1}$, a_n.

The notations are similar, but there is a fundamental difference between the polynomial function $\overline{p}$ and the polynomial $p(x) = a_n x^n + a_{n-1} x^{n-1} + \cdots a_1 x + a_0$ in $\mathbb{R}[x]$. The polynomial $p(x)$ is a formal expression in which x is an indeterminate, essentially serving as a placeholder for the real numbers a_0, a_1, $\ldots a_{n-1}$, a_n. In this way, polynomials are really just sequences of numbers (or ring elements in general). The notation we use to represent polynomials, which uses powers of the indeterminate x to keep track of the coefficients, allows us to add and multiply polynomials in a more intuitive way. (Can you imagine how much harder it would be to add and multiply polynomials if we only were able to use the formal definitions of addition and multiplication, without the algebraic intuition behind them?)

On the other hand, when working with a polynomial function $\overline{p}$, the symbol x is a variable, which means that real numbers can be substituted for x, making $\overline{p}(x)$ a real number. So there is a fundamental difference—albeit a subtle one—between a polynomial and a polynomial function. In spite of this fundamental difference, however, we can still use a polynomial over a commutative ring to define a polynomial function. The next definition describes how we can do this.

Definition 11.12. Let R be a commutative ring and let

$$p(x) = a_n x^n + a_{n-1} x^{n-1} + \cdots a_1 x + a_0$$

be a polynomial in $R[x]$. The **polynomial function induced by $\boldsymbol{p(x)}$** is the function $\overline{p} : R \to R$, where for each r in R,

$$\overline{p}(r) = a_n r^n + a_{n-1} r^{n-1} + \cdots a_1 r + a_0.$$

For simplicity, we often just say that $\overline{p}$ is a **polynomial function**.

The next activity will help illustrate the difference between polynomials and polynomial functions.

Activity 11.13.

(a) Let $p(x) = x^4$ and $q(x) = x^2$ be polynomials in $\mathbb{Z}_5[x]$.

(i) Is $p(x) = q(x)$ in $\mathbb{Z}_5[x]$?

(ii) For the polynomial function $\overline{p} : \mathbb{Z}_5 \to \mathbb{Z}_5$, determine $\overline{p}([0])$, $\overline{p}([1])$, $\overline{p}([2])$, $\overline{p}([3])$, and $\overline{p}([4])$.

(iii) For the polynomial function $\overline{q} : \mathbb{Z}_5 \to \mathbb{Z}_5$, determine $\overline{q}([0])$, $\overline{q}([1])$, $\overline{q}([2])$, $\overline{q}([3])$, and $\overline{q}([4])$.

(iv) Is the function $\overline{p}$ equal to the function $\overline{q}$? Explain.

(b) Let $p(x) = x^4$ and $q(x) = x^2$ be polynomials in $\mathbb{Z}_4[x]$.

(i) Is $p(x) = q(x)$ in $\mathbb{Z}_4[x]$?

(ii) For the polynomial function $\overline{p} : \mathbb{Z}_4 \to \mathbb{Z}_4$, determine $\overline{p}([0])$, $\overline{p}([1])$, $\overline{p}([2])$, and $\overline{p}([3])$.

(iii) For the polynomial function $\overline{q} : \mathbb{Z}_4 \to \mathbb{Z}_4$, determine $\overline{q}([0])$, $\overline{q}([1])$, $\overline{q}([2])$, and $\overline{q}([3])$.

(iv) Is the function $\overline{p}$ equal to the function $\overline{q}$? Explain.

Concluding Activities

Activity 11.14. The formal definition of a polynomial over a commutative ring R includes a term of the form a_0x^0, where a_0 is an element of R. In order to simplify notation a bit, we decided to write a_0 in place of a_0x^0. Although this replacement may seem natural, there are a few subtle points that must be considered in order for it to be valid. First, although it may be tempting to simply say that $x^0 = 1$ and so a_0x^0 is in fact *equal* to a_0, we must remember that x does not denote a ring element, but rather an indeterminate. In addition, even if we did view x as a ring element, the ring R may not have an identity, and so even in this case it might not make sense to say that $x^0 = 1$. What we *can* say is that the set of polynomials of the form a_0x^0 is a subring of $R[x]$ and is always isomorphic to R. Thus, the substitution of a_0 for a_0x^0 is legitimate, and this substitution behaves exactly as we would expect it to with respect to polynomial addition and multiplication. In parts (a) and (b) below, you are asked to prove that this is in fact the case.

(a) Let R be a commutative ring and let $S = \{a_0x^0 : a_0 \in R\}$. Prove that S is a subring of $R[x]$.

(b) Define a function $h : R \to S$ by $h(r) = rx^0$, for each r in R. Prove that h is a ring isomorphism.

Activity 11.15. We know that if R is a commutative ring, then $R[x]$ is a commutative ring, and that if R is a commutative ring with identity, then $R[x]$ is a commutative ring with identity. (See Theorem 11.6.) We also know that if D is an integral domain, then $D[x]$ is an integral domain. (See Corollary 11.11.) So it seems reasonable to ask, "If F is a field, then is $F[x]$ a field?"

(a) Is $\mathbb{R}[x]$ a field? (Hint: Consider the nonzero polynomial $p(x) = x$ in $\mathbb{R}[x]$.)

(b) Is $\mathbb{Z}_3[x]$ a field?

(c) If F is a field, is $F[x]$ always, sometimes, or never a field? Give a proof or a pair of examples (whichever is most appropriate) to justify your answer.

Activity 11.16. In Activity 11.5, we used special examples to illustrate the properties of addition for the ring $R[x]$ when R is a commutative ring. We will now explore the associative property of multiplication in $R[x]$. The formal proofs of this property can be notationally complex and somewhat difficult to follow. As such, here we will look at cases that are a bit simpler than the general forms. (A formal proof of the associative property in $R[x]$ is provided in the appendix to this investigation, and Exercise 8 explores the distributive property for $R[x]$ from a similar perspective as this activity.)

To begin, let R be a commutative ring, and let

$$p(x) = a_2x^2 + a_1x + a_0, \quad q(x) = b_2x^2 + b_1x + b_0, \quad \text{and} \quad r(x) = c_1x + c_0$$

be polynomials in $R[x]$.

(a) Show that

$$\begin{aligned} p(x)q(x) = a_2b_2x^4 &+ (a_2b_1 + a_1b_2)\,x^3 + (a_2b_0 + a_1b_1 + a_0b_2)\,x^2 \\ &+ (a_1b_0 + a_0b_1)\,x + a_0b_0 \end{aligned}$$

(b) Use your answer to part (a) to calculate $[p(x)q(x)]r(x)$. Then rewrite your result to show that

$$\begin{aligned}[p(x)q(x)]r(x) &= (a_2b_2)\, c_1x^5 + [(a_2b_2)\, c_0 + (a_2b_1)\, c_1 + (a_1b_2)\, c_1]\, x^4 \\ &+ [(a_2b_1)\, c_0 + (a_1b_2)\, c_0 + (a_2b_0)\, c_1 + (a_1b_1)\, c_1 + (a_0b_2)\, c_1]\, x^3 \\ &+ [(a_2b_0)\, c_0 + (a_1b_1)\, c_0 + (a_0b_2)\, c_0 + (a_1b_0)\, c_1 + (a_0b_1)\, c_1]\, x^2 \\ &+ [(a_1b_0)\, c_0 + (a_0b_1)\, c_0 + (a_0b_0)\, c_1]\, x + (a_0b_0)\, c_0\end{aligned}$$

(Notice that the sum of the subscripts in each term of the coefficient of x^j is equal to j.)

(c) Now use a process similar to that in parts (a) and (b) to write the product $p(x)[q(x)r(x)]$ in descending powers of x.

(d) Use the results of parts (a) – (c) to prove that

$$[p(x)q(x)]r(x) = p(x)[q(x)r(x)].$$

Activity 11.17. Write a short summary that describes the important concepts, and the relationships between these concepts, that were introduced in this investigation. Explain how the material in this investigation is connected to the material in Investigations 1 and 7.

Exercises

(1) For each of the following polynomials, state its degree, its leading coefficient, and its constant term, and write the polynomial in ascending powers of x.

(a) $5x^3 + 2x^2 - x + 7$ in $\mathbb{Z}[x]$.

(b) $7x - \sqrt{2}$ in $\mathbb{R}[x]$.

(c) $(1+i)x^4 - (2i)x^2 + x^3 + (5+2i)$ in $\mathbb{C}[x]$.

(d) $[7]x^8 + [3]x^4 + [9]$ in $\mathbb{Z}_{10}[x]$.

(2) In each of the following, perform the indicated operations and write the result in descending powers of x.

(a) $\left([3]x^2 + x + [2]\right) + \left(x^3 + x^2 + [3]x + [2]\right)$ in $\mathbb{Z}_4[x]$.

(b) $([2]x + [1])^2$ in $\mathbb{Z}_4[x]$.

(c) $([3]x + [1])^2$ in $\mathbb{Z}_6[x]$.

(d) $([4]x + [1])^2$ in $\mathbb{Z}_8[x]$.

(e) $([2]x + [5])\,([5]x + [8])$ in $\mathbb{Z}_{10}[x]$.

(f) $([2]x + [5])\,([5]x + [8])$ in $\mathbb{Z}_{11}[x]$.

(3) Give an example of each of the following or explain why no such example exists.

(a) Two polynomials of degree 1 in $\mathbb{Z}_4[x]$ whose product is a polynomial of degree 1.

(b) Two polynomials of degree 3 in $\mathbb{R}[x]$ whose sum is a polynomial of degree 1.

(c) Two polynomials of degree 2 in $\mathbb{Z}_6[x]$ whose product is a polynomial of degree 5.

$\star$ (4) Let R be a commutative ring, and let a and b be nonzero elements in R with $ab = 0$. (So R contains zero divisors.) Prove that $R[x]$ is not an integral domain by constructing two polynomials of degree 0 or 1 in $R[x]$ whose product is equal to the zero polynomial.

$\star$ (5) Let R be a commutative ring, and let $p(x), q(x) \in R[x]$ with $\deg(p(x)) = m$ and $\deg(q(x)) = n$, where m and n are nonnegative integers.

(a) Prove that the degree of $p(x) + q(x)$ is less than or equal to the maximum value of m and n.

(b) Prove that $\deg(p(x)q(x)) \leq m + n$.

(6) (a) List all polynomials of degree 2 in $\mathbb{Z}_2[x]$.

(b) List all polynomials of degree 3 in $\mathbb{Z}_2[x]$.

(c) List all polynomials of degree 2 in $\mathbb{Z}_3[x]$.

(d) List all polynomials of degree 3 in $\mathbb{Z}_3[x]$.

(7) Let m be an integer with $m \geq 1$.

(a) How many polynomials of degree m are there in $\mathbb{Z}_2[x]$?

(b) How many polynomials of degree m are there in $\mathbb{Z}_3[x]$?

(c) Let n be an integer with $n \geq 2$. How many polynomials of degree m are there in $\mathbb{Z}_n[x]$?

$\star$ (8) Let R be a commutative ring, and let

$$p(x) = a_2x^2 + a_1x + a_0, \quad q(x) = b_2x^2 + b_1x + b_0, \quad \text{and} \quad r(x) = c_1x + c_0$$

be polynomials in $R[x]$. Verify that $p(x)\left(q(x) + r(x)\right) = p(x)q(x) + p(x)r(x)$, thereby illustrating the distributive property in $R[x]$.

(9) Let R be a commutative ring, and let

$$S = \left\{a_nx^n + a_{n-1}x^{n-1} + \cdots a_1x + a_0 : a_k = 0 \text{ if } k \text{ is odd}\right\}.$$

Is S a subring of $R[x]$? Justify your conclusion.

(10) Let R be a commutative ring, and let

$$S = \left\{a_nx^n + a_{n-1}x^{n-1} + \cdots a_1x + a_0 : a_k = 0 \text{ if } k \text{ is even}\right\}.$$

Is S a subring of $R[x]$? Justify your conclusion.

(11) Let R be a commutative ring, and let

$$S = \left\{a_nx^n + a_{n-1}x^{n-1} + \cdots a_1x + a_0 : a_0 = 0_R\right\}.$$

Is S a subring of $R[x]$? Justify your conclusion.

(12) Let R be a commutative ring, and let

$$S = \{p(x) \in R[x] : \deg(p(x)) = 2\}.$$

Is S a subring of $R[x]$? Justify your conclusion.

(13) Let R and S be commutative rings, and let $h : R \to S$ be an isomorphism. Define the function $H : R[x] \to S[x]$ by

$$H\left(a_n x^n + \cdots a_1 x + a_0\right) = h\left(a_n\right) x^n + \cdots h\left(a_1\right) x + h\left(a_0\right),$$

for each $a_n x^n + a_{n-1}x^{n-1} + \cdots a_1 x + a_0$ in $R[x]$.

(a) Prove that the function H is an isomorphism.

(b) Deduce from part (a) that if R is isomorphic to S, then $R[x]$ is isomorphic to $S[x]$.

(14) In each of the following, a ring R is given and two polynomials $p(x)$ and $q(x)$ in $R[x]$ are given. In each case, determine if the polynomial functions $\overline{p} : R \to R$ and $\overline{q} : R \to R$ are equal.

(a) $\mathbb{Z}_4$ and $p(x) = [2]x^2 + [2]x$ and $q(x) = [0]$.

(b) $\mathbb{Z}_3$ and $p(x) = x^3 + [2]x$ and $q(x) = [2]x + [2]$.

(c) $\mathbb{Z}_4$ and $p(x) = x^3 + [2]x^2 + [2]x$ and $q(x) = [3]x + [2]$.

(d) $\mathbb{Z}_3$ and $p(x) = x^3 + x$ and $q(x) = [2]x$.

Connections

This investigation introduced polynomial rings. An important idea to remember from this investigation is that a polynomial ring with coefficients in an integral domain behaves much like the ring of integers from Investigation 1. We will see more of this behavior in subsequent investigations where we discuss the division algorithm, greatest common divisors, and unique factorization in polynomial rings.

Appendix – Proof that $R[x]$ Is a Commutative Ring

In this appendix, we will give a formal proof that $R[x]$ is a commutative ring when R is a commutative ring. Before we give the proof, we will show how to write the sum and product of two polynomials using summation notation. Let $f(x), g(x) \in R[x]$ with

$$\begin{aligned} f(x) &= a_n x^n + a_{n-1}x^{n-1} + \cdots + a_2 x^2 + a_1 x + a_0 \text{ with } a_n \neq 0, \text{ and} \\ g(x) &= b_m x^m + b_{m-1}x^{m-1} + \cdots + b_2 x^2 + b_1 x + b_0 \text{ with } b_m \neq 0. \end{aligned}$$

Since it must be true that $m \leq n$ or $n \leq m$, we can assume that $m \leq n$ without loss of generality. We will then use the fact that $b_{m+1} = b_{m+2} = \cdots = b_n = 0$, and so we can write

$$\begin{aligned} f(x) &= a_n x^n + a_{n-1}x^{n-1} + \cdots + a_2 x^2 + a_1 x + a_0, \text{ and} \\ g(x) &= b_n x^n + b_{n-1}x^{n-1} + \cdots + b_2 x^2 + b_1 x + b_0. \end{aligned}$$

Using summation notation, we can write the sum and product of these two polynomials as follows:

$$f(x)+g(x)=\sum_{i=0}^{n} a_i x^i+\sum_{i=0}^{n} b_i x^i=\sum_{i=0}^{n}(a_i+b_i)\,x^i, \quad \text{and}$$

$$f(x)g(x)=\left(\sum_{i=0}^{n} a_i x^i\right)\left(\sum_{i=0}^{n} b_i x^i\right)=\sum_{j=0}^{n+m}\left(\sum_{i=0}^{j}(a_{j-i}b_i)\right)x^j, \quad \text{or equivalently}$$

$$f(x)g(x)=\sum_{j=0}^{n+m}\left(\sum_{r+s=j}(a_r b_s)\right)x^j.$$

Theorem 11.16. *If R is a commutative ring, then $R[x]$ is a commutative ring. In addition, if the ring R has an identity, then the ring $R[x]$ has an identity.*

Proof. Let R be a commutative ring and let

$$\begin{aligned} f(x)&=a_n x^n+a_{n-1}x^{n-1}+\cdots+a_1 x+a_0,\\ g(x)&=b_m x^m+b_{m-1}x^{m-1}+\cdots+b_1 x+b_0, \text{ and}\\ h(x)&=c_k x^k+c_{k-1}x^{k-1}+\cdots+c_1 x+c_0 \end{aligned}$$

be elements of $R[x]$. In proving the ring properties for $R[x]$, we will assume (without loss of generality) that $k \le m \le n$ and extend the polynomials $g(x)$ and $h(x)$ to have n coefficients. This means that for $m < i \le n$, $b_i = 0$, and for $k < i \le n$, $c_i = 0$.

The definitions of addition and multiplication of polynomials show that $f(x) + g(x)$ and $f(x)g(x)$ are polynomials in $R[x]$, so $R[x]$ is closed under addition and multiplication.

To show that addition is commutative in $R[x]$, we use the commutativity of addition in R and obtain

$$f(x)+g(x)=\sum_{i=0}^{n}(a_i+b_i)x^i=\sum_{i=0}^{n}(b_i+a_i)x^i=g(x)+f(x).$$

Thus, addition is commutative in $R[x]$.

For associativity of addition in $R[x]$, the associativity of addition in R gives us

$$\begin{aligned} [f(x)+g(x)]+h(x)&=\left(\sum_{i=0}^{n}(a_i+b_i)x^i\right)+\sum_{i=0}^{n}c_i x^k\\ &=\sum_{i=0}^{n}[(a_i+b_i)+c_i]x^i\\ &=\sum_{i=0}^{n}[a_i+(b_i+c_i)]x^i\\ &=\sum_{i=0}^{n}a_i x^i+\sum_{i=0}^{n}(b_i+c_i)x^i\\ &=f(x)+[g(x)+h(x)]. \end{aligned}$$

Therefore, addition is associative in $R[x]$.

We can show that an additive identity in $R[x]$ is the polynomial $z(x) = 0_R$, the polynomial all of whose coefficients are 0_R, as follows:

$$f(x) + z(x) = \sum_{i=1}^{n} a_i x^i + z(x) = \sum_{i=1}^{n} (a_i + 0_R)\, x^i = \sum_{i=0}^{n} a_i x^i = f(x).$$

Therefore, $z(x)$ is an additive identity in $R[x]$.

We will now show that $R[x]$ contains an additive inverse for $f(x)$ (and hence for any element of $R[x]$). For each i with $0 \le i \le n$, we know that $a_i \in R$ and hence, a_i has an additive inverse, $-a_i \in R$. Let

$$q(x) = \sum_{i=0}^{n} (-a_i) x^i = (-a_n)x^n + (-a_{n-1})x^{n-1} + \cdots + (-a_1)x + (-a_0).$$

Then $q(x) \in R[x]$ and

$$f(x) + q(x) = \sum_{i=0}^{n} \left[a_i + (-a_i)\right] x^i = \sum_{i=0}^{n} 0_R x^i = z(x).$$

Therefore, $R[x]$ contains an additive inverse for each of its elements.

We now turn our attention to multiplication in $R[x]$. We will first show that multiplication is commutative. Using the definition of multiplication of polynomials, we see that

$$\begin{aligned} f(x)g(x) &= \left(\sum_{i=0}^{n} a_i x^i\right)\left(\sum_{i=0}^{m} b_i x^i\right) \\ &= \left(\sum_{i=0}^{n+m} \left(\sum_{r+s=i} (a_r b_s)\right) x^i\right). \end{aligned} \tag{11.1}$$

Since $s + r = r + s = i$ and multiplication in R is commutative, we see that $a_r b_s = b_s a_r$ and hence, we can rewrite equation (11.1) as follows:

$$\begin{aligned} f(x)g(x) &= \left(\sum_{i=0}^{n+m} \left(\sum_{s+r=i} (b_s a_r)\right) x^i\right) \\ &= \left(\sum_{i=0}^{n} a_i x^i\right)\left(\sum_{i=0}^{n} b_i x^i\right) \\ &= g(x)f(x). \end{aligned}$$

This shows that multiplication in $R[x]$ is commutative.

We will now show that multiplication is an associative operation in $R[x]$. (Note that the notation in this part of the proof can be a bit overwhelming; this is why we explored a special case of the associative property of multiplication in Activity 11.16. It will help in understanding the notation in this proof to have the work from this activity to refer to while reading the proof.) We will start with the formula for $f(x)g(x)$ in equation (11.1). To simplify the notation, for $0 \le i \le n+m$, we let $u_i = \sum_{r+s=i} a_r b_s$. We then obtain

$$\begin{aligned} [f(x)g(x)]h(x) &= \left(\sum_{i=0}^{n+m} u_i x^i\right)\left(\sum_{i=0}^{k} c_i x^i\right) \\ &= \sum_{j=0}^{(n+m)+k} \left(\sum_{p+q=j} u_p c_q\right) x^j. \end{aligned}$$

We can now substitute for u_p and then use the fact that

$$\left(\sum_{r+s=p} a_r b_s\right) c_q = \sum_{r+s=p} (a_r b_s)\, c_q,$$

which is true by the distributive property in R. This gives

$$\begin{aligned}[f(x)g(x)]h(x) &= \sum_{j=0}^{(n+m)+k} \left[\sum_{p+q=j} \left(\sum_{r+s=p} a_r b_s\right) c_q\right] x^j \\ &= \sum_{j=0}^{(n+m)+k} \left[\sum_{p+q=j} \left(\sum_{r+s=p} (a_r b_s)\, c_q\right)\right] x^j. \qquad (11.2)\end{aligned}$$

In equation (11.2), notice that $r+s+q=p+q=j$, and so equation (11.2) shows that the coefficient of x^j in $[f(x)g(x)]h(x)$ is the sum of all products of the form $(a_r b_s)\, c_q$ where $r+s+q=j$. This means that we can rewrite equation (11.2) as follows:

$$[f(x)g(x)]h(x) = \sum_{j=0}^{(n+m)+k} \left[\sum_{r+s+q=j} (a_r b_s)\, c_q\right] x^j. \qquad (11.3)$$

Using a similar procedure for $f(x)[g(x)h(x)]$, we see that

$$\begin{aligned}f(x)[g(x)h(x)] &= \left(\sum_{i=0}^{n} a_i x^i\right) \left[\sum_{w=0}^{m+k} \left(\sum_{s+q=w} b_s c_q\right) x^w\right] \\ &= \sum_{j=0}^{n+(m+k)} \left[\sum_{r+w=j} \left(a_r \sum_{s+q=w} b_s c_q\right)\right] x^j \\ &= \sum_{j=0}^{n+(m+k)} \left[\sum_{r+w=j} \left(\sum_{s+q=w} a_r\, (b_s c_q)\right)\right] x^j. \qquad (11.4)\end{aligned}$$

In equation (11.4), $r+s+q=r+w=j$, and so equation (11.4) shows that the coefficient of x^j in $f(x)[g(x)h(x)]$ is the sum of all products of the form $a_r\, (b_s c_q)$ where $r+s+q=j$. Therefore, we can write

$$f(x)[g(x)h(x)] = \sum_{j=0}^{n+(m+k)} \left[a_r\, (b_s c_q)\right] x^j. \qquad (11.5)$$

Since multiplication in R is associative, we know that $a_r\, (b_s c_q) = (a_r b_s)\, c_q$, so using this and equation (11.5), we can conclude that

$$f(x)[g(x)h(x)] = \sum_{j=0}^{n+(m+k)} \left[(a_r b_s)\, c_q\right] x^j. \qquad (11.6)$$

Comparing equations (11.3) and (11.6), we see that $[f(x)g(x)]h(x) = f(x)[g(x)h(x)]$, which proves that multiplication in $R[x]$ is associative.

We will now prove the distributive law. (Since we have proved that multiplication in $R[x]$ is commutative, we only have to prove one of the distributive laws.) For this, recall that we have assumed that $k \leq m \leq n$, and so

$$\begin{aligned}
f(x)[g(x)+h(x)] &= \left(\sum_{i=0}^{n} a_i x^i\right)\left(\sum_{j=0}^{m} (b_j + c_j) x^j\right) \\
&= \sum_{j=0}^{n+m} \left[\sum_{r+s=j} a_r(b_s + c_s)\right] x^j \\
&= \sum_{j=0}^{n+m} \left[\sum_{r+s=j} (a_r b_s + a_r c_s)\right] x^j \\
&= \sum_{j=0}^{n+m} \left(\sum_{r+s=j} a_r b_s\right) x^j + \sum_{j=0}^{n+m} \left(\sum_{r+s=j} a_r c_s\right) x^j \\
&= f(x)g(x) + f(x)h(x).
\end{aligned}$$

Therefore, multiplication distributes over addition in $R[x]$, and we conclude that $R[x]$ is a ring.

Finally, we will prove that if R has an identity, 1_R, then $R[x]$ also contains an identity. Let $u(x) = 1_R$. Then $u(x) \in R[x]$ and

$$\begin{aligned}
f(x)u(x) &= \left(\sum_{i=0}^{n} a_i x^i\right)(1_R) \\
&= f(x).
\end{aligned}$$

Therefore, if R is a commutative ring with identity, then $R[x]$ is also a commutative ring with identity. ■

Investigation 12

Divisibility in Polynomial Rings

Focus Questions

By the end of this investigation, you should be able to give precise and thorough answers to the questions listed below. You may want to keep these questions in mind to focus your thoughts as you complete the investigation.

- What does it mean to say that one polynomial divides another polynomial? What are some of the important properties of divisibility in $F[x]$, where F is a field?
- What is the Division Algorithm in $F[x]$, where F is a field? How is this similar to the Division Algorithm for integers, and how is it different?
- What is the greatest common divisor of two polynomials in $F[x]$, where F is a field? What results about greatest common divisors of integers also hold in polynomial rings?
- How can the Euclidean Algorithm be used to find the greatest common divisor of two polynomials in $F[x]$, where F is a field?

Preview Activity 12.1. In this investigation, we will explore the Division Algorithm and greatest common divisors within the context of polynomial rings. Let's begin with a review of the analogous ideas in the ring of integers.

(a) Write the definition of what it means to say that an integer a divides an integer b.

(b) Give a precise statement of the Division Algorithm in the integers.

(c) Write the definition for the greatest common divisor of two integers a and b, at least one of which is nonzero.

(d) What is a prime number? Give a precise definition.

(e) State the Fundamental Theorem of Arithmetic.

Introduction

In the next two investigations, we will follow a process similar to what we did for the integers in order to prove several results about divisibility within the integral domain $F[x]$, where F is a field. Our work in the integers used the Division Algorithm, greatest common divisors, and prime numbers to eventually prove the Fundamental Theorem of Arithmetic. Our work here will also lead to a unique factorization theorem in $F[x]$ and will set the stage for us to investigate roots of polynomials. We will begin by defining divisibility within $R[x]$, where R is a commutative ring. Notice the similarity between the next definition and the definition of divides in $\mathbb{Z}$.

Definition 12.2. Let R be a commutative ring and let $u(x)$ and $v(x)$ be polynomials in $R[x]$. The polynomial $u(x)$ **divides** the polynomial $v(x)$ provided that there exists a polynomial $q(x) \in R[x]$ such that $v(x) = u(x)q(x)$. In this case, we also say that $u(x)$ is a **factor** of $v(x)$ and sometimes write $u(x) \mid v(x)$.

For example, in $\mathbb{R}[x]$, let $f(x) = 3x+2$, $g(x) = x^2+4x-7$, and $h(x) = 3x^3+14x^2-13x-14$. it is easy to verify by multiplication that

$$f(x)g(x) = (3x+2)\left(x^2+4x-7\right) = 3x^3+14x^2-13x-14 = h(x),$$

and so we can say that in $\mathbb{R}[x]$, $f(x)$ divides $h(x)$ and $g(x)$ divides $h(x)$.

The use of Definition 12.2 in proofs is very similar to the use of the definition of divides in the integers. This will be illustrated with proofs of the various parts of Theorem 12.3. Notice that this theorem deals with polynomials over a field F. We will see later in the investigation why this restriction is necessary.

Theorem 12.3. *Let F be a field and let $f(x), g(x) \in F[x]$.*

(i) *If $f(x)$ divides $g(x)$ and $c \in F$ and $c \neq 0$, then $cf(x)$ divides $g(x)$.*

(ii) *If $f(x) \neq 0$, $g(x) \neq 0$, and $f(x)$ divides $g(x)$, then $\deg(f(x)) \leq \deg(g(x))$.*

(iii) *If $f(x) \neq 0$ and a_n is the leading coefficient of $f(x)$, then $a_n^{-1}f(x)$ is a monic polynomial.*

(iv) *If $f(x)$ divides $g(x)$ and $g(x)$ divides $f(x)$, then there exists $c \in F$ with $c \neq 0$ such that $f(x) = cg(x)$.*

(v) *Let $f(x)$ and $g(x)$ be monic polynomials in $F[x]$. If $f(x)$ divides $g(x)$ and $g(x)$ divides $f(x)$, then $f(x) = g(x)$.*

Proof. We will prove parts (i) and (iii). The other parts will be proved in Activity 12.4. For part (i), we assume that $f(x)$ divides $g(x)$ and $c \in F$ with $c \neq 0$. So there exists a polynomial $q(x) \in F[x]$ such that

$$g(x) = f(x)q(x).$$

We multiply both sides of this equation by c and obtain $cg(x) = cf(x)q(x)$. Since $c \neq 0$, c has a multiplicative inverse in F and so if we now multiply both sides of the equation by c^{-1}, we obtain

$$c^{-1}[cg(x)] = c^{-1}\left[cf(x)q(x)\right].$$

We know that $c^{-1}c = 1$ on the left side, and on the right side, we can write $c^{-1}\left[cf(x)q(x)\right] = \left[cf(x)\right]\left[c^{-1}q(x)\right]$. So we have

$$g(x) = \left[cf(x)\right]\left[c^{-1}q(x)\right].$$

This proves that $cf(x)$ divides $g(x)$.

For part (iii), we write $f(x) = a_nx^n + a_{n-1}x^{n-1} + \cdots + a_1x + a_0$, where $a_0, a_1, \ldots, a_n$ are in the field F and $a_n \neq 0$. (Recall that a monic polynomial is a polynomial whose leading coefficient is equal to 1.) So if we multiply both sides of the equation by a_n^{-1}, we obtain

$$\begin{aligned} a_n^{-1} f(x) &= a_n^{-1}\left(a_n x^n + a_{n-1}x^{n-1} + \cdots + a_1 x + a_0\right) \\ &= a_n^{-1} a_n x^n + a_n^{-1} a_{n-1} x^{n-1} + \cdots + a_n^{-1} a_1 x + a_n^{-1} a_0 \\ &= x^n + a_n^{-1} a_{n-1} x^{n-1} + \cdots + a_n^{-1} a_1 x + a_n^{-1} a_0 \end{aligned}$$

and this proves that $a_n^{-1} f(x)$ is a monic polynomial. ■

Activity 12.4. Prove parts (ii), (iv), and (v) of Theorem 12.3. Here are some hints that might help you:

- For part (ii), use the definition of divides and Theorem 11.10. (See page 143.)
- For part (iv), use the definition of divides two times and then make a substitution in one of the resulting equations.
- Part (v) is a corollary of part (iv).

The Division Algorithm in $F[x]$

Now that we have a notion of divisibility within polynomial rings, our goal is to develop a Division Algorithm for polynomials much like the Division Algorithm for integers. However, in order to do this, we will restrict our attention to polynomial rings of the form $F[x]$, where F is a field (such as $\mathbb{Q}$, $\mathbb{R}$, or $\mathbb{Z}_p$, where p is a prime). Recall that when F is a field, we know that $F[x]$ is an integral domain. (See Corollary 11.11 on page 143.)

The Division Algorithm for $F[x]$ is stated below. You may notice the similarities between it and the Division Algorithm for the integers. (See Preview Activity 12.1.)

The Division Algorithm. *Let F be a field and let $f(x)$ and $g(x)$ be polynomials in $F[x]$ with $g(x) \neq 0$. Then there exist unique polynomials $q(x)$ and $r(x)$ in $F[x]$ such that*

$$f(x) = g(x)q(x) + r(x) \quad \textit{and}$$

$r(x) = 0$ *or* $\deg(r(x)) < \deg(g(x))$.

As in the integers, the Division Algorithm for polynomials guarantees the existence and uniqueness of a quotient and remainder, but says nothing about how to find these polynomials. Thus, before proving the Division Algorithm, we will illustrate a long division process for polynomials that can be used to find the quotient $q(x)$ and the remainder $r(x)$. This process is similar to long division in the integers.

In $\mathbb{R}[x]$, let

$$f(x) = 4x^4 + 10x^3 + 7x^2 - 4x + 7 \qquad \text{and} \qquad g(x) = 2x^2 + 2x - 1.$$

The long division process for $f(x)$ divided by $g(x)$ is shown below, and this process produces a quotient and a remainder. When using long division, the polynomial $f(x)$ is called the *dividend* and the polynomial $g(x)$ is called the *divisor*. In the example below, the quotient, $q(x) = 2x^2 + 3x + \frac{3}{2}$, is on the top line; the remainder, $r(x) = -4x + \frac{17}{2}$, is the last line.

Long division provides a way to obtain the quotient $q(x)$ in a step-by-step manner. The first term of the quotient, $2x^2$, is determined by finding the term that when multiplied by the first term of the divisor, $2x^2$, is equal to the first term of the dividend, $4x^4$. This first term can be found by noting that $\frac{4x^4}{2x^2} = 2x^2$. Notice that when we complete the subtraction in the next step and obtain $f(x) - 2x^2g(x)$, the result is a polynomial whose degree is less than that of the dividend. We then repeat this process to find the remaining terms of the quotient. (Note that, in order for long division to work correctly, both $f(x)$ and $g(x)$ must be written in descending powers of x.)

$$
\begin{array}{rrl}
 & 2x^2+\ 3x+\frac{3}{2} & \leftarrow \text{quotient } q(x) \\
\cline{2-2}
2x^2+2x-1\,\big) & 4x^4+10x^3+7x^2-4x+7 & \\
 & 4x^4+4x^3-2x^2 \qquad\qquad & \leftarrow 2x^2g(x) \\
\cline{2-2}
 & 6x^3+9x^2-4x+7 & \leftarrow f(x)-2x^2g(x) \\
 & 6x^3+6x^2-3x \quad\ \ & \leftarrow 3xg(x) \\
\cline{2-2}
 & 3x^2-x+7 & \leftarrow \left[f(x)-2x^2g(x)\right]-3xg(x) \\
 & 3x^2+3x-\frac{3}{2} & \leftarrow \frac{3}{2}g(x) \\
\cline{2-2}
 & -4x+\frac{17}{2} & \leftarrow \text{remainder } r(x)
\end{array}
$$

Activity 12.5.

(a) In the division process just illustrated, use polynomial multiplication and addition to verify that $f(x) = g(x)q(x) + r(x)$ in $\mathbb{R}[x]$, where

$$f(x) = 4x^4 + 10x^3 + 7x^2 - 4x + 7 \qquad q(x) = 2x^2 + 3x + \frac{3}{2}$$

$$g(x) = 2x^2 + 2x - 1 \qquad r(x) = -4x + \frac{17}{2}$$

(b) In the Division Algorithm, if $r(x) = 0$, then we can conclude that $g(x)$ divides $f(x)$. In $\mathbb{R}[x]$, let $f(x) = 2x^4 + 9x^3 - x^2 + x - 3$ and $g(x) = 2x^2 + x + 1$. Use long division to show that $g(x)$ divides $f(x)$. What is the quotient $q(x)$ for which $f(x) = g(x)q(x)$?

The proof of the Division Algorithm will show us why we need to assume that F is a field; in particular, we will need to use multiplicative inverses of some nonzero elements. The long division process illustrated prior to Activity 12.5 gives a clue as to how to prove the Division Algorithm. Notice that after the first subtraction, we obtained a polynomial

$$f(x) - 2x^2g(x) = 6x^3 + 9x^2 - 4x + 7,$$

which is a polynomial whose degree is less than the degree of $f(x)$. The next step in the process was a similar step using $f(x) - 2x^2g(x)$ instead of $f(x)$. Notice that after the next subtraction, we obtained a polynomial of degree 2, which is less than the degree of $f(x) - 2x^2g(x)$. This process of reducing the degree at each step suggests a proof by mathematical induction. Note that this technique is different than our approach in the integers, which relied on the Well-Ordering Principle. In fact, we could use the Well-Ordering Principle to prove the Division Algorithm for $F[x]$, and we could have used induction back in Investigation 2 to prove the Division Algorithm for $\mathbb{Z}$. We are using a different technique here to illustrate the various strategies for proving results of this type.

Proof of the Division Algorithm. Let F be a field and let $f(x)$ and $g(x)$ be polynomials in $F[x]$ with $g(x) \neq 0$. We will prove that there exist unique polynomials $q(x)$ and $r(x)$ in $F[x]$ such that

$$f(x) = g(x)q(x) + r(x) \quad \text{and} \quad r(x) = 0 \text{ or } \deg(r(x)) < \deg(g(x)).$$

We will first prove the existence of the polynomials $q(x)$ and $r(x)$. We will consider three cases for $f(x)$: (1) $f(x) = 0$; (2) $f(x) \neq 0$ and $\deg(f(x)) < \deg(g(x))$; or (3) $f(x) \neq 0$ and $\deg(f(x)) \geq \deg(g(x))$.

Case 1: If $f(x) = 0$, then $0 = f(x) = g(x) \cdot 0 + f(x)$. That is, we can use $q(x) = 0$ and $r(x) = f(x) = 0$.

Case 2: If $f(x) \neq 0$ and $\deg(f(x)) < \deg(g(x))$, then

$$f(x) = g(x) \cdot 0 + f(x).$$

So once again we use $q(x) = 0$ and $r(x) = f(x)$. Notice that since $r(x) = f(x)$, $\deg(r(x)) < \deg(g(x))$.

Case 3: If $f(x) \neq 0$ and $\deg(f(x)) \geq \deg(g(x))$, we will use induction on the degree of $f(x)$ to prove the existence of $q(x)$ and $r(x)$. For the basis step, we assume $\deg(f(x)) = 0$. Since $\deg(f(x)) \geq \deg(g(x))$ and $g(x) \neq 0$, we then also know that $\deg(g(x)) = 0$. This means that there exist nonzero elements a and b in F such that $f(x) = b$ and $g(x) = a$, and we see that

$$f(x) = b = a\left(a^{-1}b\right) + 0 = g(x)\left(a^{-1}b\right) + 0.$$

So we can use $q(x) = a^{-1}b$ and $r(x) = 0$. This proves the basis step.

For the inductive step, we let $n \in \mathbb{N}$ and assume that a quotient and remainder exist whenever the dividend is a polynomial of degree less than n. We will now assume $\deg(f(x)) = n$ and use this inductive assumption to prove that there exist polynomials $q(x)$ and $r(x)$ in $F[x]$ such that $f(x) = g(x)q(x) + r(x)$, and $r(x) = 0$ or $\deg(r(x)) < \deg(g(x))$. We will write

$$\begin{aligned} f(x) &= a_n x^n + a_{n-1}x^{n-1} + \cdots + a_1 x + a_0 \\ g(x) &= b_m x^m + b_{m-1}x^{m-1} + \cdots + b_1 x + b_0 \end{aligned}$$

where $a_n \neq 0$, $b_m \neq 0$, and $n \geq m$. We will then use long division (as illustrated in Activity 12.5) to divide $f(x)$ by $g(x)$. Although it would be very formidable to write the entire process, we can complete the first step as follows:

$$\begin{array}{r|lll} & b_m^{-1}a_n x^{n-m} & & \\ \hline b_m x^m + b_{m-1}x^{m-1} + \cdots + b_0 & a_n x^n & +a_{n-1}x^{n-1} & +\cdots \\ & a_n x^n & +b_{m-1}b_m^{-1}a_n x^{n-1} & +\cdots \\ \hline \end{array} \qquad (12.1)$$

Notice that we chose the first term in the quotient so that when we multiplied $g(x)$ by this term, the leading term would be $a_n x^n$. Subtracting, we obtain

$$f(x) - b_m^{-1}a_n x^{n-m} g(x) = \left(a_{n-1} - b_{m-1}b_m^{-1}a_n\right) x^{n-1} + \cdots,$$

which is a polynomial in $F[x]$ that is either 0 or has degree less than n. We can now apply the inductive assumption with this polynomial $(f(x) - b_m^{-1}a_nx^{n-m}g(x))$ as the dividend and $g(x)$ as the divisor. By induction, there exist polynomials $q_1(x)$ and $r_1(x)$ in $F[x]$ such that

$$f(x) - b_m^{-1}a_nx^{n-m}g(x) = g(x)q_1(x) + r_1(x),$$

and $r_1(x) = 0$ or $\deg(r_1(x)) < \deg(g(x))$. We can now rewrite this equation by adding $b_m^{-1}a_nx^{n-m}g(x)$ to both sides of the equation. This gives

$$f(x) = b_m^{-1}a_nx^{n-m}g(x) + g(x)q_1(x) + r_1(x)$$
$$f(x) = \left[b_m^{-1}a_nx^{n-m} + q_1(x)\right]g(x) + r_1(x)$$

The last equation shows the existence of a quotient (with $q(x) = b_m^{-1}a_nx^{n-m} + q_1(x)$) and a remainder (with $r(x) = r_1(x)$) since we already know that $r_1(x) = 0$ or $\deg(r_1(x)) < \deg(g(x))$. This completes the inductive step for the induction proof, and we can conclude that there exist polynomials $q(x)$ and $r(x)$ in $F[x]$ such that

$$f(x) = g(x)q(x) + r(x) \quad \text{and} \quad r(x) = 0 \text{ or } \deg(r(x)) < \deg(g(x)). \tag{12.2}$$

We must now prove the uniqueness of $q(x)$ and $r(x)$. To do so, we assume that there exist polynomials $q_2(x)$ and $r_2(x)$ in $F[x]$ such that

$$f(x) = g(x)q_2(x) + r_2(x) \quad \text{and} \quad r_2(x) = 0 \text{ or } \deg(r_2(x)) < \deg(g(x)). \tag{12.3}$$

Using Equations 12.2 and 12.3, we can conclude that

$$g(x)q(x) + r(x) = g(x)q_2(x) + r_2(x).$$

We can then rewrite this equation as

$$g(x)\left[q(x) - q_2(x)\right] = r_2(x) - r(x). \tag{12.4}$$

Since both $r(x)$ and $r_2(x)$ are either 0 or have a degree less than the degree of $g(x)$, we can conclude that the right side of equation (12.4) is 0 or a polynomial whose degree is less than $\deg(g(x))$. (See Exercise 5 in Investigation 11.) On the other hand, if $q(x) - q_2(x) \neq 0$, then by Theorem 11.10, the left side of equation (12.4) is a polynomial whose degree is greater than or equal to $\deg(g(x))$. This is a contradiction and so we conclude that $q(x) - q_2(x) = 0$ or $q(x) = q_2(x)$. This means that the left side of equation (12.4) is equal to 0 and so we conclude that $r(x) - r_2(x)$ is also equal to 0. Hence, $r(x) = r_2(x)$. This completes the proof of the uniqueness of the polynomials $q(x)$ and $r(x)$ and thus completes the proof of the Division Algorithm. ■

Activity 12.6. Up to this point, we have only used long division for polynomials over the real numbers $\mathbb{R}$. However, we can also use the same process for dividing polynomials over any field F. In this case, all computations with the coefficients must be done in the field F. For example, in $\mathbb{Z}_3[x]$, we could ask the question, "Does $g(x) = [2]x^2 + x + [1]$ divide $f(x) = x^4 + x^3 + [2]x^2 + x + [2]$?" We can start the long division process as follows:

$$\begin{array}{r|l}
 & [2]x^2 \\
\hline
[2]x^2 + x + [1] & x^4 + x^3 + [2]x^2 + x + [2] \\
 & x^4 + [2]x^3 + [2]x^2 \\
\hline
 & \quad [2]x^3 \qquad\qquad + x + [2]
\end{array}$$

Remember that all of the above calculations are being performed in $\mathbb{Z}_3[x]$, and so

$$[2]x^2\left([2]x^2 + x + [1]\right) = [4]x^4 + [2]x^3 + [2]x^2 = x^4 + [2]x^3 + [2]x^2,$$

and

$$\begin{aligned} x^4 + x^3 + [2]x^2 + x + [2] &- \left(x^4 + [2]x^3 + [2]x^2\right) \\ &= ([1] - [1])\,x^4 + ([1] - [2])\,x^3 + ([2] - [2])\,x^2 + x + [2] \\ &= [2]x^3 + x + [2]. \end{aligned}$$

(a) Complete this long division process to find polynomials $q(x)$ and $r(x)$ in $\mathbb{Z}_3[x]$ such that $f(x) = g(x)q(x) + r(x)$ and $0 \leq \deg(r(x)) < 2$. Does $g(x)$ divide $f(x)$ in $\mathbb{Z}_3[x]$?

(b) In $\mathbb{Z}_5[x]$, let $g(x) = [2]x^2 + x + [1]$ and $f(x) = x^4 + x^3 + [2]x^2 + x + [2]$. Use long division to find polynomials $q(x)$ and $r(x)$ in $\mathbb{Z}_5[x]$ such that $f(x) = g(x)q(x) + r(x)$ and $0 \leq \deg(r(x)) < 2$. Does $g(x)$ divide $f(x)$ in $\mathbb{Z}_5[x]$?

Greatest Common Divisors of Polynomials

When we were working in $\mathbb{Z}$, the next step (after studying the Division Algorithm) was to define the greatest common divisor of two integers. To continue the analogy between $\mathbb{Z}$ and $F[x]$, where F is a field, we will soon define the greatest common divisor of two polynomials in $F[x]$. The following preview activity is intended to explore some of the subtleties that will be involved with this definition.

Preview Activity 12.7. In $\mathbb{R}[x]$, let $f(x) = 2x^2 + x - 6$ and let $g(x) = 2x^2 - 7x + 6$.

(a) Verify that $f(x) = (x + 2)(2x - 3)$ and $g(x) = (x - 2)(2x - 3)$. Using these factorizations as guides, what do you think is the greatest common divisor of $f(x)$ and $g(x)$?

(b) Verify that $f(x) = (2x + 4)\left(x - \frac{3}{2}\right)$ and $g(x) = (2x - 4)\left(x - \frac{3}{2}\right)$. Does this alter your opinion as to what you think is the greatest common divisor of $f(x)$ and $g(x)$?

(c) Verify that $f(x) = \left(\frac{1}{2}x + 1\right)(4x - 6)$ and $g(x) = \left(\frac{1}{2}x - 1\right)(4x - 6)$. Does this alter your opinion as to what you think is the greatest common divisor of $f(x)$ and $g(x)$?

The point of Preview Activity 12.7 is that care must be taken in defining the greatest common divisor of two polynomials. In effect, the preview activity shows that we could define the greatest common divisor of $f(x)$ and $g(x)$ to be any one of the following (as well as others):

$$\gcd(f(x), g(x)) = 2x - 3 \quad \text{or} \quad \gcd(f(x), g(x)) = x - \frac{3}{2} \quad \text{or}$$
$$\gcd(f(x), g(x)) = 4x - 6.$$

Because we want greatest common divisors of polynomials to be unique, we will adopt the following definition. Note the similarities between this definition and the one introduced by Theorem 3.15 in Investigation 3. (See page 31.)

Definition 12.8. Let F be a field and let $f(x), g(x) \in F[x]$ that are not both zero. The **greatest common divisor** of $f(x)$ and $g(x)$ is the polynomial $d(x) \in F[x]$ that satisfes the following three conditions:

(i) $d(x)$ divides $f(x)$ and $d(x)$ divides $g(x)$.

(ii) If $h(x) \in F[x]$ and $h(x)$ divides both $f(x)$ and $g(x)$, then $h(x)$ divides $d(x)$.

(iii) $d(x)$ is a monic polynomial. (Recall that this means that the leading coefficient of $d(x)$ is equal to 1_F.)

Although we have stated a reasonable definition of the greatest common divisor of two polynomials, there is still a question as to whether a polynomial $d(x)$ satisfying all three conditions from Definition 12.8 actually exists. We will first prove that if $d(x)$ exists, then it must be unique. We will then show that $d(x)$ exists whenever $f(x)$ and $g(x)$ are not both zero.

Activity 12.9. Let F be a field and let $f(x)$ and $g(x)$ be elements of $F[x]$ that are not both zero. Let $d(x)$ be the greatest divisor of $f(x)$ and $g(x)$, as in Definition 12.8. Assume that $c(x)$ is another polynomial in $F[x]$ that satisfies the three conditions of the greatest common divisor. That is: (i) $c(x)$ divides $f(x)$ and $c(x)$ divides $g(x)$; (ii) if $h(x) \in F[x]$ and $h(x)$ divides both $f(x)$ and $g(x)$, then $h(x)$ divides $c(x)$; and (iii) $c(x)$ is a monic polynomial.

We will now prove that $c(x) = d(x)$, which will prove that if $d(x)$ in Definition 12.8 exists, then it must be unique.

(a) Explain why $d(x)$ divides $c(x)$.

(b) Explain why $c(x)$ divides $d(x)$.

(c) Use one of the parts of Theorem 12.3 to conclude that $c(x) = d(x)$.

Because the greatest common divisor as defined in Definition 12.8 is unique (if it exists), we can use the notation $\gcd(f(x), g(x))$ for the greatest common divisor of $f(x)$ and $g(x)$ in $F[x]$. The next theorem will establish the fact that, indeed, $\gcd(f(x), g(x))$ exists whenever $f(x)$ and $g(x)$ are not both zero. This theorem is similar to the results we proved in Investigation 3 about the greatest common divisor of two integers. (See Proposition 3.9 and Theorem 3.10 on pages 28 and 29, respectively.)

Theorem 12.10. *Let F be a field and let $f(x)$ and $g(x)$ be polynomials in $F[x]$ that are not both zero. There exists a unique monic polynomial $d(x)$ in $F[x]$ of smallest degree that can be written in the form*

$$d(x) = f(x)u(x) + g(x)v(x)$$

for some polynomials $u(x)$ and $v(x)$ in $F[x]$. In addition, $d(x) = \gcd(f(x), g(x))$.

Proof. Using the notation in the theorem, let S be the set of all monic polynomials $h(x)$ in $F[x]$ such that there exist polynomials $m(x), n(x) \in F[x]$ with $h(x) = f(x)m(x) + g(x)n(x)$.

Since $f(x)$ and $g(x)$ are not both zero, we may assume without loss of generality that $f(x) \neq 0$. Let c be the leading coefficient of $f(x)$. Then $c \neq 0$ and by Theorem 12.3, $c^{-1}f(x)$ is a monic polynomial in $F[x]$. Since

$$c^{-1}f(x) = 1 \cdot c^{-1}f(x) + 0 \cdot g(x),$$

we see that $c^{-1}f(x) \in S$, and so $S \neq \emptyset$.

This conclusion implies that the degrees of the polynomials in S form a nonempty subset of the whole numbers, and hence, by the Well-Ordering Principle, there must be a polynomial in S of smallest degree. Let $d(x)$ be a polynomial of smallest degree in S. Then $d(x)$ is a monic polynomial, and there exist $u(x), v(x) \in F[x]$ such that

$$d(x) = f(x)u(x) + g(x)v(x). \tag{12.5}$$

We will first use the Division Algorithm for $F[x]$ to prove that $d(x)$ divides $f(x)$. We know that

$$f(x) = d(x)q(x) + r(x), \tag{12.6}$$

for some $q(x), r(x) \in F[x]$ with $r(x) = 0$ or $\deg(r(x)) < \deg(d(x))$. We will show that $r(x) = 0$ by showing that we get a contradiction if $r(x) \neq 0$.

If $r(x) \neq 0$, then we can use equations (12.5) and (12.6) to write

$$\begin{aligned} r(x) &= f(x) - d(x)q(x) \\ &= f(x) - [f(x)u(x) + g(x)v(x)]q(x) \\ &= f(x) - f(x)u(x)q(x) - g(x)v(x)q(x) \\ &= f(x)[1 - u(x)q(x)] + g(x)[-v(x)q(x)]. \end{aligned}$$

So if a is the leading coefficient of $r(x)$, then

$$a^{-1}r(x) = f(x)\left[a^{-1} - a^{-1}u(x)q(x)\right] + g(x)\left[-a^{-1}v(x)q(x)\right]$$

is a monic polynomial and therefore must be in S. However, $\deg(a^{-1}r(x)) < \deg(d(x))$, which is a contradiction to the assumption that $d(x)$ is the smallest degree polynomial in S. Therefore, $r(x) = 0$ and then equation (12.6) implies that $d(x)$ divides $f(x)$.

We can use a similar argument to prove that $d(x)$ divides $g(x)$, and this establishes that $d(x)$ is a common divisor of $f(x)$ and $g(x)$.

We now assume that $s(x) \in F[x]$ and $s(x)$ divides both $f(x)$ and $g(x)$. From equation (12.5), we see that $d(x) = f(x)u(x) + g(x)v(x)$. We can then use the result in Exercise 2 to conclude that $s(x)$ divides $f(x)u(x) + g(x)v(x)$. Thus, $s(x)$ divides $d(x)$. Because $d(x)$ is a monic polynomial, it then follows that $d(x)$ is the greatest common divisor of $f(x)$ and $g(x)$. ■

Because we now know that if F is a field, then the greatest common divisor of two polynomials $f(x)$ and $g(x)$ in $F[x]$ actually exists and is unique (whenever $f(x)$ and $g(x)$ are not both zero), we can use the notation $\gcd(f(x), g(x))$ to represent the greatest common divisor of $f(x)$ and $g(x)$.

Relatively Prime Polynomials

In Investigation 3, we proved that the greatest common divisor of two integers (not both zero), is the smallest positive linear combination of these two integers. Theorem 12.10 is a similar result for $F[x]$, where F is a field. The following definitions are similar to the corresponding definitions in Investigation 3.

Definition 12.11. Let F be a field and let $f(x)$ and $g(x)$ be polynomials in $F[x]$. A **linear combination** of $f(x)$ and $g(x)$ is a polynomial in $F[x]$ that can be written as $f(x)u(x) + g(x)v(x)$ for some polynomials $u(x)$ and $v(x)$ in $F[x]$.

So using Theorem 12.10, we can say that the greatest common divisor of $f(x)$ and $g(x)$ is the monic polynomial of least degree that is a linear combination of $f(x)$ and $g(x)$. As in the integers, a particularly important case is when the greatest common divisor of two polynomials is 1.

Definition 12.12. Let F be a field. Two polynomials $f(x)$ and $g(x)$ in $F[x]$ are **relatively prime** provided that $\gcd(f(x), g(x)) = 1$.

Activity 12.13. The following results can be proved in much the same manner as their corresponding results in the integers. For this activity, let F be a field, and let $p(x)$ and $q(x)$ be polynomials in $F[x]$ that are not both zero.

(a) Prove that $p(x)$ and $q(x)$ are relatively prime if and only if

$$p(x)u(x) + q(x)v(x) = 1$$

for some polynomials $u(x), v(x) \in F[x]$. (Hint: Use Theorem 12.10.)

(b) Prove that if $a, b \in F$ with $a \neq b$, then $\gcd(x + a, x + b) = 1$. (Hint: Notice that $(x + a) - (x + b) = a - b$. Use this observation to write 1 as a linear combination of $(x + a)$ and $(x + b)$.)

(c) Let $h(x) \in F[x]$. Prove that if $p(x)$ divides $h(x)$, $q(x)$ divides $h(x)$, and $\gcd(p(x), q(x)) = 1$, then $p(x)q(x)$ divides $h(x)$. (Hint: Use Theorem 12.10 to write $p(x)u(x) + q(x)v(x) = 1$ for some $u(x), v(x) \in F[x]$. Then multiply both sides of this equation by $h(x)$ and make appropriate substitutions.)

The Euclidean Algorithm for Polynomials

Given two integers a and b (not both zero), we learned how to use the Euclidean Algorithm to find the greatest common divisor of a and b. (See Investigation 3.) The basic idea was to repeatedly use the Division Algorithm until a remainder of zero was obtained. If F is a field, then we have a Division Algorithm in $F[x]$, and so it is not surprising that we also have a Euclidean Algorithm to find the greatest common divisor of two polynomials in $F[x]$. The Euclidean Algorithm in $F[x]$ can be described as follows:

Let $f(x)$ and $g(x)$ be polynomials over the field F, with $g(x) \neq 0$. Successively apply the Division Algorithm until the first zero remainder is obtained:

$$\begin{array}{rlcl}
f(x) &= g(x)q_1(x) + r_1(x) & \text{and} & \deg(r_1(x)) < \deg(g(x)); \\
g(x) &= r_1(x)q_2(x) + r_2(x) & \text{and} & \deg(r_2(x)) < \deg(r_1(x)); \\
r_1(x) &= r_2(x)q_3(x) + r_3(x) & \text{and} & \deg(r_3(x)) < \deg(r_2(x)); \\
\vdots & & \vdots & \vdots \\
r_{n-2}(x) &= r_{n-1}(x)q_n(x) + r_n(x) & \text{and} & \deg(r_n(x)) < \deg(r_{n-1}(x)); \\
r_{n-1}(x) &= r_n(x)q_{n+1}(x) & &
\end{array}$$

If c is the leading coefficient of the last nonzero remainder $r_n(x)$, then $c^{-1}r_n(x)$ is the greatest common divisor of $f(x)$ and $g(x)$—that is, $\gcd(f(x), g(x)) = c^{-1}r_n(x)$. For example, using the polynomials from Activity 12.5, in $\mathbb{R}[x]$, let

$$f(x) = 4x^4 + 10x^3 + 7x^2 - 4x + 7 \quad \text{and} \quad g(x) = 2x^2 + 2x - 1.$$

We have already used long division to obtain

$$f(x) = g(x)\left(2x^2 + 3x + \frac{3}{2}\right) + \left(-4x + \frac{17}{2}\right). \tag{12.7}$$

So we have $q_1(x) = 2x^2 + 3x + \frac{3}{2}$ and $r_1(x) = -4x + \frac{17}{2}$. We can now divide $g(x)$ by the remainder $r_1(x) = -4x + \frac{17}{2}$ and obtain

$$g(x) = \left(-4x + \frac{17}{2}\right)\left(-\frac{1}{2}x - \frac{25}{16}\right) + \frac{393}{32}. \tag{12.8}$$

So in the Euclidean Algorithm, $q_2(x) = -\frac{1}{2}x - \frac{25}{16}$ and $r_2(x) = \frac{393}{32}$. If we now divide $r_1(x) = -4x + \frac{17}{2}$ by $r_2(x) = \frac{393}{32}$, we will obtain a remainder of 0. This means that the greatest common divisor of $f(x)$ and $g(x)$ is

$$d(x) = \frac{32}{393}r_2(x) = 1.$$

As with the integers, we can also use the Euclidean Algorithm in reverse to write the greatest common divisor as a linear combination of $f(x)$ and $g(x)$. It will make the computations easier if we first write $r_2(x) = \frac{393}{32}$ as a linear combination of $f(x)$ and $g(x)$. However, the algebra and computations can still be quite complicated.

We first use equation (12.8) to write

$$r_2(x) = g(x) - r_1(x)q_2(x).$$

We can now use equation (12.7) and substitute for $r_1(x) = -4x + \frac{17}{2}$. This gives

$$\begin{aligned} r_2(x) &= g(x) - [f(x) - g(x)q_1(x)]\, q_2(x) \\ &= -f(x)q_2(x) + g(x)\,[1 + q_1(x)q_2(x)] \end{aligned}$$

Since $r_2(x) = \frac{393}{32}$, we see that

$$\begin{aligned} 1 &= \gcd(f(x), g(x)) \\ &= \frac{32}{393}\left[-f(x)q_2(x)\right] + g(x)\left[1 + q_1(x)q_2(x)\right] \\ &= f(x)\left[-\frac{32}{393}q_2(x)\right] + g(x)\left[\frac{32}{393}\left(1 + q_1(x)q_2(x)\right)\right] \\ &= f(x)\left[\frac{16}{393}x + \frac{50}{393}\right] + g(x)\left[-x^3 - \frac{37}{8}x^2 - \frac{87}{16}x - \frac{43}{32}\right]. \end{aligned}$$

Activity 12.14.

(a) In $\mathbb{R}[x]$, let $f(x) = 3x^5 + 7x^4 + 11x^3 + 15x^2 + 10x + 2$ and $g(x) = 3x^2 + 7x + 2$. Use the Euclidean Algorithm to determine $\gcd(f(x), g(x))$ and write $\gcd(f(x), g(x))$ as a linear combination of $f(x)$ and $g(x)$.

(b) In $\mathbb{Z}_3[x]$, let $f(x) = x^4 + [2]x^3 + x + [2]$ and $g(x) = [2]x^2 + [1]$. Determine $\gcd(f(x), g(x))$.

Concluding Activities

Activity 12.15. Although we stated the Division Algorithm for polynomials over a field, there is also a modified Division Algorithm for $D[x]$, where D is an integral domain. To explore this modification, first note that, in the proof of the Division Algorithm, we used the fact that F was a field to conclude that the leading coefficient of the divisor $g(x)$ has a multiplicative inverse. In particular, we let

$$f(x) = a_n x^n + a_{n-1}x^{n-1} + \cdots + a_1 x + a_0 \text{ and}$$
$$g(x) = b_m x^m + b_{m-1}x^{m-1} + \cdots + b_1 x + b_0,$$

where $a_n \neq 0$ and $b_m \neq 0$. In the proof, we made use of b_m^{-1} in the division process in equation (12.1). This is really the only place where we used the assumption that we were working with polynomials over a field. So if we assume that the polynomials are in $D[x]$, where D is an integral domain, we can use the same proof as long as we also assume that the leading coefficient of $g(x)$, namely b_m, is a unit in D. This means that b_m^{-1} exists in D and we can proceed with the division process. The following theorem formalizes these observations, and the problems after it provide an example of polynomial division in $\mathbb{Z}[x]$.

Theorem 12.16. *Let D be an integral domain and let $f(x)$ and $g(x)$ be polynomials in $D[x]$ with $g(x) \neq 0$. If the leading coefficient of $g(x)$ is a unit in D, then there exist unique polynomials $q(x)$ and $r(x)$ in $D[x]$ such that*

$$f(x) = g(x)q(x) + r(x) \quad \textit{and}$$

$r(x) = 0$ or $\deg(r(x)) < \deg(g(x))$.

For the problems below, assume that all polynomials are in $\mathbb{Z}[x]$.

(a) Let $f(x) = 2x^3 + 2x^2 - 5x + 3$ and $g(x) = x + 4$. Find the quotient and remainder when $f(x)$ is divided by $g(x)$.

(b) Let $f(x) = 2x^3 + 2x^2 - 5x + 3$ and $g(x) = 2x + 3$. Show that the division process does not give a unique quotient $q(x)$ and remainder $r(x)$ with $r(x) = 0$ or $0 \leq \deg(r(x)) < \deg(g(x))$.

Activity 12.17. Write a short summary that describes the important concepts, and the relationships between these concepts, that were introduced in this investigation. Explain how the material in this investigation is connected to the material in Investigations 2 and 3.

Exercises

(1) Let R be a commutative ring, and let $f(x), g(x), h(x) \in R[x]$. Prove each of the following:

   (a) If $f(x)$ divides $g(x)$ and $g(x)$ divides $h(x)$, then $f(x)$ divides $h(x)$.

   (b) If $f(x)$ divides $g(x)$ and $f(x)$ divides $h(x)$, then $f(x)$ divides $g(x) + h(x)$.

   (c) If $f(x)$ divides $g(x)$, then $f(x)$ divides $g(x)h(x)$

⋆ (2) Let R be a commutative ring, and let $p(x)$, $f(x)$, and $g(x)$ be polynomials in $R[x]$. Prove that if $p(x)$ divides both $f(x)$ and $g(x)$ in $R[x]$, then for any polynomials $u(x)$ and $v(x)$ in $R[x]$, $p(x)$ divides $f(x)u(x) + g(x)v(x)$.

(3) In each of the following, a field F is specified and two polynomials $f(x)$ and $g(x)$ in $F[x]$ are given. Find $q(x), r(x) \in F[x]$ so that $f(x) = g(x)q(x)+r(x)$ with $r(x) = 0$ or $\deg(r(x)) < \deg(g(x))$.

(a) In $\mathbb{Q}[x]$, $f(x) = x^3 + 2x^2 + 2x + 1$ and $g(x) = x + 2$.

(b) In $\mathbb{Z}_3[x]$, $f(x) = x^3 + [2]x^2 + [2]x + [1]$ and $g(x) = x + [2]$.

(c) In $\mathbb{R}[x]$, $f(x) = 2x^4 + 4x^3 + 2x^2 + 2x + 1$ and $g(x) = x^2 + 4$.

(d) In $\mathbb{Z}_5[x]$, $f(x) = [2]x^4 + [4]x^3 + [2]x^2 + [2]x + [1]$ and $g(x) = x^2 + [4]$.

(4) Let F be a field, and let $p(x), f(x), g(x) \in F[x]$. Prove that if $p(x)$ and $f(x)$ are relatively prime and $p(x)$ and $g(x)$ are relatively prime, then $p(x)$ and $f(x)g(x)$ are relatively prime.

(5) Let F be a field, and let $p(x), f(x), g(x) \in F[x]$. Prove that if $p(x)$ divides $f(x)g(x)$ and $\gcd(p(x), f(x)) = 1$, then $p(x)$ divides $g(x)$.

(6) Let F be a field, and let $p(x), q(x), f(x) \in F[x]$. Assume that $\gcd(p(x), q(x)) = 1$ and both $p(x)$ and $q(x)$ divide $f(x)$. Prove or disprove: $p(x)q(x)$ divides $f(x)$.

(7) Let F and K be fields such that F is a subfield of K.

(a) Assume that $p(x), q(x) \in F[x]$ and that $p(x)$ and $q(x)$ are relatively prime in $F[x]$. Are $p(x)$ and $q(x)$ relatively prime in $K[x]$? Justify your conclusion.

(b) Assume that $p(x), q(x) \in F[x]$ and that $p(x)$ and $q(x)$ are relatively prime in $K[x]$. Are $p(x)$ and $q(x)$ relatively prime in $F[x]$? Justify your conclusion.

(8) Use the Euclidean Algorithm to find $\gcd(f(x), g(x))$ in the indicated polynomial ring. Then write $\gcd(f(x), g(x))$ as a linear combination of $f(x)$ and $g(x)$.

(a) $f(x) = x^4 + 2x^3 + x^2 + 1$ and $g(x) = x^2 + 2x + 4$ in $\mathbb{Q}[x]$

(b) $f(x) = x^6 + x^5 + x + [1]$ and $g(x) = x^3 + [2]x^2 + [2]x + [1]$ in $\mathbb{Z}_3[x]$.

(c) $f(x) = x^4 + [4]x^3 + [2]x^2 + [4]x + [4]$ and $g(x) = x^3 + [4]x^2 + x + [4]$ in $\mathbb{Z}_5[x]$.

Connections

In Investigation 11, we saw that a polynomial ring with coefficients from an integral domain has much in common with the ring of integers. This investigation explored that connection in more detail for polynomials whose coefficients come from a field. In particular, the ideas of divisibility of polynomials and a division algorithm for such polynomials are essentially the same as divisibility and the division algorithm for integers from Investigation 2. In addition, we can define what it means for two polynomials to be relatively prime just as we did for integers in Investigation 3. We will take this connection between polynomial rings over fields and the integers a step farther in Investigation 13 when we discuss unique factorization in polynomial rings.

Investigation 13

Roots, Factors, and Irreducible Polynomials

Focus Questions

By the end of this investigation, you should be able to give precise and thorough answers to the questions listed below. You may want to keep these questions in mind to focus your thoughts as you complete the investigation.

- What does the Remainder Theorem say when a polynomial over a field F is divided by a polynomial of the form $(x - c)$, for some $c \in F$?
- What does it mean to say that an element of a field F is a root of a polynomial in $F[x]$?
- What does the Factor Theorem say about the roots of a polynomial over a field F and certain factors of that polynomial?
- How many roots can a polynomial of degree n over a field have?
- If F is a field, what does it mean to say that a polynomial in $F[x]$ is irreducible? What does it mean to say a polynomial in $F[x]$ is reducible?
- How can the Factor Theorem be used to determine if a polynomial of degree 2 or 3 over a field is irreducible or reducible?
- If F is a field, what does it mean to say that a polynomial in $F[x]$ can be factored into irreducible polynomials in a unique way?

Preview Activity 13.1. In Definition 11.12, we defined the concept of a polynomial function. If R is a commutative ring and $p(x) = a_nx^n + a_{n-1}x^{n-1} + \cdots a_1x + a_0$ is a polynomial in $R[x]$, then the **polynomial function** induced by $p(x)$ is the function $\overline{p} : R \to R$, where for each r in R,

$$\overline{p}(r) = a_nr^n + a_{n-1}r^{n-1} + \cdots a_1r + a_0.$$

To simplify notation, we will write $p(r)$ for $\overline{p}(r)$ and write $p : R \to R$ for the polynomial function induced by the polynomial $p(x)$.

In each of the following, the term "remainder" means the remainder according to the Division Algorithm, which can be determined by using long division.

(a) Let $f(x) = x^2 - 3x - 2$ be in $\mathbb{R}[x]$. Notice that $f(5) = 8$. Determine the remainder when $f(x)$ is divided by $(x - 5)$.

(b) Let $g(x) = x^2 - 3x - 10$ be in $\mathbb{R}[x]$. Determine $g(5)$ and determine the remainder when $g(x)$ is divided by $(x - 5)$.

(c) Let $h(x) = x^2 - 3x - 10$ be in $\mathbb{Q}[x]$. Determine $h(-2)$ and determine the remainder when $h(x)$ is divided by $(x - (-2)) = (x + 2)$.

(d) Let $p(x) = x^2 + [2]x + [2]$ be in $\mathbb{Z}_3[x]$. Determine $p([1])$ and determine the remainder when $p(x)$ is divided by $(x - [1])$.

Polynomial Functions and Remainders

In our comparison of the development of $\mathbb{Z}$ and $F[x]$ (where F is a field), the next natural step is to develop what is meant by unique factorization in $F[x]$. In this investigation, we will consider whether we can obtain something like the Fundamental Theorem of Arithmetic, but for polynomials rather than integers. The language for polynomials will be somewhat different. As we shall see, instead of using the term *prime*, we will use the term *irreducible* to describe polynomials that, loosely speaking, cannot be factored. We will give a precise meaning to this term later in this investigation.

Questions about whether or not a polynomial can be factored can sometimes be answered by using the corresponding induced polynomial function. Some care must be used when taking this approach. For example, in $\mathbb{R}[x]$, if we let $p(x) = x^2 + 4x - 7$, then:

- Since x is an indeterminate in $\mathbb{R}[x]$, the statement $p(x) = x^2 + 4x - 7 = 0$ is false because this is stating that $p(x)$ is equal to the zero polynomial, which it is not.
- However, if we work with the induced polynomial function $p : \mathbb{R} \to \mathbb{R}$, then it is reasonable to ask for which values $x \in \mathbb{R}$ is $p(x) = 0$. In this case, we are treating x as a variable over $\mathbb{R}$.

If R is a commutative ring, then it may be helpful to realize that statements about the indeterminate x occur in the polynomial ring $R[x]$, whereas questions about the variable x occur in the ring R.

The work in Preview Activity 13.1 was meant to illustrate an important relation between the value $p(r)$ of a polynomial function and the remainder when the polynomial is divided by $(x - r)$. This relationship is explored further in the next activity.

Activity 13.2. Let F be a field, let $p(x) \in F[x]$, and let $c \in F$.

(a) Suppose $f(x) \in F[x]$ and $f(x) = 0$ or $\deg(f(x)) = 0$. Explain why $f(x)$ must be of the form $f(x) = s$, for some $s \in F$.

(b) Write down the result of the Division Algorithm when $p(x)$ is divided by $(x - c)$. Pay particular attention to the remainder, and explain why the remainder must be a constant polynomial.

(c) Evaluate both sides of the equation from part (b) using $x = c$, and use your work to complete the following statement:

The remainder when $p(x)$ is divided by $(x - c)$ is equal to ______________.

The work in Activity 13.2 outlines of the proof of the Remainder Theorem, which can be stated formally as follows:

Theorem 13.3 (The Remainder Theorem). *Let F be a field, let $p(x) \in F[x]$, and let $c \in F$. The remainder when $p(x)$ is divided by $(x - c)$ is equal to $p(c)$. That is, there exists a unique polynomial $q(x) \in F[x]$ such that*

$$p(x) = (x - c)q(x) + p(c).$$

Proof. Let F be a commutative ring, let $p(x) \in F[x]$, and let $c \in F$. By the Division Algorithm, there exist unique $q(x)$ and $r(x)$ in $F[x]$ such that

$$p(x) = (x - c)q(x) + r(x) \quad \text{with} \quad r(x) = 0 \text{ or } \deg(r(x)) < \deg(x - c).$$

Since the degree of $(x - c)$ is 1, it follows that $r(x) = 0$ or $\deg(r(x)) = 0$. In either case, $r(x) = a$ for some $a \in F$. This then means that

$$p(x) = (x - c)q(x) + a.$$

We can now treat both sides of the last equation as polynomial functions and evaluate each at $x = c$. Doing so, we obtain:

$$p(c) = (c - c)q(c) + a = 0_F + a = a.$$

This proves that the remainder when $p(x)$ is divided by $(x - c)$ is equal to $p(c)$. ■

Roots of Polynomials and the Factor Theorem

The study of the roots of a polynomial has been an important part of the history of algebra. As it turns out, the Remainder Theorem can be used to establish an important relationship between the roots and the factors of a polynomial. To explore this relationship more, we must first define what a root is.

Definition 13.4. Let R be a commutative ring and let $p(x) \in R[x]$. An element c of R is called a **root of the polynomial $p(x)$** provided that $p(c) = 0_R$.

The Factor Theorem, stated formally below, describes precisely the relationship between the roots of a polynomial and its monic, degree 1 factors.

Theorem 13.5 (The Factor Theorem). *Let F be a field, let $p(x) \in F[x]$, and let $c \in F$. Then c is a root of the polynomial $p(x)$ if and only if $(x - c)$ is a factor of $p(x)$.*

The following activity will help guide us through a proof of the Factor Theorem.

Activity 13.6. Let F be a field, let $p(x) \in F[x]$, and let $c \in F$. Using the Remainder Theorem, there exists a unique polynomial $q(x)$ in $F[x]$ such that

$$p(x) = (x - c)q(x) + p(c). \tag{13.1}$$

(a) Assume that c is root of $p(x)$. Use equation (13.1) to prove that $(x - c)$ is a factor of $p(x)$.

(b) Now assume that $(x-c)$ is a factor of $p(x)$. Use the definition of divides to write $p(x)$ in the form of a product, and then use this product to show that $p(c) = 0$.

(c) Use your work from parts (a) and (b) to write a formal proof of the Factor Theorem.

The next activity shows how the Factor Theorem can be used to help factor polynomials.

Activity 13.7.

(a) Let $p(x) = x^4 - x^3 - 5x^2 + 3x + 6$ be a polynomial in $\mathbb{R}[x]$. Verify that $p(-1) = 0$ and $p(2) = 0$. What does the Factor Theorem allow you to conclude from these observations?

(b) Explain how Activity 12.13 (see page 162) allows us to conclude that $\gcd(x+1, x-2) = 1$ and that

$$(x+1)(x-2) = x^2 - x - 2$$

divides $p(x)$. Use this information and long division to factor $p(x)$.

(c) Let $p(x) = x^4 + [3]x^3 + [2]x^2 + [4]x + [4]$ be a polynomial in $\mathbb{Z}_5[x]$. Evaluate $p([x])$ for each $[x] \in \mathbb{Z}_5$. Then identify the roots of $p(x)$, and use the Factor Theorem and long division to factor $p(x)$ into a product of monic, degree 1 factors. (Note that some of these factors may be repeated.)

As illustrated in Activity 13.7, one way to find the roots of a polynomial $p(x)$ in $R[x]$, where R is a commutative ring, is to simply evaluate $p(c)$ for each $c \in R$ and find the values of c for which $p(c) = 0_R$. For example, let $p(x) = [3]x^2 + [3]$ be a polynomial in $\mathbb{Z}_6[x]$. We can easily observe the following:

$$p([0]) = [3] \qquad p([1]) = [0] \qquad p([2]) = [3]$$
$$p([3]) = [0] \qquad p([4]) = [3] \qquad p([5]) = [0]$$

So in this case, we have a polynomial $p(x)$ of degree 2 that has 3 roots ($[1], [3]$, and $[5]$). This may seem contrary to a well-known result from your previous algebra courses—namely, that if a polynomial has degree n, then that polynomial has at most n roots. To resolve this apparent discrepancy, it may help to remember that your prior studies of polynomials most likely dealt only with polynomials over the *field* of real numbers, $\mathbb{R}$. In this example, however, notice that we are working with $\mathbb{Z}_6$, which is not a field. In fact, $\mathbb{Z}_6$ is not even an integral domain. This distinction turns out to be very important, and we will explore it more in the next activity. Eventually, we will prove Theorem 13.9, which states that a polynomial of degree n over a field has at most n roots.

Activity 13.8. Let F be a field.

(a) Let $p(x)$ be a polynomial in $F[x]$ with $\deg(p(x)) = 0$. (So $p(x) = k$ for some $k \in F$ with $k \neq 0$.) Explain why $p(x)$ has no roots in F.

(b) Let $p(x)$ be a polynomial in $F[x]$ with $\deg(p(x)) = 1$. (So $p(x) = ax + b$ with $a, b \in F$ and $a \neq 0$.) Verify that $-a^{-1}b$ is a root of $p(x)$. Then let $r \in F$ be any root of $p(x)$. Prove that $r = -a^{-1}b$, and explain why this fact establishes that $-a^{-1}b$ is the only root of $p(x)$ in F.

(c) Let $f(x) = x^2 - 3x - 10 \in \mathbb{R}[x]$, and let $g(x) = x^2 + x + 1 \in \mathbb{R}[x]$. Notice that both of these polynomials have degree 2. How many roots do $f(x)$ and $g(x)$ have in $\mathbb{R}$? (For this part, you may use the quadratic formula.)

(d) Let $f(x) = [3]x^2 + [3] \in \mathbb{Z}_5[x]$. Notice that $\deg(f(x)) = 2$. How many roots does $f(x)$ have in $\mathbb{Z}_5$?

Theorem 13.9, which we will state shortly, formalizes the ideas from Activity 13.8. To prove it, we will use induction on the degree of the polynomial. Notice that in part (a) of Activity 13.8, we proved that a polynomial of degree 0 in $F[x]$ has no roots. In part (b), we proved that a polynomial of degree 1 has 1 root. These results establish the basis step for the induction proof. The basic idea of the induction step is that if we have a root c of a polynomial of degree $k+1$, then we can use the Factor Theorem (Theorem 13.5) to write the polynomial as a product of $(x-c)$ and a polynomial whose degree is k. This step also makes use of Theorem 11.10. (See page 143.)

Theorem 13.9. *If F is a field and $n \in \mathbb{N}$, then a polynomial of degree n in $F[x]$ has at most n roots in F. That is, a nonzero polynomial of degree n over a field has at most n roots in the field.*

Proof. Let F be a field. We will proceed by induction on the degree of $p(x)$. In Activity 13.8, we proved that a polynomial of degree 0 in $F[x]$ has no roots and a polynomial of degree 1 in $F[x]$ has exactly one root. So the basis for the induction has been established.

Now let $k \in \mathbb{Z}$ with $k \geq 0$, and suppose that any polynomial of degree k in $F[x]$ has at most k roots. Now let $p(x) \in F[x]$ with $\deg(p(x)) = k+1$. Then either $p(x)$ has no roots in F, or $p(x)$ has at least one root in F. We will consider each of these cases.

In the case where $p(x)$ has no roots, it is clear that $p(x)$ has at most $k+1$ roots. So assume that $p(x)$ has at least one root in F, and let c be such a root. By the Factor Theorem, there exists a polynomial $q(x)$ in $F[x]$ such that

$$p(x) = (x-c)q(x).$$

By Theorem 11.10, $\deg(p(x)) = \deg(x-c) + \deg(q(x))$. Since $\deg(x-c) = 1$ and $\deg(p(x)) = k+1$, it follows that $\deg(q(x)) = k$. Hence, by the induction hypothesis, we can conclude that $q(x)$ has at most k roots in F. Now notice that if r is a root of $p(x)$, then

$$0 = p(r) = (r-c)q(r).$$

Since F is a field (and an integral domain), we know that F has no zero divisors. Therefore, we know that $r - c = 0$ or $q(r) = 0$. That is, the only roots of $p(x)$ are c and the roots of $q(x)$. Since $q(x)$ has at most k roots, we conclude that $p(x)$ has at most $k+1$ roots. This completes the proof of the inductive step, and by mathematical induction, we have proved that a polynomial in $F[x]$ of degree n has at most n roots. ■

Irreducible Polynomials

In our study of the integers, the prime numbers were very important. In fact, we proved that every natural number (except 1) is either prime or a product of prime numbers. We will obtain a similar result for $F[x]$, where F is a field. However, when working with polynomials, we use the term *irreducible* rather than *prime*. In the following definition, the term *non-constant polynomial* refers to a polynomial with positive degree.

Definition 13.10. Let F be a field. A non-constant polynomial $p(x)$ in $F[x]$ is **irreducible in $F[x]$** provided that $p(x)$ cannot be factored as a product of two polynomials in $F[x]$, both of which have a positive degree. Otherwise, the polynomial $p(x)$ is called **reducible in $F[x]$**.

Basically, Definition 13.10 means that a non-constant polynomial is irreducible if and only if the only way it can be factored is by having at least one of the factors have degree 0. That is, $p(x)$ is irreducible if and only if any factorization of $p(x)$ must be of the form $cp(x)$, for some $c \in F$.

This also means that if F is a field, then a non-constant polynomial $p(x)$ in $F[x]$ is reducible if and only if there exist polynomials $f(x)$ and $g(x)$ such that

- $p(x) = f(x)g(x)$, and
- $\deg(f(x)) \geq 1$ and $\deg(g(x)) \geq 1$.

For example, in $\mathbb{R}[x]$:

- Even though we can write $\frac{3}{2}x + 4 = \frac{3}{2}(x + 6)$, the polynomial $\left(\frac{3}{2}x + 4\right)$ is irreducible in $\mathbb{R}[x]$ since it cannot be written as a product of two polynomials with positive degree.
- Since $x^4 + 5x^2 + 6 = (x^2 + 2)(x^2 + 3)$, the polynomial $(x^2 + 5x + 6)$ is reducible in $\mathbb{R}[x]$.

The next theorem states a result that is important but not terribly surprising in light of the definition of irreducibility. It can be proved easily using Theorem 11.10, and we leave this proof as an exercise.

Theorem 13.11. *Let F be a field. Any polynomial in $F[x]$ that has degree 1 is irreducible in $F[x]$.*

As you might suspect after studying the Factor Theorem, there is a relationship between the reducibility (or irreducibility) of a polynomial over a field and the roots of the polynomial. However, there are some subtleties to this relationship and some care must be taken when using it. To start, let F be a field, and let $p(x) \in F[x]$ with $\deg(p(x)) \geq 2$. If $p(x)$ has a root c in F, then by the Factor Theorem, $(x - c)$ is a factor of $p(x)$. So we know that there exists a polynomial $q(x) \in F[x]$ such that

$$p(x) = (x - c)q(x). \tag{13.2}$$

By Theorem 11.10, we can conclude that

$$\deg(p(x)) = \deg(x - c) + \deg(q(x)).$$

Since $\deg(p(x)) \geq 2$ and $\deg(x - c) = 1$, it follows that $\deg(q(x)) \geq 1$. This fact and equation (13.2) allow us to conclude that $p(x)$ is reducible. Thus, we have proved the following theorem.

Theorem 13.12. *Let F be a field and let $p(x) \in F[x]$ with $\deg(p(x)) \geq 2$. If $p(x)$ has a root in F, then $p(x)$ is reducible in $F[x]$.*

If we write the contrapositive of the conditional statement in Theorem 13.12, we obtain the following corollary:

Corollary 13.13. *Let F be a field and let $p(x) \in F[x]$ with $\deg(p(x)) \geq 2$. If $p(x)$ is irreducible in $F[x]$, then $p(x)$ has no roots in F.*

Activity 13.14. In this activity, we will explore some of the subtleties of the relationship between roots and reducibility.

(a) Let $p(x) = x^4 + 3x^2 + 2$ be a polynomial in $\mathbb{R}[x]$. Show that $p(x)$ can be factored in $\mathbb{R}[x]$ and hence, $p(x)$ is reducible in $\mathbb{R}[x]$. Also explain why $p(x)$ has no roots in $\mathbb{R}$.

(b) Use part (a) to explain why the following converse of Corollary 13.13 is false.

> *Let F be a field and let $p(x) \in F[x]$ with $\deg(p(x)) \geq 2$. If $p(x)$ has no roots in F, then $p(x)$ is irreducible in $F[x]$.*

Although we have shown that the converse of Corollary 13.13 is false, if we restrict ourselves to polynomials of degree 2 or 3, we can show that if the polynomial has no roots, then the polynomial is irreducible. Notice that the counterexample in part (b) of Activity 13.8 used a polynomial of degree 4.

Theorem 13.15. *Let F be a field and let $p(x) \in F[x]$. If $\deg(p(x)) = 2$ or $\deg(p(x)) = 3$ and if $p(x)$ has no roots in F, then $p(x)$ is irreducible in $F[x]$.*

Proof. We will use a proof by contradiction. So we assume that $\deg(p(x)) = 2$ or $\deg(p(x)) = 3$, $p(x)$ has no roots in F, and $p(x)$ is reducible in $F[x]$. This means that there exist polynomials $f(x), g(x) \in F[x]$ such that

$$\begin{aligned} p(x) &= f(x)g(x), \\ \deg(f(x)) &\geq 1, \text{ and} \\ \deg(g(x)) &\geq 1. \end{aligned}$$

In addition, by Theorem 11.10, $\deg(p(x)) = \deg(f(x)) + \deg(g(x))$. Since $\deg(p(x))$ is 2 or 3, we must have $\deg(f(x)) = 1$ or $\deg(g(x)) = 1$. Without loss of generality, we may assume that $\deg(f(x)) = 1$. This means that there exist $a, b \in F$ with $a \neq 0$ and $f(x) = ax + b$. We then see that

$$\begin{aligned} f\left(-a^{-1}b\right) &= a\left(-a^{-1}b\right) + b \\ &= -\left(a^{-1}a\right)b + b \\ &= -b + b \\ &= 0 \end{aligned}$$

Hence, $p\left(-a^{-1}b\right) = 0$ and $p(x)$ has a root in F. This contradicts the assumption that $p(x)$ has no roots in F. Therefore, $p(x)$ is irreducible in $F[x]$. ■

Unique Factorization in $F[x]$

In Investigation 4, we proved the Fundamental Theorem of Arithmetic, which states that every integer greater than 1 is either prime or a product of primes, and this factorization is unique up to the order of the factors. An important result that was needed to prove the Fundamental Theorem was Euclid's Lemma. (See page 36.) Here we will prove an analog of Euclid's Lemma for polynomials, which will then be used to prove a unique factorization theorem for polynomials over a field F. In this unique factorization theorem, irreducible polynomials will play a role similar to that of prime numbers in the integers.

Theorem 13.16. *Let F be a field and let $p(x), f(x), g(x) \in F[x]$. If $p(x)$ divides $f(x)g(x)$ and $p(x)$ is irreducible, then $p(x)$ divides $f(x)$ or $p(x)$ divides $g(x)$.*

The next activity will guide us through one way to prove Theorem 13.16. The method of proof is similar to the one outlined in Activity 4.6 (see page 37) for the integers.

Activity 13.17. Let F be a field, and let $p(x), f(x), g(x) \in F[x]$. Assume that $p(x)$ divides $f(x)g(x)$ and that $p(x)$ is irreducible. If $p(x)$ divides $f(x)$, then we are done. So assume that $p(x)$ does not divide $f(x)$.

(a) Explain why it must be the case that $\gcd(p(x), f(x)) = 1$.

(b) Use Theorem 12.10 (see page 160) to translate the observation from part (a) into an equation involving a linear combination of $p(x)$ and $f(x)$.

(c) Multiply both sides of your equation from part (b) by an appropriate polynomial in order to obtain an equation of the form

$$g(x) = \_\_\_\_\_\_ + \_\_\_\_\_\_.$$

(d) Explain why each of the terms on the right-hand side of the equation from part (c) are divisible by $p(x)$. Then conclude that $p(x)$ divides $g(x)$. (This proves that $p(x)$ divides $f(x)$ or $p(x)$ divides $g(x)$.)

As with Euclid's Lemma for the integers, Theorem 13.16 can be generalized as follows:

Corollary 13.18. *Let F be a field and let $p(x), f_1(x), f_2(x), \ldots, f_n(x) \in F[x]$. If $p(x)$ divides $f_1(x)f_2(x)\cdots f_n(x)$ and $p(x)$ is irreducible, then $p(x)$ divides $f_i(x)$, for some $i \in \mathbb{N}$ with $1 \leq i \leq n$.*

One way to prove this generalization is to use mathematical induction. This proof is left as an exercise. (See Exercise 12.)

Before we prove the unique factorization theorem for polynomials over a field, there is one more idea that we need to introduce.

Definition 13.19. Let F be a field. A polynomial $g(x)$ in $F[x]$ is said to be an **associate** of the polynomial $f(x)$ in $F[x]$ provided that there exists a nonzero $c \in F$ such that $g(x) = cf(x)$.

For example:

- In $\mathbb{R}[x]$, $g(x) = -2x + 6$ is an associate of $f(x) = x - 3$ since $g(x) = -2f(x)$. Notice that both $f(x)$ and $g(x)$ are irreducible in $\mathbb{R}[x]$ since both have degree 1.
- In $\mathbb{Q}[x]$, $g(x) = 3x^2 + x + 4$ is an associate of $f(x) = x^2 + \frac{1}{3}x + \frac{4}{3}$ since $g(x) = 3f(x)$.
- In $\mathbb{Z}_7[x]$, $q(x) = [4]x^2 + [5]x$ is an associate of $p(x) = x^2 + [3]x$ since $g(x) = [4]p(x)$

In the next activity, we will explore some results about associates that will be helpful in proving the unique factorization theorem.

Activity 13.20. Let F be a field, and let $p(x)$ and $q(x)$ be polynomials in $F[x]$ with $q(x)$ irreducible. We know that if $\deg(q(x)) \geq 1$, then $q(x)$ cannot be factored as a product of two polynomials of positive degree. Use this fact to explain why if $p(x)$ divides $q(x)$, then either $p(x) = c$ for some $c \in F$, or $p(x)$ is an associate of $q(x)$.

We are now ready to prove the unique factorization theorem for polynomials.

Theorem 13.21 (Unique Factorization in $\boldsymbol{F[x]}$). *Let F be a field. If $f(x) \in F[x]$ is a polynomial with positive degree, then $f(x)$ is either irreducible or a product of irreducible polynomials in $F[x]$. This factorization is unique in the following sense: If*

$$f(x) = p_1(x)p_2(x)\cdots p_m(x) \qquad \text{and} \qquad f(x) = q_1(x)q_2(x)\cdots q_k(x) \tag{13.3}$$

with $p_1(x), p_2(x), \ldots p_m(x)$ and $q_1(x), q_2(x), \ldots q_k(x)$ irreducible in $F[x]$, then $m = k$ and after the $q_j(x)$ are reordered and relabeled, if necessary, $p_j(x)$ is an associate of $q_j(x)$, for each j with $1 \leq j \leq m$.

Proof. We will first prove the existence of a factorization into irreducibles using induction on the degree of $f(x)$. If $\deg(f(x)) = 1$, then we know that $f(x)$ is irreducible (by Theorem 13.11), and this establishes the basis step.

For the induction step, we assume that $n \in \mathbb{N}$, that $\deg(f(x)) = n$, and that all polynomials in $F[x]$ of degree 1, 2, ..., $n-1$ are either irreducible or the product of irreducible polynomials. We will consider two cases: either $f(x)$ is irreducible or $f(x)$ is reducible.

There is nothing to prove in the case that $f(x)$ is irreducible. In the case that $f(x)$ is reducible, we have

$$f(x) = g(x)h(x) \tag{13.4}$$

for some polynomials $g(x)$ and $h(x)$ of positive degree in $F[x]$. We can use Theorem 11.10 to conclude that $\deg(f(x)) = \deg(g(x)) + \deg(h(x))$, and hence that the degrees of both $g(x)$ and $h(x)$ are less than n. The inductive hypothesis then implies that both $g(x)$ and $h(x)$ are either irreducible or a product of irreducible polynomials. We can then use equation (13.4) to conclude that $f(x)$ is irreducible or a product of irreducible polynomials. This completes the induction proof that any polynomial of positive degree in $F[x]$ is either irreducible or a product of irreducible polynomials.

To prove that the factorization is unique in the sense stated in the theorem, we will again use induction on the degree of $f(x)$. If $\deg(f(x)) = 1$, then $f(x)$ is irreducible, and the only way that equation (13.3) can be true is if $f(x) = p_1(x) = q_1(x)$. This establishes the basis step.

For the inductive step, assume that $n \in \mathbb{N}$, that $\deg(f(x)) = n$, and that the factorization into irreducible polynomials is unique (as stated in the theorem) for all polynomials in $F[x]$ of degree 1, 2, ..., $n-1$. Furthermore, assume that

$$f(x) = p_1(x)p_2(x)\cdots p_m(x) \qquad \text{and} \qquad f(x) = q_1(x)q_2(x)\cdots q_k(x) \tag{13.5}$$

for some irreducible $p_1(x), p_2(x), \ldots p_m(x)$ and $q_1(x), q_2(x), \ldots q_k(x)$ in $F[x]$. Since

$$p_1(x)p_2(x)\ldots p_m(x) = q_1(x)q_2(x)\ldots q_k(x),$$

it follows that $p_1(x)$ divides $q_1(x)q_2(x)\cdots q_k(x)$. We can then use Corollary 13.18 to conclude that $p_1(x)$ divides $q_j(x)$ for some j. After rearranging and relabeling, if necessary, we may assume that $p_1(x)$ divides $q_1(x)$. Since $p_1(x)$ and $q_1(x)$ are irreducible, $p_1(x)$ must be an associate of $q_1(x)$. So $p_1(x) = cq_1(x)$ for some $c \in F$. Using this and the equations in (13.5), we see that

$$f(x) = cq_1(x)p_2(x)\cdots p_m(x) = q_1(x)q_2(x)\cdots q_k(x).$$

Since $F[x]$ is an integral domain, we can cancel the $q_1(x)$ on both sides of this equation and obtain

$$cp_2(x)\cdots p_m(x) = q_2(x)\cdots q_k(x).$$

Since $\deg(f(x)) = n$ and $\deg q_1(x) \geq 1$, we can conclude that the polynomials on each side of the last equation have degree less than n. After rearranging and relabeling the factors, if necessary, we can now use the inductive hypothesis to conclude that $m = k$ and that each $p_j(x)$ is an associate of $q_j(x)$. This completes the induction proof that the factorization of $f(x)$ into irreducibles is unique. ■

Concluding Activities

Activity 13.22. In this activity, we will explore some different ways to write the factorization of a polynomial into irreducibles. Let F be a field, and let $f(x) \in F[x]$.

(a) Use Theorem 13.21 to explain why there exist distinct irreducible polynomials $p_1(x)$, $p_2(x)$, $\ldots$, $p_n(x)$ and positive integers $r_1, r_2, \ldots, r_n$ such that

$$f(x) = p_1(x)^{r_1} p_2(x)^{r_2} \cdots p_n(x)^{r_n}.$$

(b) Let $c \in F$ be the leading coefficient of $f(x)$. Explain why there exist unique monic irreducible polynomials $\hat{p}_1(x)$, $\hat{p}_2(x)$, $\ldots$, $\hat{p}_n(x)$ and positive integers $r_1, r_2, \ldots, r_n$ such that

$$f(x) = c\hat{p}_1(x)^{r_1} \hat{p}_2(x)^{r_2} \cdots \hat{p}_n(x)^{r_n}.$$

Activity 13.23. In geometry, we learned that 2 points determine a straight line. We can view this fact in terms of linear functions with real coefficients. A linear function can be considered a polynomial of degree one. Let $f(x) = ax + b$ and $g(x) = cx + d$ be degree 1 polynomials in $\mathbb{R}[x]$. Assume that $f(1) = g(1)$ and that $f(3) = g(3)$. (This means that the straight lines that are the graphs of these two functions both pass through the points $(1, f(1))$ and $(3, f(3))$, so the graphs are the same straight line.)

(a) Can we conclude that $f(x) = g(x)$ in $\mathbb{R}[x]$? To answer this question, first notice that:

- $f(1) = g(1)$ implies that $a + b = c + d$; and
- $f(3) = g(3)$ implies that $3a + b = 3c + d$.

Use these equations to show that $a = c$ and $b = d$.

(b) Although the method from part (a) could be used to prove a similar result for degree 2 polynomials, another approach involves using Theorem 13.9. Let $f(x) = ax^2 + bx + c$ and $g(x) = px^2 + qx + r$ be degree 2 polynomials in $\mathbb{R}[x]$. Assume further that u, v, and w are three distinct real numbers, and that $f(u) = g(u)$, $f(v) = g(v)$, and $f(w) = g(w)$. Construct the polynomial $h(x) = f(x) - g(x)$ in $\mathbb{R}[x]$. Explain why $h(x)$ has three roots, and then use Theorem 13.9 to explain why $h(x)$ has to be the zero polynomial, and thus $f(x) = g(x)$.

(c) Use an argument similar to the argument in part (b) to prove the following theorem. (A proof by contradiction could be helpful.)

Theorem 13.24. *Let F be a field, let n be a natural number, and let $u(x)$ and $v(x)$ be polynomials in $F[x]$.*

- *If* $\deg(u(x)) < n$ *and* $\deg(v(x)) < n$, *and*
- *if there exist n distinct elements $a_1, a_2, \ldots, a_n$ in F such that $u(a_1) = v(a_1)$, $u(a_2) = v(a_2)$, $\ldots$, $u(a_n) = v(a_n)$,*

then $u(x) = v(x)$ in $F[x]$.

Activity 13.25. Write a short summary that describes the important concepts, and the relationships between these concepts, that were introduced in this investigation. Explain how the material in this investigation is connected to the material in Investigation 4.

Exercises

(1) Determine all of the roots of $f(x) = x^2 + [3]x + [2]$ in $\mathbb{Z}_6[x]$.

(2) Let $f(x) = x^2 + [7]x$ be a polynomial in $\mathbb{Z}_{10}[x]$. Find two different factorizations of $f(x)$ into a product of two monic polynomials of degree 1. Does this contradict Theorem 13.21? Explain.

(3) Let $p(x) = x^4 - 3x^2 - 10$. Completely factor $p(x)$ in:

(a) $\mathbb{C}[x]$ (b) $\mathbb{R}[x]$ (c) $\mathbb{Q}[x]$

(4) (a) Find all real numbers c so that $(x - c)$ is a factor of the polynomial $p(x) = 3x^3 + 4x^2 - 5x + c$ in $\mathbb{R}[x]$.

(b) Find all real numbers c so that $(x - c)$ is a factor of the polynomial $p(x) = 3x^3 + 4x^2 + cx - 8$ in $\mathbb{R}[x]$.

(5) Prove Theorem 13.11. (Hint: Let $p(x)$ be a polynomial of degree 1 in $F[x]$, and write $p(x) = f(x)g(x)$. Use Theorem 11.10 to conclude that $\deg(p(x)) = \deg(f(x)) + \deg(g(x))$.)

(6) Theorem 13.15 can be useful in determining if polynomials of degree 2 or degree 3 are irreducible. In addition, the quadratic formula can often be used to find the roots of a degree 2 polynomial.

(a) Let $f(x) = 2x^2 - 5x - 7$ be a polynomial in $\mathbb{R}[x]$. Is $f(x)$ irreducible or reducible in $\mathbb{R}[x]$? Justify your conclusion.

(b) Let $f(x) = x^2 + x + 1$ be a polynomial in $\mathbb{R}[x]$. Is $f(x)$ irreducible or reducible in $\mathbb{R}[x]$? Justify your conclusion.

(c) Let $f(x) = x^2 + x + 1$ be a polynomial in $\mathbb{C}[x]$. Is $f(x)$ irreducible or reducible in $\mathbb{C}[x]$? Justify your conclusion.

(7) Let $p(x) = ax^2 + bx + c$ be a polynomial in $\mathbb{C}[x]$ with $a, b, c \in \mathbb{C}$ and $a \neq 0$. Prove that $p(x)$ is reducible in $\mathbb{C}[x]$.

(8) Let $p(x) = ax^2 + bx + c$ be a polynomial in $\mathbb{R}[x]$ with $a, b, c \in \mathbb{R}$ and $a \neq 0$. Use the quadratic formula to determine a necessary and sufficient condition on the coefficients for $p(x)$ to be irreducible in $\mathbb{R}[x]$.

(9) Let F be a field. In Activity 13.8, we proved that every polynomial of degree 1 in $F[x]$ is irreducible and has exactly one root in F. Now let $p(x) \in F[x]$ with $\deg(p(x)) = 2$. Prove that $p(x)$ either has no roots in F or has two roots in F. In the latter case, prove that $p(x) = c(x - a)(x - b)$, for some $a, b, c \in F$ with $c \neq 0$.

(10) Let F be a field and let $q(x) = a_n x^{2n} + a_{n-1} x^{2(n-1)} + \cdots + a_1 x^2 + a_0$ be a polynomial in $F[x]$. Prove that if c is a root of $q(x)$ in F, then c^2 is a root of the polynomial $p(x) = a_n x^n + a_{n-1} x^{n-1} + \cdots + a_1 x + a_0$ in $F[x]$.

Suggestion: If you are having a difficult time with the notation, you might first try to prove the result for a polynomial of degree 2 or 3. For example, you could prove the result for $q(x) = a_2 x^4 + a_1 x^2 + a_0$, and then try the general case.

(11) Let F be a field and let $p(x) = a_n x^n + a_{n-1} x^{n-1} + \cdots + a_1 x + a_0$ be a polynomial in $F[x]$. Prove that if c is a nonzero root of $p(x)$ in F, then c^{-1} is a root of the polynomial $q(x) = a_0 x^n + a_1 x^{n-1} + \cdots + a_{n-1} x + a_n$ in $F[x]$.

Suggestion: If you are having a difficult time with the notation, you might first try to prove the result for a polynomial of degree 2 or 3. For example, you could prove the result for $p(x) = a_2 x^2 + a_1 x + a_0$, and then try the general case.

⋆ (12) Use induction to prove Corollary 13.18, which is a generalization of Theorem 13.16.

Connections

In Investigations 11 and 12, we saw that a polynomial ring with coefficients from a field has much in common with the ring of integers, including: a divides relation, a division algorithm, and the idea of relatively prime polynomials. This investigation made an additional connection regarding unique factorization. The Fundamental Theorem of Arithmetic from Investigation 4 showed that the prime numbers form the building blocks of the ring of integers. In the same way, Theorem 13.21 establishes that the irreducible polynomials in $F[x]$ play the same role as the primes do in the integers, forming the building blocks for all non-constant polynomials. Furthermore, just like in the integers, factorization into irreducibles is unique (up to order and associates).

Investigation 14

Irreducible Polynomials

Focus Questions

By the end of this investigation, you should be able to give precise and thorough answers to the questions listed below. You may want to keep these questions in mind to focus your thoughts as you complete the investigation.

- What is an irreducible polynomial?
- What does the Fundamental Theorem of Algebra say about irreducible polynomials in $\mathbb{C}[x]$?
- How can we determine if a polynomial in $\mathbb{R}[x]$ is irreducible? What are some important results related to irreducibility and factorization of polynomials in $\mathbb{R}[x]$?
- How can do we determine if a polynomial in $\mathbb{Q}[x]$ is irreducible? What are some important results related to irreducibility and factorization of polynomials in $\mathbb{Q}[x]$?
- What are the main differences between factorization and irreducibility in $\mathbb{C}[x]$, $\mathbb{R}[x]$, and $\mathbb{Q}[x]$?
- What is the cubic formula? How can the cubic formula be proved, and what steps are involved in using it to find roots of cubic polynomials in $\mathbb{C}[x]$?

Preview Activity 14.1. In previous investigations, we defined irreducible polynomials and showed that irreducible polynomials in polynomial rings over fields play the same role as primes play in $\mathbb{Z}$. In this investigation we will explore some methods to determine when a polynomial is irreducible, with a special emphasis on polynomials with coefficients in $\mathbb{C}$, $\mathbb{R}$, and $\mathbb{Q}$. To begin, we will review the definition and a simple case. Let F be a field.

(a) Give a formal definition of what it means for a polynomial $f(x) \in F[x]$ to be irreducible in $F[x]$.

(b) Is the polynomial $x^2 - 2$ irreducible in $\mathbb{C}[x]$? Explain.

(c) Is the polynomial $x^2 - 2$ irreducible in $\mathbb{R}[x]$? Explain.

(d) Is the polynomial $x^2 - 2$ irreducible in $\mathbb{Q}[x]$? Explain.

(e) The simplest non-constant polynomials are the linear (degree 1) polynomials. Show that any degree 1 polynomial in $F[x]$ is irreducible.

Introduction

We have seen that every non-constant polynomial in a polynomial ring over a field can be factored into a product of irreducible polynomials. While there is no general technique to find the factorization of an arbitrary polynomial into a product of irreducible polynomials, there are some tools that are helpful in this endeavor. As we will see, the factorization of a polynomial into a product of irreducible polynomials, and exactly which polynomials are irreducible, depends heavily on the coefficient field. In this investigation, we will discuss which polynomials are irreducible over $\mathbb{C}$, $\mathbb{R}$, and $\mathbb{Q}$, and we will discover some techniques for determining whether polynomials in $\mathbb{C}[x]$, $\mathbb{R}[x]$, and $\mathbb{Q}[x]$ are irreducible.

Factorization in $\mathbb{C}[x]$

We will begin our work in this investigation by considering polynomials in $\mathbb{C}[x]$. To determine which polynomials in $\mathbb{C}[x]$ are irreducible, we will invoke a very important result known as The Fundamental Theorem of Algebra.

Theorem 14.2 (The Fundamental Theorem of Algebra). *Every polynomial of degree 1 or greater in $\mathbb{C}[x]$ has a root in $\mathbb{C}$.*

There are many different proofs of the Fundamental Theorem of Algebra, and most of them require some knowledge of complex analysis that is beyond the scope of this text. However, it is still instructive to have some idea of how this important theorem is proved. In the appendix to this investigation, we provide a proof that glosses over some of the points from complex analysis. While the proof is not as complete and rigorous as you would see in a complex analysis text, it does give the general idea behind the argument.

As its name would suggest, the importance of the Fundamental Theorem of Algebra cannot to be overstated. Because every non-constant polynomial in $\mathbb{C}[x]$ has a root in $\mathbb{C}$, we say that $\mathbb{C}$ is an *algebraically closed* field. This is a fundamental difference between $\mathbb{C}$ and $\mathbb{R}$. Furthermore, since every root of a polynomial corresponds to a linear factor of the polynomial, the Fundamental Theorem of Algebra tells us the following:

Corollary 14.3. *Every polynomial of degree 1 or greater in $\mathbb{C}[x]$ can be factored as a product of linear polynomials in $\mathbb{C}[x]$.*

In other words, the only irreducible polynomials in $\mathbb{C}[x]$ are the linear polynomials. Unfortunately, although in theory we can factor any polynomial in $\mathbb{C}[x]$ as a product of linear polynomials, there are no general techniques that will perform the factorization. Even so, if a polynomial in $\mathbb{C}[x]$ is of small enough degree or has a particularly convenient form, we may be able to find its factors quite easily.

As an example, consider the polynomial $x^3 - x^2 + x \in \mathbb{C}[x]$. We can factor out a common factor of x to obtain

$$x^3 - x^2 + x = x(x^2 - x + 1).$$

The quadratic formula then tells us that the roots of $x^2 - x + 1$ are $\frac{1 \pm \sqrt{3}i}{2}$. So the Factor Theorem shows that

$$x^2 - x + 1 = \left[x - \left(\frac{1+\sqrt{3}i}{2}\right)\right]\left[x - \left(\frac{1-\sqrt{3}i}{2}\right)\right].$$

Therefore,

$$x^3 - x^2 + x = x\left[x - \left(\frac{1+\sqrt{3}i}{2}\right)\right]\left[x - \left(\frac{1-\sqrt{3}i}{2}\right)\right]. \qquad (14.1)$$

Since every degree 1 polynomial is irreducible, the factorization from equation (14.1) is the unique factorization of $x^3 - x^2 + x$ into a product of irreducible polynomials in $\mathbb{C}[x]$.

Activity 14.4. Factor $f(x) = x^4 - 1$ in $\mathbb{C}[x]$ into a product of irreducible polynomials in $\mathbb{C}[x]$.

In addition to what Corollary 14.3 tells us about irreducible polynomials in $\mathbb{C}[x]$, it also tells us something about the number of roots that a polynomial of degree n in $\mathbb{C}[x]$ must have. You may recall that Theorem 13.9 (see page 171) states that a polynomial of degree n over a field F can have at most n roots in F. Corollary 14.3 allows us to take this result one step further for polynomials in $\mathbb{C}[x]$. In particular, since every polynomial in $\mathbb{C}[x]$ can be factored as a product of linear polynomials in $\mathbb{C}[x]$, and since the Factor Theorem tells us that every linear factor corresponds to a root, it follows that every polynomial in $\mathbb{C}[x]$ has exactly n roots in $\mathbb{C}$. There is, however, one small caveat to this conclusion. Since the factorization of a polynomial in $\mathbb{C}[x]$ may include repeated linear factors, the n roots guaranteed by the Factor Theorem and Corollary 14.3 may not be distinct. So, for instance, the polynomial $p(x) = x^4 - 6x^3 + 13x^2 - 24x + 36$ can be factored as follows:

$$p(x) = (x - 2i)(x + 2i)(x - 3)^2.$$

In this case, 3 is a repeated root—that is, a root of *multiplicity* 2. In general, the multiplicity of a root c of a polynomial $f(x)$ is the largest integer k for which $(x - c)^k$ divides $f(x)$. Using this language, we obtain the following result:

Corollary 14.5. *Every polynomial of degree* $n \geq 1$ *in* $\mathbb{C}[x]$ *has exactly* n *roots in* $\mathbb{C}$ *(with repeated roots counted according to their multiplicities).*

Factorization in $\mathbb{R}[x]$

The irreducible polynomials in $\mathbb{R}[x]$ are closely related to the irreducible polynomials in $\mathbb{C}[x]$. As an example, consider again the polynomial $f(x) = x^3 - x^2 + x \in \mathbb{R}[x]$. In equation (14.1), we factored $f(x)$ into a product of irreducible linear factors in $\mathbb{C}[x]$. Each linear factor corresponds to a root of $f(x)$ (in $\mathbb{C}$), and two of the roots of $f(x)$ are *complex conjugates* of each other.* Furthermore, the product of the factors corresponding to these two roots, $\left[x - \left(\frac{1+\sqrt{3}i}{2}\right)\right]\left[x - \left(\frac{1-\sqrt{3}i}{2}\right)\right]$, is a polynomial with real coefficients. This observation is the key idea in understanding which polynomials are irreducible in $\mathbb{R}[x]$.

Recall that the roots of a quadratic $ax^2 + bx + c \in \mathbb{R}[x]$ are $-\frac{b}{2a} \pm \frac{\sqrt{b^2-4ac}}{2a}$. If either root is complex, then its complex conjugate is also a root. That this always happens is the subject of the next theorem.

*The *complex conjugate* of a complex number $z = a + bi$ is the complex number $\overline{z} = a - bi$. Note that $\overline{\overline{z}} = z$, $z + \overline{z} = 2a$, and $z\overline{z} = a^2 + b^2$.

Theorem 14.6. *Let $f(x) \in \mathbb{R}[x]$. If $z \in \mathbb{C}$ is a root of $f(x)$, then so is $\overline{z}$.*

Proof. Let $f(x) = a_n x^n + a_{n-1}x^{n-1} + \cdots + a_1 x + a_0$ with $a_n, a_{n-1}, \ldots, a_1, a_0 \in \mathbb{R}$. Let $z \in \mathbb{C}$ be a root of $f(x)$. Then, using the facts that $\overline{u+w} = \overline{u} + \overline{w}$ and $\overline{uw} = \overline{u}\,\overline{w}$ for any complex numbers u and w, and the fact that $\overline{y} = y$ for any real number y, we see that

$$\begin{aligned} f(\overline{z}) &= a_n \overline{z}^n + a_{n-1}\overline{z}^{n-1} + \cdots + a_1 \overline{z} + a_0 \\ &= \overline{a_n}\,\overline{z^n} + \overline{a_{n-1}}\,\overline{z^{n-1}} + \cdots + \overline{a_1}\,\overline{z} + \overline{a_0} \\ &= \overline{a_n z^n} + \overline{a_{n-1} z^{n-1}} + \cdots + \overline{a_1 z} + \overline{a_0} \\ &= \overline{a_n z^n + a_{n-1} z^{n-1} + \cdots + a_1 z + a_0} \\ &= \overline{f(z)} \\ &= \overline{0} \\ &= 0. \end{aligned}$$

Thus, $\overline{z}$ is a root of $f(x)$. ∎

We can use Theorem 14.6 in the following way. By the Fundamental Theorem of Algebra, we can factor any polynomial $f(x) \in \mathbb{R}[x] \subset \mathbb{C}[x]$ into a product of linear factors in $\mathbb{C}[x]$. Suppose $x - z$ is one such factor. Theorem 14.6 and the Factor Theorem tell us that $x - \overline{z}$ is also a factor of $f(x)$. Let $z = a + bi$. Then

$$(x - z)(x - \overline{z}) = x^2 - (z + \overline{z})x + z\overline{z} = x^2 - 2ax + (a^2 + b^2) \in \mathbb{R}[x].$$

So if we factor a polynomial $f(x) \in \mathbb{R}[x]$ into a product of irreducibles in $\mathbb{C}[x]$ and collect the linear factors corresponding to complex conjugate roots of $f(x)$, then we can write $f(x)$ as a product of irreducible linear and quadratic factors. Therefore, any polynomial in $\mathbb{R}[x]$ can be factored into a product of linear and quadratic polynomials in $\mathbb{R}[x]$. This proves the following theorem.

Theorem 14.7. *If $f(x) \in \mathbb{R}[x]$ is an irreducible polynomial, then $\deg(f(x))$ is either 1 or 2.*

We can determine which quadratic polynomials in $\mathbb{R}[x]$ are irreducible by using the quadratic formula and checking for real roots.

Activity 14.8. Factor $f(x) = x^5 - 4x$ in $\mathbb{R}[x]$ into a product of irreducible polynomials in $\mathbb{R}[x]$.

Factorization in $\mathbb{Q}[x]$

Unlike in $\mathbb{R}[x]$ and $\mathbb{C}[x]$, we will see that there are irreducible polynomials of every degree in $\mathbb{Q}[x]$, and there is no general theory to tell us which polynomials are irreducible in $\mathbb{Q}[x]$. However, there are some tools we can use in addition to the Factor Theorem in order to determine if a given polynomial in $\mathbb{Q}[x]$ has roots in $\mathbb{Q}$.

Let

$$f(x) = \frac{a_n}{b_n}x^n + \frac{a_{n-1}}{b_{n-1}}x^{n-1} + \cdots + \frac{a_1}{b_1}x + \frac{a_0}{b_0} \in \mathbb{Q}[x],$$

with $a_n, a_{n-1}, \ldots, a_1, a_0, b_n, b_{n-1}, \ldots, b_1, b_0 \in \mathbb{Z}$. Let $B = b_n b_{n-1} \cdots b_1 b_0$ and $B_i = \frac{B}{b_i}$ for each i between 0 and n. We can then express $f(x)$ in the form

$$f(x) = \frac{1}{B}\left((a_n B_n)x^n + (a_{n-1}B_{n-1})x^{n-1} + \cdots + (a_1 B_1)x + (a_0 B_0)\right),$$

which is a nonzero rational number times a polynomial in $\mathbb{Z}[x]$. Hence, when looking for irreducible polynomials in $\mathbb{Q}[x]$, it suffices to study polynomials with integer coefficients.

Our first result in factoring polynomials in $\mathbb{Q}[x]$ helps us determine all of the possible rational roots.

Activity 14.9. Consider the polynomial $f(x) = 2x^3 + 7x^2 + 2x - 3$ in $\mathbb{Q}[x]$. Assume $\frac{p}{q} \in \mathbb{Q}$ is a root of $f(x)$, with $\frac{p}{q}$ in reduced form (that is, $\gcd(p,q) = 1$).

(a) Substitute $\frac{p}{q}$ for x in $f(x)$ and show that the equation

$$2p^3 + 7p^2q + 2pq^2 - 3q^3 = 0 \tag{14.2}$$

results.

(b) Add $3q^3$ to both sides of equation (14.2), and factor all common factors from $2p^3 + 7p^2q + 2pq^2$. What does this tell us about p? (Hint: $\gcd(p,q) = 1$. What are the possible values for p?)

(c) Now subtract $2p^3$ from both sides of equation (14.2), and factor all common factors from $7p^2q + 2pq^2 - 3q^3$. What does this tell us about q? (Hint: $\gcd(p,q) = 1$. What are the possible values for q?)

(d) What are the possible roots of $f(x)$ in $\mathbb{Q}$?

(e) Find all of the rational roots of $f(x)$.

Activity 14.9 is a specific case of the next theorem, which tells us how to find all of the rational roots of a polynomial in $\mathbb{Z}[x]$.

Theorem 14.10 (Rational Root Theorem). *Let*

$$f(x) = a_n x^n + a_{n-1}x^{n-1} + \cdots + a_1 x + a_0 \in \mathbb{Z}[x]$$

with $a_n \neq 0$. *If* $r = \frac{p}{q} \in \mathbb{Q}$ *is a root of* $f(x)$ *with* $\gcd(p,q) = 1$, *then* $p \mid a_0$ *and* $q \mid a_n$.

Proof. Let $f(x) = a_n x^n + a_{n-1}x^{n-1} + \cdots + a_1 x + a_0 \in \mathbb{Z}[x]$ and suppose $r = \frac{p}{q} \in \mathbb{Q}$ is a root of $f(x)$ with $\gcd(p,q) = 1$. Then

$$\begin{aligned} 0 &= f\left(\frac{p}{q}\right) \\ &= a_n\left(\frac{p}{q}\right)^n + a_{n-1}\left(\frac{p}{q}\right)^{n-1} + \cdots + a_1\left(\frac{p}{q}\right) + a_0 \\ &= \frac{1}{q^n}\left(a_n p^n + a_{n-1}p^{n-1}q + a_{n-2}p^{n-2}q^2 + \cdots + a_2 p^2 q^{n-2} + a_1 p q^{n-1} + a_0 q^n\right). \end{aligned}$$

Therefore, we must have

$$0 = a_n p^n + a_{n-1}p^{n-1}q + a_{n-2}p^{n-2}q^2 + \cdots + a_2 p^2 q^{n-2} + a_1 p q^{n-1} + a_0 q^n. \tag{14.3}$$

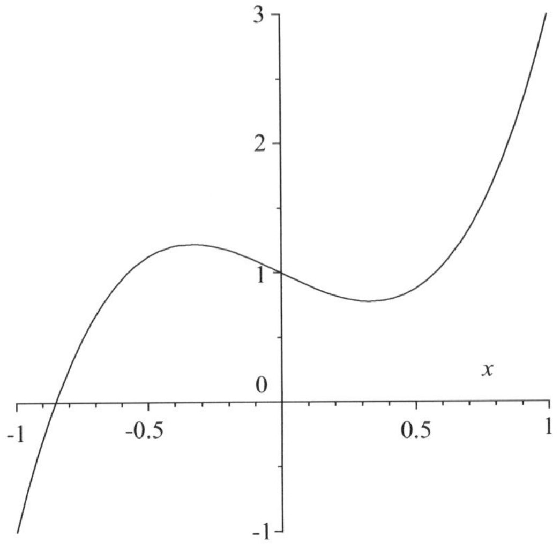

Figure 14.1
The graph of $x^3 - x + 1$ over $\mathbb{R}$.

Subtracting $a_n p^n$ from both sides of equation (14.3) gives us

$$\begin{aligned} -a_n p^n &= a_{n-1}p^{n-1}q + a_{n-2}p^{n-2}q^2 + \cdots + a_2p^2q^{n-2} + a_1pq^{n-1} + a_0q^n \\ &= q\left(a_{n-1}p^{n-1} + a_{n-2}p^{n-2}q + \cdots + a_2p^2q^{n-3} + a_1pq^{n-2} + a_0q^{n-1}\right). \end{aligned}$$

Thus, q divides $a_n p^n$. Since $\gcd(p, q) = 1$, we must have $q \mid a_n$.

Similarly, subtracting $a_0 q^n$ from both sides of equation (14.3) gives us

$$\begin{aligned} -a_0 q^n &= a_n p^n + a_{n-1}p^{n-1}q + a_{n-2}p^{n-2}q^2 + \cdots + a_2p^2q^{n-2} + a_1pq^{n-1} \\ &= p\left(a_n p^{n-1} + a_{n-1}p^{n-2} + a_{n-2}p^{n-3}q + \cdots + a_2pq^{n-3} + a_1q^{n-2}\right). \end{aligned}$$

Thus p divides $a_0 q^n$. Since $\gcd(p, q) = 1$, we must have $p \mid a_0$. ■

We can use Theorem 14.10 to help us determine if certain polynomials in $\mathbb{Q}[x]$ are irreducible. For example, Theorem 14.10 tells us that the possible rational roots of the polynomial $f(x) = 3x^3 - x + 1 \in \mathbb{Z}[x]$ are of the form $\frac{p}{q}$, where p divides 1 and q divides 3. So the possible rational roots of $f(x)$ are ± 1 or $\pm\frac{1}{3}$. The graph of $f(x)$, shown in Figure 14.1 indicates that these potential rational roots are not roots of $f(x)$ at all. (We could also verify this by simply evaluating $f(x)$ at $x = \pm 1$ and $x = \pm\frac{1}{3}$.) Since $\deg(f(x)) = 3$ and $f(x)$ has no roots in $\mathbb{Q}$, we can conclude that $f(x)$ is irreducible in $\mathbb{Q}[x]$.

Activity 14.11.

(a) Factor $f(x) = 18x^3 + 9x^2 - 5x - 2$ in $\mathbb{Q}[x]$.

(b) Is $f(x) = 6x^3 - 7x^2 + x - 1$ irreducible in $\mathbb{Q}[x]$?

Polynomials with No Linear Factors in $\mathbb{Q}[x]$

The Rational Root Theorem helps us find all the linear factors of a polynomial with integer coefficients. But what happens if a polynomial $f(x)$ in $\mathbb{Q}[x]$ has no linear factors? Can we still determine if $f(x)$ is irreducible in $\mathbb{Q}[x]$? The answer is yes, but the problem becomes more difficult.

Consider, for example, the polynomial $f(x) = x^4 + 2x^3 + 5x^2 + 4x + 3$ in $\mathbb{Q}[x]$. A quick use of the Rational Root Theorem shows us that $f(x)$ has no linear factors in $\mathbb{Q}[x]$. It is still possible, however, that $f(x)$ could factor into a product of two quadratic polynomials in $\mathbb{Q}[x]$. The problem would then be to determine if there are any quadratic polynomials $ax^2 + bx + c$ and $rx^2 + sx + t$ in $\mathbb{Q}[x]$ such that $f(x) = \left(ax^2 + bx + c\right)\left(rx^2 + sx + t\right)$. This would be a very hard task, since there are infinitely many combinations of coefficients that might yield such a factorization, and we can't possibly check them all. Fortunately, there is a result that makes the problem much easier.

Activity 14.12. Let $f(x) = 2x^2 + 3x + 1 \in \mathbb{Z}[x]$.

(a) Show that $f(x) = g(x)h(x)$ with $g(x) = x + \frac{1}{2}$ and $h(x) = 2x + 2$ in $\mathbb{Q}[x]$.

(b) From this factorization of $f(x)$ as $g(x)h(x)$, find associates $g_0(x)$ and $h_0(x)$ of $g(x)$ and $h(x)$ so that $g_0(x), h_0(x) \in \mathbb{Z}[x]$ and $f(x) = g_0(x)h_0(x)$.

Activity 14.12 showed how a factorization of $f(x)$ in $\mathbb{Q}[x]$ gave rise to a corresponding factorization of $f(x)$ in $\mathbb{Z}[x]$. This powerful result is the subject of the next lemma.

Lemma 14.13 (Gauss' Lemma)**.** *Let $f(x) \in \mathbb{Z}[x]$. If there are polynomials $g(x)$ and $h(x)$ in $\mathbb{Q}[x]$ such that $f(x) = g(x)h(x)$, then there are also polynomials $g_0(x), h_0(x) \in \mathbb{Z}[x]$ so that $f(x) = g_0(x)h_0(x)$.*

Proof. Let

$$f(x) = a_n x^n + a_{n-1}x^{n-1} + \cdots + a_1 x + a_0 \in \mathbb{Z}[x].$$

Let

$$g(x) = \frac{b_m}{c_m}x^m + \frac{b_{m-1}}{c_{m-1}}x^{m-1} + \cdots + \frac{b_1}{c_1}x + \frac{b_0}{c_0}$$

and

$$h(x) = \frac{r_k}{s_k}x^k + \frac{r_{k-1}}{s_{k-1}}x^{k-1} + \cdots + \frac{r_1}{s_1}x + \frac{r_0}{s_0}$$

be polynomials in $\mathbb{Q}[x]$ so that $f(x) = g(x)h(x)$. We can factor out $c = c_m c_{m-1} \cdots c_1 c_0$ from $g(x)$ and $s = s_k s_{k-1} \cdots s_1 s_0$ from $h(x)$ to obtain polynomials

$$\hat{g}(x) = u_m x^m + u_{m-1}x^{m-1} + \cdots + u_1 x + u_0$$

and

$$\hat{h}(x) = v_k x^k + v_{k-1}x^{k-1} + \cdots + v_1 x + v_0$$

in $\mathbb{Z}[x]$ so that

$$g(x) = \frac{1}{c}\hat{g}(x) \text{ and } h(x) = \frac{1}{s}\hat{h}(x).$$

Then

$$f(x) = g(x)h(x) = \frac{1}{cs}\hat{g}(x)\hat{h}(x)$$

and

$$csf(x) = \hat{g}(x)\hat{h}(x). \tag{14.4}$$

If $cs = 1$, then we can let $g_0(x) = \hat{g}(x)$ and $h_0(x) = \hat{h}(x)$, and we are done. If $cs \neq 1$, then let $|cs| = p_1 p_2 \cdots p_l$ for prime integers $p_1, p_2, \ldots p_l$. Let t be between 1 and l, inclusive. Then p_t divides every coefficient of the polynomial $csf(x)$. Thus p_t divides every coefficient of the product $\hat{g}(x)\hat{h}(x)$.

To continue the proof, we will first need to prove the following claim:

Claim. Either p_t divides every coefficient of $\hat{g}(x)$ or p_t divides every coefficient of $\hat{h}(x)$.

Proof of Claim:. We will prove the claim by contradiction. Let $p = p_t$. Assume that there exist coefficients of both $\hat{g}(x)$ and $\hat{h}(x)$ that are not divisible by p. Let r and s be the smallest integers such that u_r and v_s are not divisible by p. Then p divides u_i for each $i < r$ and p divides v_j for each $j < s$. The coefficient of x^{r+s} in $\hat{g}(x)\hat{h}(x)$ is

$$\begin{aligned} &u_{r+s}v_0 + u_{r+s-1}v_1 + u_{r+s-2}v_2 \\ &\qquad + \cdots + u_{r+1}v_{s-1} + u_r v_s + u_{r-1}v_{s+1} \\ &\qquad + \cdots + u_1 v_{r+s-1} + u_0 v_{r+s}. \end{aligned} \tag{14.5}$$

Now p divides every coefficient in $\hat{g}(x)\hat{h}(x)$, so p divides the expression in (14.5) above. Then

$$\begin{aligned} u_r v_s \equiv &-(u_{r+s}v_0 + u_{r+s-1}v_1 + u_{r+s-2}v_2 + \cdots + u_{r+1}v_{s-1}) \\ &- (u_{r-1}v_{s+1} + \cdots + u_1 v_{r+s-1} + u_0 v_{r+s}) \pmod{p}. \end{aligned}$$

Since p divides v_j for each $j < s$, we know p divides $u_{r+s}v_0 + u_{r+s-1}v_1 + u_{r+s-2}v_2 + \cdots + u_{r+1}v_{s-1}$. Similarly, since p divides u_i for $i < r$, we know p divides $u_{r-1}v_{s+1} + \cdots + u_1 v_{r+s-1} + u_0 v_{r+s}$. Therefore, p divides $u_r v_s$. Since p is a prime, p must divide u_r or p must divide v_s. Either conclusion is a contradiction to our assumption that both u_r and v_s are not divisible by p. Therefore, it must be the case that either p divides every coefficient of $\hat{g}(x)$ or p divides every coefficient of $\hat{h}(x)$. ■

Having established that p_t divides every coefficient of $\hat{g}(x)$ or p divides every coefficient of $\hat{h}(x)$, we may assume, without loss of generality, that p_t divides every coefficient of $\hat{g}(x)$. Then we can factor out p_t from each coefficient and write $\hat{g}(x) = p_t g'(x)$ for some $g'(x) \in \mathbb{Z}[x]$. Then we can cancel the factor of p_t from both sides of equation (14.4) to obtain

$$p_1 p_2 \cdots p_{t-1} p_{t+1} \cdots p_l f(x) = g'(x)\hat{h}(x).$$

We can continue to do this for each value of t to ultimately obtain

$$f(x) = g_0(x)h_0(x)$$

for some $g_0(x), h_0(x) \in \mathbb{Z}[x]$. ■

The beauty of Gauss' Lemma is that it allows us to consider the problem of factoring with integer coefficients rather than rational ones.

As an example, let's return to the polynomial $f(x) = x^4 + 2x^3 + 5x^2 + 4x + 3$ in $\mathbb{Q}[x]$. To determine if $f(x)$ is irreducible, it suffices to show that there are no polynomials $g(x) = ax^2 + bx + c$

and $h(x) = rx^2 + sx + t$ in $\mathbb{Z}[x]$ with $a, r \neq 0$ so that $f(x) = g(x)h(x)$. By working with integer coefficients, we reduce the possible values of these coefficients to a manageable amount.

To see this in action, suppose there exist $g(x) = ax^2 + bx + c$ and $h(x) = rx^2 + sx + t \in \mathbb{Z}[x]$ such that

$$\begin{aligned} x^4 &+ 2x^3 + 5x^2 + 4x + 3 \\ &= g(x)h(x) \\ &= (ar)x^4 + (as + br)x^3 + (at + bs + cr)x^2 + (bt + cs)x + ct. \end{aligned}$$

By equating the coefficients, we obtain the following system of equations:

$$\begin{aligned} ar &= 1 \\ as + br &= 2 \\ at + bs + cr &= 5 \\ bt + cs &= 4 \\ ct &= 3. \end{aligned}$$

Note that we only need to find integer solutions to these equations. This leaves us with only two possibilities for a and r: either $a = r = 1$ or $a = r = -1$. Since we can always factor out a factor of -1 from $g(x)$ and $h(x)$, we can assume $a = r = 1$. The fact that $ct = 3$ gives us four possibilities: $c = 1, t = 3$; $c = -1, t = -3$; $c = 3, t = 1$; or $c = -3, t = -1$. The first case ($c = 1$, $t = 3$, and $a = r = 1$) leaves us with the following equations:

$$\begin{aligned} s + b &= 2 \\ 3 + bs + 1 &= 5 \\ 3b + s &= 4. \end{aligned}$$

Note that $b = s = 1$ gives a complete solution. Therefore,

$$f(x) = x^4 + 2x^3 + 5x^2 + 4x + 3 = (x^2 + x + 1)(x^2 + x + 3)$$

is reducible in $\mathbb{Q}[x]$.

Activity 14.14. Factor $x^4 + 2x^3 + 5x^2 + 6x + 6$ into a product of two quadratic polynomials in $\mathbb{Z}[x]$.

Of course, the problem of determining whether a polynomial in $\mathbb{Q}[x]$ is irreducible becomes much more involved if the degree of the polynomial is larger than 4. For example, to show that a degree 8 polynomial $f(x)$ is irreducible using the previous method, we would have to consider all of the ways that $f(x)$ could be written as a product $g(x)h(x)$, with $\deg(g(x)) + \deg((h(x)) = 8$. So we would need to consider the cases where $\deg(g(x)) = 1$ and $\deg(h(x)) = 7$; $\deg(g(x)) = 2$ and $\deg(h(x)) = 6$; $\deg(g(x)) = 3$ and $\deg(h(x)) = 5$; and $\deg(g(x)) = 4$ and $\deg(h(x)) = 4$. This is a lot of work!

Reducing Polynomials in $\mathbb{Z}[x]$ Modulo Primes

Another tool in determining reducibility of polynomials with integer coefficients in $\mathbb{Q}[x]$ is reducing modulo a prime p. For example, consider the polynomial $f(x) = x^3 + x + 1 \in \mathbb{Z}[x]$. We can reduce this polynomial to the polynomial $\overline{f(x)} = [1]x^3 + [1]x + [1]$ in $\mathbb{Z}_2[x]$. Note that $\overline{f(x)}$ is irreducible in $\mathbb{Z}_2[x]$. (This follows from the Factor Theorem, since $f(x)$ is a degree 2 polynomial and has no roots in $\mathbb{Z}_2$.) So if $f(x) = g(x)h(x) \in \mathbb{Z}[x]$, then we should have $\overline{f(x)} = \overline{g(x)}\ \overline{h(x)} \in \mathbb{Z}_2[x]$. So it would appear that the irreducibility of $\overline{f(x)}$ in $\mathbb{Z}_2[x]$ implies the irreducibility of $f(x) \in \mathbb{Z}[x]$. To generalize this argument to $\mathbb{Z}_p$, we need only to require that the leading coefficient of $f(x)$ is not divisible by p.

Theorem 14.15. *Let $f(x) = a_nx^n + a_{n-1}x^{n-1} + \cdots + a_1x + a_0$ be a polynomial with integer coefficients and $a_n \neq 0$. If there is a prime p so that $[a_n] \neq [0]$ in $\mathbb{Z}_p$ and the polynomial $\overline{f(x)} = [a_n]x^n + [a_{n-1}]x^{n-1} + \cdots + [a_1]x + [a_0]$ is irreducible in $\mathbb{Z}_p[x]$, then $f(x)$ is irreducible in $\mathbb{Q}[x]$.*

Proof. Let $f(x) = a_nx^n + a_{n-1}x^{n-1} + \cdots + a_1x + a_0$ be a polynomial with integer coefficients. We will prove the contrapositive. That is, we assume $f(x)$ is not irreducible in $\mathbb{Q}[x]$. Then Gauss' Lemma tells us there are polynomials $g(x), h(x) \in \mathbb{Z}[x]$ such that $f(x) = g(x)h(x)$ with $1 \leq \deg(g(x)), \deg(h(x)) < \deg(f(x))$. Let b_m be the leading coefficient of $g(x)$ and c_k the leading coefficient of $h(x)$. Then $b_mc_k = a_n$. Suppose p is a prime so that $[a_n] \neq [0]$ in $\mathbb{Z}_p$. Let $\overline{f(x)}$ be the polynomial obtained by reducing the coefficients of $f(x)$ modulo p. Then $\deg(\overline{f(x)}) = \deg(f(x))$. Since $[a_n] = [b_m][c_k]$ and $\mathbb{Z}_p$ is an integral domain, $[b_m] \neq [0]$ and $[c_k] \neq [0]$ in $\mathbb{Z}_p$. Thus, $\deg(\overline{g(x)}) = \deg(g(x))$ and $\deg(\overline{h(x)}) = \deg(h(x))$. Also, $\overline{f(x)} = \overline{g(x)}\ \overline{h(x)}$, with $1 \leq \deg(\overline{g(x)}), \deg(\overline{h(x)}) < \deg(\overline{f(x)})$. So $\overline{f(x)}$ is reducible in $\mathbb{Z}_p[x]$. ■

Note that we only need to find one prime p so that $\overline{f(x)}$ is irreducible in $\mathbb{Z}_p[x]$ to conclude that $f(x)$ is irreducible in $\mathbb{Q}[x]$. However, the relationship does not work in the other direction. For example, the polynomial $x^2 + [1]$ is reducible in $\mathbb{Z}_2[x]$, but is irreducible in $\mathbb{Q}[x]$.

Activity 14.16. Is $x^3 + 2x + 2$ irreducible in $\mathbb{Q}[x]$? Check by reducing modulo a prime.

Eisenstein's Criterion

There is one other well-known tool that is often used to determine the irreducibility of polynomials in $\mathbb{Q}[x]$.

Theorem 14.17 (Eisenstein's Criterion). *Let $f(x) = a_nx^n + a_{n-1}x^{n-1} + \cdots + a_1x + a_0 \in \mathbb{Z}[x]$ be of positive degree. If there is a prime p so that $p \mid a_0$, $p \mid a_1$, $p \mid a_2$, $\ldots$, $p \mid a_{n-1}$, but p does not divide a_n and p^2 does not divide a_0, then $f(x)$ is irreducible in $\mathbb{Q}[x]$.*

Proof. Let $f(x) = a_nx^n + a_{n-1}x^{n-1} + \cdots + a_1x + a_0 \in \mathbb{Z}[x]$ be of positive degree. Let p be a prime so that $p \mid a_0, p \mid a_1, p \mid a_2, \ldots, p \mid a_{n-1}$, p does not divide a_n and p^2 does not divide a_0. To prove that $f(x)$ is irreducible in $\mathbb{Q}[x]$, we proceed by contradiction. Assume there are non-constant polynomials $g(x), h(x) \in \mathbb{Z}[x]$ (by Gauss's Lemma) so that $f(x) = g(x)h(x)$. Let

$$g(x) = b_mx^m + b_{m-1}x^{m-1} + \cdots + b_1x + b_0$$

and

$$h(x) = c_kx^k + c_{k-1}x^{k-1} + \cdots + c_1x + c_0,$$

with $b_m, c_k \neq 0$. Since $a_0 = b_0c_0$, and p divides a_0, we know $p \mid b_0c_0$. So $p \mid b_0$ or $p \mid c_0$. However, since p^2 does not divide a_0, we know that p cannot divide both b_0 and c_0. Assume without loss of generality that p does not divide b_0. We also know $a_n = b_m c_k$. Since p does not divide a_n, we cannot have p as a factor of b_m or c_k. Let l be the smallest integer so that c_l is not divisible by p. In other words, p divides c_i for each $i < l$. Note that we have $l \leq k < n$. Therefore, we know p divides a_l by hypothesis. Now

$$a_l = b_l c_0 + b_{l-1}c_1 + b_{l-2}c_2 + \cdots b_1 c_{l-1} + b_0 c_l,$$

and so

$$b_0 c_l = a_l - (b_l c_0 + b_{l-1}c_1 + b_{l-2}c_2 + \cdots b_1 c_{l-1}). \tag{14.6}$$

Notice that since p divides a_l and c_i for each $i < l$, we have p as a factor of each term on the right side of (14.6). Therefore, p must divide $b_0 c_l$. But neither b_0 nor c_l is divisible by p, so this is impossible. We conclude that no such polynomials $g(x), h(x)$ exist, and $f(x)$ is irreducible in $\mathbb{Q}[x]$. ■

Activity 14.18.

(a) Show that $2x^5 + 27x^2 + 9x + 3$ is irreducible in $\mathbb{Q}[x]$.

(b) Show that for any prime p and any positive integer n, the polynomial $f(x) = x^n + p$ is irreducible in $\mathbb{Q}[x]$. Conclude that there are irreducible polynomials in $\mathbb{Q}[x]$ of any degree.

Factorization in $F[x]$ for Other Fields F

To factor polynomials over fields other than $\mathbb{Q}$, $\mathbb{R}$, or $\mathbb{C}$, trial and error is often the best option. We can first look for roots, which (by the Factor Theorem) give us linear factors. If we find roots, we can then divide by the corresponding linear factor to reduce the degree and look for roots again.

Activity 14.19. Factor $f(x) = [3]x^3 + [4]x^2 + [3]$ in $\mathbb{Z}_5[x]$.

Summary

Let $f(x) \in F[x]$. To factor $f(x)$ into a product of irreducible polynomials, we can always begin by testing for roots.

- If $f(x)$ can be written with integer coefficients, then we can use the Rational Root Theorem to check for roots in $\mathbb{Q}$.
- If $f(x) \in \mathbb{Z}_p[x]$ for some prime p, then we can just test each element in $\mathbb{Z}_p$ to see if $f(x)$ has any roots.
- If $r \in F$ is a root of $f(x)$, then we can divide $f(x)$ by $x - r$ and obtain a quotient $q(x)$ of degree $\deg(f(x)) - 1$. We can then repeat this process with $q(x)$.
- If $f(x)$ is of degree 2 or 3 and has no roots in F, then we know that $f(x)$ is irreducible in $F[x]$.

- If we can reduce $f(x)$ down to a polynomial of degree 2 in $\mathbb{Q}[x]$, $\mathbb{R}[x]$ or $\mathbb{C}[x]$, then we can use the quadratic formula to find the roots of $f(x)$ in $\mathbb{C}$.

Remember that the polynomial rings $\mathbb{Q}[x]$, $\mathbb{R}[x]$, and $\mathbb{C}[x]$ are fundamentally different with regard to the types of irreducible polynomials they contain. In particular:

- The Fundamental Theorem of Algebra tells us that the only irreducible polynomials in $\mathbb{C}[x]$ are the linear polynomials.
- The Fundamental Theorem of Algebra and Theorem 14.6 show us that there are no irreducible polynomials in $\mathbb{R}[x]$ of degree higher than 2.
- There are irreducible polynomials of any degree in $\mathbb{Q}[x]$.

The Cubic Formula

The quadratic formula tells us how to find all solutions to quadratic equations in $\mathbb{C}[x]$. As it turns out, there is also a general formula for solving cubic equations, although it is much more complicated than the quadratic formula.

First note that when we want to find roots of polynomials, it suffices to work with only monic polynomials. To see why, consider the general polynomial $f(x) = \sum_{i=0}^{n} a_i x^i$ (with $a_n \neq 0$) in $F[x]$, where F is a field. Let $g(x) = x^n + \sum_{i=0}^{n+1} a_n^{-1} a_i x^i$, and assume $r \in R$ is a root of $f(x)$. Then $f(r) = 0$, and so

$$a_n r^n + a_{n-1} r^{n-1} + \cdots + a_1 r + a_0 = 0.$$

Multiplying both sides by a_n^{-1} gives us

$$r^n + a_n^{-1} a_{n-1} r^{n-1} + \cdots + a_n^{-1} a_1 r + a_n^{-1} a_0 = a_n^{-1} 0 = 0.$$

But the left hand side of the last equation is just $g(r)$. Thus, $g(r) = 0$.

Now assume $r \in R$ is a root of $g(x)$. Then $g(r) = 0$. So

$$r^n + a_n^{-1} a_{n-1} r^{n-1} + \cdots + a_n^{-1} a_1 r + a_n^{-1} a_0 = 0.$$

Multiplying both sides by a_n yields

$$a_n r^n + a_{n-1} r^{n-1} + \cdots + a_1 r + a_0 = a_n 0 = 0.$$

But the left hand side of the last equation is just $f(r)$. Thus, $f(r) = 0$. What we have shown is that r is a root of $f(x)$ if and only if r is a root of $g(x)$.

One conclusion we can draw from this is that when looking for roots of polynomials over a field, it is enough to consider roots of monic polynomials. So when finding roots of cubics, it suffices to consider only cubics of the form $x^3 + ax^2 + bx + c$. Let $a, b, c \in \mathbb{C}$, and consider the cubic equation

$$x^3 + ax^2 + bx + c = 0. \tag{14.7}$$

Our first step in solving this cubic equation is to reduce the cubic polynomial $x^3 + ax^2 + bx + c$ to what is often called a *depressed* cubic—that is, a cubic of the form $x^3 + px + q$. One way to do this is to make the change of variable $x = z - \frac{a}{3}$.

Activity 14.20.

(a) Evaluate the cubic polynomial x^3+6x^2+x+3 at $x=z-\frac{a}{3}$ and show that the result is the depressed cubic $z^3-11z+17$.

(b) Evaluate the general cubic x^3+ax^2+bx+c at $x=z-\frac{a}{3}$ and show that

$$x^3+ax^2+bx+c=z^3+\left(\frac{3b-a^2}{3}\right)z+\left(\frac{2a^3-9ab+27c}{27}\right).$$

Activity 14.20 shows that the substitution $x=z-\frac{a}{3}$ transforms our general cubic x^3+ax^2+bx+c to the depressed cubic $z^3+pz+q=0$ with $p=\frac{3b-a^2}{3}$ and $q=\frac{2a^3-9ab+27c}{27}$. Now we just need to solve the depressed cubic.

Theorem 14.21. *Let $p\in\mathbb{C}$ be nonzero, and let $q\in\mathbb{C}$. The roots of*

$$z^3+pz+q=0 \tag{14.8}$$

are given by $z=\sqrt[3]{A}-\frac{p}{3\sqrt[3]{A}}$, where $A=-\frac{q}{2}+\sqrt{\frac{q^2}{4}+\frac{p^3}{27}}$ and $\sqrt[3]{A}$ can be any one of the three cube roots of A in $\mathbb{C}$.

Proof. Our first step will be to reduce equation (14.8) to a quadratic. We do this by substituting $y-\frac{p}{3y}$ for z to obtain

$$\begin{aligned}
\left(y-\frac{p}{3y}\right)^3+p\left(y-\frac{p}{3y}\right)+q&=0\\
y^3-3\left(\frac{p}{3y}\right)y^2+3\left(\frac{p}{3y}\right)^2y-\left(\frac{p}{3y}\right)^3+py-p\left(\frac{p}{3y}\right)+q&=0\\
y^3-py+\left(\frac{p^2}{3y}\right)-\left(\frac{p^3}{27y^3}\right)+py-\left(\frac{p^2}{3y}\right)+q&=0\\
y^3-\left(\frac{p^3}{27}\right)\left(\frac{1}{y^3}\right)+q&=0\\
\left(y^3\right)^2+qy^3-\left(\frac{p^3}{27}\right)&=0.
\end{aligned} \tag{14.9}$$

Setting $v=y^3$ transforms the last equation into the quadratic equation

$$v^2+qv-\left(\frac{p^3}{27}\right)=0. \tag{14.10}$$

We can solve equation (14.10) with the quadratic formula, which yields the following two roots:

$$A=\frac{-q+\sqrt{q^2+4\left(\frac{p^3}{27}\right)}}{2} \quad\text{and}\quad B=\frac{-q-\sqrt{q^2+4\left(\frac{p^3}{27}\right)}}{2}.$$

After some simplifying, we see that

$$A=-\frac{q}{2}+\sqrt{\frac{q^2}{4}+\frac{p^3}{27}} \quad\text{and}\quad B=-\frac{q}{2}-\sqrt{\frac{q^2}{4}+\frac{p^3}{27}}.$$

So our solutions to equation (14.9) are the cube roots of A and B. At first glance it might appear that there are then 6 solutions to equation (14.10) and therefore 6 solutions to equation (14.8). However, if r is a cube root of A, then

$$\left(-\frac{p}{3r}\right)^3 = -\frac{p^3}{27r^3} = \frac{p^3}{27A}.$$

Since

$$AB = \frac{q^2}{4} - \left(\frac{q^2}{4} + \frac{p^3}{27}\right) = \frac{p^3}{27},$$

we see that

$$\left(-\frac{p}{3r}\right)^3 = -\frac{p^3}{27r^3} = \frac{p^3}{27A} = B,$$

and so $-\frac{p}{3r}$ is a cube root of B. So if we let r_1, r_2, and r_3 be the cube roots of A, then $s_1 = -\frac{p}{3r_1}$, $s_2 = -\frac{p}{3r_2}$, and $s_3 = -\frac{p}{3r_3}$ are the cube roots of B. Since $z = y - \frac{p}{3y}$ is equal to $r_i + s_i$ whenever $y = r_i$ or $y = s_i$, it follows that the solutions to equation (14.8) are r_1+s_1, r_2+s_2, and r_3+s_3. ■

To illustrate the cubic formula, we will find the solutions to the cubic equation $x^3+x^2+x+1 = 0$. We begin by reducing this equation to the depressed cubic

$$z^3 + \frac{2}{3}z + \frac{20}{27} = 0$$

by substituting $x = z - \frac{1}{3}$, where $p = \frac{3(1)-(1)^2}{3} = \frac{2}{3}$ and $q = \frac{2(1)^3-9(1)(1)+27(1)}{27} = \frac{20}{27}$. To solve the depressed cubic, we need to find the cube roots of

$$A = -\frac{q}{2} + \sqrt{\frac{q^2}{4} + \frac{p^3}{27}} = -\frac{10}{27} + \frac{2\sqrt{3}}{9}.$$

Let $r = \sqrt[3]{A}$, and let ω be the primitive cube root† of 1 given by $\omega = -\frac{1}{2} + \frac{\sqrt{3}}{2}i$. Then, using the same notation as in the proof of Theorem 14.21, we obtain:

$$\begin{aligned} r_1 &= r\omega & s_1 &= -\frac{p}{3r_1} = -\frac{2}{9r\omega} = -\frac{2}{9r}\omega^2 \\ r_2 &= r\omega^2 & s_2 &= -\frac{p}{3r_2} = -\frac{2}{9r}\omega \\ r_3 &= r & s_3 &= -\frac{p}{3r_3} = -\frac{2}{9r}. \end{aligned}$$

So the roots of the depressed polynomial $z^3 + \frac{2}{3}z + \frac{20}{27}$ are

$$\begin{aligned} z_1 &= r\omega - \frac{2}{9r}\omega^2 \\ z_2 &= r\omega^2 - \frac{2}{9r}\omega \\ z_3 &= r - \frac{2}{9r}. \end{aligned}$$

We can simplify these roots a bit. First, note that $\left(-\frac{1}{3} + \frac{\sqrt{3}}{3}\right)^3 = -\frac{10}{27} + \frac{2\sqrt{3}}{9}$, so $r = -\frac{1}{3} + \frac{\sqrt{3}}{3}$.

†If A is a real number and $\omega = \cos(\frac{2\pi}{n}) + i\sin(\frac{2\pi}{n})$, then ω is called a *primitive* n^{th} root of 1, and the n complex n^{th} roots of A are given by $\sqrt[n]{A}, \omega\sqrt[n]{A}, \omega^2\sqrt[n]{A}, \ldots, \omega^{n-1}\sqrt[n]{A}$.

Then

$$\begin{aligned} z_3 &= r - \frac{2}{9r} \\ &= -\frac{1}{3} + \frac{\sqrt{3}}{3} - \frac{2}{9\left(-\frac{1}{3} + \frac{\sqrt{3}}{3}\right)} \\ &= -\frac{1}{3} + \frac{\sqrt{3}}{3} - \frac{2}{9\left(-\frac{1}{3} + \frac{\sqrt{3}}{3}\right)} \left(\frac{-\frac{1}{3} - \frac{\sqrt{3}}{3}}{-\frac{1}{3} - \frac{\sqrt{3}}{3}}\right) \\ &= -\frac{1}{3} + \frac{\sqrt{3}}{3} - \frac{2\left(-\frac{1}{3} - \frac{\sqrt{3}}{3}\right)}{9\left(\frac{1}{9} - \frac{3}{9}\right)} \\ &= -\frac{1}{3} + \frac{\sqrt{3}}{3} + \left(-\frac{1}{3} - \frac{\sqrt{3}}{3}\right) \\ &= -\frac{2}{3}. \end{aligned}$$

Similar simplification gives $z_1 = \frac{1}{3} + i$ and $z_2 = \frac{1}{3} - i$. The solutions to our original equation $x^3 + x^2 + x + 1 = 0$ have the form $x_i = z_i - \frac{a}{3} = z_i - \frac{1}{3}$. So the solutions to $x^3 + x^2 + x + 1 = 0$ are

$$x_1 = i, \quad x_2 = -i, \text{ and } \quad x_3 = -1.$$

Theorem 14.21 shows us how to find the roots of cubic polynomials in $\mathbb{C}[x]$. There is a corresponding formula for finding roots of quartic (degree 4) polynomials as well, but we won't consider that formula here. When we use the quadratic and cubic formulas to solve equations, we are finding the solutions in a form that only depends on the sums, differences, products, or quotients of the coefficients of the polynomial along with roots (square, cube, etc.) of such combinations of the coefficients. When we do this, we say we are *solving an equation by radicals*. Some of the best mathematicians throughout history, including Euler and Lagrange, attempted to find solutions by radicals of general quintic (degree 5) polynomial equations over $\mathbb{C}$. It wasn't until 1826 that the first generally accepted proof of the insolvability of quintic polynomials was published by Abel. Galois later developed a theory of solvability of equations involving groups and fields, and he used this theory to show that polynomial equations of degree 5 or higher over $\mathbb{C}$ are not solvable by radicals.

Concluding Activities

Activity 14.22. Let $f(x) = ax^2 + bx + c \in \mathbb{C}[x]$ with $a \neq 0$. The discriminant of $f(x)$ is the number $D = b^2 - 4ac$.

(a) Show that the roots of $f(x)$ in $\mathbb{C}$ are $\frac{-b+\sqrt{D}}{2a}$ and $\frac{-b-\sqrt{D}}{2a}$.

(b) We can use the discriminant to completely characterize the irreducible quadratic polynomials in $\mathbb{R}[x]$. If $f(x) = ax^2 + bx + c \in \mathbb{R}[x]$ with $a \neq 0$, show that $f(x)$ is irreducible in $\mathbb{R}[x]$ if and only if $D < 0$.

Activity 14.23. We have used the quadratic formula to find roots of polynomials with complex coefficients, and we have seen that there is a cubic formula as well. As this activity will show, there is a quadratic formula for fields other than $\mathbb{C}$. Let F be any field, and let $a, b, c \in F$ with $a \neq 0$. Suppose also, that for each $t \neq 0$ in F, we have $t + t = 2t \neq 0$ as well. (Note that this assumption implies that F does not have characteristic 2.)

(a) Show that if there is an element $q \in F$ so that $q^2 = b^2 - 4ac$, then the polynomial $f(x) = ax^2 + bx + c$ has the two roots $r_1 = (2a)^{-1}(-b + q)$ and $r_2 = (2a)^{-1}(-b - q)$. Explain why these are all of the roots of $f(x)$.

(b) Prove the converse of the previous problem. That is, show that if there is no element $q \in F$ so that $q^2 = b^2 - 4ac$, then the polynomial $f(x) = ax^2 + bx + c$ has no roots in F. (Hint: Complete the square and perform a little algebra.)

(c) Apply the previous results to find the roots of the polynomial $x^2 + [2]x + [2]$ in $\mathbb{Z}_5$. Hint: Compute q^2 for each element $q \in \mathbb{Z}_5$. Use the method above instead of computing the roots directly.

(d) Apply the previous results to show that the polynomial $[3]x^2 + x + [2]$ is irreducible in $\mathbb{Z}_5[x]$.

(e) Compare the results of the first two parts of this problem to the well-known quadratic formula.

(f) Why did we insist that $t + t = 2t \neq 0$ for each nonzero $t \in F$?

Activity 14.24. Write a short summary that describes the important concepts, and the relationships between these concepts, that were introduced in this investigation. Explain how the material in this investigation is connected to the material in Investigation 13.

Exercises

(1) Factor $f(x) = 18x^3 + 9x^2 - 5x - 2$ into a product of irreducible polynomials in $\mathbb{Q}[x]$.

(2) Factor $f(x) = 2x^4 + x^3 + 4x^2 + x + 2$ into a product of irreducible polynomials in $\mathbb{Q}[x]$.

(3) Factor $f(x) = x^3 - 2x^2 - 3x + 6$ into a product of irreducible polynomials in $\mathbb{R}[x]$.

(4) Factor $f(x) = x^3 - 4x^2 + 6x - 4$ into a product of irreducible polynomials in $\mathbb{C}[x]$.

(5) Factor $f(x) = [3]x^3 + [4]x^2 + [3]$ into a product of irreducible polynomials in $\mathbb{Z}_5[x]$.

(6) Is $f(x) = 235x^3 + 110x + 59$ irreducible in $\mathbb{Q}[x]$?

(7) Is $f(x) = 6x^3 - 7x^2 + x - 6$ irreducible in $\mathbb{Q}[x]$?

(8) Is $f(x) = x^4 + 3x^2 + 2$ irreducible in $\mathbb{Q}[x]$?

(9) Is $f(x) = 2x^8 + 15x^5 - 21x^2 + 9x + 3$ irreducible in $\mathbb{Q}[x]$?

(10) Is $f(x) = [6]x^3 + x + [2]$ irreducible in $\mathbb{Z}_7[x]$?

(11) Find all the roots in $\mathbb{C}$ of the given polynomial.

(a) $x^3 + 3x^2 + 3x + 12$

(b) $x^3 + 6x^2 + 18x + 22$

(12) Let $f(x) \in \mathbb{R}[x]$.

(a) If $\deg(f(x)) = 2$, then how many real roots can $f(x)$ have? Explain.

(b) If $\deg(f(x)) = 3$, then how many real roots can $f(x)$ have? Explain.

(c) If $\deg(f(x)) = 4$, then how many real roots can $f(x)$ have? Explain.

(d) If $\deg(f(x)) = 5$, then how many real roots can $f(x)$ have? Explain.

(13) Can an odd degree polynomial of degree greater than one in $\mathbb{R}[x]$ be irreducible? If yes, find one. If no, explain why.

(14) Let u and w be complex numbers and y a real number. Prove each of the following.

(a) $\overline{u+w} = \overline{u} + \overline{w}$

(b) Extend part (a) to show that if n is a positive integer and $w_1, w_2, \ldots, w_n$ are complex numbers, then

$$\overline{w_1 + w_2 + \cdots + w_n} = \overline{w_1} + \overline{w_2} + \cdots \overline{w_n}.$$

(c) $\overline{uw} = \overline{u}\,\overline{w}$

(d) Extend part (c) to show that if n is a positive integer and $w_1, w_2, \ldots, w_n$ are complex numbers, then

$$\overline{w_1 w_2 \cdots w_n} = \overline{w_1}\;\overline{w_2}\;\cdots\;\overline{w_n}.$$

(e) $\overline{uw} = \overline{u}\,\overline{w}$

(f) $\overline{y} = y$

(15) Let F be a field and $c \in F$. Show that a polynomial $f(x) \in F[x]$ is irreducible in $F[x]$ if and only if the polynomial $f(x+c)$ is irreducible in $F[x]$.

(16) Prove that for any prime number p, the polynomial $f(x) = x^{p-1} + x^{p-2} + \cdots + x + 1$ is irreducible in $\mathbb{Q}[x]$. (Hint: Consider the polynomial $g(x) = f(x+1)$, and use the fact that $\sum_{i=0}^{n} x^i = \frac{1-x^{n+1}}{1-x}$.)

(17) We showed that the factorization of a polynomial into a product of irreducible polynomials is unique when the quotient ring is a field. This need not be true if the quotient ring is not a field. For example, find a polynomial $f(x) \in \mathbb{Z}_8[x]$ that can be factored in at least two different ways as a product of linear polynomials.

(18) Show that for any prime integer p and any integer $n \geq 2$, the number $\sqrt[n]{p}$ is not a rational number. This will show that there are infinitely many irrational numbers. (Hint: Consider an appropriate polynomial equation with integer coefficients.)

(19) You might wonder if the converse of Theorem 14.15 is true. That is, if $f(x) \in \mathbb{Z}[x]$ and the polynomial $\overline{f(x)} \in \mathbb{Z} - p[x]$ is reducible for each prime p, must it be the case that $f(x)$ is reducible in $\mathbb{Z}[x]$? We will answer that question in this exercise. Let $f(x) = x^4 + 1 \in \mathbb{Z}[x]$.

(a) Show that $\overline{f(x)}$ is reducible in $\mathbb{Z}_2[x]$.

(b) Now we will see how to show that $\overline{f(x)}$ is reducible in $\mathbb{Z}_p[x]$ for any odd prime p. Let p be an odd prime and let U_p be the set of units in $\mathbb{Z}_p$. Let $U_p^2 = \{[a]^2 : [a] \in U_p\}$ be the set of squares in U_p.

(i) Why is $p = 2k+1$ for some integer k?

(ii) Explain why $U_p^2 = \{[1]^2, [2]^2, \ldots, [k-1]^2, [k]^2\}$.

(iii) Show that if $a^2 \equiv b^2 \pmod{p}$, then $[a] = [b]$ or $[a] = -[b]$ in $\mathbb{Z}_p$. Use this result to conclude that U_p^2 contains exactly k elements.

(iv) Let $[a]$ be in $U_p \setminus U_p^2$ and define $[a]U_p^2$ as

$$[a]u_p^2 = \{[a][b] : [b] \in U_p^2\}.$$

Prove that $[a]U_p^2 \cap U_p^2 = \emptyset$ and $[a]U_p^2 \cup U_p^2 = U_p$.

(v) Show that the product of two non-squares in U_p is a square. (Hint: Based on the previous part of this exercise, what form do all of the non-squares have?)

(vi) Explain why at least one of $[-1]$, $[2]$, or $[2]$ is in U_p^2. Show that in each case, $\overline{f(x)}$ is reducible in $\mathbb{Z}_p[x]$.

(vii) Show that $f(x)$ is irreducible in $\mathbb{Q}[x]$. Conclude that the converse of Theorem 14.15 is false.

(c) Do some research and determine if there are other polynomials in $\mathbb{Z}[x]$ that are irreducible in $\mathbb{Q}[x]$ but reducible in $\mathbb{Z}_p[x]$ for every prime p. If there are other such polynomials, list at least 3, if possible.

Appendix – Proof of the Fundamental Theorem of Algebra

This particular proof comes from the paper "The Fundamental Theorem of Algebra" by Frode Terkelsen in *The American Mathematical Monthly*, Vol. 83, No. 8, (Oct. 1976), p. 647. As is true with most proofs of this theorem, some complex analysis is required, and we will gloss over those points. While our proof will not be complete and rigorous, it is instructive to have some idea of how this important theorem is proved. For complete details, consult a text on complex analysis.

Proof. Let $f(z) \in \mathbb{C}[z]$ be a non-constant polynomial. We know that $|f(z)| \to \infty$ as $|z| \to \infty$, and along with the continuity of the polynomial $f(z)$, this implies the existence of a complex number z_0 such that $|f(z_0)| \leq |f(z)|$ for all $z \in \mathbb{C}$. In other words, the function $|f(z)|$ attains a minimum value.[‡] (Recall that the norm $|f(z)|$ of the complex number $f(z)$ is a nonnegative real number.) We can always translate f so that z_0 is at the origin (by considering the polynomial $f(z + z_0)$), so we will assume $z_0 = 0$ without loss of generality. Our job is then to show that $f(0) = 0$. We will proceed by contradiction and assume $f(0) \neq 0$.

First, we will rewrite $f(z)$ in a more useful form. By subtracting out the constant term of $f(z)$, we obtain a polynomial whose smallest degree term is n for some $n \geq 1$. Thus,

$$f(z) = a_0 + a_n z^n + z^{n+1} Q(z),$$

where $a_n \neq 0$ and $Q(z)$ is a polynomial. (As an illustration, consider the polynomial $f(z) = 2 + 3z^3 + 4z^4 + 5z^6$. Note that $n = 3$ and $Q(z) = 4 + 5z^2$ in this case.) For ease of notation, we

[‡]The formal proof of this result requires some more sophisticated results from analysis and topology, but the idea is similar to why every even-degree polynomial $p(x)$ in $\mathbb{R}[x]$ attains a minimum value. For any such $p(x)$, the Extreme Value Theorem guarantees that $p(x)$ will attain a minimum value m_a on each closed interval of the form $[-a, a]$. But $p(x) \to \infty$ as $|x| \to \infty$, and so there must be some $a \in \mathbb{R}^+$ for which $p(x) > m_a$ for all $x < -a$ and all $x > a$. It follows that this m_a is the (global) minimum value of $p(x)$.

let $a = a_0$ and $b = a_n$. Note that $f(0) \neq 0$ implies $a \neq 0$. We will now use the fact that every complex number has n^{th} roots. In particular, let ω be an nth root of $-\frac{a}{b}$—that is $\omega^n = -\frac{a}{b}$. For each real number x, we have

$$\begin{aligned} f(x\omega) &= a + b(x\omega)^n + (x\omega)^{n+1}Q(x\omega) \\ &= a + bx^n\omega^n + (x\omega)^{n+1}Q(x\omega) \\ &= a + bx^n\left(-\frac{a}{b}\right) + (x\omega)^{n+1}Q(x\omega) \\ &= a(1 - x^n) + (x\omega)^{n+1}Q(x\omega). \end{aligned}$$

Suppose $Q(z) = q_0 + q_1 z + q_2 z^2 + \cdots + q_m z^m$. If $0 < t < 1$, then the triangle inequality shows that

$$|Q(t\omega)| \leq \sum_{k=0}^{m} |q_k(t\omega)^k| = \sum_{k=0}^{m} |t|^k |q_k\omega^k| < \sum_{k=0}^{m} |q_k\omega^k|.$$

Note that the quantity $\sum_{k=0}^{m} |q_k\omega^k|$ does not depend on t. Therefore, there exists $t \in \mathbb{R}$ with $0 < t < 1$ such that

$$t|\omega^{n+1}Q(t\,\omega)| < |a|.$$

Thus,

$$\begin{aligned} |f(t\omega)| &\leq |a|(1 - t^n) + t^n\left(t|\omega^{n+1}Q(t\omega)|\right) \\ &< |a|(1 - t^n) + t^n|a| \\ &= |a| \\ &= |f(0)|, \end{aligned}$$

which contradicts the fact that $|f(z)|$ attains its minimum value at 0. Therefore, it must be that $f(0) = 0$, and $f(z)$ has a root. ■

Investigation 15

Quotients of Polynomial Rings

Focus Questions

By the end of this investigation, you should be able to give precise and thorough answers to the questions listed below. You may want to keep these questions in mind to focus your thoughts as you complete the investigation.

- How is congruence modulo a polynomial defined? How is this definition similar to congruence modulo a positive integer n in the integers?
- What is a quotient of a polynomial ring, and are such quotients constructed? How are quotient rings of polynomials similar to $\mathbb{Z}_n$?
- When is a quotient of a polynomial ring a field?
- How can quotients of polynomial rings be used to find roots of irreducible polynomials?
- What is an algebraic number, and what is the structure of the set of all algebraic numbers?

Preview Activity 15.1. Recall that we defined the ring $\mathbb{Z}_n$ to be the set of congruence classes of integers modulo the positive integer n, where congruence on $\mathbb{Z}$ is defined by $a \equiv b \pmod{n}$ if and only if $n \mid (b - a)$. We then defined addition and multiplication on the set of congruence classes to make $\mathbb{Z}_n$ into a ring. This construction only depended on having a notion of divisibility, so we can attempt the same construction in other sets in which we have a divides relation—for example, in polynomial rings.

Consider the polynomial $x^2 + 1$ in $\mathbb{Q}[x]$. Define a relation on $\mathbb{Q}[x]$ as follows:

> The polynomial $f(x)$ is congruent to the polynomial $g(x)$ modulo $x^2 + 1$ if $x^2 + 1$ divides $f(x) - g(x)$ in $\mathbb{Q}[x]$.

(a) Find 3 distinct polynomials congruent to $x - 1$ modulo $x^2 + 1$ in $\mathbb{Q}[x]$.

(b) Characterize (that is, describe in a precise way) all polynomials congruent to $x - 1$ modulo $x^2 + 1$ in $\mathbb{Q}[x]$.

Introduction

In previous investigations, we have seen different ways in which polynomial rings over fields are similar to the ring of integers. The similarities have included a notion of divisibility and a Division Algorithm, and a decomposition of elements into products of irreducible elements. We will now take another step and use the divides relation to define congruence in polynomial rings just like we did in $\mathbb{Z}$. We will then study the resulting equivalence class structures, which will allow us to better understand Kronecker's Theorem (see page 113) and roots of polynomials in general.

Congruence Modulo a Polynomial

We have a definition of divides in a polynomial ring over a field, and so we can use this idea (just as we did in $\mathbb{Z}$) to define congruence modulo a polynomial.

Definition 15.2. Let F be a field and $f(x) \in F[x]$ a non-constant polynomial. The polynomial $g(x) \in F[x]$ is **congruent to the polynomial** $h(x) \in F[x]$ **modulo** $f(x)$ if and only if $f(x) \mid (g(x) - h(x))$.

If $g(x)$ is congruent to $h(x)$ modulo $f(x)$, we will denote this relationship by writing

$$g(x) \equiv h(x) \ (\text{mod} f(x)).$$

The polynomial $f(x)$ is called the *modulus*. Note that congruence modulo a non-constant polynomial is a relation on the set $F[x]$, just as congruence modulo n (where n is a positive integer) is a relation on $\mathbb{Z}$. The next activity asks some natural questions pertaining to the congruence relation on $F[x]$.

Activity 15.3. Let F be a field and $f(x)$ a non-constant polynomial in $F[x]$.

(a) Is congruence modulo $f(x)$ a reflexive relation on $F[x]$? Prove your answer.

(b) Is congruence modulo $f(x)$ a symmetric relation on $F[x]$? Prove your answer.

(c) Is congruence modulo $f(x)$ a transitive relation on $F[x]$? Prove your answer.

(d) Why do we assume $f(x)$ is a non-constant polynomial in Definition 15.2?

Congruence Classes of Polynomials

In Activity 15.3, we saw that congruence of polynomials is an equivalence relation on $F[x]$. The equivalence classes corresponding to the congruence relation on $F[x]$ are called congruence classes, defined formally as follows:

Definition 15.4. Let F be a field and $f(x)$ a non-constant polynomial in $F[x]$. The **congruence class of the polynomial** $g(x)$ **modulo** $f(x)$ is the set

$$\{h(x) \in F[x] : h(x) \equiv g(x) \ (\text{mod} f(x))\}.$$

We will use the notation $\overline{g(x)}_{\langle f(x)\rangle}$ to represent the congruence class of the polynomial $g(x)$ modulo the polynomial $f(x)$. Note the change in notation from $\mathbb{Z}_n$, where we used brackets (for instance, $[1]$) to denote congruence classes. The difference in notation is intended to avoid confusion. For example, if we choose $\mathbb{Z}_5$ as our coefficient field, we want to be able to distinguish between the coefficients of our polynomials in $\mathbb{Z}_5[x]$ and the congruence class of a polynomial in $\mathbb{Z}_5[x]$. When the modulus $f(x)$ is clear from the context, we usually drop the subscript and denote the congruence class of $g(x)$ as $\overline{g(x)}$. As with any equivalence relation, Theorem 5.6 (see page 49) tells us that two congruence classes are either equal or disjoint and that the union of all of the congruence classes is the entire polynomial ring $F[x]$. Or, stated differently, each element of $F[x]$ belongs to exactly one congruence class modulo $f(x)$.

Activity 15.5.

(a) Find all of the distinct congruence classes modulo $f(x) = x^2 - [1]$ in $\mathbb{Z}_3[x]$. (Hint: The Division Algorithm guarantees that the remainder when a polynomial is divided by $f(x)$ will either be zero or have a degree less than $\deg(f(x))$.)

(b) Find all of the distinct congruence classes modulo $p(x) = x^3 + [1]$ in $\mathbb{Z}_2[x]$.

(c) Based on your work in parts (a) and (b), make a conjecture to answer the following question: How many distinct congruence classes are there in $\mathbb{Z}_n[x]$ modulo a polynomial of degree m? Explain your conjecture and how you might prove it.

The Set $F[x]/\langle f(x)\rangle$

After we defined the set of congruence classes in $\mathbb{Z}$ modulo a positive integer n, we then defined addition and multiplication on those congruence classes in order to ultimately define the ring $\mathbb{Z}_n$ of integers modulo n. Now that we have a congruence relation on polynomials in a polynomial ring over a field, it seems reasonable to attempt the same constructions here. If F is a field and $f(x) \in F[x]$ is a non-constant polynomial, we let $F[x]/\langle f(x)\rangle$ represent the set of distinct congruence classes modulo $f(x)$. That is,

$$F[x]/\langle f(x)\rangle = \{\overline{g(x)} : g(x) \in F[x]\}.$$

We read $F[x]/\langle f(x)\rangle$ as "F bracket x mod $f(x)$."

Analogous to our work with congruence in $\mathbb{Z}$, we can define addition and multiplication on $F[x]/\langle f(x)\rangle$ using the addition and multiplication from $F[x]$. In particular, if $\overline{g(x)}, \overline{h(x)} \in F[x]/\langle f(x)\rangle$, then

$$\overline{g(x)} + \overline{h(x)} = \overline{g(x) + h(x)} \text{ and } \overline{g(x)}\ \overline{h(x)} = \overline{g(x)h(x)}. \quad (15.1)$$

Note that these definitions involve calculating $g(x)+h(x)$ and $g(x)h(x)$, using the standard addition and multiplication operations in $F[x]$. The next activity provides a few examples and poses an important question.

Activity 15.6. For parts (a) and (b), write the result of each operation in the form $\overline{h(x)}$, where $h(x)$ has as small a degree as possible.

(a) $\overline{x^2 + [2]x + [1]} + \overline{[2]x^3 + x + [3]}$ in $\mathbb{Z}_5[x]/\langle x^2 + x\rangle$

(b) $\left(\overline{x^2 + [2]x + [1]}\right)\left(\overline{[2]x^3 + x + [3]}\right)$ in $\mathbb{Z}_5[x]/\langle x^2 + x\rangle$

(c) In general, is addition as we have defined it in $F[x]/\langle f(x)\rangle$ well-defined? Prove your answer. (Hint: You may want to refer back to Definition 5.16 on page 54 and the subsequent discussion.)

(d) In general, is multiplication as we have defined it in $F[x]/\langle f(x)\rangle$ well-defined? Prove your answer.

(e) Why are these last two questions important? Why do we need to consider whether addition and multiplication in $F[x]/\langle f(x)\rangle$ are well-defined?

We will now look at the structure of $F[x]/\langle f(x)\rangle$ with a specific example.

Activity 15.7. In this activity, we will work in the set $\mathbb{Z}_3[x]/\langle x^2 - x\rangle$.

(a) Why is $\overline{x^2} = \overline{x}$ in $\mathbb{Z}_3[x]/\langle x^2 - x\rangle$?

(b) Show that the element $\overline{x^3 + [3]x^2 + x + [2]}$ can be written as $\overline{h(x)}$ in $\mathbb{Z}_3[x]/\langle x^2 - x\rangle$ for some polynomial $h(x) \in \mathbb{Z}_3[x]$ of degree 1.

(c) Explain why $\mathbb{Z}_3[x]/\langle x^2 - x\rangle$ contains exactly 9 elements. Find all of these elements.

(d) Write the addition table for $\mathbb{Z}_3[x]/\langle x^2 - x\rangle$.

(e) Write the multiplication table for $\mathbb{Z}_3[x]/\langle x^2 - x\rangle$.

(f) What properties of addition and multiplication properties appear to hold in $\mathbb{Z}_3[x]/\langle x^2 - x\rangle$? What algebraic structure do you believe $\mathbb{Z}_3[x]/\langle x^2 - x\rangle$ has?

Now that we have some experience working with elements of $\mathbb{Z}_3[x]/\langle x^2 - x\rangle$, we will move on to the general context and examine the algebraic structure of the set $F[x]/\langle f(x)\rangle$ for an arbitrary field F and an arbitrary non-constant polynomial $f(x) \in F[x]$.

Activity 15.8. Let F be a field and $f(x)$ a non-constant polynomial in $F[x]$.

(a) Why is $F[x]/\langle f(x)\rangle$ closed under the addition and multiplication defined in (15.1)?

(b) Rewrite all of the properties of integer arithmetic listed on page 5 in the context of $F[x]/\langle f(x)\rangle$.

(c) Verifying the properties of integer arithmetic in the context of $F[x]/\langle f(x)\rangle$ is really no different than verifying them in $\mathbb{Z}_n$. For example, to prove that left multiplication distributes over addition in $F[x]/\langle f(x)\rangle$, we can use the distributive laws from $F[x]$. To illustrate, let

$\overline{g(x)}, \overline{h(x)}$, and $\overline{k(x)}$ be elements in $F[x]/\langle f(x)\rangle$. Then

$$\begin{aligned}\overline{g(x)}\left(\overline{h(x)}+\overline{k(x)}\right) &= \overline{g(x)}\left(\overline{h(x)+k(x)}\right)\\ &= \overline{g(x)(h(x)+k(x))}\\ &= \overline{g(x)h(x)+g(x)k(x)}\\ &= \overline{g(x)h(x)}+\overline{g(x)k(x)}\\ &= \overline{g(x)}\,\overline{h(x)}+\overline{g(x)}\,\overline{k(x)}.\end{aligned}$$

Prove the remaining properties of integer arithmetic are valid in the context of $F[x]/\langle f(x)\rangle$. (Note the similarity between Activity 5.12 and your work here.)

The next theorem is the consequence of Activity 15.8.

Theorem 15.9. *Let F be a field and $f(x)$ a non-constant polynomial in $F[x]$. Then $F[x]/\langle f(x)\rangle$ is a commutative ring with identity under the operations defined in (15.1).*

The ring $F[x]/\langle f(x)\rangle$ is called a quotient of $F[x]$ by $\langle f(x)\rangle$, or simply a quotient ring. Quotient rings can be defined in general (that is, for rings other than polynomial rings), and we will do so in Investigation 16.

Special Quotients of Polynomial Rings

Recall that if p is prime, then $\mathbb{Z}_p$ is a field, but if n is composite, then $\mathbb{Z}_n$ is not even an integral domain. So $\mathbb{Z}_n$ has a nicer structure for some values of n than for others. Given the similarities between $\mathbb{Z}_n$ and $F[x]/\langle f(x)\rangle$, it is natural to ask if the structure of $F[x]/\langle f(x)\rangle$ depends on certain properties of $f(x)$.

Activity 15.10. In Activity 15.7 we investigated the quotient ring $\mathbb{Z}_3[x]/\langle x^2 - x\rangle$. In this activity, we will do the same with $\mathbb{Z}_3[x]/\langle x^2 - [1]\rangle$ and $\mathbb{Z}_3[x]/\langle x^2 + x + [2]\rangle$. We will then compare the two rings.

(a) Find all the elements in $\mathbb{Z}_3[x]/\langle x^2 - [1]\rangle$.

(b) Find all the elements in $\mathbb{Z}_3[x]/\langle x^2 + x + [2]\rangle$.

(c) Explain why the addition tables for the rings $\mathbb{Z}_3[x]/\langle x^2 - x\rangle$, $\mathbb{Z}_3[x]/\langle x^2 - [1]\rangle$, and $\mathbb{Z}_3[x]/\langle x^2 + x + [2]\rangle$ have the same structure.

(d) The multiplication tables for the rings $\mathbb{Z}_3[x]/\langle x^2 - [1]\rangle$ and $\mathbb{Z}_3[x]/\langle x^2 + x + [2]\rangle$ are given in Tables 15.1 and 15.2. Explain in detail how the entry for $\left(\overline{[2]x+[1]}\right)\left(\overline{[2]x+[2]}\right)$ is obtained in each.

(e) There is a significant difference between the rings $\mathbb{Z}_3[x]/\langle x^2 - [1]\rangle$ and $\mathbb{Z}_3[x]/\langle x^2 + x + [2]\rangle$. What is the difference, and why do you think it happens?

The result of Activity 15.10 shows that $F[x]/\langle f(x)\rangle$ has a nicer structure for some polynomials $f(x)$ than for others. You might notice the similarity between this observation and Activity 7.28 on page 87, where we saw that $\mathbb{Z}_p$ is a field if and only if p is prime. The next theorem is an analogous result for polynomials over a field.

Theorem 15.11. *Let F be a field and $f(x)$ a non-constant polynomial in $F[x]$. The ring $F[x]/\langle f(x)\rangle$ is a field if and only if $f(x)$ is irreducible in $F[x]$.*

Proof. Let F be a field and $f(x)$ a non-constant polynomial in $F[x]$. We will prove the forward implication first. Suppose $F[x]/\langle f(x)\rangle$ is a field. To show that $f(x)$ is irreducible in $F[x]$, we will proceed by contradiction and assume $f(x)$ is reducible in $F[x]$. So we can find polynomials $g(x), h(x)$ with $1 \leq \deg(g(x), \deg(h(x)) < \deg(f(x))$, so that $f(x) = g(x)h(x)$. Since $1 \leq \deg(g(x)) < \deg(f(x))$ and $1 < \deg(h(x)) < \deg(f(x))$, we know $\overline{g(x)}$ and $\overline{h(x)}$ are nonzero in $F[x]/\langle f(x)\rangle$. But $\overline{g(x)}\ \overline{h(x)} = \overline{g(x)h(x)} = \overline{f(x)} = \overline{0}$, and so $\overline{g(x)}$ is a zero divisor in $F[x]/\langle f(x)\rangle$. However, this cannot happen in a field, so we conclude $f(x)$ is irreducible in $F[x]$.

To prove the reverse implication, we assume $f(x)$ is irreducible in $F[x]$. Let $\overline{g(x)} \in F[x]$ with $\overline{g(x)} \neq \overline{0}$. So $f(x)$ does not divide $g(x)$. Since $f(x)$ is irreducible in $F[x]$, we know $\gcd(f(x), g(x)) = 1$. Thus, there exist polynomials $s(x), t(x)$ in $F[x]$ so that

$$1 = s(x)f(x) + t(x)g(x).$$

Then

$$\overline{1} = \overline{s(x)f(x) + t(x)g(x)} = \overline{s(x)}\ \overline{f(x)} + \overline{t(x)}\ \overline{g(x)} = \overline{t(x)}\ \overline{g(x)}.$$

Therefore, $\overline{g(x)}$ is a unit in $F[x]/\langle f(x)\rangle$, and $F[x]/\langle f(x)\rangle$ is a field. ■

Theorem 15.11 gives us many ways to construct new types of fields. Also, note that any field of the form $\mathbb{Z}_p[x]/\langle q(x)\rangle$, where $q(x)$ is an irreducible polynomial of degree n in $\mathbb{Z}_p[x]$, has p^n elements. It turns out that if F is a finite field, then $|F| = p^n$ for some prime p and positive integer n, although we won't prove that fact here.

There is a critically important application of quotient rings to the problem of finding roots of polynomials. To illustrate, let's look at an example. The polynomial $f(x) = x^2 + 1$ is irreducible in $\mathbb{Q}[x]$. Thus, the quotient $E = \mathbb{Q}[x]/\langle x^2 + 1\rangle$ is a field. We can consider $\mathbb{Q}$ as a subset of E in a natural way, so $f(x) = x^2 + 1$ is also a polynomial in $E[x]$. In E, we have $f(\overline{x}) = \overline{x}^2 + 1 = \overline{x^2 + 1}$. But in E, we know that $\overline{x^2 + 1} = 0$, and so $f(\overline{x}) = 0$, and $\overline{x}$ is a root of $f(x)$ in E. As a matter of convention, we typically refer to $\overline{x}$ as i. Note that since $\overline{x}^2 + 1 = 0$, the familiar property that $i^2 = -1$ is satisfied.

In general, suppose we have a polynomial $f(x)$ with coefficients in a field F. The argument in the previous paragraph shows how we can always find a field in which $f(x)$ has a root. The process works by first factoring $f(x)$ into a product of irreducible polynomials. Let $p(x)$ be one of the irreducible factors of $f(x)$. Theorem 15.11 shows that $E = F[x]/\langle p(x)\rangle$ is a field. Consider the element $\overline{x} \in E$. In E we have

$$p(\overline{x}) = \overline{p(x)} = \overline{0}.$$

So $\overline{x}$ is a root of $p(x)$ in E and, consequently, a root of $f(x)$ in E. This proves the following important theorem referred to in Investigation 9.

Theorem 15.12 (Kronecker's Theorem). *Let F be a field and $p(x)$ an irreducible polynomial in $F[x]$. There exists a field E containing F such that $p(x)$ has a root in E.*

We can continue to apply this idea repeatedly to ultimately factor any polynomial into a product of linear factors in some large field.

Activity 15.13. Let $p(x) = x^2 + x + [2]$ in $\mathbb{Z}_3[x]$

(a) Show that $p(x)$ is an irreducible polynomial in $\mathbb{Z}_3[x]$.

(b) Let $E = \mathbb{Z}_3[x]/\langle x^2 + x + [2]\rangle$. Show explicitly that $\overline{x}^2 + \overline{x} + \overline{[2]} = \overline{[0]}$ in E. Then use this observation to determine exactly which element of E is guaranteed to be a root of $p(x)$.

Algebraic Numbers

Kronecker's Theorem shows that every polynomial with coefficients in a field has a root in some larger field. If we restrict ourselves to polynomials with rational coefficients, then the Fundamental Theorem of Algebra shows that all such polynomials can be completely factored into linear factors with coefficients from the field $\mathbb{C}$ of complex numbers. We will now use this idea to define algebraic numbers and find another field that is an extension of $\mathbb{Q}$.

First, we need to review a relationship between polynomials in $\mathbb{Q}[x]$ and polynomials in $\mathbb{Z}[x]$. (This relationship is closely related to Gauss' Lemma (see page 185), and you may notice similarities between the following argument and parts of the proof of Gauss' Lemma.) Suppose

$$p(x) = \left(\frac{r_n}{s_n}\right) x^n + \left(\frac{r_{n-1}}{s_{n-1}}\right) x^{n-1} + \cdots + \left(\frac{r_1}{s_1}\right) x + \frac{r_0}{s_0}$$

is a polynomial in $\mathbb{Q}[x]$. Then, with a little bit of algebra, we can rewrite $p(x)$ as a rational number times a polynomial $q(x) \in \mathbb{Z}[x]$. In particular, let $S = s_n s_{n-1} \cdots s_1 s_0$ and $S_i = \frac{S}{s_i} = s_n s_{n-1} \cdots s_{i+1} s_{i-1} s_{i-2} \cdots s_1 s_0$ for each i between 0 and n, inclusive. Then

$$p(x) = \left(\frac{1}{S}\right) \left(S_n r_n x^n + S_{n-1} r_{n-1} x^{n-1} + \cdots + S_1 r_1 x + S_0 r_0\right),$$

with $q(x) = S_n r_n x^n + S_{n-1} r_{n-1} x^{n-1} + \cdots + S_1 r_1 x + S_0 r_0$. Now if $p(r) = 0$—that is, if r is a root of $p(x)$—then $q(r) = 0$ as well. Also, if $q(r) = 0$, then $p(r) = 0$. So roots of polynomials with rational coefficients correspond to roots of polynomials with integer coefficients. In other words, when considering roots of polynomials in $\mathbb{Q}[x]$, it suffices to consider roots of polynomials in $\mathbb{Z}[x]$. The advantage of this perspective is that polynomials with integer coefficients are easier to work with than polynomials with rational coefficients.

As we saw in Investigation 9, we can attach roots of polynomials to $\mathbb{Q}$ to build extension fields of $\mathbb{Q}$. We can extend this idea by considering all of the roots of polynomials with integer coefficients, which leads us to the definition of an algebraic number.

Definition 15.14. A real number a is an **algebraic** number if $p(a) = 0$ for some polynomial $p(x) \in \mathbb{Z}[x]$.

A real number that is not algebraic is said to be transcendental. Examples of algebraic numbers include any rational number $\frac{r}{s}$ (using $p(x) = sx - r$), $\sqrt{7}$ (using $p(x) = x^2 - 7$), and $\sqrt[3]{2}$ (using $p(x) = x^3 - 2$). Examples of transcendental numbers include π, e, and $2^{\sqrt{2}}$.

Let $\mathcal{A}$ denote the set of algebraic numbers. Once we have defined a set of numbers like this, it is natural to ask what kind of structure this set possesses. For example, is the sum of two algebraic numbers algebraic? What about the product of two algebraic numbers?

One way to answer these questions about combinations of algebraic numbers is with linear algebra. Recall that a set of vectors spans a space if every vector in the space can be written as a linear combination of vectors in the set. Also, the dimension of a space is the number of vectors in a minimal spanning set. With that in mind, consider the question of whether the number $z = \sqrt{2} + \sqrt{7}$ is algebraic. Since $\sqrt{2}$ is a zero of the polynomial $x^2 - 2$ and $\sqrt{7}$ is a zero of $x^2 - 7$, we see that both $\sqrt{2}$ and $\sqrt{7}$ are algebraic. To determine if z is algebraic, we need to look for a polynomial with z as a zero. The next activity demonstrates how we might do so.

Activity 15.15.

(a) Explain why $z = \sqrt{2} + \sqrt{7}$ cannot be the root of any linear polynomial with integer coefficients.

(b) Calculate z^2, and simplify your answer as much as possible.

(c) Use your answer to part (b) to find an integer k such that $(z^2 - k)^2 \in \mathbb{Z}$.

(d) Use your answer to part (c) to show that z is a root of a degree 4 polynomial with integer coefficients.

The argument from Activity 15.15 will be difficult to repeat for every sum of algebraic numbers, but the general idea is not. Since z^4 could be written in terms of linear combinations of lower powers of z, we were able to combine those powers to make a polynomial that had z as a zero. In other words, since z^4 can be written as a linear combination of $1, z, z^2, z^3$, the elements $1, z, z^2, z^3$ form a spanning set for a four-dimensional vector space V over $\mathbb{Q}$. We will exploit this idea in the following theorem.

Theorem 15.16. *Let a and b be algebraic numbers. Then $a + b$, ab, and $-a$ are also algebraic. In addition, if $a \neq 0$, then $\frac{1}{a}$ is algebraic.*

Proof. Let a and b be algebraic numbers. If either a or b is zero, then it is clear that $a + b$ and ab are algebraic. So suppose that a and b are both nonzero. Then there are polynomials

$$p(x) = a_k x^k + a_{k-1} x^{k-1} + \cdots a_1 x + a_0$$

and

$$q(x) = b_m x^m + b_{m-1} x^{m-1} + \cdots b_1 x + b_0$$

in $\mathbb{Z}[x]$ so that $a_k, b_m \neq 0$ and $p(a) = q(b) = 0$. So

$$a^k = \frac{-1}{a_k}\left(a_{k-1}x^{k-1} + \cdots a_1 x + a_0\right) \quad \text{and} \tag{15.2}$$

$$b^m = \frac{-1}{b_m}\left(b_{m-1}x^{m-1} + \cdots b_1 x + b_0\right). \tag{15.3}$$

First we will show that $-a \in \mathcal{A}$. Notice that

$$\begin{aligned} a_0 - a_1(-a) &+ a_2(a^2) - a_3(-a)^3 + \cdots + (-1)^k a_k(-a)^k \\ &= a_0 + a_1(a) + a_2(a^2) + \cdots + a_k(a^k) \\ &= p(a) \\ &= 0, \end{aligned}$$

so $-a$ is a root of the polynomial $a_0 - a_1 x + a_2 x^2 - a_3 x^3 + \cdots + (-1)^k a_k x^k$. Thus, $-a$ is an algebraic number and the set A contains the additive inverse of each of its elements.

Now we will show that $a^{-1} \in \mathcal{A}$. Note that

$$\begin{aligned} a_0\left(\frac{1}{a}\right)^k &+ a_1\left(\frac{1}{a}\right)^{k-1} + a_2\left(\frac{1}{a}\right)^{k-2} + \cdots + a_k \\ &= \left(\frac{1}{a}\right)^k \left(a_0 + a_1(a) + a_2(a^2) + a_3(a^3) + \cdots + a_k(a^k)\right) \\ &= \left(\frac{1}{a}\right)^k p(a) = 0, \end{aligned}$$

so a^{-1} is a root of the polynomial $a_0x^k + a_1x^{k-1} + a_2x^{k-2} + \cdots + a_{k-1}x + a_k$. Thus, a^{-1} is an algebraic number and $\mathcal{A}$ contains the inverse of each of its nonzero elements.

Next we will show that $a + b$ is algebraic. Let $n = \max\{k, m\}$.

Let $z = a + b$. By the Binomial Theorem (see Exercise 9 on page 103 of Investigation 8), we know that

$$\begin{aligned} z^n &= (a+b)^n \\ &= a^n + na^{n-1}b + \binom{n}{2}a^{n-2}b^2 + \cdots + \binom{n}{n-2}a^2b^{n-2} + nab^{n-1} + b^n \\ &= \sum_{i=0}^{n}\binom{n}{i}a^{n-i}b^i. \end{aligned}$$

By (15.2), all the powers a^r for $r \geq k$ can be written as linear combinations of the powers $1, a, a^2, a^3, \ldots, a^{k-1}$. Similarly, (15.3) shows that all the powers b^s for $s \geq m$ can be written as linear combinations of the powers $1, b, b^2, b^3, \ldots, b^{m-1}$. Therefore, each monomial of the form a^ib^j in the expansion of z^n can be written as a product of linear combinations of $1, a, a^2, a^3, \ldots, a^{k-1}$ and $1, b, b^2, b^3, \ldots, b^{m-1}$. We can conclude that the elements of the form a^ib^j for $0 \leq i \leq k-1$ and $0 \leq j \leq m-1$ form a spanning set of a vector space V over $\mathbb{Q}$ of which z is a member. So the elements $1, z, z^2, z^3, \ldots, z^{km}$ have to be linearly dependent and satisfy some polynomial relationship over $\mathbb{Z}$. This shows that $a + b$ is algebraic.

The proof that ab is algebraic is similar and left as an exercise. (See Exercise 6.) ■

Activity 15.17. What does Theorem 15.16 tell us about the structure of the set $\mathcal{A}$ of algebraic numbers? Be specific and explain why.

In a sense, the proof given for Theorem 15.16 that $a + b$ is algebraic is constructive. Let's illustrate with an example. Let $a = \sqrt{2}$, $b = \sqrt[3]{3}$, and $z = a + b$. We know a is a zero of $x^2 - 2$ and b is a zero of $x^3 - 3$. In this case, $k = 2$ and $m = 3$, so z must satisfy some polynomial equation of degree no more than $km = 6$. Now we know that

$$\begin{aligned} z^0 &= 1 \\ z^1 &= a + b \\ z^2 &= (a+b)^2 = 2 + b^2 + 2ab \\ z^3 &= (a+b)^3 = 3 + 2a + 6b + 3ab^2 \\ z^4 &= (a+b)^4 = 4 + 12a + 3b + 12b^2 + 8ab \\ z^5 &= (a+b)^5 = 60 + 4a + 20b + 3b^2 + 15ab + 20ab^2 \\ z^6 &= (a+b)^6 = 17 + 120a + 90b + 60b^2 + 24ab + 18ab^2. \end{aligned}$$

If z satisfies a polynomial equation of the form

$$c_6x^6 + c_5x^5 + c_4x^4 + c_3x^3 + c_2x^2 + c_1x + c_0 = 0,$$

then we have

$$\begin{aligned}
0 = {} & c_6\left(17 + 24ab + 60b^2 + 120a + 90b + 18ab^2\right) \\
& + c_5\left(60 + 4a + 20b + 20ab^2 + 15ab + 3b^2\right) \\
& + c_4\left(4 + 8ab + 12b^2 + 12a + 3b\right) \\
& + c_3\left(3 + 2a + 6b + 3ab^2\right) \\
& + c_2\left(2 + 2ab + b^2\right) \\
& + c_1(a + b) \\
& + c_0.
\end{aligned}$$

Equating like terms gives us the system of equations

$$\begin{aligned}
17c_6 + 60c_5 + 4c_4 + 3c_3 + 2c_2 + c_0 &= 0 \\
120c_6 + 4c_5 + 12c_4 + 2c_3 + c_1 &= 0 \\
90c_6 + 20c_5 + 3c_4 + 6c_3 + c_1 &= 0 \\
60c_6 + 3c_5 + 12c_4 + c_2 &= 0 \\
24c_6 + 15c_5 + 8c_4 + 2c_2 &= 0 \\
18c_6 + 20c_5 + 3c_3 &= 0.
\end{aligned}$$

We can solve this system using row reduction to obtain the solution $c_0 = 1, c_1 = -36, c_2 = 12, c_3 = -6, c_4 = -6, c_5 = 0, c_6 = 1$. So $\sqrt{2} + \sqrt[3]{3}$ is a zero of the polynomial $p(x) = x^6 - 6x^4 - 6x^3 + 12x^2 - 36x + 1$. Check it if you dare!

Concluding Activities

Activity 15.18. Let F be a field and $f(x)$ a non-constant polynomial in $F[x]$.

(a) Show that the congruence class $\overline{f(x)}_{\langle f(x)\rangle}$ is a subring of $F[x]$.

(b) Show that if $f_0(x) \in \overline{f(x)}_{\langle f(x)\rangle}$ and $h(x) \in F[x]$, then both $h(x)f_0(x)$ and $f_0(x)h(x)$ are in $\overline{f(x)}_{\langle f(x)\rangle}$. (Any subset of a ring that satisfies the properties in this activity is called an *ideal*.)

Activity 15.19. Find a field E over which the polynomial $f(x) = x^3 + x + [1] \in \mathbb{Z}_3[x]$ factors completely. Give the complete factorization of $f(x)$ into a product of linear factors in $E[x]$.

Activity 15.20. Use the methods of this investigation to find a polynomial in $\mathbb{Z}[x]$ that has $\sqrt{3} + \sqrt{6}$ as a root.

Activity 15.21. Write a short summary that describes the important concepts, and the relationships between these concepts, that were introduced in this investigation. Explain how the material in this investigation is connected to the material in Investigations 5 and 9.

Exercises

(1) In this problem, we will work within the field $\mathbb{Z}_2$.

(a) Find an irreducible polynomial in $\mathbb{Z}_2[x]$ of degree 3.

(b) Use the polynomial from part (a) to construct a field E with 8 elements. List the elements in E and then create the operations tables for E.

(c) Find a field with 16 elements and list its elements.

(2) Find a field with 25 elements and list its elements.

(3) Let F be a field and $p(x)$ an irreducible polynomial in $F[x]$. In this investigation we showed that $E = F[x]/\langle p(x)\rangle$ is a field, and we implied that F is a subfield of E. Now we will examine what we mean by that statement.

(a) There is a natural mapping ι from F to E. Identify this mapping (ι is called the *inclusion* mapping). Show that ι preserves the structure of F. Is ι an isomorphism? Explain.

(b) Explain how E contains an isomorphic copy of F. (It is in this sense that we say F is a subfield of E. This subfield of E that is isomorphic to F is called an *embedding* of F in E.)

(4) In a previous investigation, we saw that the field $\mathbb{Q}(\sqrt{2}) = \{a + b\sqrt{2} : a, b \in \mathbb{Q}\}$ is an extension of $\mathbb{Q}$ that contains a root of the irreducible polynomial $x^2 - 2 \in \mathbb{Q}[x]$. In this investigation, we saw that $\mathbb{Q}[x]/\langle x^2 - 2\rangle$ is also an extension of $\mathbb{Q}$ that contains a root of the irreducible polynomial $x^2 - 2$. In this problem, we will explore the connection between the fields $\mathbb{Q}[x]/\langle x^2 - 2\rangle$ and $\mathbb{Q}(\sqrt{2})$.

(a) Since $\mathbb{Q}[x]/\langle x^2 - 2\rangle$ contains a root of $x^2 - 2$, there is an element in $\mathbb{Q}[x]/\langle x^2 - 2\rangle$ that behaves like $\sqrt{2}$. What element is this? Why?

(b) Explain why every element in $\mathbb{Q}[x]/\langle x^2 - 2\rangle$ has the form $\overline{a + bx}$ for some $a, b \in \mathbb{Q}$.

(c) Use the previous parts of this exercise to define a function φ from $\mathbb{Q}[x]/\langle x^2 - 2\rangle$ to $\mathbb{Q}(\sqrt{2})$. Show that φ is an isomorphism. What is the relationship between $\mathbb{Q}[x]/\langle x^2 - 2\rangle$ and $\mathbb{Q}(\sqrt{2})$?

(5) Use the methods of this investigation to find a polynomial in $\mathbb{Z}[x]$ that has $\sqrt[3]{2} + \sqrt{6}$ as a root.

$\star$ (6) **Closure of algebraic numbers under multiplication.** Prove that if a and b are algebraic numbers, then ab is an algebraic number.

(7) **Constructible numbers.** We are familiar with the classification of real numbers as rational and irrational numbers, and in this investigation we learned about algebraic numbers. There are other important classifications that are also of interest. Much of the mathematics of ancient times, even pre-dating the Greek mathematicians of Euclid's age, was very geometric in nature. The number theory of the time dealt with geometric constructions of numbers. The constructions were performed with straightedge (unmarked) and compass. In fact, the compass of the time was a collapsible one; it would not hold its position when lifted from the page.

With the tools of an unmarked ruler and collapsible compass, every integer can be constructed. It is also possible to build a rich variety of construction techniques, which allow one to see that many other objects are constructible. A sample of such techniques includes:

- Given points P and Q, we can construct a ray emanating from P that passes through Q.
- Given points P and Q, we can construct a circle with center P and radius PQ.
- Given a point P and a constructible length r, we can construct a circle with center P and radius r.
- Given a point P and a line l, we can construct a line through P perpendicular to l.
- Given a point P and a line l, we can construct a line through P parallel to l.

We now call numbers that can be constructed with these tools *constructible numbers*. It is not difficult to show that every rational number is constructible. In fact, as we will see in this exercise, the quotient of any two constructible numbers is a constructible number.

Suppose a and b are constructible numbers. We can then construct a from an origin point O. Let A be the other endpoint of this constructed segment. We can extend the constructible segment to a line by constructing the line through A parallel to the segment. At A, construct a circle with radius b. Let B be the point of intersection of this circle with the ray $\overrightarrow{OA}$. Then $\overline{OB}$ has length $a+b$, and $a+b$ is constructible. (See Figure 15.1.) So the sum of two constructible numbers is constructible.

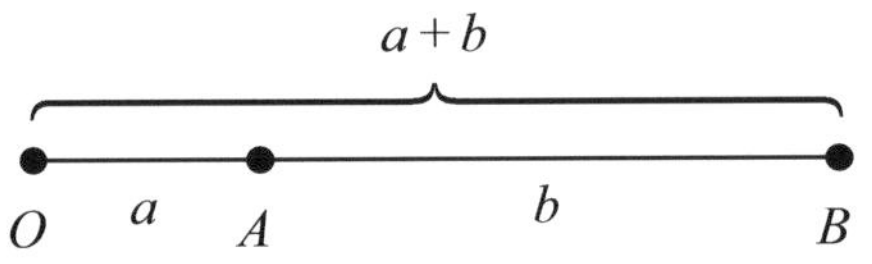

Figure 15.1
Constructing the sum of constructible a and b.

(a) Figure 15.2 illustrates how the difference of two constructible numbers is constructed. Explain how this construction works.

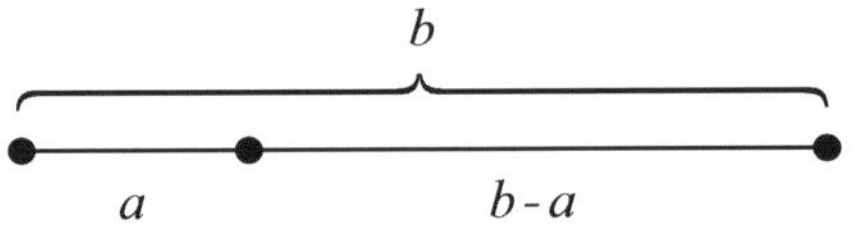

Figure 15.2
Constructing the difference of constructible a and b.

(b) Figure 15.3 illustrates how the product of two constructible numbers is constructed. Explain how this construction works.

(c) Figure 15.4 illustrates how the quotient of two constructible numbers is constructed. Explain how this construction works.

(d) Let S be the set of constructible numbers. Explain why S is a subfield of $\mathbb{R}$ that contains $\mathbb{Q}$.

(e) The Greeks were very interested in geometric constructions. There are three problems

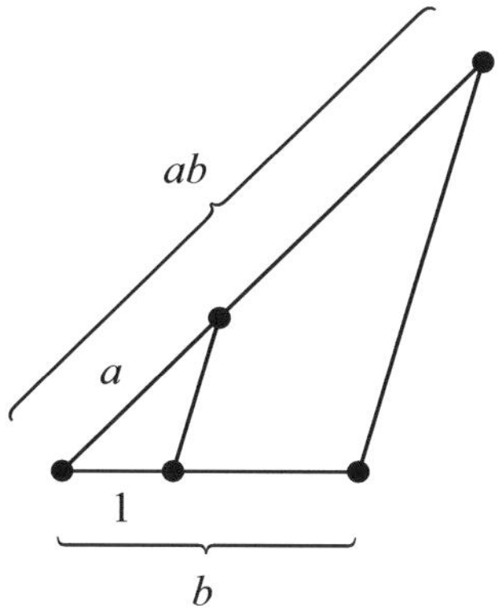

Figure 15.3
Constructing the product of constructible a and b.

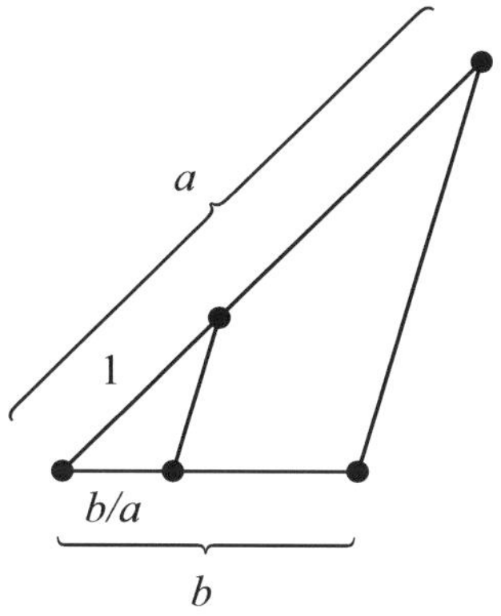

Figure 15.4
Constructing the quotient of constructible a and b.

from this time that were extremely influential in the subsequent development of geometry and algebra. These problems are labeled as: squaring the circle, trisecting an angle, and doubling the cube. To square the circle means to construct a square of the same area as that of a given circle. To trisect an angle means to subdivide a given angle into three congruent angles. To double the cube means to construct a cube whose volume is twice that of a given cube. All of these constructions are to be performed using only a straightedge and compass. Whether these constructions could actually be performed with the given tools went unanswered for a very long time. The answers to these problem (squaring the circle, trisecting an angle, and doubling the cube) depend on the constructibility of the numbers π, $\cos(20°)$, and $\sqrt[3]{2}$. Research this problem to find sources that prove that π, $\cos(20°)$, and $\sqrt[3]{2}$ are not constructible. Explain then why it is impossible to square the circle, trisect an angle, or double the cube using only an unmarked straightedge and collapsible compass.

Connections

Given a polynomial in a polynomial ring $F[x]$, we can define congruence modulo a polynomial and form the quotient ring $F[x]/\langle f(x)\rangle$. Quotient structures are useful in mathematics in that they often have a simpler structure than the original set and can therefore provide important information about the original set. We are familiar with congruence and quotient structures in the integers, namely $\mathbb{Z}_n$ (from Investigation 5). If you studied group theory before ring theory, you should notice connections between the topics in this investigation and those in Investigation 27. In particular, the set $F[x]/\langle f(x)\rangle$ represents the set of distinct congruence classes modulo $f(x)$ in the same way that the set G/N represents the set of left cosets of N in G. Although G/N is a group only if N is normal in G, we don't have that problem in $F[x]$. Since $F[x]$ is an Abelian group under addition, any subgroup is normal.

$\times$	$\overline{[0]}$	$\overline{[1]}$	$\overline{[2]}$	$\overline{x}$	$\overline{x+[1]}$	$\overline{x+[2]}$	$\overline{[2]x}$	$\overline{[2]x+[1]}$	$\overline{[2]x+[2]}$
$\overline{[0]}$	$\overline{[0]}$	$\overline{[0]}$	$\overline{[0]}$	$\overline{[0]}$	$\overline{[0]}$	$\overline{[0]}$	$\overline{[0]}$	$\overline{[0]}$	$\overline{[0]}$
$\overline{[1]}$	$\overline{[0]}$	$\overline{[1]}$	$\overline{[2]}$	$\overline{x}$	$\overline{x+[1]}$	$\overline{x+[2]}$	$\overline{[2]x}$	$\overline{[2]x+[1]}$	$\overline{[2]x+[2]}$
$\overline{[2]}$	$\overline{[0]}$	$\overline{[2]}$	$\overline{[1]}$	$\overline{[2]x}$	$\overline{[2]x+[2]}$	$\overline{[2]x+[1]}$	$\overline{x}$	$\overline{x+[2]}$	$\overline{x+[1]}$
$\overline{x}$	$\overline{[0]}$	$\overline{x}$	$\overline{[2]x}$	$\overline{[1]}$	$\overline{x+[1]}$	$\overline{[2]x+[1]}$	$\overline{[2]}$	$\overline{x+[2]}$	$\overline{[2]x+[2]}$
$\overline{x+[1]}$	$\overline{[0]}$	$\overline{x+[1]}$	$\overline{[2]x+[2]}$	$\overline{x+[1]}$	$\overline{[2]x+[2]}$	$\overline{[0]}$	$\overline{[2]x+[2]}$	$\overline{[0]}$	$\overline{x+[1]}$
$\overline{x+[2]}$	$\overline{[0]}$	$\overline{x+[2]}$	$\overline{[2]x+[1]}$	$\overline{[2]x+[1]}$	$\overline{[0]}$	$\overline{x+[2]}$	$\overline{x+[2]}$	$\overline{[2]x+[1]}$	$\overline{[0]}$
$\overline{[2]x}$	$\overline{[0]}$	$\overline{[2]x}$	$\overline{x}$	$\overline{[2]}$	$\overline{[2]x+[2]}$	$\overline{x+[2]}$	$\overline{[1]}$	$\overline{[2]x+[1]}$	$\overline{x+[1]}$
$\overline{[2]x+[1]}$	$\overline{[0]}$	$\overline{[2]x+[1]}$	$\overline{x+[2]}$	$\overline{x+[2]}$	$\overline{[0]}$	$\overline{[2]x+[1]}$	$\overline{[2]x+[1]}$	$\overline{x+[2]}$	$\overline{[0]}$
$\overline{[2]x+[2]}$	$\overline{[0]}$	$\overline{[2]x+[2]}$	$\overline{x+[1]}$	$\overline{[2]x+[2]}$	$\overline{x+[1]}$	$\overline{[0]}$	$\overline{x+[1]}$	$\overline{[0]}$	$\overline{[2]x+[2]}$

table for $\mathbb{Z}_3[x]/\langle x^2 - [1]\rangle$.

$\times$	$\overline{[0]}$	$\overline{[1]}$	$\overline{[2]}$	$\overline{x}$	$\overline{x+[1]}$	$\overline{x+[2]}$	$\overline{[2]x}$	$\overline{[2]x+[1]}$	$\overline{[2]x+[2}$
$\overline{[0]}$	$\overline{[0]}$	$\overline{[0]}$	$\overline{[0]}$	$\overline{[0]}$	$\overline{[0]}$	$\overline{[0]}$	$\overline{[0]}$	$\overline{[0]}$	$\overline{[0]}$
$\overline{[1]}$	$\overline{[0]}$	$\overline{[1]}$	$\overline{[2]}$	$\overline{x}$	$\overline{x+[1]}$	$\overline{x+[2]}$	$\overline{[2]x}$	$\overline{[2]x+[1]}$	$\overline{[2]x+[2}$
$\overline{[2]}$	$\overline{[0]}$	$\overline{[2]}$	$\overline{[1]}$	$\overline{[2]x}$	$\overline{[2]x+[2]}$	$\overline{[2]x+[1]}$	$\overline{x}$	$\overline{x+[2]}$	$\overline{x+[1]}$
$\overline{x}$	$\overline{[0]}$	$\overline{x}$	$\overline{[2]x}$	$\overline{[2]x+[1]}$	$\overline{[1]}$	$\overline{x+[1]}$	$\overline{x+[2]}$	$\overline{[2]x+[2]}$	$\overline{[2]}$
$\overline{x+[1]}$	$\overline{[0]}$	$\overline{x+[1]}$	$\overline{[2]x+[2]}$	$\overline{[1]}$	$\overline{x+[2]}$	$\overline{[2]x}$	$\overline{[2]}$	$\overline{x}$	$\overline{[2]x+[}$
$\overline{x+[2]}$	$\overline{[0]}$	$\overline{x+[2]}$	$\overline{[2]x+[1]}$	$\overline{x+[1]}$	$\overline{[2]x}$	$\overline{[2]}$	$\overline{[2]x+[2]}$	$\overline{[1]}$	$\overline{x}$
$\overline{[2]x}$	$\overline{[0]}$	$\overline{[2]x}$	$\overline{x}$	$\overline{x+[2]}$	$\overline{[2]}$	$\overline{[2]x+[2]}$	$\overline{[2]x+[1]}$	$\overline{x+[1]}$	$\overline{[1]}$
$\overline{[2]x+[1]}$	$\overline{[0]}$	$\overline{[2]x+[1]}$	$\overline{x+[2]}$	$\overline{[2]x+[2]}$	$\overline{x}$	$\overline{[1]}$	$\overline{x+[1]}$	$\overline{[2]}$	$\overline{[2]x}$
$\overline{[2]x+[2]}$	$\overline{[0]}$	$\overline{[2]x+[2]}$	$\overline{x+[1]}$	$\overline{[2]}$	$\overline{[2]x+[1]}$	$\overline{x}$	$\overline{[1]}$	$\overline{[2]x}$	$\overline{x+[2]}$

ion table for $\mathbb{Z}_3[x]/\langle x^2+x+[2]\rangle$.

Part V

More Ring Theory

Investigation 16

Ideals and Homomorphisms

Focus Questions

By the end of this investigation, you should be able to give precise and thorough answers to the questions listed below. You may want to keep these questions in mind to focus your thoughts as you complete the investigation.

- What is an ideal of a ring?
- What is a principal ideal domain? What is a Euclidean domain? What is the relationship between principal ideal domains and Euclidean domains?
- How is congruence modulo an ideal defined? How can congruence modulo an ideal be used to construct quotient rings?
- What are prime and maximal ideals, and how are they related to certain properties of quotient rings?
- What is a homomorphism of rings? What are the similarities and differences between homomorphisms and isomorphisms?
- What are the kernel and image of a homomorphism? Why are the kernel and image important?
- What important relationship is established by the First Isomorphism Theorem for rings?

In Investigation 15, we considered quotients of polynomial rings. In this investigation, we will expand on these ideas to construct quotients of other rings. To begin, let's consider an example.

Preview Activity 16.1. Let I be the subset of $\mathbb{Z}_{12}$ defined by $I = \{[0], [3], [6], [9]\} = \{k[3] : k \in \mathbb{Z}\}$.

(a) Show that I is a subring of $\mathbb{Z}_{12}$.

(b) Show that if $[a] \in \mathbb{Z}_{12}$ and $[b] \in I$, then $[a][b] \in I$ and $[b][a] \in I$.

(c) Define a relation $\equiv_I$ on $\mathbb{Z}_{12}$ by

$$[a] \equiv_I [b] \text{ if and only if } [b] - [a] \in I.$$

(i) Describe all the elements in $\mathbb{Z}_{12}$ that are related to $[0]$.

(ii) Describe all the elements in $\mathbb{Z}_{12}$ that are related to $[1]$.

(iii) Describe all the elements in $\mathbb{Z}_{12}$ that are related to $[2]$.

(iv) Describe all the elements in $\mathbb{Z}_{12}$ that are related to $[7]$.

(v) It can be shown that $\equiv_I$ is an equivalence relation. Find all of the disjoint equivalence classes corresponding to $\equiv_I$.

Introduction

In Investigation 15, we saw that for a polynomial ring $F[x]$ over a field F, there is a quotient ring $F[x]/\langle f(x)\rangle$ for any non-constant polynomial $f(x) \in F[x]$. We showed in Activity 15.18 (see page 208) that the congruence class $\overline{f(x)}_{\langle f(x)\rangle}$ is a subring of $F[x]$ with the property that both $h(x)f_0(x)$ and $f_0(x)h(x)$ are in $\overline{f(x)}_{\langle f(x)\rangle}$ for any $f_0(x) \in \overline{f(x)}_{\langle f(x)\rangle}$ and any $h(x) \in F[x]$. Any subset of a ring that satisfies these properties is called an *ideal*. For any ideal of R, we can construct a quotient structure analogous to $F[x]/\langle f(x)\rangle$. In this investigation, we will study ideals and the corresponding quotient structures. We will also introduce a special type of function called a homomorphism of rings, and we will study the relationship between homomorphisms and ideals.

Ideals

In an arbitrary ring R, the subsets that allow us to create a quotient structure are called *ideals*.

Definition 16.2. An **ideal** I in a ring R is a subring of R such that $rx \in I$ and $xr \in I$ for all $r \in R$ and $x \in I$.

As an example, we saw in Preview Activity 16.1 that $I = \{[0],[3],[6],[9]\}$ is an ideal of $\mathbb{Z}_{12}$. Notice that an ideal I is not only closed under multiplication by elements of I, but also is closed under multiplication by elements of the larger ring R. This property (that rx and xr are in I for any ring $r \in R$ and any $x \in I$), is called the *absorbing property*, or *closure under outside multiplication*.

Recall that the Subring Test (Theorem 9.4 on page 108) shows that we only need to establish three conditions to show that a subset of a ring is a subring. We can use the Subring Test, along with the absorbing property, to develop a relatively simple test to determine if a subset I of a ring R is an ideal.

Activity 16.3. Let R be a ring and I a subset of R. Assume that I is nonempty, closed under subtraction, and that $ra \in I$ and $ar \in I$ for every $r \in R$ and $a \in I$.

(a) Explain why I must be closed under multiplication.

(b) A formal statement of the Ideal Test is the following:

Theorem 16.4 (The Ideal Test). *Let R be a ring. A subset I of R is an ideal of R if and only if:*

(i) *I is nonempty;*

(ii) $a - b \in I$ *for every* $a, b \in I$*; and*

(iii) $ra \in I$ *and* $ar \in I$ *for every* $r \in R$ *and* $a \in I$.

Explain how we have proved this theorem.

Now some applications of this test are in order.

Activity 16.5. For each of the following parts, determine if the set I is an ideal of the ring R. Use the Ideal Test to justify your answer.

(a) $R = \mathbb{Z}$ and $I = \{0\}$.

(b) $R = \mathbb{Z}$ and $I = \{2n : n \in \mathbb{Z}\}$.

(c) $R = \mathbb{R}$ and $I = \mathbb{Q}$.

(d) R is a commutative ring, $a \in R$, and $I = \{ra : r \in R\}$.

If R is any ring with additive identity 0_R, then the set $\{0_R\}$ is always an ideal of R. This ideal is called the *trivial ideal* and is the simplest of all ideals. An ideal I in R is a *proper ideal* if $I \neq R$. If an ideal I in a commutative ring R contains a non-identity element a, then the absorbing property of I shows that ra and ar are in I for any $r \in R$. So the ideal $\{ra : r \in R\}$ is the smallest ideal of a commutative ring R containing the element a. These ideals are the next simplest ideals (compared to the trivial ideal) and are given a special name.

Definition 16.6. An ideal I in a commutative ring R is a **principal ideal** if $I = \{ra : r \in R\}$ for some $a \in R$.

We say that the ideal $I = \{ra : r \in R\}$ is *generated by* a and denote this ideal as $\langle a \rangle$. For example, the ideal $\langle 3 \rangle$ in $\mathbb{Z}$ is a principal ideal. This ideal consists of all the integer multiples of 3, which we have denoted by $3\mathbb{Z}$ in previous investigations. In general, the principal ideal $\langle k \rangle$ in $\mathbb{Z}$ consisting of all integer multiples of k is also denoted by $k\mathbb{Z}$. (See Exercise 5 on page 88 of Investigation 7.)

For some familiar rings, the only ideals are the principal ones.

Activity 16.7. Let I be an ideal of $\mathbb{Z}$.

(a) If I is the trivial ideal, explain why I is a principal ideal.

(b) Now assume I is a non-trivial ideal. Use the fact that I contains a nonzero element b to show that I contains a positive integer. (Hint: The integer b is either positive or negative.)

(c) Let $S = \{x \in I : x > 0\}$. Explain why S contains a smallest element a.

(d) Explain why it must be the case that $\langle a \rangle \subseteq I$.

(e) Let $y \in I$. Use the Division Algorithm to divide y by a, and then use the fact that I is an ideal to show that the remainder r after this division must be an element of I. What can we conclude about r, and why does this imply that $I \subseteq \langle a \rangle$?

(f) Explain how we have proved the following theorem:

Theorem 16.8. *Every ideal of* $\mathbb{Z}$ *is a principal ideal.*

Integral domains like $\mathbb{Z}$ in which every ideal is a principal ideal are given a special name.

Definition 16.9. An integral domain R is a **principal ideal domain** (PID) if every ideal of R is a principal ideal.

The ring of integers is not the only principal ideal domain. We have seen that the polynomial ring $F[x]$ is like $\mathbb{Z}$ in many ways, and so it is natural to ask if $F[x]$ is also a PID. The next theorem provides the answer.

Theorem 16.10. *Let F be a field. Then $F[x]$ is a principal ideal domain.*

Proof. Let F be a field and I an ideal of $F[x]$. Let 0 denote the additive identity in F and 1 the multiplicative identity. If $I = \{0\}$, then I is generated by 0. Suppose $I \neq \{0\}$. Then I contains a nonzero element $b(x)$. Let $S = \{\deg(p(x)) : p(x) \in I\}$. Since I contains a nonzero element, we know that S is not empty. The Well-Ordering Principle tells us that S contains a smallest element n. Let $a(x)$ be a polynomial in I of degree n. Note that if the leading coefficient a_n of $a(x)$ is not equal to 1, then the polynomial $\frac{1}{a_n}a(x)$ is in I. So we can assume $a(x)$ is a monic polynomial. We will show that I is generated by $a(x)$—that is, I is equal to the ideal $J = \{f(x)a(x) : f(x) \in F[x]\}$. First we will demonstrate that $J \subseteq I$. Let $p(x) \in J$. Then $p(x) = h(x)a(x)$ for some $h(x) \in F[x]$. Since $a(x) \in I$ we know $h(x)a(x) \in I$ and $p(x) \in I$. Thus, $J \subseteq I$. For the reverse containment, let $g(x) \in I$. By the Division Algorithm, there are polynomials $q(x), r(x)$ such that $g(x) = q(x)a(x) + r(x)$ with $0 \leq \deg(r(x)) < \deg(a(x))$ or $r(x) = 0$. Now $g(x) \in I$ and $q(x)a(x) \in I$, so $r(x) = g(x) - q(x)a(x) \in I$. But $a(x)$ is a polynomial in S of smallest degree, so $\deg(r(x)) < \deg(a(x))$ would be a contradiction. Therefore, we must have $r(x) = 0$ and $g(x) = q(x)a(x) \in J$. So $I = J$, and I is generated by a single element. ■

One more example of a principal ideal domain is the ring of Gaussian integers $\mathbb{Z}[i]$. (See Exercise 2 on page 75 of Investigation 6.) Recall that $\mathbb{Z}[i]$ is the subring of the field of complex numbers consisting of all elements of the form $a + bi$, where $a, b \in \mathbb{Z}$ and $i^2 = -1$.

Lemma 16.11. *The ring $\mathbb{Z}[i]$ is a principal ideal domain.*

Proof. Let $I \subset \mathbb{Z}[i]$ be an ideal. We need to show that there is an element $a + bi \in \mathbb{Z}[i]$ such that $I = \langle a + bi \rangle$. If $I = \{0\}$, then $I = \langle 0 \rangle$. So assume $I \neq \langle 0 \rangle$. As in the proofs of Theorems 16.8 and 16.10, we will look for an element in I that is smallest in some sense. In this case, we will use the complex norm of elements in $\mathbb{Z}[i]$ as our measure of size. In other words, we will define a function $\delta : \mathbb{Z}[i] \to \mathbb{Z}^+ \cup \{0\}$ by $\delta(a + bi) = a^2 + b^2$. It is left to the reader to show that $\delta((a + bi)(c + di)) = \delta(a + bi)\delta(c + di)$. Now let $S = \{\delta(u + vi) > 0 : u + vi \in I\}$. By definition, S is bounded below. We know S is not empty because I contains a nonzero element. Thus, by the Well-Ordering Principle, we know that S contains a smallest positive integer of the form $\delta(a + bi) = a^2 + b^2$ for some $a + bi \in I$.

To show that $I = \langle a + bi \rangle$, we need to show that every element in I is a multiple of $a + bi$. Let $c + di \in I$. We will show that when $c + di$ is divided by $a + bi$, the remainder is 0. But how do we divide $a + bi$ into $c + di$ in $\mathbb{Z}[i]$? First, we will do the division in $\mathbb{C}$:

$$\frac{c + di}{a + bi} = \frac{(c + di)(a - bi)}{a^2 + b^2} = \left(\frac{ca + db}{a^2 + b^2}\right) + \left(\frac{da - cb}{a^2 + b^2}\right) i.$$

Let $x = \frac{ca+db}{a^2+b^2}$ and $y = \frac{da-cb}{a^2+b^2}$, so that $\frac{c+di}{a+bi} = x + yi$. Now x and y are rational numbers, so there exist integers m and n such that

$$|m - x| \leq \frac{1}{2} \text{ and } |n - y| \leq \frac{1}{2}.$$

Now $\frac{c+di}{a+bi} = x + yi$ implies

$$\begin{aligned} c + di &= (a + bi)(x + yi) \\ &= (a + bi)[(x - m + m) + (y - n + n)i] \\ &= (a + bi)[(m + ni) + ((x - m) + (y - n)i)] \\ &= (a + bi)(m + ni) + (a + bi)\left((x - m) + (y - n)i\right)]. \end{aligned}$$

Let $q = m + ni$ and $r = (a + bi)\left((x - m) + (y - n)i\right)$. We know $q \in \mathbb{Z}[i]$ and $r = (c + di) - q(a + bi)$, so $r \in \mathbb{Z}[i]$. Also, we know $a + bi, c + di \in I$, so $r \in I$. However, we have

$$\begin{aligned} \delta(r) &= \delta(a + bi)\delta((x - m) + (y - n)i) \\ &= (a^2 + b^2)\left((x - m)^2 + (y - n)^2\right) \\ &\leq (a^2 + b^2)\left(\frac{1}{4} + \frac{1}{4}\right) \\ &< a^2 + b^2, \end{aligned}$$

which means that r is in I but has a smaller δ value than $a + bi$. This is a contradiction unless $r = 0$. Thus, $(a + bi)q = c + di$, and so $c + di \in \langle a + bi \rangle$. It follows that $I = \langle a + bi \rangle$ is a principal ideal. ■

The three examples of principal ideal domains ($\mathbb{Z}$, $F[x]$, and $\mathbb{Z}[i]$) all have something in common. In the proof of Lemma 16.11 we essentially showed that there is a division algorithm in $\mathbb{Z}[i]$ just like in $\mathbb{Z}$ and $F[x]$. Integral domains in which there is a division algorithm are called *Euclidean domains*. In a Euclidean domain, we can define greatest common divisors, and the function δ helps establish a Euclidean algorithm to find greatest common divisors. The next definition formalizes these ideas.

Definition 16.12. Let D be an integral domain with additive identity 0_D. Then D is a **Euclidean domain** if there is a function $\delta : D - \{0_D\} \to \mathbb{Z}^+ \cup \{0\}$ such that the following conditions hold:

(i) If $a, b \in D$ are nonzero, then $\delta(a) \leq \delta(ab)$.

(ii) If $a, b \in D$ with a nonzero, then there exist elements $q, r \in D$ such that

$$b = aq + r \text{ with either } r = 0_D \text{ or } \delta(r) < \delta(a).$$

The function δ in Definition 16.12 is called a **norm** for D.

Some of our previous examples can be used to illustrate Definition 16.12. In particular, note the following:

- In $\mathbb{Z}$, $\delta(x) = |x|$.
- In $F[x]$, where F is a field, $\delta(f(x)) = \deg(f(x))$.
- In $\mathbb{Z}[i]$, $\delta(a + bi) = a^2 + b^2$.

As we have seen, all of these rings are principal ideal domains. The arguments we used to prove this fact can be generalized to establish the following theorem:

Theorem 16.13. *Every Euclidean domain is a principal ideal domain.*

There are principal ideal domains that are not Euclidean domains, but none that we have encountered in our investigations up to this point.* It is also important to note that not every integral domain is a principal ideal domain. Consider the ring $\mathbb{Z}[x]$ and the ideal I generated by 2 and x—that is,

$$I = \langle 2, x\rangle = \{2f(x) + xg(x) : f(x), g(x) \in \mathbb{Z}[x].\}$$

(As an exercise, you may want to verify that I is in fact an ideal.) Suppose that I was a principal ideal with generator $a(x)$. Since $2 \in I$, we must have $2 = c(x)a(x)$ for some $c(x) \in \mathbb{Z}[x]$. It follows that $\deg(a(x)) = \deg(c(x)) = 1$. So $a(x) = a \in \mathbb{Z}$. We also have $x \in I$, so $x = b(x)a$ for some $b(x) \in \mathbb{Z}[x]$. Comparing degrees gives us $\deg(b(x)) = 1$, so $b(x) = b_1x + b_0$ for some $b_1, b_0 \in \mathbb{Z}$. Then $x = (b_1x + b_0)a$ implies $b_1a = 1$. Therefore, we can conclude that $a = 1$ or $a = -1$. In each of these cases, a is a unit in $\mathbb{Z}[x]$, which implies that $I = \langle a\rangle = \mathbb{Z}[x]$, and so $1 \in I$. By the definition of I, there must then exist polynomials $p(x), q(x) \in \mathbb{Z}[x]$ such that

$$1 = 2p(x) + xq(x).$$

If p_0 is the constant term of $p(x)$, it follows that the constant term of $2p(x) + xq(x)$ is $2p_0$. Thus, we have $2p_0 = 1$ for some $p_0 \in \mathbb{Z}$, a contradiction. Therefore, I is not a principal ideal of $\mathbb{Z}[x]$.

Given two principal ideals, one might ask if they can be equal without having their generators be equal.

Activity 16.14.

(a) How are the two principal ideals $\langle 2\rangle$ and $\langle -2\rangle$ in $\mathbb{Z}$ related?

(b) There is an integer k so that $2 = k(-2)$. What is this integer and what special property does it have in $\mathbb{Z}$?

(c) Let R be any commutative ring with identity, and let $a, b \in R$ with $a = ub$ for some unit $u \in R$. What conclusion can you draw about $\langle a\rangle$ and $\langle b\rangle$? Explain.

Elements a and b that are related as in part (c) of Activity 16.14 are given a special name.

Definition 16.15. Let R be a commutative ring with identity. Elements $a, b \in R$ are said to be **associates** if $a = ub$ for some unit $u \in R$.

The result of the previous activity suggests the following lemma, which is almost—but not quite—a biconditional statement.

Lemma 16.16. *Let R be a commutative ring with identity, and let $a, b \in R$.*

(i) *If a and b are associates, then $\langle a\rangle = \langle b\rangle$.*

(ii) *If R is an integral domain and $\langle a\rangle = \langle b\rangle$, then a and b are associates.*

Proof. Let R be a commutative ring with additive identity 0_R and multiplicative identity 1_R. Let $a, b \in R$.

To prove (i), assume a and b are associates. Then $a = ub$ for some unit $u \in R$. To show $\langle a\rangle = \langle b\rangle$, we will prove that $\langle a\rangle \subseteq \langle b\rangle$ and $\langle b\rangle \subseteq \langle a\rangle$. Let $x \in \langle a\rangle$. Then $x = ra$ for some $r \in R$. So $x = r(ub) = (ru)b \in \langle b\rangle$, and $\langle a\rangle \subseteq \langle b\rangle$. Now let $y \in \langle b\rangle$. Then $y = sb$ for some $s \in R$. So $y = s\left(u^{-1}a\right) = \left(su^{-1}\right)a \in \langle a\rangle$, and $\langle b\rangle \subseteq \langle a\rangle$. The two containments show that $\langle a\rangle = \langle b\rangle$.

*For an example, see the paper "The Euclidean Algorithm" by T. Motzkin, *Bulletin of the American Mathematical Society*, **55**(12), 1949, pp. 1142–1146.

To prove the forward implication, assume $\langle a\rangle = \langle b\rangle$. If $\langle a\rangle = \langle b\rangle = \{0_R\}$, then we must have $a = b = 0_R$, and so $a = 1_R b$. Thus, a and b are associates. Now assume a and b are nonzero elements in R. Since $a \in \langle a\rangle$, it follows that $a \in \langle b\rangle$. Similarly, since $b \in \langle b\rangle$, we know $b \in \langle a\rangle$. Thus, $a = ub$ and $b = va$ for some $u, v \in R$. Then $a = u(va) = (uv)a$. Therefore, $a(1_R - uv) = 0_R$. The fact that R is an integral domain and $a \neq 0_R$ implies that $1_R - uv = 0_R$ or $uv = 1_R$. We can therefore conclude that u and v are units and that a and b are associates. ■

That Lemma 16.16 is not a biconditional is not so easy to see. However, Exercise 14 provides an example.

Congruence Modulo an Ideal

In Investigation 15, we defined congruence modulo a polynomial and constructed quotients of polynomial rings based on this relation. The congruence class of a polynomial is an ideal, so now we extend that construction to congruence modulo an arbitrary ideal of a ring R. The definition is analogous to Definition 15.2.

Definition 16.17. Let R be a ring and I an ideal of R. The element $a \in R$ is **congruent modulo** I to $b \in R$ if $b - a \in I$.

If a is congruent to b modulo I, we denote this relation by writing

$$a \equiv b \pmod{I}.$$

The next activity provides an example.

Activity 16.18. Let $R = \mathbb{Z}_{12}$ and $I = \langle [3]\rangle$ as in Preview Activity 16.1.

(a) Find all the elements of R that are congruent to $[0]$ modulo I.

(b) Find all the elements of R that are congruent to $[1]$ modulo I.

(c) Find all the elements of R that are congruent to $[2]$ modulo I.

(d) What can you say about the three sets of elements you have just found? What might we expect to be true about the relation of congruence modulo an ideal?

As with any relation, there are three natural questions to ask.

Activity 16.19. Let R be a ring and I an ideal of R.

(a) Is congruence modulo I a reflexive relation? Prove your answer.

(b) Is congruence modulo I a symmetric relation? Prove your answer.

(c) Is congruence modulo I a transitive relation? Prove your answer.

The collection of distinct equivalence classes of R under the congruence modulo I relation is denoted R/I (read as "R mod I"). As we did with congruence classes of polynomials, we can make R/I into a ring. To do so, we first need to understand some properties of the congruence relation.

Activity 16.20. Let R be a ring and I an ideal. Let $a, b, c, d \in R$ such that $a \equiv c \pmod{I}$ and $b \equiv d \pmod{I}$. (Note the similarities between this activity, Activity 5.25 on page 57, and Activity 15.6 on page 202.)

(a) Show that $(a+b) \equiv (c+d) \pmod{I}$.

(b) Show that $(ab) \equiv (cd) \pmod{I}$.

We can make the set R/I into a ring by defining addition and multiplication on the set of equivalence classes. We denote by $r+I$ the equivalence class of r modulo I, so that $R/I = \{a+I : a \in R\}$. We can then define addition and multiplication on R/I in a natural way; in particular, for $r+I, s+I \in R/I$, we define

$$(r+I)+(s+I) = (r+s)+I \tag{16.1}$$
$$(r+I)(s+I) = rs+I. \tag{16.2}$$

We will now investigate the structure that addition and multiplication impose on the set R/I.

Activity 16.21. Let $R = \mathbb{Z}_{12}$ and $I = \langle [3] \rangle$, as in Preview Activity 16.1.

(a) Find all the elements in R/I.

(b) Construct the addition table for R/I.

(c) Construct the multiplication table for R/I.

(d) What kind of structure does R/I appear to have? Explain.

Of course, one question we must ask about R/I is whether the operations defined in (16.1) and (16.2) are well-defined. This is a question we have asked many times by now, so answering it should entail a familiar process.

Activity 16.22. Let R be a ring and I an ideal of R.

(a) Formally state what it means for the operation defined in (16.1) to be well-defined. Then prove that this operation is well-defined.

(b) Formally state what it means for the operation defined in (16.2) to be well-defined. Then prove that this operation is well-defined.

As we might expect from our work in this section, the set R/I of congruence classes is a ring under the operations defined in (16.1) and (16.2).

Activity 16.23. Let R be a ring and I an ideal of R.

(a) Why is R/I closed under the operation in (16.1)?

(b) Why is R/I closed under the operation in (16.2)?

(c) What is the additive identity in R/I? Prove your conjecture.

(d) What is the additive inverse of the element $a+I$ in R/I? Prove your conjecture.

(e) Complete the proof that R/I is a ring. (Note the similarity between this proof and that from Activity 15.8 on page 202.)

(f) If R is a commutative ring, must R/I be a commutative ring? Prove your conjecture.

(g) If R contains an identity, must R/I also be a ring with identity? Prove your conjecture.

The result of Activity 16.23 is the following theorem.

Theorem 16.24. *Let R be a ring and I an ideal of R. The set R/I of congruence classes modulo I is a ring under the operations (16.1) and (16.2). Moreover, If R is a commutative ring, then R/I is also a commutative ring. If R is a ring with identity, then R/I is also a ring with identity.*

In Investigation 9, we saw how field extensions and direct sums can be used to construct new rings from old ones. Theorem 16.24 shows that congruence modulo an ideal can be used for the same purpose. As in Investigation 15, the resulting rings are called *quotient rings*.

Maximal and Prime Ideals

Recall that the ring $\mathbb{Z}_n$ is a field for prime values of n, but is not even an integral domain when n is composite. A similar type of behavior happens with quotient rings.

Activity 16.25. In Activity 16.21, we created the operation tables for the quotient ring $\mathbb{Z}_{12}/I$, where $I = \langle [3] \rangle$. We will now compare that quotient ring to another.

(a) Let $J = \langle [6] \rangle \subset \mathbb{Z}_{12}$. Construct the addition and multiplication tables for $\mathbb{Z}_{12}/J$.

(b) The ring $\mathbb{Z}_{12}/I$ is isomorphic to a familiar ring R_1. Identify this familiar ring and exhibit an isomorphism between $\mathbb{Z}_{12}/I$ and R_1.

(c) The ring $\mathbb{Z}_{12}/J$ is isomorphic to a familiar ring R_2. Identify this familiar ring and exhibit an isomorphism between $\mathbb{Z}_{12}/J$ and R_2.

(d) The rings R_1 and R_2 are fundamentally different. Explain the differences and why you think they might occur. (Note that $J \subset I$.)

Activity 16.25 shows that some quotient rings have more algebraic structure than others. Of the various types of rings we have considered, fields have the most structure. This being the case, it would be useful to know when a quotient ring is a field. In Activity 16.25, we saw that because J was contained (properly) in another proper ideal I, the ring $\mathbb{Z}_{12}/J$ was not even an integral domain. This observation suggests that proper ideals that are not contained in any larger ideals may play an important role in determining when a quotient ring is a field.

Definition 16.26. An ideal I in a commutative ring R is a **maximal ideal** if $I \neq R$ and there is no proper ideal J in R such that $I \subset J$.

In fact, maximal ideals are important because of the following theorem.

Theorem 16.27. *Let R be a commutative ring with identity and I an ideal of R. Then R/I is a field if and only if I is a maximal ideal.*

Proof. Let R be a commutative ring with identity and I a proper ideal of R. To prove this biconditional statement, first assume that R/I is a field. We will proceed by contradiction and assume that there is a proper ideal J in R with $I \subset J$. Let $b \in J$ such that $b \notin I$. Then $b + I \neq I$ and $b + I$ is a unit in R/I. So there exists $c \in R$ such that $c+I = (b+I)^{-1}$. Thus, $1+I = (b+I)(c+I) = bc+I$ and $1 - bc \in I$. Now $I \subset J$, so $1 - bc \in J$. Let $1 - bc = j$. Recall that $b \in J$, so we also have $bc \in J$. This gives us $1 = bc + j \in J$. But if $1 \in J$, then $J = R$, a contradiction. Therefore, we cannot have a proper ideal of R that contains I as a proper subset.

To prove the reverse implication, assume I is a maximal ideal of R. We know that R/I is a commutative ring with identity, so it remains to show that every nonzero element in R/I is a unit. Let $a+I$ be a nonzero element in R/I. Let $\overline{J} = \langle a+I \rangle$. Since $a+I \neq I$, we know $\overline{J}$ is a non-trivial ideal of R/I. Let $J = \{r \in R : r + I \in \overline{J}\}$. Note that $a \in J$, so J is nonempty. Also, if $r, s \in J$, then

$$(r + I) - (s + I) = (r - s) + I \in \overline{J}$$

because $\overline{J}$ is an ideal. Thus, we have $r - s \in J$. Finally, if $x \in R$, then

$$(x + I)(r + I) = xr + I \in \overline{J}$$

and $xr \in J$. Therefore, J is an ideal of R. If $i \in I$, then $i + I = I \in \overline{J}$, so $i \in J$. Thus, we have $I \subseteq J$. However, we know that $a \in J$ and $a \notin I$, so $I \subset J$. Since I is a maximal ideal, we conclude that $J = R$. Thus, $1 \in J$ and $1 + I \in \overline{J}$. This implies that there is an element $c \in R$ such that $1 + I = (c + I)(a + I)$, which makes $a + I$ a unit in R/I. Therefore, we conclude that R/I is a field if I is a maximal ideal of R. ■

The next activity shows that maximal ideals of $\mathbb{Z}$ and $F[x]$ (for a field F) are easy to recognize.

Activity 16.28.

(a) Let $m \in \mathbb{Z}^+$ be a composite integer. Show that there is a proper ideal I in $\mathbb{Z}$ that contains $\langle m \rangle$. Use this fact to classify the maximal ideals of $\mathbb{Z}$.

(b) Let F be a field. We have seen many instances where $F[x]$ shares the same properties as $\mathbb{Z}$. With that in mind, write a problem analogous to that in part (a), but for $F[x]$ instead of $\mathbb{Z}$. Provide a solution that includes a classification of the maximal ideals of $F[x]$.

Theorem 16.27 provides us with a method of constructing new fields, similar to what we saw in Investigation 15.

Activity 16.29. Let $I = \langle 3 \rangle$ in $\mathbb{Z}[i]$.

(a) Find all the elements in $\mathbb{Z}[i]/I$. (Hint: Remember that $i^2 = -1$.)

(b) We know that $\mathbb{Z}[i]/I$ is a commutative ring with identity. Is it true that every nonzero element in $\mathbb{Z}[i]/I$ is a unit? If so, explain why. If not, verify your statement.

(c) Is I a maximal ideal of $\mathbb{Z}[i]$? Why or why not?

(d) Do you know of any fields with 9 elements? Explain.

At this point, it might be natural to ask if it is necessary that I be a maximal ideal in order for R/I to be an integral domain. The next activity provides an answer to this question.

Activity 16.30. Let $R = \mathbb{Z}[x]$ and let $I = \langle 3 \rangle$.

(a) Describe the elements in R/I.

(b) The ring R/I is isomorphic to a familiar ring S. What ring is S? Why is S an integral domain?

(c) Explain why I is not a maximal ideal of R.

Activity 16.30 shows that we can have R/I be an integral domain without I being a maximal ideal. The next activity explores conditions on the ideal I under which this happens in $\mathbb{Z}[x]$.

Activity 16.31. Let $R = \mathbb{Z}[x]$ and let $I = \langle 3 \rangle$.

(a) Suppose $f(x), g(x) \in R$ and $f(x)g(x) \in I$. How does this relate 3 to the product $f(x)g(x)$?

(b) What conclusion can you draw from the previous question?

The property illustrated in Activity 16.31 suggests the following definition:

Definition 16.32. A proper ideal I in a commutative ring R is a **prime ideal** if for any $a, b \in R$, whenever $ab \in I$ then $a \in I$ or $b \in I$.

You may notice a similarity between the definition of a prime ideal and the statement of Euclid's Lemma from Investigation 4. (See page 36.) In fact, Euclid's Lemma can be restated in terms of prime ideals as follows:

Euclid's Lemma. *Let p be any prime number. Then $\langle p \rangle$ is a prime ideal of $\mathbb{Z}$.*

As discussed above, we defined prime ideals with the goal of classifying the ideals I for which R/I is an integral domain. The next activity accomplishes this goal.

Activity 16.33. Let R be a commutative ring with identity and I an ideal of R.

(a) Give a formal statement of what a zero divisor looks like in R/I.

(b) Assume I is a prime ideal of R. Use the fact that I is a prime ideal to show that a nonzero element $a + I \in R/I$ cannot be a zero divisor in R/I.

(c) Assume R/I is an integral domain and let $a, b \in R$. Use the fact that R/I is an integral domain to show that if $ab \in I$, then $a \in I$ or $b \in I$.

(d) Explain how we have proven the following theorem:

Theorem 16.34. *Let R be a commutative ring with identity and I a proper ideal of R. Then R/I is an integral domain if and only if I is a prime ideal.*

Note that a prime ideal need not be maximal, as our example of the ideal $\langle 3 \rangle$ in $\mathbb{Z}[x]$ (from Activity 16.30) shows. However, Theorems 16.27 and 16.34 together establish the following result:

Theorem 16.35. *Let R be a commutative ring with identity. Then every maximal ideal of R is a prime ideal.*

Homomorphisms

Investigation 10 introduced us to isomorphisms of rings. Recall that a ring isomorphism is a bijective function that preserves the additive and multiplicative structure of a ring, thereby identifying isomorphic rings as being essentially the same. It is also worth considering functions that preserve the structure of a ring but are not necessarily bijections. Such functions are called *homomorphisms* and will be the focus of the remainder of this investigation.†

The formal definition of a homomorphism is as follows:

Definition 16.36. Let R and S be rings. A function $\varphi : R \to S$ is a **homomorphism** of rings if

$$\varphi(a+b) = \varphi(a) + \varphi(b) \text{ and } \varphi(ab) = \varphi(a)\varphi(b)$$

for all $a, b \in R$.

Certain types of homomorphisms are given special names. In particular, suppose R and S are rings and let $\varphi : R \to S$ be a function.

- If φ is a homomorphism and an injection, then φ is called a *monomorphism*.
- If φ is a homomorphism and a surjection, then φ is called an *epimorphism*.
- If φ is a homomorphism and a bijection, then φ is called an *isomorphism*. In this case we say that the rings R and S are isomorphic rings.

If φ maps the ring R onto the ring S, then S inherits the additive and multiplicative structure of R. We call S a *homomorphic image* of R. The next activity provides some examples.

Activity 16.37. Determine if the given function is a homomorphism. If the function is a homomorphism, is it a monomorphism, an epimorphism, an isomorphism, or none of the above?

(a) Let R and S be any rings, and define $\varphi : R \to S$ by $\varphi(r) = 0_S$ (where 0_S is the identity in S)

(b) Let R be any ring, and define $id_R : R \to R$ by $id_R(r) = r$.

(c) Let $n \in \mathbb{Z}^+$, $n > 1$, and let $\varphi : \mathbb{Z} \to \mathbb{Z}_n$ be defined by $\varphi(k) = [k]$.

(d) Let R be a field and $r \in R$. Let $ev_r : R[x] \to R$ be defined by $ev_r(f(x)) = f(r)$. (This map is called the *evaluation map*.)

(e) Let $\varphi : \mathbb{Z}_{12} \to \mathbb{Z}_6$ be defined by $\varphi([k]_{12}) = [4k]_6$, where $[k]_n$ denotes the congruence class of k modulo n.

In Investigation 10, we saw that isomorphisms preserve certain properties of rings called *invariants*. Homomorphisms also preserve some properties of rings, as the next activity shows.

Activity 16.38. Let $\varphi : R \to S$ be a ring homomorphism, and let 0_R and 0_S be the additive identities in R and S, respectively.

(a) Isomorphisms preserve the additive identity. Use the fact that $0_R = 0_R + 0_R$ to show that $\varphi(0_R) = 0_S$. Conclude that homomorphisms also preserve additive identities.

(b) Isomorphisms preserve additive inverses. Let $a \in R$. Use the fact that φ is a homomorphism to show that $\varphi(-a) = -\varphi(a)$. Conclude that homomorphisms also preserve additive inverses.

†This section assumes an understanding of injective, surjective, and bijective functions. For a review of these topics, see Appendix A.

(c) Isomorphisms preserve differences. Let $a, b \in R$. Use the result from part (b) and the fact that φ is a homomorphism to show that $\varphi(a-b) = \varphi(a) - \varphi(b)$. Conclude that homomorphisms also preserve differences.

(d) If A is a ring with identity 1_A and $\psi : A \to B$ is an isomorphism, then B is a ring with identity $1_B = \psi(1_A)$. Is it true that if R is a ring with identity 1_R, then S is also a ring with identity $1_S = \varphi(1_R)$? Explain. (Hint: Consider $R = \mathbb{Z}_3$, $S = \mathbb{Z}_6$, and $\varphi([k]_3) = [3k]_6$.)

(e) If A is a ring with identity 1_A, $\psi : A \to B$ is an isomorphism, and $u \in A$ is a unit, then $\psi(u)$ is a unit in B and $\psi(u)^{-1} = \psi\left(u^{-1}\right)$. Is it true that if R is a ring with identity 1_R and u is a unit in R, then $\varphi(u)$ is a unit in S and $\psi(u)^{-1} = \psi\left(u^{-1}\right)$? Explain. (Hint: Again consider $R = \mathbb{Z}_3$, $S = \mathbb{Z}_6$, and $\varphi([k]_3) = [3k]_6$.)

Activity 16.38 shows that homomorphisms don't preserve all the invariants that isomorphisms do. However, under certain conditions, homomorphisms do preserve these invariants, as demonstrated in the next lemma.

Theorem 16.39. *Let $\varphi : R \to S$ be a ring homomorphism. Let 0_R and 0_S be the additive identities in R and S, respectively.*

(i) $\varphi(0_R) = 0_S$.

(ii) *If $a \in R$, then $-\varphi(a) = \varphi(-a)$.*

(iii) *If $a, b \in R$, then $\varphi(a-b) = \varphi(a) - \varphi(b)$.*

(iv) *If R has identity 1_R, φ is an epimorphism, and S is not the trivial ring (that is, $S \neq \{0_S\}$), then S has identity 1_S and $\varphi(1_R) = 1_S$.*

(v) *If R has identity 1_R, φ is an epimorphism, and S is not the trivial ring (that is, $S \neq \{0_S\}$), then $\varphi(u)$ is a unit in S for any unit u in R and $(\varphi(u))^{-1} = \varphi\left(u^{-1}\right)$.*

Proof. Let $\varphi : R \to S$ be a ring homomorphism and let 0_R and 0_S be the additive identities in R and S, respectively. The proofs of the first three properties were part of Activity 16.38. Thus, we will focus on the latter properties.

For property (iv), let $s \in S$. Since φ is an epimorphism, there is an element $r \in R$ such that $\varphi(r) = s$. Then

$$\varphi(1_R)s = \varphi(1_R)\varphi(r) = \varphi(1_R r) = \varphi(r) = \varphi(r1_R) = \varphi(r)\varphi(1_R) = s\varphi(1_R),$$

and so $\varphi(1_R)$ is an identity in S.

Finally, to verify property (v), choose u to be a unit in R. So there exists $u^{-1} \in R$ such that $uu^{-1} = u^{-1}u = 1_R$. Then

$$\varphi(u)\varphi\left(u^{-1}\right) = \varphi\left(uu^{-1}\right) = \varphi(1_R) = \varphi\left(u^{-1}u\right) = \varphi\left(u^{-1}\right)\varphi(u).$$

Therefore, we have $(\varphi(u))^{-1} = \varphi\left(u^{-1}\right)$. ■

The Kernel and Image of a Homomorphism

There are two important sets that are related to a ring homomorphism: the kernel and image. To understand why the kernel is important, consider the homomorphism $\varphi : \mathbb{Z}_{12} \to \mathbb{Z}_6$ defined by $\varphi([k]_{12}) = [4k]_6$ (from Activity 16.37). As with all homomorphisms, we have $\varphi([0]_{12}) = [0]_6$. If φ was a monomorphism, then $[0]_{12}$ would be the only element in $\mathbb{Z}_{12}$ that φ sends to $[0]_6$. However, if we look at the preimage $\varphi^{-1}([0]_6)$ of $[0]_6$ (that is, all the elements in $\mathbb{Z}_{12}$ that map to $[0]_6$), we see that $\varphi^{-1}([0]_6) = \{[0]_{12}, [3]_{12}, [6]_{12}, [9]_{12}\}$. As we saw in our preview activity, this pre-image is an ideal of $\mathbb{Z}_{12}$. For convenience, let $K = \varphi^{-1}([0]_6)$. Recall that whenever I is an ideal of a ring R and $a \in R$, $a + I = \{a + r : r \in I\}$. Now notice that

- $\varphi^{-1}([0]_6) = \{[0]_{12}, [3]_{12}, [6]_{12}, [9]_{12}\} = K$;
- $\varphi^{-1}([1]_6) = \emptyset$;
- $\varphi^{-1}([2]_6) = \{[2]_{12}, [5]_{12}, [8]_{12}, [11]_{12}\} = [2]_{12} + K$;
- $\varphi^{-1}([3]_6) = \emptyset$;
- $\varphi^{-1}([4]_6) = \{[1]_{12}, [4]_{12}, [7]_{12}, [10]_{12}\} = [1]_{12} + K$; and
- $\varphi^{-1}([5]_6) = \emptyset$.

Thus, every nonempty preimage is just a translation of the ideal K. So if $K = \{[0]_{12}\}$, then each preimage will be a one element set, and φ will be a monomorphism. In this sense, the set $\varphi^{-1}([0]_6)$ tells us how close φ is to being a monomorphism. This particular preimage is called the *kernel* of φ, denoted Ker(φ).

Definition 16.40. Let $\varphi : R \to S$ be a homomorphism of rings. The **kernel** of φ is the set

$$\text{Ker}(\varphi) = \{r \in R : \varphi(r) = 0_S\}.$$

In our example of $\varphi : \mathbb{Z}_{12} \to \mathbb{Z}_6$ defined by $\varphi([k]_{12}) = [4k]_6$, we have $\text{Ker}(\varphi) = \{[0]_{12}, [3]_{12}, [6]_{12}, [9]_{12}\}$. As we saw in this example, the kernel was an ideal of R. It is natural to ask if this is always the case.

Activity 16.41. Let $\varphi : R \to S$ be a ring homomorphism. Let 0_R be the additive identity in R and 0_S the additive identity in S.

(a) Theorem 16.39 shows that $0_R \in \text{Ker}(\varphi)$. Use Theorem 16.39 and the fact that φ is a homomorphism to show that Ker(φ) is closed under subtraction.

(b) Let $a \in R$ and $k \in \text{Ker}(\varphi)$. Use the fact that φ is a homomorphism to explain why $ak \in \text{Ker}(\varphi)$ and $ka \in \text{Ker}(\varphi)$.

(c) Explain how we have proved the following result:

Theorem 16.42. *If $\varphi : R \to S$ is a ring homomorphism, then Ker(φ) is an ideal of R.*

As mentioned earlier, the kernel of a homomorphism φ tells us how close φ is to being a monomorphism. The next theorem formalizes this statement.

Theorem 16.43. *Let $\varphi : R \to S$ be a ring homomorphism. Then φ is a monomorphism if and only if Ker(φ) $= \{0_R\}$, where 0_R is the additive identity in R.*

Proof. Let $\varphi : R \to S$ be a ring homomorphism. Let 0_R and 0_S be the additive identities in R and S, respectively. We will prove the forward implication first. Assume φ is a monomorphism. Theorem 16.39 shows that $0_R \in \text{Ker}(\varphi)$. Suppose $k \in \text{Ker}(\varphi)$. Then $\varphi(k) = 0_S = \varphi(0_R)$. Since φ is a monomorphism, we must have $k = 0_R$. Thus, $\text{Ker}(\varphi) = \{0_R\}$.

For the reverse implication, assume $\text{Ker}(\varphi) = \{0_R\}$. To show that φ is a monomorphism, let $a, b \in R$ such that $\varphi(a) = \varphi(b)$. Then $\varphi(a) - \varphi(b) = 0_S$, which implies $\varphi(a - b) = 0_S$. Thus, we have $a - b \in \text{Ker}(\varphi)$. Since $\text{Ker}(\varphi) = \{0_R\}$, we can conclude that $a - b = 0_R$, or $a = b$. This proves that φ is a monomorphism. ■

The second important set related to a homomorphism is its image. In our example of $\varphi : \mathbb{Z}_{12} \to \mathbb{Z}_6$ defined by $\varphi([k]_{12}) = [4k]_6$, note that

- $\varphi([0]_{12}) = \varphi([3]_{12}) = \varphi([6]_{12}) = \varphi([9]_{12}) = [0]_6$,
- $\varphi([1]_{12}) = \varphi([4]_{12}) = \varphi([7]_{12}) = \varphi([10]_{12}) = [4]_6$, and
- $\varphi([2]_{12}) = \varphi([5]_{12}) = \varphi([8]_{12}) = \varphi([11]_{12}) = [2]_6$.

In this case, φ maps onto the set $\{[0]_6, [4]_6, [2]_6\}$. It is not difficult to see that this set is in fact a subring of $\mathbb{Z}_6$. The set of all images of elements under the homomorphism φ is called the *image* (or range) of φ and is denoted as Im(φ).

Definition 16.44. Let $\varphi : R \to S$ be a ring homomorphism. The **image** of φ is the set

$$\text{Im}(\varphi) = \{\varphi(r) : r \in R\}.$$

In our example, we saw that the image of φ was a subring of $\mathbb{Z}_6$. Is it always the case that the image of a homomorphism is a ring?

Activity 16.45. Let $\varphi : R \to S$ be a ring homomorphism. Let 0_R be the additive identity in R and 0_S the additive identity in S.

(a) Theorem 16.39 shows that $0_S \in \text{Im}(\varphi)$. Use Theorem 16.39 and the fact that φ is a homomorphism to show that Im(φ) is closed under subtraction.

(b) Show that Im(φ) is closed under multiplication.

(c) Explain how we have proved the following result:

Theorem 16.46. *If $\varphi : R \to S$ is a ring homomorphism, then Im(φ) is a subring of S.*

The First Isomorphism Theorem for Rings

If $\varphi : R \to S$ is a ring homomorphism, then there is an important relationship between R, Ker(φ), and Im(φ). We will investigate this relationship in the next activity.

Activity 16.47. Let $R = \mathbb{Z}_{12}$, $S = \mathbb{Z}_6$, and $\varphi : R \to S$ defined by $\varphi([k]_{12}) = [4k]_6$. Recall that $\text{Ker}(\varphi) = \{[0]_{12}, [3]_{12}, [6]_{12}, [9]_{12}\}$ and $\text{Im}(\varphi) = \{[0]_6, [4]_6, [2]_6\}$.

(a) Construct the addition and multiplication tables for $\text{Im}(\varphi)$.

(b) In Activity 16.21, we constructed the addition and multiplication tables for R/I, where $I = \langle[3]\rangle$. Show in detail that R/I is isomorphic to $\text{Im}(\varphi)$.

The important general relationship uncovered in Activity 16.47 between R, $\text{Ker}(\varphi)$, and $\text{Im}(\varphi)$ is given in the next theorem.

Theorem 16.48 (First Isomorphism Theorem for Rings). *Let $\varphi : R \to S$ be a ring homomorphism. Then $R/\text{Ker}(\varphi) \cong \text{Im}(\varphi)$.*

Proof. Let $\varphi : R \to S$ be a ring homomorphism. For ease of notation, let $K = \text{Ker}(\varphi)$. Define a function $\Phi : R/K \to \text{Im}(\varphi)$ by $\Phi(a + K) = \varphi(a)$. We will show that Φ is a well-defined isomorphism. First we will show that Φ is well-defined. Suppose $a + K = a' + K$ for some $a, a' \in R$. Thus, $a = a' + k$ for some $k \in K$. Then

$$\Phi(a' + K) = \varphi(a') = \varphi(a') + 0_s = \varphi(a') + \varphi(k) = \varphi(a' + k) = \varphi(a) = \Phi(a + K).$$

Therefore, Φ is well-defined.

Next we will show that Φ is a ring homomorphism. Let $a + K, b + K \in R/K$. Then

$$\begin{aligned}\Phi((a + K) + (b + K)) &= \Phi((a + b) + K)\\ &= \varphi(a + b)\\ &= \varphi(a) + \varphi(b)\\ &= \Phi(a + K) + \Phi(b + K).\end{aligned}$$

Furthermore,

$$\begin{aligned}\Phi((a + K)(b + K)) &= \Phi((ab) + K)\\ &= \varphi(ab)\\ &= \varphi(a)\varphi(b)\\ &= \Phi(a + K)\Phi(b + K).\end{aligned}$$

Thus, Φ is a homomorphism.

It remains to show that Φ is an isomorphism. To show that Φ is an injection, suppose $\Phi(a+K) = \Phi(b + K)$. Then $\varphi(a) = \varphi(b)$. So $\varphi(a) + (-\varphi(b)) = 0_S$. Recall that $-\varphi(b) = \varphi(-b)$, so we have $\varphi(a - b) = 0_S$. Therefore, $a - b \in K$, and so $a + K = b + K$. Thus, Φ is an injection. To conclude, we will show that Φ is a surjection. Let $s \in \text{Im}(\varphi)$. So $s = \varphi(r)$ for some $r \in R$. Then $\Phi(r + K) = \varphi(r) = s$, which proves that Φ is a surjection.

Since Φ is a bijective ring homomorphism, it follows that Φ is an isomorphism and $R/\text{Ker}(\varphi) \cong \text{Im}(\varphi)$. ■

In addition to Theorem 16.48, there are two other ring isomorphism theorems that are introduced in Exercises 21 and 22.

Concluding Activities

Activity 16.49. Let $n \in \mathbb{Z}^+$, and define $\varphi : \mathbb{Z} \to \mathbb{Z}_n$ by $\varphi(k) = [k]$.

(a) Show that $\text{Ker}(\varphi) = \langle n \rangle$. We also denote this ideal as $n\mathbb{Z}$.

(b) Find $\text{Im}(\varphi)$.

(c) Explain why $\mathbb{Z}_n \cong \mathbb{Z}/n\mathbb{Z}$. (Note that this gives us another way of constructing the ring of integers modulo n.)

Activity 16.50. Let R be a commutative ring with identity, and let 0 denote the additive identity in R. Let $ev_0 : R[x] \to R$ be the evaluation map (see Activity 16.37) defined by $ev_0(f(x)) = f(0)$.

(a) Explain why $\text{Ker}(ev_0) = \langle x \rangle$.

(b) What is $\text{Im}(ev_0)$? Explain.

(c) What is $R[x]/\langle x \rangle$? Explain.

(d) Under what conditions is $\langle x \rangle$ a prime ideal of $R[x]$? Give a precise answer with justification.

(e) Under what conditions is $\langle x \rangle$ a maximal ideal of $R[x]$? Give a precise answer with justification.

(f) Can $\langle x \rangle$ ever be a prime ideal of $R[x]$ but not a maximal ideal? Explain.

Activity 16.51. Write a short summary that describes the important concepts, and the relationships between these concepts, that were introduced in this investigation. Explain how the material in this investigation is connected to the material in Investigations 5 and 15.

Exercises

(1) If an ideal I in a ring R with identity contains a unit, what must I be? Prove your answer.

(2) Find all of the ideals of the following rings.

(a) $\mathbb{Z}_3$

(b) $\mathbb{Z}_6$

(c) $\mathbb{Z}_{12}$

(d) $\mathbb{Z}$

(e) $\mathbb{R}$

(3) Determine which of the following sets I is an ideal of the indicated ring.

(a) Let R be any ring, and let I be the collection of all polynomials in $R[x]$ with a constant term equal to 0_R, the additive identity in R.

(b) Let $R = M_{2\times 2}(\mathbb{Z})$ and $I = \left\{ \begin{pmatrix} a & 0 \\ b & 0 \end{pmatrix} : a, b \in \mathbb{Z} \right\}$.

(c) Let R be a commutative ring and $a_1, a_2, \ldots, a_k \in R$ for some $k \in \mathbb{Z}^+$. Let $I = \{r_1a_1 + r_2a_2 + \cdots + r_ka_k : r_1, r_2, \ldots, r_k \in R\}$.

(d) Let $R = \mathbb{Z}[x]$, and let I be the collection of all polynomials in $R[x]$ with even coefficients.

(4) Determine if the given function is a homomorphism. If the function is a homomorphism, is it a monomorphism, epimorphism, isomorphism, or none of the above?

(a) Let $n \in \mathbb{Z}^+$, $n > 1$, and let $\varphi : \mathbb{Z} \to \mathbb{Z}$ be defined by $\varphi(k) = nk$.

(b) Let $\varphi : \mathbb{Z}_3 \to \mathbb{Z}_{12}$ be defined by $\varphi([k]_3) = [4k]_{12}$.

(c) Let $\varphi : \mathbb{Z}_{12} \to \mathbb{Z}_3$ be defined by $\varphi([k]_{12}) = [k]_3$.

(d) Let $\varphi : \mathbb{R} \to M_{2\times 2}(\mathbb{R})$ be defined by $\varphi(x) = \begin{bmatrix} x & 0 \\ -x & 0 \end{bmatrix}$.

(e) Let $\varphi : \mathbb{Z}_3 \to \mathbb{Z}_3$ be defined by $\varphi([n]) = [n^3]$.

(5) Let R be a ring.

(a) Is every ideal of R a subring of R? Explain.

(b) Is every subring of R an ideal of R? Explain.

(6) Let $R = \left\{ \begin{pmatrix} a & b \\ 0 & a \end{pmatrix} : a, b \in \mathbb{R} \right\}$ and let $I = \left\{ \begin{pmatrix} 0 & c \\ 0 & 0 \end{pmatrix} : c \in \mathbb{R} \right\}$.

(a) Prove that I is an ideal of R.

(b) Show that $R/I \cong \mathbb{R}$ by exhibiting a specific homomorphism from R onto $\mathbb{R}$ with kernel I. Be sure to verify that your map is a homomorphism.

(7) Let $\varphi : R \to S$ be a ring homomorphism.

(a) Is it the case that $\text{Im}(\varphi)$ is always an ideal of S? Prove your answer.

(b) Show that if φ is a surjection, then $\text{Im}(\varphi)$ is an ideal of S. Is it possible for $\text{Im}(\varphi)$ to be an ideal of S even if φ is not a surjection?

(8) Let $R = \mathbb{Z}_5[x]$.

(a) Show that the polynomial $f(x) = x^2 + [2]x + [3]$ is irreducible in R.

(b) Let I be the ideal generated by $f(x)$. What kind of structure must R/I have? Explain.

(c) Is the element $([2]x + [3]) + I$ in R/I a unit? If yes, find its multiplicative inverse. If no, explain why.

(9) Let $R = \mathbb{Z}_{24}$. For parts (a) – (d), construct the addition and multiplication tables for R/I for the given ideal I.

(a) $I = \langle [3] \rangle$

(b) $I = \langle [4] \rangle$

(c) $I = \langle [6] \rangle$

(d) $I = \langle [8] \rangle$

(e) Which, if any, of the ideals from parts (a) – (d) is a prime/maximal ideal? Explain.

(f) Show that the only principal ideals of $\mathbb{Z}_{24}$ are the ideals generated by $[0]$, $[1]$, $[2]$, $[3]$, $[4]$, $[6]$, $[8]$, and $[12]$. (You can do so by proving the more general result that if $n \in \mathbb{Z}^+$, then the only non-trivial principal ideals of $\mathbb{Z}_n$ are those of the form $\langle [a] \rangle$, where a is a divisor of n.)

(10) Construct the addition and multiplication tables for $\mathbb{Z}_2[x]/\langle x^3 \rangle$.

(11) Let $R = \mathbb{Z}_3[x]$ and $I = \langle x^2 \rangle$.

(a) Construct the addition and multiplication tables for R/I.

(b) We know of two rings with 9 elements: $\mathbb{Z}_9$ and $\mathbb{Z}_3 \oplus \mathbb{Z}_3$. Is R/I isomorphic to either of these rings? Justify your answer.

(c) Is I a prime ideal of R? Why or why not?

(d) Explain why I is not a maximal ideal of R and find an ideal J of R that is between I and R.

(12) Show that the quotient ring $\mathbb{Z}[x]/\langle 2x - 1 \rangle$ is isomorphic to the ring $\mathbb{Z}\left[\frac{1}{2}\right] = \left\{ \frac{p}{q} : p \in \mathbb{Z} \text{ and } q = 2^m \text{ for some nonnegative integer } m \right\}$. (Hint: Consider the evaluation homomorphism $ev_{1/2}$.)

(13) Is every ideal of $\mathbb{Z}_n$ a principal ideal? Verify your conjecture.

⋆ (14) Prove that Lemma 16.16 is not a biconditional with the following example.[‡] Let R be the ring of all continuous real-valued functions on the interval $[0, 3]$. Let

$$a(t) = \begin{cases} 1 - t, & \text{if } 0 \leq t \leq 1, \\ 0, & \text{if } 1 < t \leq 2, \\ t - 2, & \text{if } 2 < t \leq 3. \end{cases}$$

and let

$$b(t) = a(t) \text{ on } [0, 2] \text{ and } b(t) = -a(t) \text{ on } (2, 3].$$

(a) Find a continuous function $k(t)$ defined on $[0, 3]$ so that $a(t) = k(t)b(t)$. Explain how this shows that $\langle a(t) \rangle \subseteq \langle b(t) \rangle$.

(b) Find a continuous function $m(t)$ defined on $[0, 3]$ so that $b(t) = m(t)a(t)$. Explain how this shows that $\langle a(t) \rangle = \langle b(t) \rangle$.

(c) Prove that $a(t)$ and $b(t)$ are not associates in R. (Hint: Show that no unit in R can ever attain the value 0, and then use the Intermediate Value Theorem from Calculus.)

(15) **Ideals of fields.** Prove that if R is a field, then the only ideals of R are $\{0_R\}$ and R. Is the converse true if R is a commutative ring with identity? Prove your answer.

[‡]Based on the paper "Elementary divisors and modules" by I. Kaplansky, *Transactions of the American Mathematical Society*, **66**(2), 1949, pp. 464–491, available at `http://www.ams.org/journals/tran/1949-066-02/S0002-9947-1949-0031470-3/S0002-9947-1949-0031470-3.pdf`.

(16) **Simple rings.** As Exercise 15 shows, a commutative ring R with identity is field if and only if the only ideals of R are $\{0_R\}$ and R. As this exercise will demonstrate, this result is not true for non-commutative rings. In general, we say that a ring R is **simple** if R is nonzero and the only ideals of R are $\{0_R\}$ and R. So every field is a simple ring. In this exercise, we illustrate a non-commutative simple ring.

Recall that H (the ring of quaternions from Exercise 3 on page 117 of Investigation 9) is a division ring—that is, H is a non-commutative ring in which every nonzero element is a unit. There are other examples of division rings, but H is the only one we have encountered so far. Let D be any division ring. In this exercise we will prove that $R = M_{n\times n}(D)$ is a non-commutative simple ring.

(a) Let E_{ij} be the elementary matrix in R whose entries are all 0 except the (i, j) entry, which is 1. Note that if $A = [a_{ij}] \in R$, then

$$A = \sum_{ij} a_{ij}E_{ij}$$

and every element in R can be written as a linear combination of these elementary matrices. Show that

$$E_{ij}E_{rs} = \begin{cases} 0 & \text{if } j \neq r \\ E_{is} & \text{if } j = r. \end{cases}$$

(b) Now we will show that if I is a nonzero ideal of R, then $I = R$. So let I be a nonzero ideal of R and let $A = [a_{ij}]$ be a nonzero matrix in I with nonzero entry a_{rs}.

(i) Let u and v be integers between 1 and n. Explain why $E_{ur}AE_{sv} \in I$.

(ii) Use the fact that $A = \sum_{ij} a_{ij}E_{ij}$ and the product of elementary matrices to show that

$$E_{ur}AE_{sv} = a_{rs}E_{uv}.$$

Explain how this shows that $E_{uv} \in I$. Why does this complete our proof that R is a non-commutative simple ring?

(17) Prove that if R is a ring and I is an ideal of R, then $I[x]$ is an ideal of $R[x]$.

(18) **Intersections and unions of ideals.** Ideals are subsets of the rings that contain them, so we can form the intersection and union of two ideals of the same ring.

(a) Let $I = \{k[4] : k \in \mathbb{Z}\}$ and $J = \{m[6] : m \in \mathbb{Z}\}$ be ideals of $\mathbb{Z}_{48}$. Find the elements in $I \cap J$ and $I \cup J$. Is either an ideal of $\mathbb{Z}_{48}$?

(b) In general, if I and J are ideals of a ring R, must it be true that $I \cap J$ is an ideal of R? Prove your answer.

(c) In general, if I and J are ideals of a ring R, must it be true that $I \cup J$ is an ideal of R? Is $I \cup J$ ever an ideal of R? If yes, find necessary and sufficient conditions so that $I \cup J$ is an ideal of R. Prove your answer.

(19) Ideals are subsets of the rings that contain them, so we can add and multiply elements in ideals of the same ring. If I and J are ideals of a ring R, define the sum $I + J$ and product IJ of I and J as follows:

$$I + J = \{i + j : i \in I, j \in J\}$$
$$IJ = \{i_1j_1 + i_2j_2 + \cdots + i_kj_k : i_k \in I, j_k \in J, k \in \mathbb{Z}^+\}.$$

(a) Let $I = \{k[4] : k \in \mathbb{Z}\}$ and $J = \{m[6] : m \in \mathbb{Z}\}$ be ideals of $\mathbb{Z}_{48}$. Find the elements in $I + J$ and IJ. Is either an ideal of $\mathbb{Z}_{48}$?

(b) In general, if I and J are ideals of a commutative ring R, must it be true that $I + J$ is an ideal of R? Prove your answer.

(c) In general, if I and J are ideals of a commutative ring R, must it be true that IJ is an ideal of R? Prove your answer.

(20) Characterize all of the ring homomorphisms $\varphi : \mathbb{Z} \to \mathbb{Z}$.

⋆ (21) There is a **Second Isomorphism Theorem** that we will investigate in this problem. Let R be a ring, I a subring of R, and J an ideal of R. For each of the results you are asked to prove, also illustrate the result with the example $R = \mathbb{Z}_{24}$, $I = \{4[a] : [a] \in \mathbb{Z}_{24}\}$, and $J = \{[0], [6], [12], [18]\}$.

(a) Prove that $I \cap J$ is an ideal of I.

(b) Define $I + J$ to be the set $\{i + j : i \in I, j \in J\}$. Prove that $I + J$ is a ring and that J is an ideal of $I + J$.

(c) Define $\varphi : I \to (I + J)/J$ by $\varphi(a) = a + J$. Prove that φ is a homomorphism.

(d) Find $\text{Ker}(\varphi)$.

(e) Prove the following theorem:

Theorem 16.52 (The Second Isomorphism Theorem)**.** *Let R be a ring, I a subring of R, and J an ideal of R. Then*

$$I/(I \cap J) \cong (I + J)/J.$$

⋆ (22) There is a **Third Isomorphism Theorem** that we will investigate in this problem. Let R be a ring, and let I and J be ideals of R with $J \subset I$. For each of the results you are asked to prove, also illustrate the result with the example $R = \mathbb{Z}_{24}$, $I = \langle [3] \rangle$, and $J = \langle [6] \rangle$.

(a) Show that J is an ideal of I.

(b) Explain how I/J is a subset of R/J. Prove that I/J is an ideal of R/J.

(c) Define $\varphi : R \to \frac{R/J}{I/J}$ by $\varphi(a) = (a + J) + I/J$. Prove that φ is a homomorphism.

(d) Find $\text{Ker}(\varphi)$.

(e) Prove the following theorem:

Theorem 16.53 (The Third Isomorphism Theorem)**.** *Let R be a ring, and let I and J be ideals of R with $J \subset I$. Then*

$$R/I \cong \frac{R/J}{I/J}.$$

Connections

Given an ideal I in a ring R, we can define congruence modulo I and form the quotient ring R/I. Quotient structures are useful in mathematics in that they often have a simpler structure than the original set and can therefore provide important information about the original set. We are familiar with congruence and quotient structures in the integers—namely $\mathbb{Z}_n$ (from Investigation 5)—and quotient rings of the form $F[x]/\langle f(x)\rangle$ in a polynomial ring (from Investigation 15). If you studied group theory before ring theory, you should notice connections between the topics in this investigation and those in Investigation 27. In particular, the set R/I represents the set of distinct congruence classes modulo the ideal I in the same way that the set G/N represents the set of left cosets of N in G. Although G/N is a group only if N is normal in G, we don't have that problem in a ring R. Since R is an Abelian group under addition, and any ideal I is a subgroup under addition, I is also a normal subgroup of R under addition.

A homomorphism from a ring R to a ring S is a structure preserving map. The isomorphism theorems for rings show us different ways that homomorphisms can determine isomorphisms. If you studied group theory before ring theory, you should notice the connections between ring homomorphisms in this investigation and group homomorphisms in Investigation 30. A group homomorphism preserves the structure of a group, but since groups have only one operation, there is a bit less structure to preserve. In addition, a careful perusal of the ring isomorphism theorems shows that they are essentially the same as the corresponding group isomorphism theorems.

Investigation 17

Divisibility and Factorization in Integral Domains

Focus Questions

By the end of this investigation, you should be able to give precise and thorough answers to the questions listed below. You may want to keep these questions in mind to focus your thoughts as you complete the investigation.

- What is a Euclidean domain, and what are some examples of Euclidean domains?
- What is the difference between prime and irreducible elements in an integral domain?
- What is a unique factorization domain?
- What is the relationship between Euclidean domains, principal ideal domains, and unique factorization domains?
- What are two different ways of proving that every Euclidean domain is a unique factorization domain?

Preview Activity 17.1. Look back to the proof of the Fundamental Theorem of Arithmetic on page 38 of Investigation 4. List all of the results from Investigations 2 – 4 that were needed for the proof. Your list should include one theorem that is central to all the others. Identify this theorem, and explain why it is so important.

Introduction

We began our investigations of abstract algebra by studying arithmetic, divisibility, and factorization in the integers. Later, we generalized these ideas to polynomial rings. In this investigation, we will take our work one step further by showing that a result analogous to the Fundamental Theorem of Arithmetic holds in any integral domain that admits a division algorithm. This shouldn't come as too much of a surprise; after all, it was the Division Algorithm in $\mathbb{Z}$ that gave us the Euclidean Algorithm. We used the Euclidean Algorithm to prove Bezout's Identity, which in turn allowed us to prove Euclid's Lemma—the key ingredient in our proof of the uniqueness of prime factorizations. What we will see in this activity is that each of these steps can be generalized. In particular, we will argue that every *Euclidean domain*—that is, every integral domain that admits a division algorithm—is a *unique factorization domain*. We will consider two different proofs of this important result: one that mirrors our work in the integers and does not require the machinery of ideals, and one that uses ideals to prove a more general result. The former approach relies on material from Investigations 1 – 7 and references a few results from Investigations 12 and 13. The latter additionally relies on concepts from Investigation 16. For those who have not yet studied ideals, some of the relevant definitions from Investigation 16 are repeated here; these definitions are sufficient to complete most of the investigation, with the exception of the second proof of our main result.

Divisibility and Euclidean Domains

To begin our study of divisibility in integral domains, we must first specify more generally what it means for one element of an integral domain to divide another. The next definition does just that and is analogous to those we used in the integers and in polynomial rings.

Definition 17.2. Let D be an integral domain, and let $a, b \in D$. Then a **divides** b, denoted $a \mid b$, if there is an element $q \in D$ such that $b = aq$.

Next, we must define precisely what it means for an integral domain to "admit a division algorithm."

Definition 17.3. Let D be an integral domain with additive identity 0_D. Then D is a **Euclidean domain** if there is a function $\delta : D - \{0_D\} \to \mathbb{Z}^+ \cup \{0\}$ such that the following conditions hold:

(i) If $a, b \in D$ are nonzero, then $\delta(a) \leq \delta(ab)$.

(ii) If $a, b \in D$ with a nonzero, then there exist elements $q, r \in D$ such that

$$b = aq + r \text{ with either } r = 0_D \text{ or } \delta(r) < \delta(a).$$

The function δ in Definition 17.3 is called a **norm** for D.

Activity 17.4.

(a) How is the second condition in Definition 17.3 related to the Division Algorithm for $\mathbb{Z}$? What are the similarities and differences?

(b) Is $\mathbb{Z}$ a Euclidean domain? If so, what function serves as a norm for $\mathbb{Z}$, and why does this function satisfy the conditions listed in Definition 17.3?

(c) Let F be a field. Is $F[x]$ a Euclidean domain? If so, what function serves as a norm for $F[x]$, and why does this function satisfy the conditions listed in Definition 17.3?

Primes and Irreducibles

In order to generalize prime factorization, we will need to first revisit the way we defined prime numbers in our study of the integers. Recall that we defined a prime number in $\mathbb{Z}$ to be an integer

$p > 1$ whose only positive divisors are 1 and p. In general, the word *prime* is defined slightly differently, and what we've called prime in the past is actually closer to what is called *irreducible* in other contexts. It turns out that in the integers, the notions of prime and irreducible are equivalent. They are not, however, equivalent in all settings, as we will see shortly.

Definition 17.5. Let D be an integral domain.

- A nonzero element $a \in D$ is said to be **irreducible** provided that a is not a unit, and if $a = bc$ for some b, $c \in D$, then b or c is a unit.
- A nonzero element $a \in D$ is said to be **prime** provided that a is not a unit, and for all b, $c \in D$, if $a \mid bc$, then $a \mid b$ or $a \mid c$.
- Two elements a, $b \in D$ are said to be **associates** if $a = ub$ for some unit $u \in D$.

Activity 17.6.

(a) Explain why all of the integers that we called prime in Investigation 4 are in fact irreducible according to Definition 17.5.

(b) Explain why all of the integers that we called prime in previous activities are also prime according to Definition 17.5.

(c) List at least three irreducible integers that we did not previously consider to be prime. Would each of these integers be considered prime according to Definition 17.5?

(d) Which term defined in Definition 17.5 is closest to our definition of prime in $\mathbb{E}$? (See page 39.)

(e) The definition of prime should look familiar to you. How is this definition related to a result we used in Investigation 4?

(f) Find two integers that are associates and two that are not.

(g) Find two real numbers that are associates and two that are not.

It is important to note that the definition of prime from Definition 17.5 is more general than the definition we considered for $\mathbb{Z}$ in Investigation 4. In fact, as Activity 17.6 demonstrates, there are many integers that we did not consider prime in Investigation 4 that are considered prime according to Definition 17.5. Also, as noted earlier, the notions of prime and irreducible are equivalent in $\mathbb{Z}$. The next theorem formalizes this result.

Theorem 17.7. *An integer a is prime if and only if a is irreducible.*

To prove Theorem 17.7, we will first argue that if an integer a is prime, then a is irreducible. In fact, we will actually prove the more general result described in the next activity.

Activity 17.8. Let D be an integral domain, and let a be a prime element of D. Use the definition of prime to show that if $a = bc$ for some elements b, $c \in D$, then either b or c is a unit. Explain why this proves the following theorem:

Theorem 17.9. *Let D be an integral domain, and let $a \in D$. If a is prime, then a is irreducible.*

To prove the reverse implication of Theorem 17.7, first note that the statement we need to prove is nearly identical to Euclid's Lemma. (See page 36.) Thus, our proof of Euclid's Lemma should generalize nicely. The main difference is that we will need to replace our prior assumption of primality (in the sense that we defined prime in $\mathbb{Z}$) with the new notion of irreducibility.

Activity 17.10. Modify the proof of Euclid's Lemma from Investigation 4 to complete the proof of Theorem 17.7.

So now we have shown that the notions of prime and irreducible are equivalent in the integers, and that Euclid's Lemma works even with the more general assumption of irreducibility. The next natural question for us to consider is whether there are integral domains in which the notions of prime and irreducible are not equivalent. Since we have already shown that in *any* integral domain, every prime element is irreducible, we will be looking for an element of an integral domain that is irreducible, but not prime.

Activity 17.11. Let D be the set of all polynomials in $\mathbb{Z}[x]$ for which the coefficient on the linear term is zero. That is,

$$D = \{a_n x^n + a_{n-1}x^{n-1} + \cdots + a_2 x^2 + a_0\} \subseteq \mathbb{Z}[x].$$

Prove that D is an integral domain, and find a polynomial of the form $p(x) = x^n$ (for some $n \in \mathbb{Z}^+$) that is irreducible, but not prime.*

Activity 17.11 demonstrates that there are integral domains in which not all irreducible elements are prime. However, we will see later in the investigation that the notions of prime and irreducible are equivalent in certain types of integral domains—specifically, principal ideal domains.

Unique Factorization Domains

We are now ready to generalize our prior results about unique factorization—namely, the Fundamental Theorem of Arithmetic (see page 35) and the unique factorization theorem for polynomials over a field (Theorem 13.21 on page 175). We'll begin with a definition.

Definition 17.12. An integral domain D is said to be a **unique factorization domain** if the following conditions hold:

(i) Every nonzero, non-unit element of D is either irreducible or can be written as a finite product of irreducible elements of D.

(ii) Factorization into irreducibles is unique up to associates and the order of the factors. In particular, if $x \in D$ can be factored as a product of irreducibles in two different ways, say

$$x = p_1 p_2 \cdots p_m \text{ and } x = q_1 q_2 \cdots q_k$$

for some irreducible elements $p_1, p_2, \ldots, p_m, q_1, q_2, \ldots, q_k$ of D, then $m = k$ and the factors of x can be reordered so that p_i and q_i are associates for each i.

Condition (ii) of Definition 17.12 warrants some clarification. The idea behind it can be illustrated by a simple example.

Note that the integer 30 can be factored in several different ways. For instance,

$$30 = 2 \cdot 3 \cdot 5 \text{ and } 30 = -3 \cdot 5 \cdot -2.$$

*Gerald Wildenberg, "An integral domain lacking unique factorization into irreducibles," *Mathematics Magazine*, **80**, 2007, pp. 75–76.

Note also that each factor in the first product can be paired with an associate factor in the second product, and vice versa. In particular, 2 can be paired with -2, 3 with -3, and 5 with 5. Since each pair consists of two associate elements of $\mathbb{Z}$, the two factorizations are considered to be the same up to associates and the order of the factors.

Activity 17.13. Look back at the proof of the Fundamental Theorem of Arithmetic. (See page 38.) Can this proof be modified to show that $\mathbb{Z}$ is a unique factorization domain? If so, what modifications would need to be made, and why is the modified proof still valid? If not, why does the proof not generalize? Is $\mathbb{Z}$ a unique factorization domain?

And now, for the punchline—the theorem that generalizes our prior investigations of divisibility and factorization:

Theorem 17.14. *Every Euclidean domain is a unique factorization domain.*

Proof 1: Generalizing Greatest Common Divisors

One way to prove Theorem 17.14 is to recognize the observation from Preview Activity 17.1 that all of the work leading up to the Fundamental Theorem of Arithmetic (in Investigations 2 – 4) relied on the Division Algorithm. The same could be said for factorization of polynomials: Theorem 13.21 (see page 175) ultimately rests on the existence of a division algorithm for polynomials (page 155). In order to generalize our work from these two settings, we will need one more definition.

Definition 17.15. Let D be a Euclidean domain with norm function δ, and let a and b be elements of D, not both zero. Then a **greatest common divisor** of a and b is an element $d \in D$ such that:

(i) $d \mid a$ and $d \mid b$; and

(ii) for all $d' \in D$, if $d' \mid a$ and $d' \mid b$, then $\delta(d') \leq \delta(d)$.

The elements a and b are said to be **relatively prime** if 1 is a greatest common divisor of a and b.

Note that condition (i) of Definition 17.15 is what makes d a common divisor of a and b, whereas condition (ii) makes d a *greatest* common divisor in the sense that d has a norm greater than or equal to any other common divisor of a and b. Note also that we refer to *a* greatest common divisor of a and b instead of *the* greatest common divisor of a and b. This suggests that greatest common divisors, at least according to Definition 17.15, are not unique. The next activity confirms this suspicion.

Activity 17.16. Find two distinct greatest common divisors of 6 and 8 in $\mathbb{Z}$. What is the relationship between these two greatest common divisors? (Recall that the absolute value function is a norm for $\mathbb{Z}$.)

As Activity 17.16 demonstrates, it is possible for two elements a and b of a Euclidean domain to have more than one greatest common divisor. However, any two greatest common divisors of a and b must be associates. (See Exercise 3.) It can also be shown, as the next activity suggests, that greatest common divisors always exist, as long as at least one of the elements involved is nonzero.

Activity 17.17. Prove that if a and b are elements of a Euclidean domain D, not both zero, then there is at least one greatest common divisor of a and b. (Hint: 1 necessarily divides both a and b, and so there is at least one common divisor. Use the definition of the norm function to argue that at least one of these common divisors must be maximal.)

Now that we have formally defined greatest common divisors for any Euclidean domain, we are ready to proceed with the proof of Theorem 17.14. Rather than providing a complete proof, we will state a sequence of theorems that can be used to build up the proof one step at a time. These theorems should look familiar to you. In fact, each one is a generalization of an analogous result that we proved for the integers. Thus, you should be able to prove each of the theorems by making small modifications to the proofs of the analogous results from Investigations 2 – 4. Together, they establish Theorem 17.14—that every Euclidean domain is a unique factorization domain.

Theorem 17.18. *Let D be a Euclidean domain with norm function δ, and let a and b be elements of D, not both zero. Furthermore, let q and r be elements of D such that $b = aq + r$ and $\delta(r) < \delta(a)$. Then d is a greatest common divisor of a and b if and only if d is a greatest common divisor of a and r.*

Theorem 17.19. *Let D be a Euclidean domain, and let a and b be elements of D, not both zero. Then the last nonzero remainder produced by the Euclidean algorithm (starting with a and b) is a greatest common divisor of a and b.*

Theorem 17.20. *Let D be a Euclidean domain, let a and b be elements of D, not both zero, and let d be the greatest common divisor of a and b produced by the Euclidean algorithm. Then there exist $x, y \in D$ such that $ax + by = d$.*

Theorem 17.21. *Let D be a Euclidean domain, and let p and a be relatively prime elements of D. For all $b \in D$, if $p \mid ab$, then $p \mid b$.*

Theorem 17.22. *Let D be a Euclidean domain, and let $p \in D$ be irreducible. For all $a, b \in D$, if $p \mid ab$, then $p \mid a$ or $p \mid b$.*

Theorem 17.23. *Let D be a Euclidean domain. Then every nonzero, non-unit element of D is either irreducible or can be written as a finite product of irreducible elements of D.*

Theorem 17.24. *Let D be a Euclidean domain, and let $x \in D$. Suppose*

$$x = p_1 p_2 \cdots p_m \ \text{ and } \ x = q_1 q_2 \cdots q_k$$

for some irreducible elements $p_1, p_2, \ldots, p_m, q_1, q_2, \ldots, q_k$ of D. Then $m = k$, and the factors of x can be reordered so that p_i and q_i are associates for each i.

Proof 2: Principal Ideal Domains

If you have studied ideals (in Investigation 16), then a second way to approach the proof of Theorem 17.14 is to prove the following more general result:

Theorem 17.25. *Every principal ideal domain is a unique factorization domain.*

Note that, by Theorem 16.13 (see page 221), every Euclidean domain is a principal ideal domain. Therefore, if we can prove Theorem 17.25, we will have also established that every Euclidean domain is a unique factorization domain.

The first step in proving Theorem 17.25 is to show that every principal ideal domain satisfies a condition known as the *ascending chain condition*. To illustrate the ascending chain condition, recall that $\mathbb{Z}$ is a principal ideal domain. Therefore, any ideal of $\mathbb{Z}$ has the form $\langle m \rangle$ for some positive integer m. Consider, for example, the ideal $I = \langle 24 \rangle$. Note that the positive divisors of 24 are 1, 2, 3, 4, 6, 12, and 24. For each such divisor k, let $I_k = \langle k \rangle$. Then

$$I = I_{24} \subset I_{12} \subset I_6 \subset I_3 \subset I_1 = \mathbb{Z}$$

and

$$I = I_{24} \subset I_{12} \subset I_4 \subset I_2 \subset I_1 = \mathbb{Z}.$$

No matter which set of factors of 24 we use, the corresponding "chain" of ideals must always eventually terminate—that is, it must contain only a finite number of distinct ideals. Since any given integer has only a finite number of factors, the same idea applies to any ideal in $\mathbb{Z}$. Because of this, we say that $\mathbb{Z}$ satisfies the ascending chain condition on ideals, defined formally below.

Definition 17.26. A commutative ring R is said to satisfy the **ascending chain condition** on ideals[†] if there does not exist an infinite sequence of ideals of R, each of which is a proper subset of the next. In other words, R satisfies the ascending chain condition if whenever

$$I_1 \subseteq I_2 \subseteq I_3 \subseteq \cdots$$

is a chain of ideals of R, then there is some integer m for which $I_k = I_m$ for all $k \geq m$.

Are there rings that do not satisfy the ascending chain condition? The answer is yes, although we have to dig a little deeper to find them.

Activity 17.27. Let $\mathcal{C}(\mathbb{R})$ denote the set of all continuous functions from $\mathbb{R}$ to $\mathbb{R}$, with addition and multiplication defined by

$$(f+g)(x) = f(x) + g(x) \quad \text{and} \quad (fg)(x) = f(x)g(x)$$

for all $f, g \in \mathcal{C}(\mathbb{R})$. It can be shown fairly easily that $\mathcal{C}(\mathbb{R})$ is a commutative ring with identity. For every integer $n \geq 0$, let I_n denote the set of all functions f in $\mathcal{C}(\mathbb{R})$ for which $f(x) = 0$ for all $x \geq n$.

(a) Show that each I_n is an ideal of $\mathcal{C}(\mathbb{R})$.

(b) Show that I_n is a proper subset of I_{n+1} for all n.

(c) Explain how parts (a) and (b) establish that $\mathcal{C}(\mathbb{R})$ does not satisfy the ascending chain condition.

That $\mathbb{Z}$ satisfies the ascending chain conditions, while other rings, like $\mathcal{C}(\mathbb{R})$, do not is a consequence of the fact that $\mathbb{Z}$ is a principal ideal domain. The next theorem states this result formally, and the activity that follows outlines its proof.

Theorem 17.28. *Every principal ideal domain satisfies the ascending chain condition.*

Activity 17.29. Let D be a principal ideal domain, and let $I_1 \subseteq I_2 \subseteq I_3 \subseteq \cdots$ be a chain of ideals of D. Define the set I to be the union of all of these ideals—that is,

$$I = \bigcup_{i=1}^{\infty} I_i.$$

Note that $x \in I$ if and only if $x \in I_i$ for some i.

[†]Rings that satisfy the ascending chain condition on ideals are also called *Noetherian* rings, named after German mathematician Emmy Noether (1882-1935), who is known for her many contributions to abstract algebra and theoretical physics.

(a) Use the Ideal Test (see page 218) to show that I is an ideal of D.

(b) Use the fact that D is a principal ideal domain to conclude that $I = \langle a \rangle$ for some $a \in D$. Then use the definition of I to deduce that $a \in I_m$ for some m.

(c) Explain why it must be that $I_m = \langle a \rangle$, and so $I_k = I_m = I$ for all $k \geq m$. Deduce that D satisfies the ascending chain condition.

The ascending chain condition allows us to establish that every nonzero, non-unit element in a principal ideal domain can be factored into irreducibles. In particular:

Theorem 17.30. *Let D be a principal ideal domain. Then every nonzero, non-unit element of D is either irreducible or can be written as a finite product of irreducible elements of D.*

Proof. Let D be a principal ideal domain, and let a be a nonzero, non-unit element of D. If a is irreducible, then we are done. Therefore, suppose that $a = a_1 a_2$ for some non-unit elements a_1, $a_2 \in D$. If both a_1 and a_2 are irreducible, then, once again, we are done. So suppose that one of a_1 or a_2—say, a_1—is not irreducible. Then there exist $a_{1,1}, a_{1,2} \in D$, with neither $a_{1,1}$ nor $a_{1,2}$ a unit, such that $a_1 = a_{1,1}a_{1,2}$. Continue factoring in this manner for as long as possible—that is, continue to factor the factors of a, and their factors, and so on. This process will terminate if and only if a can be written as a finite product of irreducible elements. Now suppose the process does not terminate. That is, suppose there is a sequence a_1, $a_{1,1}$, $a_{1,1,1}$, $\ldots$ of non-unit elements of D such that

$$\cdots \mid a_{1,1,1} \mid a_{1,1} \mid a_1 \mid a.$$

Then we would have an infinite chain of ideals

$$\langle a \rangle \subset \langle a_1 \rangle \subset \langle a_{1,1} \rangle \subset \langle a_{1,1,1} \rangle \subset \cdots,$$

each of which is properly contained in the next. (The proof that these containments are proper is left to you in Activity 17.35.) But this is a contradiction to the ascending chain condition. Therefore, it must be that a can be written as a finite product of irreducible elements of D, as desired. ■

Now that we have established the existence of factorizations into irreducibles in any principal ideal domain, the next step in proving Theorem 17.25 is to show that these factorizations are unique. As in the integers, we will need a result analogous to Euclid's Lemma. We will prove such a result by first showing that the notions of prime and irreducible are equivalent in every principal ideal domain.

Theorem 17.31. *Let D be a principal ideal domain, and let $a \in D$. Then a is prime if and only if a is irreducible.*

Proof. Let D be a principal ideal domain, and let $a \in D$. By Theorem 17.9, if a is prime, then a is irreducible. Thus, it remains to show that if a is irreducible, then a is prime. Let $a \in D$ be irreducible. First note that a is prime if and only if $\langle a \rangle$ is a prime ideal. Furthermore, by Theorem 16.35 (see page 227), every maximal ideal is prime. Therefore, we can show that a is prime by showing that $\langle a \rangle$ is maximal. Let I be any ideal containing $\langle a \rangle$. Since D is a principal ideal domain, $I = \langle b \rangle$ for some $b \in D$. Furthermore, since $a \in I$, it must be that $a = bd$ for some $d \in D$. But a is irreducible, and so either b or d must be a unit. If b is a unit, then $I = \langle b \rangle = D$. If d is a unit, then $I = \langle b \rangle = \langle a \rangle$. Therefore, the only ideals containing $\langle a \rangle$ are D and $\langle a \rangle$. It follows that $\langle a \rangle$ is maximal, and so a is prime. ■

A corollary of Theorem 17.31 provides the analog of Theorem 17.22 for principal ideal domains. In particular:

Corollary 17.32. *Let D be a principal ideal domain, and let $p \in D$ be irreducible. For all $a, b \in D$, if $p \mid ab$, then $p \mid a$ or $p \mid b$.*

Corollary 17.32 can then be used to prove the following analog of Theorem 17.24, which establishes the uniqueness of factorizations into irreducibles in any principal ideal domain. Its proof is completely analogous to the proof of the uniqueness portion of the Fundamental Theorem of Arithmetic from Investigation 4 (see page 38) and is therefore left as an exercise. (See Exercise 4.)

Theorem 17.33. *Let D be a principal ideal domain, and let $x \in D$. Suppose*

$$x = p_1 p_2 \cdots p_m \text{ and } x = q_1 q_2 \cdots q_k$$

for some irreducible elements $p_1, p_2, \ldots, p_m, q_1, q_2, \ldots, q_k$ of D. Then $m = k$, and the factors of x can be reordered so that p_i and q_i are associates for each i.

Taken together, Theorems 17.30 and 17.33 establish that every principal ideal domain is a unique factorization domain. And, as noted earlier, since every Euclidean domain is a principal ideal domain, Theorem 17.14 then follows as a corollary.

Concluding Activities

Activity 17.34. Prove Theorems 17.18 – 17.24. By doing so, you will prove Theorem 17.14.

Activity 17.35. Let D be a principal ideal domain, and let $a \in D$ such that $a \neq 0$ and $a = a_1 a_2$ for some non-unit elements $a_1, a_2 \in D$. Prove that $\langle a \rangle \subset \langle a_1 \rangle \subset D$, where each containment is proper.

Activity 17.36. We have now seen that every principal ideal domain, and hence every Euclidean domain, is a unique factorization domain. So if we want to find an integral domain that is not a unique factorization domain, we will need to consider integral domains that are neither Euclidean domains nor principal ideal domains. How can we look for such integral domains? One way is to use Theorem 17.31.

(a) Use Theorem 17.31 to explain why the integral domain D in Activity 17.11 is not a Euclidean domain.

(b) Prove that D is not a unique factorization domain.

Activity 17.37. Write a short summary that describes the important concepts, and the relationships between these concepts, that were introduced in this investigation. Explain how the material in this investigation is connected to the material in Investigations 2, 3, 4, 12, 13, and 16.

Exercises

(1) Prove that the set $\mathbb{Z}[i]$ of Gaussian integers (see Exercise 2 on page 75) is a Euclidean domain.

(2) **Primes and irreducibles in Euclidean domains.** Generalize Theorem 17.7 by proving that in any Euclidean domain, every irreducible element is prime. Deduce that the notions of prime and irreducible are equivalent in Euclidean domains.

⋆ (3) **Uniqueness of greatest common divisors.** Let D be a Euclidean domain, and let $a, b \in D$, not both zero. Prove that if d and d' are both greatest common divisors of a and b, then d and d' are associates.

⋆ (4) Prove Theorem 17.33.

(5) (a) Show that the ring $\mathcal{C}(\mathbb{R})$ (from Activity 17.27) is an integral domain.

(b) Show that $\mathcal{C}(\mathbb{R})$ is not a principal ideal domain.

(6) Prove that every field is a Euclidean domain.

(7) In this investigation, we showed that every principal ideal domain is a unique factorization domain, but does the converse hold? That is, is every unique factorization domain a principal ideal domain? As it turns out, the answer is no, and $\mathbb{Z}[x]$ is a canonical example of a unique factorization domain that is not a principal ideal domain.

(a) Construct an example to show that $\mathbb{Z}[x]$ is not a principal ideal domain. (That is, find an ideal of $\mathbb{Z}[x]$ that is not principal.)

(b) Find a proof, either online or in a textbook, that $\mathbb{Z}[x]$ is a unique factorization domain. How do the techniques used in the proof you found compare to those in this investigation, Investigation 4, and Investigation 13?

Connections

In Investigations 1 – 4, we proved the Division Algorithm for integers and then used the Division Algorithm to establish several other important results, all leading up to the Fundamental Theorem of Arithmetic. We followed a similar progression for polynomials in Investigations 12 and 13, proving the Division Algorithm for polynomials over a field and ultimately showing that all such polynomials have unique factorizations into irreducibles. Unique factorization in each of these settings is a consequence of the more general result we explored in this investigation—namely, that every Euclidean domain is a unique factorization domain. This result can be proved by either adapting the proof of the Fundamental Theorem of Arithmetic (and the results leading up to it) or—if you studied ideals in Investigation 16—by proving the stronger result that every principal ideal domain is a unique factorization domain.

Investigation 18

From $\mathbb{Z}$ to $\mathbb{C}$

Focus Questions

By the end of this investigation, you should be able to give precise and thorough answers to the questions listed below. You may want to keep these questions in mind to focus your thoughts as you complete the investigation.

- What is the field of fractions of an integral domain, and how is this field constructed?
- What are real numbers, and how can they be constructed from the rational numbers? What special property do the real numbers have that the rational numbers do not?
- How can we formally define the set of complex numbers? What special property do the complex numbers have that the real numbers do not?
- What makes $\mathbb{Z}$ unique among all ordered integral domains?

Preview Activity 18.1. Let $Q = \{(a, b) : a, b \in \mathbb{Z} \text{ and } b \neq 0\}$. Define a relation $\sim$ on Q as follows:

$$(a, b) \sim (c, d) \text{ if } ad = bc.$$

(a) Find three elements in Q that are related to $(1, 2)$.

(b) Find all of the elements in Q that are related to $(0, 1)$.

(c) Is $\sim$ a reflexive relation? Explain.

(d) Is $\sim$ a symmetric relation? Explain.

(e) Is $\sim$ a transitive relation? Explain.

(f) Is there a natural way to define what it means for an element of Q to be positive, negative, or zero? Explain.

(g) Is there a natural way to define an ordering on Q? Explain.

(h) Compare Q (with the relation $\sim$) to the set $\mathbb{Q}$ of rational numbers. What similarities and differences do you see?

Introduction

Throughout previous investigations, we have studied many different number systems, including the integers, the rational numbers, the real numbers, and the complex numbers. We have defined these number systems with varying degrees of formality, but we have also relied on our intuition to some extent. In this investigation, we will consider some of the formal details that we have glossed over in our prior work. Preview Activity 18.1 gives a hint of what this process will look like. It suggests one way to *formally* define the rational numbers as equivalence classes of ordered pairs of integers. And just as the rational numbers can be constructed from the integers, the real numbers can be constructed from the rational numbers, and the complex numbers from the real numbers. But before we get to any of that, we must start with the most basic of all numbers: the counting numbers.

From $\mathbb{W}$ to $\mathbb{Z}$

Counting techniques have been important for ages.

- Ancient cultures used pebbles or notches on sticks to identify a number.
- The Mayas, Aztecs, and Celts adopted base 20 counting systems (using both fingers and toes).
- The Babylonians "invented" the number 0 and used a counting system with base 60. (To this we owe the division of the hour into 60 minutes and the division of the central angle of a circle into 360 angles of measure 1 degree.)

While many of the counting techniques used by these civilizations, and others, were quite sophisticated, none of them wrestled with the logical underpinnings of number systems. While this lack of formality does not present a big problem from a practical standpoint, from the mathematical point of view, it prohibits a formal development of the mathematical ideas that are based on these number systems. In this sense, much of the mathematics we take for granted rests on the logical foundation of number systems, which, as it turns out, was not developed until the late 1800s.

Giuseppe Peano (1858-1932) was an Italian mathematician whose mathematical works spanned the fields of analysis, differential equations, mathematical logic, and set theory.* In 1889, Peano published his famous axioms, now called the Peano axioms, which constructed a formal foundation upon which whole numbers could be defined in terms of sets. The Peano Axioms are as follows (where $\mathbb{W}$ is the set of whole numbers):

I. $0 \in \mathbb{W}$.

II. There is a function S from $\mathbb{W}$ to $\mathbb{W}$ (called the successor function).

III. Suppose $U \subseteq \mathbb{W}$ with $0 \in U$. If $n \in U$ implies $S(n) \in U$, then $U = \mathbb{W}$. (This property is the Principle of Mathematical Induction.)

*Peano is especially noted for his invention (or discovery) of space filling curves. These are continuous mappings from the closed interval [0,1] onto the unit square in the plane. The existence of such curves has been described as one of the most remarkable facts in set theory.

IV. If $n \in \mathbb{W}$, then $S(n) > 0$.

V. If $n, m \in \mathbb{W}$ and $S(n) = S(m)$, then $n = m$.

In these axioms, the whole number $n + 1$ is the successor of the whole number n, and the successor function S maps each whole number to its successor.

As it turns out, all of the structure of the integers can be derived from the Peano axioms.[†] These axioms created a logical foundation for our integer, real, and complex number systems and at one time were considered to be the fountainhead of all mathematical knowledge. From this point forward, we will assume the Peano axioms and the subsequent construction of the ring of integers. More specifically, we will assume the familiar properties of the set of integers $\mathbb{Z}$ under the standard addition and multiplication operations, as described below.

(1) $a + b \in \mathbb{Z}$ for every $a, b \in \mathbb{Z}$. (The set $\mathbb{Z}$ is closed under addition.)

(2) $a + b = b + a$ for every $a, b \in \mathbb{Z}$. (Addition is commutative in $\mathbb{Z}$.)

(3) $(a + b) + c = a + (b + c)$ for all $a, b, c \in \mathbb{Z}$. (Addition is associative in $\mathbb{Z}$.)

(4) $\mathbb{Z}$ contains an element 0 so that $0 + a = a$ for all $a \in \mathbb{Z}$. (0 is an additive identity in $\mathbb{Z}$.)

(5) For each $a \in \mathbb{Z}$ there is an element $-a \in \mathbb{Z}$ so that $a + (-a) = 0$. (Each element in $\mathbb{Z}$ has an additive inverse in $\mathbb{Z}$.) Using the additive inverse, we can define subtraction on $\mathbb{Z}$ by $a - b = a + (-b)$.

(6) $ab \in \mathbb{Z}$ for every $a, b \in \mathbb{Z}$. ($\mathbb{Z}$ is closed under multiplication.)

(7) $(ab)c = a(bc)$ for all $a, b, c \in \mathbb{Z}$. (Multiplication is associative in $\mathbb{Z}$.)

(8) $\mathbb{Z}$ contains an element $1 \neq 0$ so that $1a = a = a1$ for all $a \in \mathbb{Z}$. (1 is a multiplicative identity in $\mathbb{Z}$.)

(9) $a(b + c) = ab + ac$ for all $a, b, c \in \mathbb{Z}$. (Multiplication in $\mathbb{Z}$ distributes over addition in $\mathbb{Z}$.)

(10) There is a subset of $\mathbb{Z}$, denoted $\mathbb{Z}^+$ and called the positive integers, so that for each $a \in \mathbb{Z}$, exactly one of the following is true:

$$a \in \mathbb{Z}^+, \qquad a = 0, \qquad \text{or} \qquad -a \in \mathbb{Z}^+.$$

(This property is called the **trichotomy** principle.) Using the trichotomy principle, we can define a relation $<$ on $\mathbb{Z}$ by

$$a < b \qquad \text{if and only if} \qquad b - a \in \mathbb{Z}^+.$$

This inequality relation satisfies all of the standard properties with which we are familiar. Other inequalities ($>$, $\geq$, and $\leq$) are defined in a similar way.

Note that these axioms mirror those that were introduced in Investigation 1, with the exception of trichotomy, which is stated somewhat differently. As we will see shortly, this variation of trichotomy implies the version that we stated for the less than ($<$) relation on page 8.

The integers are a natural place to begin the study of algebra, in part because the integers are the smallest ring with identity and characteristic zero. (More formally, every ring with identity and

[†]The axioms were published in a pamphlet *Arithmetices principia, nova methodo exposita*, which was applauded as a landmark in the history of mathematical logic and the foundations of mathematics.

characteristic zero contains a subring isomorphic to the integers.) In addition, the foundation of algebra is the study of polynomials and solutions to polynomial equations, so to study algebra we might begin by asking what kinds of polynomial equations we can solve within the set $\mathbb{Z}$. We can certainly solve some simple polynomial equations (e.g., $x - 2 = 0$) in $\mathbb{Z}$, but there are many others that we cannot solve. For example, the polynomial equation $2x - 1 = 0$ has no integer solutions. To solve this and other polynomial equations with integer coefficients, we will need to have bigger sets from which to draw solutions. In particular, to solve equations like $ax + b = 0$, where a and b are integers with $a \neq 0$, we will need to find a set that contains the multiplicative inverse of each nonzero integer. To do so, we will formally construct what is called the *field of fractions* of $\mathbb{Z}$, which turns out to be just the rational numbers. Before we do that, however, we will need to take a short digression to talk about ordered rings.

Ordered Rings

The less than ($<$) relation on the set of integers allows us to compare integers so that the entire set of integers can be organized on a number line. In particular, the trichotomy principle implies that any integer can be compared to the integer 0; the integers that are greater than 0 are placed to the right of 0 on the number line and the integers that are less than 0 are placed to the left of 0. To determine how two different integers a and b are placed relative to each other, we just need to know whether $a - b$ is positive or $b - a$ is positive. The placement of every integer is thus determined by the positive integers. As it turns out, this idea generalizes to other rings as well. Whenever a ring contains a set of elements that behaves like the positive integers, it can be made into an *ordered ring*, defined formally as follows:

Definition 18.2. An **ordered ring** is a commutative ring R containing a subset P such that:

(i) P is not empty;

(ii) if $a \in P$ and $b \in P$, then $a + b \in P$ and $ab \in P$; and

(iii) for any $a \in R$, exactly one of following conditions holds:

$$a \in P, \quad a = 0, \quad \text{or} \quad -a \in P.$$

The elements in P are called the *positive* elements of the ordered ring. If $a \in P$, then we will also write $a > 0$. The integers form the standard example of an ordered ring. But the integers are also an integral domain, and this is not a requirement in the definition of an ordered ring. An ordered ring that is also an integral domain is called an **ordered domain**. However, it turns out that an ordered ring cannot have any zero divisors, and so an ordered ring with identity must be an ordered domain. The next theorem formalizes this result.

Theorem 18.3. *An ordered ring contains no zero divisors.*

Proof. Suppose R is an ordered ring, and let $a, b \in R$ with both a and b nonzero. If $a, b > 0$, then $ab > 0$ and so $ab \neq 0$. Suppose one of a or b is positive and the other negative. Without loss of generality, assume $a > 0$ and $b < 0$. Then $-b > 0$ and so $(a)(-b) > 0$. Thus, $-(ab) > 0$, which means that $ab \neq 0$. The final case is when $-a > 0$ and $-b > 0$. Then $ab = (-a)(-b) > 0$, and so $ab \neq 0$. ∎

We know of several ordered rings: $\mathbb{Z}$, $\mathbb{Q}$, and $\mathbb{R}$, for example. As a less familiar example, let R be an ordered ring, and define the set P to be the polynomials in $R[x]$ with positive leading coefficient. Then the ring $R[x]$ with positive elements P is an ordered ring. (See Exercise 8.)

There are some standard properties of elements in ordered rings that we might expect to be true, and a few are given in the next lemma.

Lemma 18.4. *Let R be an ordered ring.*

(i) *If $a \in R$ is not the zero element, then $a^2 > 0$.*

(ii) *If R contains an identity, then $1 > 0$.*

(iii) *If R contains an identity, then $(-1) < 0$.*

The proofs of the second and third parts of Lemma 18.4 are left for you as exercises. (See Exercise 2.) The next activity provides some guidance for proving the first part of the lemma.

Activity 18.5. Let R be an ordered ring, and let $a \in R$. We will show that $a^2 > 0$ by considering cases.

(a) If $a \neq 0$ in R, what else can we say about a? Explain.

(b) If $a > 0$, why must a^2 be positive?

(c) Complete the proof of part 1 of Lemma 18.4 by assuming that $(-a) > 0$ and proving that $a^2 > 0$.

The positive elements in an ordered ring allow us to compare elements to 0, but we know in the integers that we can compare any two elements to each other. For example, we know that $4 > 2$ since $4 - 2 > 0$. We can extend this idea to any ordered ring. If R is an ordered ring and $a, b \in R$, then we know by trichotomy that exactly one of the following must be true: $a - b > 0$, $a - b = 0$, or $-(a - b) > 0$. When $a - b > 0$ we will say that a is greater than b and write $a > b$. In the case where $a - b = 0$, we have that $a = b$. When $-(a - b) > 0$, we say that b is greater than a and write $b > a$. If $b > a$, we will also say that a is less than b and write $a < b$. In this way, we can use the set of positive elements to define a less than relation (an ordering) on the entire ring. There are many familiar properties satisfied by the less than relation $<$ on $\mathbb{Z}$ (and also by the greater than relation). We might expect these properties to hold in any ordered ring, and they in fact do. The next theorem formalizes this result. You may recognize the first four parts from the ordering axioms on page 8 of Investigation 1. Part (v) is a generalization of Exercise 8 from the same investigation. (See page 10.)

Theorem 18.6. *Let R be an ordered ring. The following conditions are satisfied.*

(i) ***Trichotomy.*** *For all a, $b \in R$, exactly one of the conditions*

$$a < b, \ a = b, \ a > b$$

is satisfied.

(ii) ***Transitivity.*** *For all a, b, $c \in R$, if $a > b$ and $b > c$, then $a > c$.*

(iii) ***Translation invariance.*** *For all a, b, $c \in R$, if $a > b$, then $a + c > b + c$.*

(iv) ***Scaling.*** *For all a, b, $c \in R$, if $a > b$ and $c > 0$, then $ac > bc$.*

(v) *For all a, b, $c \in R$, if $ac > bc$ and $c > 0$, then $a > b$.*

Proof. Let $a, b \in R$. By condition (iii) of Definition 18.2, we know that exactly one of $a - b > 0$, $a - b = 0$, or $-(a - b) > 0$ must be true. Now $a - b > 0$ if and only if $a > b$; $a - b = 0$ if and only if $a = b$; and $-(a - b) > 0$ if and only if $b - a > 0$ or, equivalently, $b > a$. This completes the proof of the trichotomy condition.

Now let $a, b, c \in R$. Assume $a > b$ and $b > c$. Then $a-b > 0$ and $b-c > 0$. So $(a-b)+(b-c) > 0$. But this means that $a - c > 0$ or, equivalently, $a > c$, proving transitivity.

For part (iii), assume $a > b$, and let c be any element in R. Then $a - b > 0$. Note that $0 < a - b = (a + c) - (b + c)$, and so $a + c > b + c$.

The proof of the last two parts is left for the exercises. (See Exercise 3.) ■

Now that we have considered ordered rings, we can return to our problem of constructing the field of rational numbers from the ring of integers.

From $\mathbb{Z}$ to $\mathbb{Q}$

In this section, we will investigate how to formally construct the field of rational numbers from the integers. It turns out that the same construction works for any integral domain, so we will make the argument in that more general context. In particular, we will show that for any integral domain D, we can construct a field that contains an isomorphic copy of D and also contains solutions to all linear equations in $D[x]$. The details of this construction are essentially contained in Preview Activity 18.1. To translate the intuition from that activity to our more general construction, it might be helpful throughout this section to think of the integral domain D as the ring of integers.

Recall that rational numbers are fractions made from pairs of integers. So we can think of a rational number as an ordered pair (a, b), where $a, b \in \mathbb{Z}$ with $b \neq 0$. Unfortunately, these ordered pairs do not give us the rational numbers because, for example, the ordered pairs $(1, 2)$ and $(2, 4)$ are different, but the rational numbers $\frac{1}{2}$ and $\frac{2}{4}$ are equal. To rectify this problem, we define a relation on the set of ordered pairs so that (a, b) is related to (c, d) if $ad = bc$. In this way, equivalent ordered pairs represent "equal" rational numbers. This same idea works in any integral domain.

Let D be an integral domain, and let

$$D' = \{(a, b) : a, b \in D \text{ with } b \neq 0\}.$$

We will define a relation $\sim$ on D' by saying that $(a, b) \sim (c, d)$ in D' if and only if $ad = bc$ in D. As with any relation, it is natural to consider what kind of relation $\sim$ is.

Activity 18.7.

(a) Prove that $\sim$ is a reflexive relation.

(b) Prove that $\sim$ is a symmetric relation.

(c) Prove that $\sim$ is a transitive relation.

(d) What do your answers to parts (a) – (c) allow you to conclude about $\sim$?

Since $\sim$ is an equivalence relation, $\sim$ partitions D' into disjoint equivalence classes. (See Theorem 5.6 on page 49.) We will denote the equivalence class of $(a, b) \in D'$ as $[(a, b)]$. Let $Q(D)$ be the collection of equivalence classes of elements in D'—that is,

$$Q(D) = \{[(a, b)] : a, b \in D \text{ with } b \neq 0\}.$$

Note that $[(a, b)] = [(c, d)]$ in $Q(D)$ if and only if $ad = bc$ in D.

To make $Q(D)$ into a field like $\mathbb{Q}$, we will need to define addition and multiplication operations on $Q(D)$ that mimic addition and multiplication in $\mathbb{Q}$. Recall that the sum $\frac{a}{b} + \frac{c}{d}$ and product $\left(\frac{a}{b}\right)\left(\frac{c}{d}\right)$ of two rational numbers $\frac{a}{b}$ and $\frac{c}{d}$ are defined as

$$\frac{a}{b} + \frac{c}{d} = \frac{ad + bc}{bd} \quad \text{and} \quad \left(\frac{a}{b}\right)\left(\frac{c}{d}\right) = \frac{ac}{bd}.$$

We can easily translate these definitions to $Q(D)$. For $[(a, b)], [(c, d)] \in Q(D)$, define the sum and product by

$$[(a, b)] + [(c, d)] = [(ad + bc, bd)] \quad \text{and} \quad [(a, b)][(c, d)] = [(ac, bd)]. \tag{18.1}$$

Activity 18.8.

(a) Let $D = \mathbb{Z}$. Compute the sums $[(1, 3)] + [(4, 5)]$ and $[(2, 6)] + [(-4, -5)]$ in $Q(\mathbb{Z})$. What do you notice about these sums? Why should we have expected this?

(b) Let $D = \mathbb{Z}$. Compute the products $[(1, 3)][(4, 5)]$ and $[(2, 6)][(-4, -5)]$ in $Q(\mathbb{Z})$. What do you notice about these products? Why should we have expected this?

(c) Describe in symbols what it would mean for addition in $Q(D)$ to be well-defined.

(d) Describe in symbols what it would mean for multiplication in $Q(D)$ to be well-defined.

(e) Verify that addition in $Q(D)$ is well-defined, and explain why it is important to do so.

(f) Verify that multiplication in $Q(D)$ is well-defined

Now that we have well-defined addition and multiplication operations on $Q(D)$, we need to determine if $Q(D)$ is a field.

Activity 18.9. Let D be an integral domain.

(a) What element would you expect to be the additive identity in $Q(D)$? Prove that you have the right element.

(b) What element would you expect to be the additive inverse of the element $[(a, b)]$ in $Q(D)$? Prove that you have the right element.

(c) What element would you expect to be the multiplicative identity in $Q(D)$? Prove that you have the right element.

(d) Now we will determine the units in $Q(D)$.

(i) In order for an element to be a unit in $Q(D)$, it first has to be nonzero. What can we say about $a \in D$ if the element $[(a, b)]$ is nonzero in $Q(D)$? Prove your answer.

(ii) If $[(a, b)]$ is nonzero in $Q(D)$, must $[(a, b)]$ be a unit? If so, determine $[(a, b)]^{-1}$, and prove your answer. If not, give an example of a nonzero element of $Q(D)$ that is not a unit.

Activity 18.9 establishes some of the details that we need to prove the next theorem.

Theorem 18.10. *The set $Q(D)$, with the operations in (18.1), is a field.*

We should have learned in Activity 18.9 that $[(0,1)]$ is the additive identity in $Q(D)$, that $[(-a,b)]$ is the additive inverse of the element $[(a,b)]$ in $Q(D)$, that $[(1,1)]$ is the multiplicative identity in $Q(D)$, and that $[(b,a)]$ is the multiplicative inverse of the nonzero element $[(a,b)]$ in $Q(D)$. To complete the proof of Theorem 18.10, we need to know that $Q(D)$ is closed under its operations, addition and multiplication are commutative and associative in $Q(D)$, and that multiplication distributes over addition in $Q(D)$. We will verify that multiplication distributes over addition in $Q(D)$ and leave the proofs of the remaining properties as an exercise. (See Exercise 6.)

To show that multiplication distributes over addition in $Q(D)$, let $[(a,b)]$, $[(c,d)]$, and $[(e,f)]$ be in $Q(D)$. The commutative, associative and distributive laws from D, along with Exercise 5, show that

$$\begin{aligned}
[(a,b)]([(c,d)]+[(e,f)]) &= [(a,b)][(cf+de,df)] \\
&= [(a(cf+de),b(df))] \\
&= [(acf+ade,bdf)] \\
&= [(b(acf+dae),b(bdf))] \\
&= [((ac)(bf)+(bd)(ae),(bd)(bf))] \\
&= [(ac,bd)]+[(ae,bf)] \\
&= [(a,b)][(c,d)]+[(a,b)][(e,f)].
\end{aligned}$$

Thus, left multiplication distributes over addition in $Q(D)$. The commutativity of multiplication in $Q(D)$ shows that right multiplication also distributes over addition in $Q(D)$.

The field $Q(D)$ is called the *field of fractions* or the *quotient field* of the integral domain D. Since we defined the elements of $Q(D)$ to be equivalence classes of *pairs* of elements of D, the ring D is not a subring of $Q(D)$. However, if we let

$$\overline{D} = \left\{\frac{a}{1} : a \in D\right\},$$

then $\overline{D}$ is a subring of $Q(D)$ that is isomorphic to D. (See Activity 18.24.) So when we say that D is contained in $Q(D)$, we mean that there is a subring of $Q(D)$ that is isomorphic to D in a natural sense. One final note about the field of fractions of an integral domain: any field that contains an integral domain D will have to contain a multiplicative inverse of each of D's elements, so $Q(D)$ is the smallest field that contains a copy of D. From this point forward, we will use the standard fraction notation $\frac{a}{b}$ to represent the element $[(a,b)]$ in $Q(D)$. It is important to remember, though, that the fraction $\frac{a}{b}$ is really an equivalence class of ordered pairs of elements from the integral domain D.

Ordering on $\mathbb{Q}$

Not every integral domain is an ordered ring. (See Exercise 13, for example.) However, the integers are an ordered ring and so we might wonder if we can extend the ordering on $\mathbb{Z}$ to an ordering on its field of fractions $\mathbb{Q}$. We can expand this question to ask if the field of fractions of an ordered integral domain can always be made into an ordered field. As we address this question in this section, it may again be helpful to think of the ordered domain D as the integers $\mathbb{Z}$ and the field of fractions $Q(D)$ as $\mathbb{Q}$.

Activity 18.11. Let D be an ordered integral domain. For $Q(D)$ to be an ordered field, there needs to be a nonempty subset P of $Q(D)$ (the positive elements) that is closed under addition and multiplication so that for any $a \in Q(D)$, exactly one of $a \in P$, $a = 0$, or $-a \in P$ is true.

(a) What subset of $Q(D)$ should we define as our set P? Note that since we are attempting to define an ordering on $Q(D)$, your definition of P cannot use such an ordering and can depend only on the ordering in D.

(b) Explain why the set P defined in part (a) is nonempty.

(c) Show that the set P from part (a) is closed under both addition and multiplication. (If you cannot prove this, then you may need to rethink how you defined P.)

(d) To prove the last condition on P, let $\frac{a}{b} \in Q(D)$. By the trichotomy principle in D, we know that exactly one of $ab < 0$, $ab = 0$, or $ab > 0$ is satisfied.

(i) Explain why $\frac{a}{b} > 0$ if $ab > 0$.

(ii) Explain why $\frac{a}{b} = 0$ if $ab = 0$.

(iii) Explain why $-\frac{a}{b} > 0$ if $ab < 0$.

The result of Activity 18.11 is the following theorem:

Theorem 18.12. *If D is an ordered integral domain, then $Q(D)$ is an ordered field.*

A specific consequence of Theorem 18.12 is that $\mathbb{Q}$ is an ordered field.

As in every ordered ring, we can use the set P to define an ordering on $Q(D)$. We say that $\frac{a}{b} > \frac{c}{d}$ in $Q(D)$ if $\left(\frac{a}{b} - \frac{c}{d}\right) \in P$. The other inequalities are defined in a similar manner, and from this point on we will treat these inequalities as the old friends they are.

One final question we should ask about the ordering we have defined on $Q(D)$ is whether it really extends the ordering on D. As noted earlier, you will be asked to show in Activity 18.24 that the set $\overline{D} = \left\{\frac{a}{1} : a \in D\right\}$ is a subring of $Q(D)$ isomorphic to D (where the natural isomorphism is given by sending $a \in D$ to $\frac{a}{1}$ in $\overline{D}$). So we want to know that if $a > 0$ in D, it must follow that $\frac{a}{1} > 0$ in $Q(D)$. But we know that $1 > 0$, so if $a > 0$, then it must be that $(a)(1) > 0$ in D. This, however, implies (by part (d) of Activity 18.11) that $\frac{a}{1} > 0$ in $Q(D)$. So the ordering we defined on $Q(D)$ really does extend the ordering on D.

Now that we have the rational numbers, we can find roots of all linear polynomials in $D[x]$. If we want to find solutions to equations like $x^2 - 2 = 0$, however, the rationals are not enough. This leads us to the field of real numbers.

From $\mathbb{Q}$ to $\mathbb{R}$

The real number system was first given a rigorous treatment by Richard Dedekind, who defined irrational numbers as certain types of sets (called *Dedekind cuts*). Dedekind was a German mathematician who lived between 1831 and 1916. His work on the real number system took place in 1858, though it was not published until 1872.

A **Dedekind cut** is a subset α of $\mathbb{Q}$ satisfying:

(1) $\alpha \neq \emptyset$ and $\alpha \neq \mathbb{Q}$;

(2) if $p \in \alpha$, $q \in \mathbb{Q}$ and $q < p$, then $q \in \alpha$; and

(3) if $p \in \alpha$, then $p < r$ for some $r \in \alpha$.

It is important to note that this construction of the real numbers depends on the fact that there is an ordering on the set of rational numbers. In this way, rational numbers can be used to define any irrational number by identifying the irrational number with the set α of all rational numbers less than that irrational number. With some formal mathematics, we can then define the entire set of real numbers and prove all of the relevant field properties. The work of Dedekind essentially filled in the holes in the number line not occupied by the rational numbers, thus creating a continuous number line.[‡] After Dedekind, we could all feel much more comfortable with the mathematics of the real number system. Thank you, Richard!

It would take us quite a ways off course to rigorously construct the real numbers from the rationals in this way, and the construction is not really algebraic in nature, so we will content ourselves with a more informal discussion of the real numbers using sequences.

Activity 18.13. In the episode *Wolf in the Fold* of the original Star Trek television series, the Enterprise's computer system was taken over by an alien entity. To force the entity out of the computer system, Mr. Spock directed the computer to calculate the exact value of the number π. As the computer chugged away at the problem it consumed more and more resources which eventually drove the alien entity from the computer.

In this activity, we will examine some of the characteristics of this number π and, as a result, understand some of the important structure of the real number system. For reference, an approximation of π to 25 decimal places is

$$\pi \approx 3.1415926535897932384626433.$$

(a) As presented in Table 18.1, the decimal representations for some approximations to π are presented along with their equivalent representations as rational numbers. For example, the decimal 3.14 is equal to the rational number $\frac{314}{100}$. Complete Table 18.1.

(b) Explain what happens to the numbers in the last two columns of the completed Table 18.1 as we allow the number, n, of digits in the decimal approximation to increase without bound. You may use whatever language from calculus you need. Be very specific.

(c) (i) Explain how we could construct a sequence $\{p_n\}$ of rational numbers whose values get as close to the value of π as we want.

(ii) Suppose our universe of numbers consisted **only** of rational numbers. Explain why, in this universe, the sequence $\{p_n\}$ would **not** have a limit.

‡Dedekind describes the essence of continuity in the principle: "If all points of the straight line fall into two classes such that every point of the first class lies to the left of every point of the second class, then there exists one and only one point which produces this division of all points into two classes, this severing of the straight line into two portions" (from *Stetigkeit und irrationale Zahlen (Continuity and Irrational Numbers)*, which can be found at various sources online as part of *Essays on the Theory of Numbers*).

Number of digits in decimal expansion of π	Decimal expansion	Fractional form of decimal expansion
2	3.14	$\frac{314}{100}$
3	3.141	$\frac{3141}{1000}$
4		
5		
6		
7		
8		
9		
10		

Table 18.1
Approaching π.

(iii) It is important to note that computers are only able to calculate with finite decimal expansions—that is, with rational numbers. How would you explain to a fellow student why the Enterprise computer would be unable to complete the task it was given by Mr. Spock (calculating the exact value of π)?

The property of sequences described in Activity 18.13 illustrates a very important property that the set of rational numbers lacks. There are convergent sequences of rational numbers whose entries appear to approach a fixed number that is **not rational**. For this reason, we say that the set of rational numbers is not *complete*. The objects that are the limits of convergent sequences of rational numbers that are not themselves rational numbers are called *irrational numbers*. Since the rational numbers all have terminating or repeating decimal expansions (see Exercise 17), the irrational numbers are those numbers (like π) with non-terminating, non-repeating decimal expansions.

Activity 18.14.

(a) Explain how any irrational number can be represented as a limit of a sequence of rational numbers. Illustrate your answer by finding the first 6 entries in a sequence of rational numbers whose limit is $\sqrt{2}$.

(b) Explain how any rational number can be represented as a limit of a sequence of rational numbers. Illustrate your answer by finding the first 6 entries in a sequence of rational numbers whose limit is $\frac{2}{5}$.

(c) Let $\mathbb{R}$ be the set of all limits of convergent sequences of rational numbers. Explain as best you can in your own words how this set $\mathbb{R}$ represents the set of all real numbers. Include in your explanation a discussion of how the set of real numbers is fundamentally different than the set of rational numbers.

So we have informally defined the set of real numbers as the collection of all limits of convergent sequences of rational numbers. This is not a rigorous definition, so we won't attempt to formally define addition and multiplication on $\mathbb{R}$. Instead, we will simply accept these operations (and their familiar properties) as we have used them in the past. Note that the set of constant sequences of rational numbers in $\mathbb{R}$ is a subset of $\mathbb{R}$ that is isomorphic to $\mathbb{Q}$, so the reals contain an isomorphic copy of the rationals, although we usually just say that $\mathbb{Q}$ itself is a subfield of $\mathbb{R}$.

We have seen that $\mathbb{Z}$ and $\mathbb{Q}$ are ordered rings, and the same is true of $\mathbb{R}$. The set $\mathbb{R}$ contains a subset of positive reals that satisfies the conditions given in Definition 18.2, which in turn defines the familiar relation $>$ that makes $\mathbb{R}$ into an ordered field. The result of Activity 18.14 is that the field of real numbers is a complete ordered field. This makes $\mathbb{R}$ a very nice field indeed, but $\mathbb{R}$ is still lacking in an important algebraic way. Recall that we began our construction of $\mathbb{Q}$ and $\mathbb{R}$ in order to solve polynomial equations. While we can solve all linear equations in $\mathbb{R}$, and we can even solve other equations like $x^3 - 2 = 0$ (which we couldn't solve in $\mathbb{Q}$), there are still some very simple equations—such as $x^2 + 1 = 0$—that do not have solutions in $\mathbb{R}$. To solve these kinds of equations, we will need the complex numbers, which we define formally in the next section.

From $\mathbb{R}$ to $\mathbb{C}$

Complex numbers are often introduced as a tool to solve the quadratic equation $x^2+1 = 0$. However, that is not how complex numbers first came to light. The story actually involves solutions to the general cubic equation, and there are many sources that discuss this interesting history.[§]

A complex number (which we usually write in the form $a + bi$) can be identified with a pair (a, b) of real numbers. So we can define the set of complex numbers $\mathbb{C}$ as

$$\mathbb{C} = \{(a, b) : a, b \in \mathbb{R}\}.$$

(Note that, unlike the construction of $\mathbb{Q}$ from $\mathbb{Z}$, we don't need to define an additional equivalence relation on this set since two complex numbers $a + bi$ and $c + di$ are equal exactly when $a = c$ and $b = d$.) To work with complex numbers, we will need to be able to add and multiply them. Fortunately, the familiar operations from our prior discussions of complex numbers carry over nicely to the formal definition of $\mathbb{C}$ (as a collection of ordered pairs of real numbers).

Activity 18.15.

(a) Explain how the operations

$$(a, b) + (c, d) = (a + c, b + d) \quad (a, b)(c, d) = (ac - bd, ad + bc) \tag{18.2}$$

on $\mathbb{C}$ mimic the familiar sum and product of complex numbers as you know them.

(b) Explain why the operations defined in (18.2) are well-defined.

As we might expect, the set $\mathbb{C}$, with the operations defined above, forms the familiar field of complex numbers.

Theorem 18.16. *The set $\mathbb{C} = \{(a, b) : a, b \in \mathbb{R}\}$, with addition and multiplication defined as*

$$(a, b) + (c, d) = (a + c, b + d) \quad \textit{and} \quad (a, b)(c, d) = (ac - bd, ad + bc),$$

is a field.

[§]See, for example, *Journey Through Genius, The Great Theorems of Mathematics* by William Dunham, John Wiley and Sons, Inc., 1990.

We will prove certain parts of Theorem 18.16 in the next activity and leave the remainder for the exercises. (See Exercise 7.)

Activity 18.17. Let $\mathbb{C} = \{(a, b) : a, b \in \mathbb{R}\}$ with operations defined as in (18.2).

(a) What element would you expect to be the additive identity in $\mathbb{C}$? Prove that you have the right element.

(b) What element would you expect to be the additive inverse of the element (a, b) in $\mathbb{C}$? Prove that you have the right element.

(c) What element would you expect to be the multiplicative identity in $\mathbb{C}$? Prove that you have the right element.

(d) Now we will determine the units in $\mathbb{C}$. What is the form of a nonzero element in $\mathbb{C}$? If (a, b) is nonzero in $\mathbb{C}$, must (a, b) be a unit? If so, determine $(a, b)^{-1}$, and prove your answer. If not, give an example of a nonzero element of $\mathbb{C}$ that is not a unit.

From this point on, we will use the standard notation $a + bi$ for the complex number (a, b), where $i^2 = -1$ or, equivalently, $i = \sqrt{-1}$. This notation is helpful in that if we treat i as though it has the properties of a real number, then addition and multiplication of complex numbers follow naturally from the properties of the corresponding operations in $\mathbb{R}$.

Since we defined the elements of $\mathbb{C}$ to be ordered pairs of real numbers, the field $\mathbb{R}$ is not a subfield of $\mathbb{C}$. However, if we let

$$\overline{R} = \{a + 0i : a \in \mathbb{R}\},$$

then $\overline{R}$ is a subfield of $\mathbb{C}$ that is isomorphic to $\mathbb{R}$. (See Activity 18.25.) So when we say that $\mathbb{R}$ is contained in $\mathbb{C}$, we mean that there is a subfield of $\mathbb{C}$ that is isomorphic to $\mathbb{R}$ in a natural sense.

Recall that we constructed the rational numbers in order to solve linear equations in $\mathbb{Z}[x]$. Then we needed the real numbers in order to solve more equations—for example, $x^2 - 2 = 0$ in $\mathbb{Q}[x]$—and to have numbers that represent limits of convergent sequences of rational numbers. As we have seen, the complex numbers take us one step farther. In $\mathbb{R}[x]$, there are polynomials that have no roots in $\mathbb{R}$. (The polynomial $x^2 + 1$ is the canonical example.) However, if $f(x) \in \mathbb{C}[x]$ is a nonconstant polynomial, then $f(x)$ must have a root in $\mathbb{C}$, as we saw in the Fundamental Theorem of Algebra (Theorem 14.2 on page 180). This property makes $\mathbb{C}$ what we call an *algebraically closed* field. However, we do give something up when we move from $\mathbb{R}$ to $\mathbb{C}$; in particular, the field $\mathbb{C}$ is no longer an ordered field. (See Exercise 11.) There are some other important attributes of complex numbers that we have seen in previous investigations (like the polar form of a complex number) and will study more in Exercise 1.

A Characterization of the Integers

We will conclude this investigation with a discussion of the uniqueness of the set of integers. As we will see, The Principle of Mathematical Induction (and the equivalent Well-Ordering Principle) makes $\mathbb{Z}$ unique among all ordered integral domains. To begin, we need to know what it means for an ordered ring to be well-ordered.

Definition 18.18. Let S be a nonempty subset of an ordered ring R. An element a in S is called a **smallest element of S** provided that $a \leq x$ for each $x \in S$.

Note that if S contains a smallest element, then that smallest element is unique. (See Exercise 15.) In the set of integers, we are used to the fact that 1 is the smallest positive integer. That leads us to ask if 1 is always the smallest element in any ordered integral domain, provided its set of positive elements contains a smallest element. We will investigate this question in the next activity.

Activity 18.19. Let D be an ordered integral domain with D^+ as its set of positive elements. Assume D^+ contains a smallest element s. To prove that $s = 1$, we will proceed by contradiction and assume $s \neq 1$.

(a) Recall that 1 is positive in D. Explain why it must be the case that $0 < 1$ and $s < 1$.

(b) Explain why $0 < s^2$ and $s^2 < s$.

(c) Explain why $s^2 < s$ provides us with the contradiction we need to conclude that $s = 1$.

We can now define what it means for a set to be *well-ordered.*

Definition 18.20. A subset S of an ordered ring R is said to be **well-ordered** provided that every nonempty subset of S contains a smallest element.

As we will show in Appendix B, the Principle of Mathematical Induction is equivalent to the set of positive integers being well-ordered. Since we are assuming the Principle of Mathematical Induction through the Peano axioms, we also take the following as an axiom.

Axiom 18.21 (The Well-Ordering Principle)**.** *The set of natural numbers is a well-ordered subset of the integers.*

The Well-Ordering Principle is not a trivial assumption, since there are many familiar sets that are not well-ordered.

Activity 18.22.

(a) Is the set of positive rational numbers a well-ordered subset of $\mathbb{Q}$? Explain.

(b) Is the set of positive real numbers a well-ordered subset of $\mathbb{R}$? Explain.

As Activity 18.22 demonstrates, not every ordered ring has a well-ordered subset of positive elements. In fact, it is exactly this property that characterizes the integers, as the next theorem demonstrates.

Theorem 18.23. *Let D be an ordered integral domain in which the set of positive elements D^+ is well-ordered. Then D is isomorphic to $\mathbb{Z}$.*

Proof. Let D be an ordered integral domain in which the set of positive elements D^+ is well-ordered. To prove that D is isomorphic to $\mathbb{Z}$, we will exhibit an isomorphism $\varphi : \mathbb{Z} \to D$. Since every integer is a multiple of 1, we can define φ from $\mathbb{Z}$ into D by

$$\varphi(n) = n \cdot 1_D,$$

where 1_D is the identity in D.

First, we will prove that φ is a homomorphism. Let m and n be integers. Using properties from Theorem 8.5, we have the following:

$$\begin{aligned} \varphi(m+n) &= (m+n) \cdot 1_D \\ &= m \cdot 1_D + n \cdot 1_D \\ &= \varphi(m) + \varphi(n) \end{aligned} \qquad \begin{aligned} \varphi(mn) &= (mn) \cdot 1_D \\ &= (mn) \cdot (1_D 1_D) \\ &= (m 1_D) \cdot (n 1_D) \\ &= \varphi(m) \cdot \varphi(n) \end{aligned}$$

Thus, φ is a homomorphism.

To show that φ is an injection, assume that $m, n \in \mathbb{Z}$ with $\varphi(m) = \varphi(n)$. Without loss of generality, we can assume that $m \geq n$. Since $\varphi(m) = \varphi(n)$ it follows that $m \cdot 1_D = n \cdot 1_D$ and

$$(m - n) \cdot 1_D = m \cdot 1_D - n \cdot 1_D = 0_D,$$

where 0_D is the additive identity in D. We know that 1_D is positive, and so if $m - n > 0$, Exercise 4 shows that $(m - n)1_D > 0$, a contradiction to trichotomy. We can thus conclude that $(m - n) = 0$, or that $m = n$. Thus, φ is an injection.

To complete the proof, we must show that φ is a surjection. To prove this result, we need to show that every element of D is of the form $m \cdot 1_D$ for some integer m. We will do so by using cases based on trichotomy.

Case 1. We will first prove that for each $y \in D^+$, there exists an integer m such that $y = m \cdot 1_D$. We will use a proof by contradiction. Assume that there exists an element y in D^+ such that y cannot be written in the form $n \cdot 1_D$ for any integer n. Let S be the set of all such elements—that is,

$$S = \{z \in D^+ \mid z \neq n \cdot 1_D \text{ for every } n \in \mathbb{Z}\}.$$

By the definition of y, we know that $y \in S$ and so S is not empty. Since D^+ is well-ordered, the set S contains a smallest element s. Now 1_D is the smallest element of D^+ and $1_D = 1 \cdot 1_D$, so $1_D \notin S$. Therefore, 1_D is not equal to s, and so $s > 1_D$. It follows that $s - 1_D > 0$, and so $s - 1_D \in D^+$. In addition,

$$s - (s - 1_D) = 1_D,$$

and so $s > (s - 1_D)$. This means that

$$(s - 1_D) \notin S,$$

and so there exists an integer k such that

$$s - 1_D = k \cdot 1_D.$$

But then

$$\begin{aligned}(k + 1) \cdot 1_D &= (k \cdot 1_D) + (1 \cdot 1_D)\\ &= (s - 1_D) + 1_D\\ &= s,\end{aligned}$$

and so $s \notin S$, a contradiction. It follows that every element in D^+ is of the form $m \cdot 1_D$ for some $m \in \mathbb{Z}$.

Case 2. If $y = 0_D$, then $y = 0 \cdot 1_D$.

Case 3. Now assume that $-y \in D^+$. By our previous work, there exists an integer k such that

$$-y = k \cdot 1_D.$$

But then

$$\begin{aligned}(-k) \cdot 1_D &= -(k \cdot 1_D)\\ &= -(-y)\\ &= y,\end{aligned}$$

and so y has the form $(-k) \cdot 1_D$.

We have shown that for each $y \in D$, there exists an integer m such that $y = m \cdot 1_D$. It follows that if $y \in D$ and $y = m \cdot 1_D$, then $\varphi(m) = m \cdot 1_D = y$, which proves that φ is a surjection. We can therefore conclude that φ is an isomorphism, and so $\mathbb{Z} \cong D$. ■

In conclusion, $\mathbb{Z}$ is the only ordered integral domain with a well-ordered set of positive elements. So the Principle of Mathematical Induction (or the Well-Ordering Principle) is a determining property of $\mathbb{Z}$.

Concluding Activities

Activity 18.24. Let D be an integral domain, and let $\overline{D} = \left\{\frac{a}{1} : a \in D\right\}$ in $Q(D)$.

(a) Prove that $\overline{D}$ is a subring of $Q(D)$.

(b) Prove that D is isomorphic to $\overline{D}$.

Activity 18.25. Let $\overline{R}$ be the subset of $\mathbb{C}$ defined by $\overline{R} = \{a + 0i : a \in \mathbb{R}\}$.

(a) Prove that $\overline{R}$ is a subfield of $\mathbb{C}$.

(b) Prove that $\mathbb{R}$ is isomorphic to $\overline{R}$.

Activity 18.26. Write a short summary that describes the important concepts, and the relationships between these concepts, that were introduced in this investigation. Explain how the material in this investigation is connected to the material in Investigations 1, 7, 10, and 15.

Exercises

⋆ (1) **Properties of complex numbers.** Let $z = a + bi$ be a complex number.

i. We define the real part of z to be the real number a.

ii. We define the imaginary part of z to be the real number b.

iii. The complex conjugate $\overline{z}$ of z is the complex number $a - bi$.

iv. The modulus (or norm, or absolute value) of z, denoted $|z|$, is the real number $\sqrt{a^2 + b^2}$.

(a) Let $w = 2 + 3i$ and $z = -1 + 5i$.

(i) Find $\overline{w}$ and $\overline{z}$.

(ii) Compute $|w|$ and $|z|$.

(iii) Compute $w\overline{w}$ and $z\overline{z}$.

(b) Let z be an arbitrary complex number. There is a relationship between $|z|$, z, and $\overline{z}$. Find and prove this relationship.

(c) What is $\overline{z}$ if $z \in \mathbb{R}$?

⋆ (2) Let R be an ordered ring with identity. Prove that $1 > 0$ and $(-1) < 0$ in R.

⋆ (3) Let R be an ordered ring.

(a) Prove that for all a, b, and c in R, if $a > b$ and $c > 0$, then $ac > bc$.

(b) Prove that for all a, b, and c in R, if $ac > bc$ and $c > 0$, then $a > b$.

⋆ (4) Prove that if x is a positive element of an ordered ring and $n \in \mathbb{Z}$, then $nx > 0$ in R if and only if $n > 0$ in $\mathbb{Z}$. (Hint: Recall how the element nx is defined, and then use induction.)

⋆ (5) One important property of rational numbers is that if $\frac{a}{b}$ is a rational number and m is any nonzero integer, then $\frac{a}{b} = \frac{am}{bm}$. We use this property often, and so it is natural to ask if it holds in the field of fractions of *any* arbitrary integral domain. Let D be an integral domain. The equivalent formulation of this statement about rational numbers in the context of the set $Q(D)$ is the following:

If $[(x, y)] \in Q(D)$ and m is a nonzero element in D, then $[(x, y)] = [(mx, my)]$.

Prove this statement. (Hint: First consider what we need to do to show that $[(x, y)] = [(mx, my)]$ in $Q(D)$.)

⋆ (6) Complete the proof of Theorem 18.10 by verifying the properties below. Throughout, let D be an integral domain.

(a) Show that $Q(D)$ is closed under addition and multiplication.

(b) Prove addition is commutative and associative in $Q(D)$.

(c) Prove that multiplication is commutative and associative in $Q(D)$.

⋆ (7) Let $\mathbb{C} = \{(a, b) : a, b \in \mathbb{R}\}$ with operations defined as in (18.2). Complete the proof of Theorem 18.16 by carrying out the following steps:

(a) Explain why $\mathbb{C}$ is closed under addition and multiplication.

(b) Prove that addition and multiplication are commutative in $\mathbb{C}$.

(c) Prove that multiplication distributes over addition in $\mathbb{C}$.

⋆ (8) (a) Let R be an ordered ring, and define a relation $>$ on $R[x]$ where $f(x) > g(x)$ if $f(x) - g(x)$ has a positive leading coefficient. Show that $R[x]$ is an ordered ring with this relation.

(b) Is the set of positive elements (the polynomials with positive leading coefficient) a well-ordered subset of $\mathbb{Z}[x]$? Explain.

(9) **Rational functions.** If R is an integral domain, then $R[x]$ is also an integral domain. Describe as best you can the elements of $Q(R[x])$, the field of quotients of $R[x]$. When $R = F$ is a field, this field of quotients of $F[x]$ is usually denoted as $F(x)$ and is called the **field of rational functions** over F.

(10) Let $\mathbb{Q}(i)$ be the subset of $\mathbb{C}$ defined by $\mathbb{Q}(i) = \{r + si : r, s \in \mathbb{Q}\}$.

(a) Show that $\mathbb{Q}(i)$ is a subfield of $\mathbb{C}$.

(b) What is the specific relationship between $\mathbb{Q}(i)$ and the ring of Gaussian integers introduced in Exercise 2 of Investigation 6? (See page 75.) Explain.

$\star$ (11) Can we define an ordering on $\mathbb{C}$ to make $\mathbb{C}$ an ordered ring? Explain.

(12) We are used to having exactly one set of positive elements in $\mathbb{Z}$, $\mathbb{Q}$, and $\mathbb{R}$. But is it possible that an ordered ring could contain more than one set of positive elements?

(a) Let r be a real number, and let

$$P_r = \{f(x) \in \mathbb{Z}[x] : f(r) > 0\}.$$

(Note that the inequality $f(r) > 0$ takes place in R.)

(i) Show that P_r satisfies the conditions of a set of positive elements in $\mathbb{Z}[x]$. You may use all of the properties of the standard ordering on $\mathbb{R}$.

(ii) Show that the sets P_r are not all the same. Conclude that a ordered ring may be ordered with more than one choice of a set of positive elements.

(b) How many subsets of positive elements can $\mathbb{Z}$ contain? Prove your answer.

$\star$ (13) **The characteristic of an ordered ring.** Let R be an ordered ring.

(a) What are the possibilities for the characteristic of R? Prove your answer.

(b) Is there a prime p for which $\mathbb{Z}_p$ is an ordered ring? Explain.

(14) Let R be an ordered ring. Under what conditions on a and b in R is $ab > 0$? Under what conditions is $ab < 0$? Prove your answers.

$\star$ (15) Let S be a subset of an ordered ring R such that S contains a smallest element. Show that this smallest element is unique.

(16) Let R be any ordered ring. For all $a \in R$, we can define the **absolute value** of a as

$$|a| = \begin{cases} a, & \text{if } a \geq 0 \\ -a, & \text{otherwise.} \end{cases}$$

Prove that this absolute value function satisfies the following properties for any $a, b \in R$.

(a) $|ab| = |a|\ |b|$

(b) $|a| \geq 0$ with $|a| = 0$ if and only if $a = 0$.

(c) $|-a| = |a|$

(d) $|a + b| \leq (|a| + |b|)$

$\star$ (17) **Decimal expansions of real numbers.** A real number x has a decimal expansion

$$x = N + \sum_{i \geq 1} 10^{-i} x_i,$$

where N is an integer and each x_i is an integer between 0 and 9. We will also write this decimal expansion of x as

$$x = N.x_1 x_2 x_3 \ldots.$$

The decimal expansion for x *terminates* if $x_i = 0$ for all i larger than some integer m. The decimal expansion for x is *repeating* or *periodic* if there are integers m and k so that $x_{k+(sm+t)} = x_{k+t}$ for all $s \geq 1$ and $1 \leq t \leq m$. In this case, we write

$$x = N.x_1 x_2 \ldots x_{k-1} \overline{x_k x_{k+1} \ldots x_{k+m}}.$$

In this exercise, we will show that a real number x is a rational number if and only if the decimal expansion for x terminates or is periodic.

(a) We will first consider the implication that every rational number has a terminating or repeating decimal expansion.

(i) Find the decimal representations of $\frac{81}{500}$ and $\frac{5}{11}$. Explain your process.

(ii) Prove the forward implication by showing that every rational number has a terminating or repeating decimal expansion.

(b) Next, we will consider the implication that every terminating or periodic decimal is a rational number.

(i) Express each of the following decimals as rational numbers: 0.213 and $0.42\overline{123}$. Explain your process. (Hint: Use the formula for the sum of a geometric series for the latter.)

(ii) Show that any real number with a terminating or repeating decimal expansion is a rational number.

Connections

In this investigation, we began with the ring of integers, as introduced in Investigation 1. The desire to solve polynomial equations led us to integral domains and fields (namely the field of fractions of an integral domain, as well as the fields $\mathbb{R}$ and $\mathbb{C}$), as introduced in Investigation 7. In each construction we saw that the previous structure could be identified with an isomorphic copy (as in Investigation 10) of the new structure. The constructions of new number systems culminated with an algebraically closed field—namely, the field of complex numbers $\mathbb{C}$. However, this field $\mathbb{C}$ contains much more than just solutions to polynomial equations over $\mathbb{R}$; it also contains algebraic, transcendental, and constructible numbers, as discussed in Investigation 15.

Part VI

Groups

Investigation 19

Symmetry

Focus Questions

By the end of this investigation, you should be able to give precise and thorough answers to the questions listed below. You may want to keep these questions in mind to focus your thoughts as you complete the investigation.

- What is a rigid motion, and what is a symmetry? How are the two related and how are they different?
- How can permutation notation be used to describe the symmetries of a regular polygon?
- What operation can be performed on a collection of symmetries? What structure does the resulting set of symmetries have?

Preview Activity 19.1. The four-petal flower at right (you may recognize this figure as the graph of $r = \cos(4\theta)$ in polar coordinates) exhibits several kinds of symmetry. Identify all of the ways in which the figure is symmetric, and describe the symmetries as best you can.

Introduction

Symmetry is a basic element of design. We also see symmetry in nature—for instance, in crystals, plants, and animals. Throughout history, different cultures have embraced the use of symmetry in their architecture and art.

The major topic in this part of the text is group theory. In a sense, group theory can be thought of as the study of symmetry, and so that is where we will begin. In subsequent investigations, we will define groups, examine a large variety of different groups, and investigate the structure of groups.

In a translation, everything is moved the same amount.	
In a rotation, there is a point (called the *rotocenter* or *center of rotation*) around which everything is spun by a fixed amount (called the *rotation angle*).	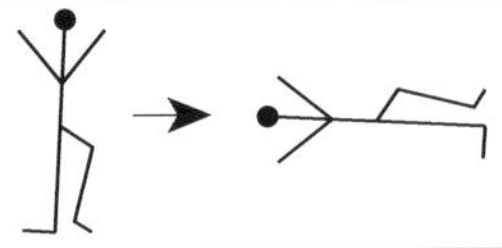
In a reflection, there is a line (called the *axis of reflection*), and the reflection consists of the mirror images of all points across this axis.	
A glide reflection consists of a reflection, followed by a translation parallel to the axis of reflection.	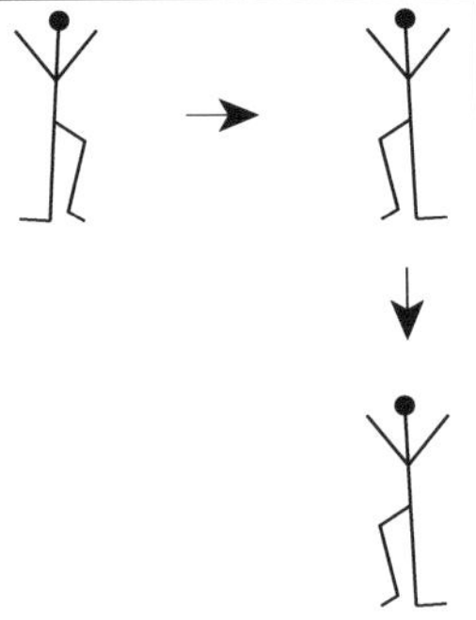

Figure 19.1
Rigid motions in the plane.

Symmetries

To introduce the idea of a group, we will first study the symmetries of certain objects. We will see that we can define an operation on the symmetries of a given object to form a set that has an interesting structure.

In order to talk about symmetries, we must first discuss rigid motions in the plane. Rigid motions preserve the shape of an object, but not necessarily its location or orientation. In other words, after performing a rigid motion on an object, the resulting object must be congruent to the original, but may be in a different position. A formal definition of a rigid motion is as follows:

Definition 19.2. A **rigid motion** in the plane is a bijective function $f : \mathbb{R}^2 \to \mathbb{R}^2$ such that, for all $x, y \in \mathbb{R}^2$, the distance between $f(x)$ and $f(y)$ is the same as the distance between x and y.

In other words, a rigid motion is a distance-preserving function, or an *isometry*. For example, if we simply move an object from one location to another, we have performed an isometry, as we have not altered the distances between points in that object. It turns out (though we won't prove it) that there are four rigid motions in the plane: translation, rotation, reflection, and glide reflection. These rigid motions are illustrated on the dancing man in Figure 19.1.

For our purposes, we will only be concerned with certain types of rigid motions—namely, symmetries. For the purposes of the next definition, a *geometric object* is simply any subset of $\mathbb{R}^2$. (For example, the octagon in Figure 19.2 is a geometric object.)

Definition 19.3. A **symmetry** of a geometric object O is a rigid motion f so that $f(O) = O$.

Note that every symmetry is either a rotation or a reflection, since translations and glide reflections change the location of the object and are therefore not symmetries. To visualize some of the symmetries of the octagon, we can label its vertices with the numbers 1, 2, ..., 8, as shown in Figure 19.2. (Note that the figure gives examples, but does not include all of the symmetries of the octagon.)

Symmetries of Regular Polygons

To more easily identify the symmetries of regular polygons (like the octagon), we will introduce some special notation. Notice that we can completely identify a symmetry by what it does to each vertex. For example, the reflection r_V permutes the vertices of the octagon as follows:

- vertex 1 is left alone;
- vertex 2 is moved to the original position of vertex 8;
- vertex 3 is moved to the original position of vertex 7;
- vertex 4 is moved to the original position of vertex 6;
- vertex 5 is left alone;
- vertex 6 is moved to the original position of vertex 4;
- vertex 7 is moved to the original position of vertex 3; and
- vertex 8 is moved to the original position of vertex 2.

Notice that r_V permutes the vertices of the octagon in a particular way. We can describe this permutation concisely by constructing an array in which the original vertices are listed in the top row, and their corresponding images are listed in the second row. Doing so, we can represent r_V as follows:

$$r_V = \begin{pmatrix} 1 & 2 & 3 & 4 & 5 & 6 & 7 & 8 \\ 1 & 8 & 7 & 6 & 5 & 4 & 3 & 2 \end{pmatrix}.$$

Using the same notation to describe I, r_H, and R_{45}, we obtain:

$$I = \begin{pmatrix} 1 & 2 & 3 & 4 & 5 & 6 & 7 & 8 \\ 1 & 2 & 3 & 4 & 5 & 6 & 7 & 8 \end{pmatrix}, \quad r_H = \begin{pmatrix} 1 & 2 & 3 & 4 & 5 & 6 & 7 & 8 \\ 5 & 4 & 3 & 2 & 1 & 8 & 7 & 6 \end{pmatrix}, \text{ and}$$

$$R_{45} = \begin{pmatrix} 1 & 2 & 3 & 4 & 5 & 6 & 7 & 8 \\ 2 & 3 & 4 & 5 & 6 & 7 & 8 & 1 \end{pmatrix}.$$

A permutation is really a function from a set to itself (the set of vertices of an octagon in our example), so we can combine permutations through composition. For example, $R_{45}(4) = 5$ and

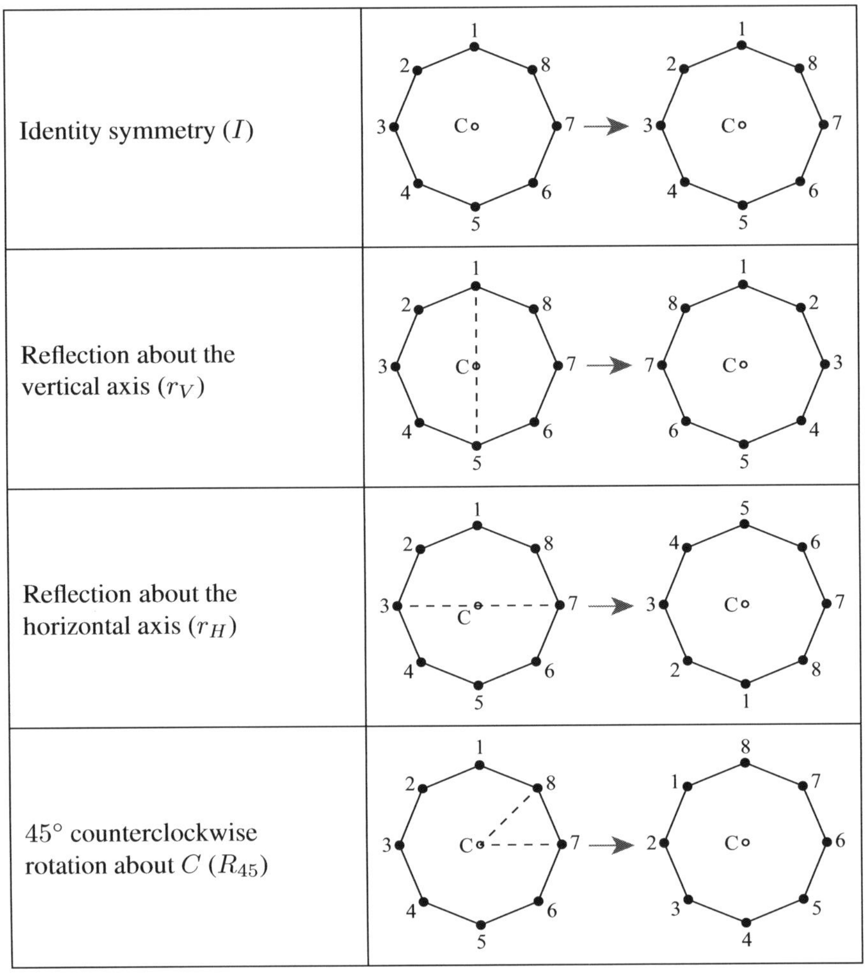

Figure 19.2
Symmetries of an octagon.

$r_H(5) = 1$, so in the composite $r_H \circ R_{45}$ we have $(r_H \circ R_{45})(4) = r_H(R_{45}(4)) = r_H(5) = 1$. Continuing with the other vertices gives

$$r_H \circ R_{45} = \begin{pmatrix} 1 & 2 & 3 & 4 & 5 & 6 & 7 & 8 \\ 4 & 3 & 2 & 1 & 8 & 7 & 6 & 5 \end{pmatrix}.$$

Activity 19.4. Is $r_H \circ R_{45}$ a symmetry of the regular octagon? Why or why not?

For any polygon, choose one vertex of the polygon to label as vertex 1. Label the remaining vertices in order proceeding counterclockwise from vertex 1 as shown in the Figure 19.3. In addition:

- let r_i denote the reflection about the line through the origin and the vertex i;
- let $\overline{r}_i$ denote the reflection about the perpendicular to the segment joining adjacent vertices i and $i+1$ through the midpoint of that segment; and
- let R_k denote a counterclockwise rotation of the polygon about its center by an angle of $\left(\frac{360}{n}\right) k$ degrees. (Note that the identity symmetry, I, is equal to R_0.)

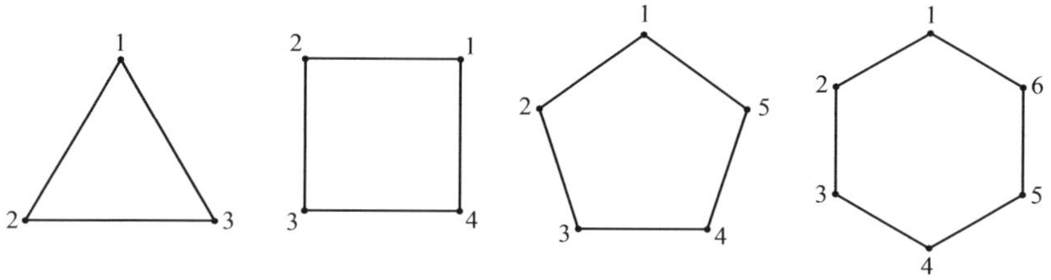

Figure 19.3
Equilateral triangle, square, regular pentagon, and regular hexagon.

As an example, the symmetries of the square (with vertices labeled as shown in Figure 19.3) can be represented as follows:

$$I = \begin{pmatrix} 1 & 2 & 3 & 4 \\ 1 & 2 & 3 & 4 \end{pmatrix} \quad r_1 = \begin{pmatrix} 1 & 2 & 3 & 4 \\ 1 & 4 & 3 & 2 \end{pmatrix}$$
$$r_2 = \begin{pmatrix} 1 & 2 & 3 & 4 \\ 3 & 2 & 1 & 4 \end{pmatrix} \quad \overline{r}_1 = \begin{pmatrix} 1 & 2 & 3 & 4 \\ 2 & 1 & 4 & 3 \end{pmatrix}$$
$$\overline{r}_2 = \begin{pmatrix} 1 & 2 & 3 & 4 \\ 4 & 3 & 2 & 1 \end{pmatrix} \quad R_1 = \begin{pmatrix} 1 & 2 & 3 & 4 \\ 2 & 3 & 4 & 1 \end{pmatrix}$$
$$R_2 = \begin{pmatrix} 1 & 2 & 3 & 4 \\ 3 & 4 & 1 & 2 \end{pmatrix} \quad R_3 = \begin{pmatrix} 1 & 2 & 3 & 4 \\ 4 & 1 & 2 & 3 \end{pmatrix}$$

The operation table (with the operation of function composition) for the set of symmetries of a square is shown in Table 19.1.

$\circ$	I	$\overline{r}_1$	$\overline{r}_2$	r_1	r_2	R_1	R_2	R_3
I	I	$\overline{r}_1$	$\overline{r}_2$	r_1	r_2	R_1	R_2	R_3
$\overline{r}_1$	$\overline{r}_1$	I	R_2	R_1	R_3	r_1	$\overline{r}_2$	r_2
$\overline{r}_2$	$\overline{r}_2$	R_2	I	R_3	R_1	r_2	$\overline{r}_1$	r_1
r_1	r_1	R_3	R_1	I	R_2	$\overline{r}_2$	r_2	$\overline{r}_1$
r_2	r_2	R_1	R_3	R_2	I	$\overline{r}_1$	r_1	$\overline{r}_2$
R_1	R_1	r_2	r_1	$\overline{r}_1$	$\overline{r}_2$	R_2	R_3	I
R_2	R_2	$\overline{r}_2$	$\overline{r}_1$	r_2	r_1	R_3	I	R_1
R_3	R_3	r_1	r_2	$\overline{r}_2$	$\overline{r}_1$	I	R_1	R_2

Table 19.1
Symmetries of a square.

In the next activity, we will study the common properties that all sets of symmetries share. In particular, we will determine the symmetries of and create the operation tables for an equilateral triangle, a regular pentagon, and a regular hexagon.

Activity 19.5.

(a) (i) Describe all of the symmetries of an equilateral triangle (as labeled in Figure 19.3) using permutation notation and the notation from the preceding example.

(ii) A partial operation table for the set of symmetries of an equilateral triangle is given in Table 19.2. Complete the table.

$\circ$	I	r_1	r_2	r_3	R_1	R_2
I	I	r_1	r_2	r_3	R_1	R_2
r_1	r_1		R_1	R_2		r_3
r_2	r_2	R_2	I			r_1
r_3	r_3		R_2	I		r_2
R_1	R_1	r_3			R_2	I
R_2	R_2	r_2		r_1		R_1

Table 19.2
Symmetries of an equilateral triangle.

(b) (i) Describe all of the symmetries of a regular pentagon (as labeled in Figure 19.3) using permutation notation and the notation from the preceding example.

(ii) A partial operation table for the set of symmetries of a regular pentagon is given in Table 19.3. Complete the table.

$\circ$	I	r_1	r_2	r_3	r_4	r_5	R_1	R_2	R_3	R_4
I	I	r_1	r_2	r_3	r_4	r_5	R_1	R_2	R_3	R_4
r_1	r_1	I	R_3	R_1	R_4	R_2	r_3	r_5	r_2	r_4
r_2	r_2	R_2	I	R_3	R_1	R_4	r_4	r_1	r_3	r_5
r_3	r_3		R_2	I		R_1	r_5	r_2	r_4	r_1
r_4	r_4	R_1	R_4	R_2	I	R_3	r_1	r_3	r_5	r_2
r_5	r_5	R_3	R_1	R_4	R_2	I	r_2	r_4	r_1	r_3
R_1	R_1		r_5	r_1	r_2		R_2	R_3	R_4	I
R_2	R_2	r_2	r_3	r_4	r_5	r_1	R_3	R_4	I	R_1
R_3	R_3	r_5	r_1	r_2	r_3			I	R_1	R_2
R_4	R_4	r_3	r_4	r_5	r_1	r_2	I	R_1	R_2	R_3

Table 19.3
Symmetries of a regular pentagon.

(c) (i) Describe all of the symmetries of a regular hexagon (as labeled in Figure 19.3) using permutation notation and the notation from the preceding example.

(ii) A partial operation table for the set of symmetries of a regular hexagon is given in Table 19.4. Complete the table.

$\circ$	I	r_1	r_2	r_3	$\overline{r}_1$	$\overline{r}_2$	$\overline{r}_3$	R_1	R_2	R_3	R_4	R_5
I	I	r_1	r_2	r_3	$\overline{r}_1$	$\overline{r}_2$	$\overline{r}_3$	R_1	R_2	R_3	R_4	R_5
r_1	r_1	I	R_4	R_2	R_5	R_3	R_1	$\overline{r}_3$	r_3		r_2	$\overline{r}_1$
r_2	r_2	R_2	I	R_4	R_1	R_5	R_3	$\overline{r}_1$	r_1	$\overline{r}_3$	r_3	$\overline{r}_2$
r_3	r_3	R_4	R_2	I	R_3	R_1		$\overline{r}_2$	r_2		r_1	$\overline{r}_3$
$\overline{r}_1$	$\overline{r}_1$	R_1	R_5	R_3	I	R_4	R_2	r_1	$\overline{r}_3$	r_3	$\overline{r}_2$	r_2
$\overline{r}_2$	$\overline{r}_2$	R_3	R_1	R_5	R_2	I	R_4	r_2	$\overline{r}_1$	r_1	$\overline{r}_3$	r_3
$\overline{r}_3$	$\overline{r}_3$	R_5	R_3	R_1	R_4	R_2	I	r_3	$\overline{r}_2$	r_2	$\overline{r}_1$	r_1
R_1	R_1	$\overline{r}_1$	$\overline{r}_2$	$\overline{r}_3$	r_2	r_3	r_1	R_2	R_3	R_4	R_5	I
R_2	R_2	r_2		r_1	$\overline{r}_2$		$\overline{r}_1$	R_3	R_4	R_5	I	R_1
R_3	R_3	$\overline{r}_2$	$\overline{r}_3$	$\overline{r}_1$	r_3	r_1	r_2	R_4	R_5	I	R_1	R_2
R_4	R_4	r_3		r_2	$\overline{r}_3$	$\overline{r}_1$	$\overline{r}_2$		I	R_1	R_2	R_3
R_5	R_5	$\overline{r}_3$	$\overline{r}_1$	$\overline{r}_2$	r_1	r_2	r_3	I	R_1	R_2	R_3	R_4

Table 19.4
Symmetries of a regular hexagon.

(d) What are some properties that all of the operation tables from parts (a) – (c) have in common? List as many properties as you can find.

The properties that you identified in Activity 19.5 describe the group structure that we will study in the next investigations. As we proceed, we will discover many other sets that have the same properties, including the sets of symmetries of all of the regular polygons.

Concluding Activities

Activity 19.6. Find, via a library or Internet search, an object (building, tiling, painting, sculpture, mosaic, fractal, rug, etc.) that has significant symmetry. Then complete the following.

(a) Identify the object and the source through which you found it. Choose something other than a simple polygon; that is, find an object that is interesting to you and that possesses at least 6 symmetries, including both rotational and reflective symmetry.

(b) Describe all of the symmetries possessed by your object. Choose 6 symmetries (including at least one non-trivial rotation and one non-trivial reflection), and make a copy of the picture of your object for each symmetry. Find a convenient way to label your object so that you can use permutation notation to represent each symmetry. Then illustrate each symmetry on one of the copies of your picture.

(c) Choose 3 of the symmetries, and find all of the compositions of these three symmetries. Is each composition a symmetry of your object? Explain.

Activity 19.7. Let G be the set of symmetries of an object.

(a) Is G closed under the operation of function composition? Prove your answer.

(b) Is there an identity element for composition in G? If yes, what is it?

(c) Does each element in G have an inverse in G under composition? If yes, what is the inverse of each element?

(d) Is the operation in G associative? If yes, prove it. If no, provide an example to illustrate. (Hint: Is there a more general argument we can use here that involves functions?)

Activity 19.8. Write a short summary that describes the important concepts, and the relationships between these concepts, that were introduced in this investigation.

Exercises

(1) Let O be the object shown in Figure 19.4.

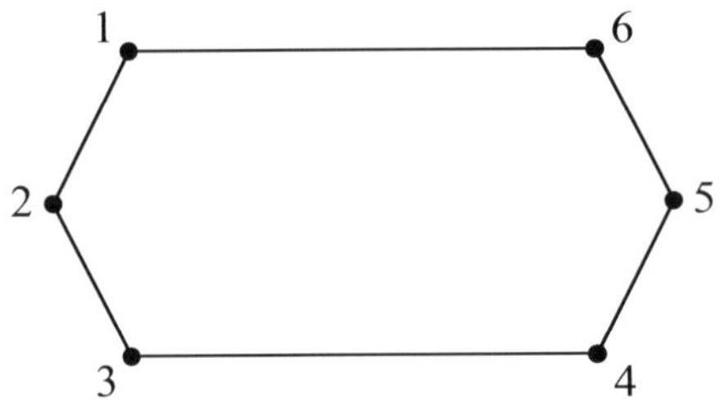

Figure 19.4
The object O.

(a) Write the permutation notation for the reflection α of O around the line through the points labeled 2 and 5.

(b) Write the permutation notation for the reflection β of O around perpendicular bisector of the segment connecting the points labeled 1 and 6.

(c) Determine $\beta\alpha$ and $\alpha\beta$.

(d) Write the permutation notation for the $180°$ counterclockwise rotation γ of O around its center.

(e) Construct the operation table for the set of symmetries of O.

(2) The permutations

$$\alpha = \begin{pmatrix} 1 & 2 & 3 & 4 & 5 & 6 & 7 & 8 & 9 & 10 \\ 3 & 4 & 5 & 6 & 7 & 8 & 9 & 10 & 1 & 2 \end{pmatrix}$$

and

$$\beta = \begin{pmatrix} 1 & 2 & 3 & 4 & 5 & 6 & 7 & 8 & 9 & 10 \\ 4 & 3 & 2 & 1 & 10 & 9 & 8 & 7 & 6 & 5 \end{pmatrix}$$

are two symmetries of a regular decagon.

(a) Identify α and β as either rotations or reflections. Explain your choices.

(b) Find $\alpha\beta$ and $\beta\alpha$, and identify them as reflections or rotations. Explain.

(c) Does α have an inverse? If so, describe the inverse of α geometrically and using permutation notation.

(d) Does β have an inverse? If so, describe the inverse of β geometrically and using permutation notation.

(3) (a) Find all of the symmetries of the letter B, and create the operation table for the set of symmetries of B.

(b) Find all of the symmetries of the letter T, and create the operation table for the set of symmetries of T.

(c) Find all of the symmetries of the letter Z, and create the operation table for the set of symmetries of Z.

(d) Compare the operation tables for B, T, and Z. Describe all of the similarities and differences you observe.

(4) Is composition of symmetries a commutative operation? Prove your answer.

(5) Let A, B, and C be the objects shown in Figure 19.5.

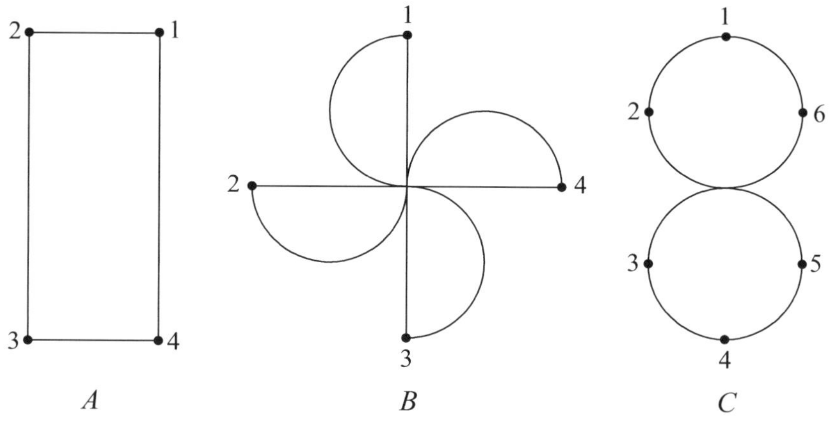

Figure 19.5
Three geometric objects.

(a) For each object, find all of the symmetries. Describe the symmetries in words and using the permutation notation introduced in this investigation.

(b) Create the operation table for the set of symmetries of each object.

(c) Describe the similarities and differences in the operation tables you made in part (b). Your description should include not only obvious attributes like the number of elements, but also how the elements interact within a given set of symmetries.

(6) **Mattress flipping.** Mattress manufacturers always recommend that users periodically flip their mattresses around in order to promote even wear. A flip of a mattress is not necessarily just a flip, but rather a symmetry of the mattress. Consequently, we should be able to describe all possible ways to rearrange a mattress using permutation notation. Label the corners of a rectangular mattress as 1, 2, 3, 4. Find all symmetries of the mattress. Then create the operation table for the set of symmetries of the mattress. What properties does this set of symmetries have?

* (7) **Symmetries of a circle.** The figures we have considered in this investigation have all possessed only finitely many symmetries. In contrast, the unit circle (or any circle) has infinitely many rotational and reflective symmetries. We will examine those symmetries in this problem.

(a) We will first consider rotational symmetries. Let θ be an angle. We can explicitly represent the counterclockwise rotation around the origin by the angle θ using linear algebra. Let $P = (x, y) = (\cos(\alpha), \sin(\alpha))$ and $Q = (w, z) = (\cos(\alpha + \theta), \sin(\alpha + \theta))$ be points on the unit circle, as illustrated in Figure 19.6.

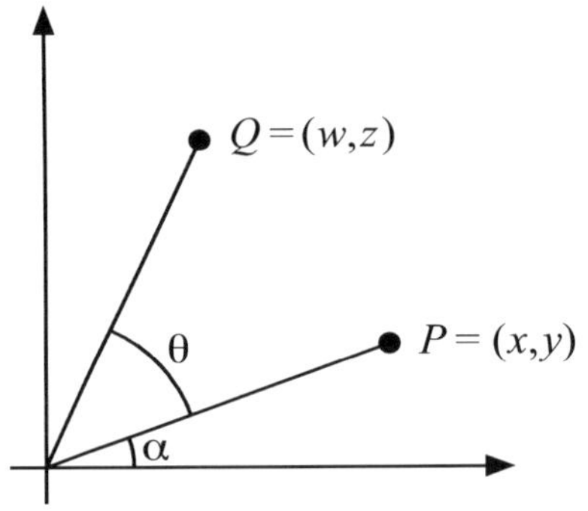

Figure 19.6
A rotation in the plane.

(i) Use suitable trigonometric identities to show that

$$\begin{aligned} w &= \cos(\theta)x - \sin(\theta)y \\ z &= \sin(\theta)x + \cos(\theta)y. \end{aligned}$$

(ii) Explain why the counterclockwise rotation around the origin by an angle θ can be represented by left multiplication by the matrix $\begin{bmatrix} \cos(\theta) & -\sin(\theta) \\ \sin(\theta) & \cos(\theta) \end{bmatrix}$.

(b) We will now examine the reflective symmetries by finding a matrix that performs a reflection across the line l specified by the parametric equations $x = at$, $y = bt$. We can use the previous result about rotations to complete this problem. Assume the line l makes an angle θ with the positive x-axis and that we want to reflect the point P across l to the point Q, as shown top left in Figure 19.7.

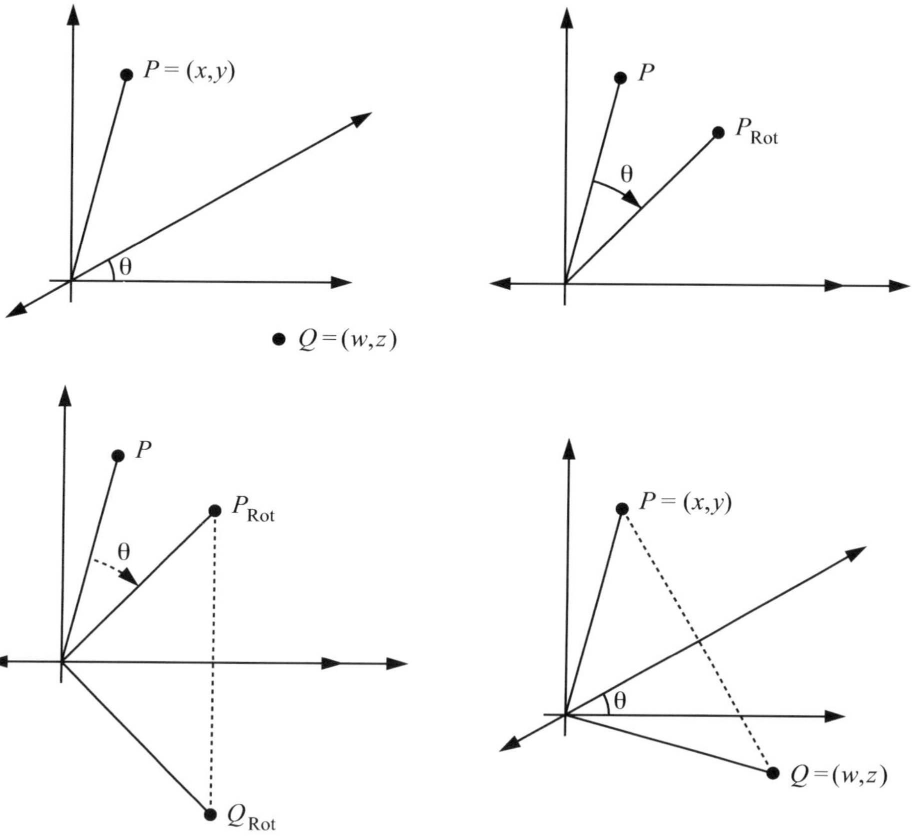

Figure 19.7
Finding a reflection matrix.

(i) We will first rotate everything clockwise around the origin by an angle of θ to make the x-axis the axis of reflection, as shown top right in Figure 19.7. Find the matrix M_1 that performs this rotation. Under this rotation, our original point P gets transformed to the point P_{Rot}.

(ii) Now we will reflect the point P_{Rot} across the x-axis (our transformed line of reflection) to the point Q_{Rot}, as shown bottom left in Figure 19.7. Show that the matrix $M_2 = \begin{bmatrix} 1 & 0 \\ 0 & -1 \end{bmatrix}$ performs this reflection across the x-axis.

(iii) Finally, we will rotate everything around the origin counterclockwise by an angle of θ to obtain the reflection Q of our original point P across the line l, as shown bottom right in Figure 19.7. Find the matrix M_3 that performs this rotation.

(iv) Put parts (i) – (iii) together to find the matrix $R_{(a,b)}$ that performs a reflection across the line with parameterization $x = at$, $y = bt$. Use appropriate trigonometric identities to show that

$$R_{(a,b)} = \begin{bmatrix} \cos(2\theta) & \sin(2\theta) \\ \sin(2\theta) & -\cos(2\theta) \end{bmatrix}.$$

(v) We can also write the matrix $R_{(a,b)}$ in terms of a and b. Show that

$$R_{(a,b)} = \frac{1}{a^2+b^2}\begin{bmatrix} a^2-b^2 & 2ab \\ 2ab & b^2-a^2 \end{bmatrix}.$$

(c) Show that the composition of two rotations is a rotation.

(d) What is the composition of a rotation and a reflection? Explain geometrically and by multiplying appropriate matrices. Does order matter? Why or why not?

(e) What is the composition of two reflections? Explain geometrically and by multiplying appropriate matrices. Does order matter? Why or why not?

Investigation 20

An Introduction to Groups

Focus Questions

By the end of this investigation, you should be able to give precise and thorough answers to the questions listed below. You may want to keep these questions in mind to focus your thoughts as you complete the investigation.

- What is a group, and what are some examples of groups?
- What is an Abelian group? What are some examples of Abelian and non-Abelian groups?
- What are some basic properties that hold in all groups and that can be proved from the group axioms?
- What uniqueness properties are satisfied by identities and inverses in groups, and why do these properties hold?
- What is the order of a group? What is the difference between finite and infinite groups?
- What is the group of units of a set, and for which types of sets can a group of units be defined?

In Investigation 6, we considered a variety of different number systems, some familiar and some not. While there were significant differences between these number systems, there were also common features that seemed to be shared by all of them. In this investigation, we will focus on some of these common features and their implications.

Preview Activity 20.1. In Investigation 19, we studied the set of symmetries of an object under the operation of function composition. Compare and contrast this set to each of the following sets with the given operation. Which properties do they share, and which do they not share?

(a) $\mathbb{Z}$, the set of integers under addition

(b) $\mathbb{E}$, the set of even integers under addition

(c) $\mathbb{Q}$, the set of rational numbers under addition

(d) $\mathbb{R}^+$, the set of all positive real numbers under multiplication

(e) $\mathbb{Z}_n$, the set of all congruence classes of integers modulo n under addition

(f) $\{-1, 1\}$, the two element subset of the integers under integer multiplication

(g) $\mathcal{M}_{n\times n}(\mathbb{R})$, the set of $n \times n$ matrices with real entries under addition

(h) $GL_n(\mathbb{R})$, the set of all invertible $n \times n$ matrices with real entries under the operation of multiplication

(i) $\mathcal{F}$, the set of all functions mapping the reals to the reals under the operation of addition of functions

Groups

In Preview Activity 20.1, we identified a set of properties that seemed to be satisfied in a variety of different sets including $\mathbb{Z}$, $\mathbb{E}$, $\mathbb{Q}$, $\mathbb{R}^+$, $\mathbb{C}$, $\mathbb{Z}_n$, $\mathcal{M}_{n\times n}(\mathbb{R})$, and the set of symmetries of an object. All of these sets are examples of a special type of algebraic structure known as a *group*,* defined formally as follows:

Definition 20.2. A **group** is a set G on which one binary operation, denoted $\cdot$, is defined such that all of the following axioms hold:

The Group Axioms

- **The set G is closed under its operation**, meaning that $a \cdot b \in G$ for all $a, b \in G$.
- **The operation $\cdot$ is associative in G**, meaning that $(a \cdot b) \cdot c = a \cdot (b \cdot c)$ for all a, b, $c \in G$.
- **The set G contains an identity element**, meaning that there exists an element $e \in G$ such that $a \cdot e = a = e \cdot a$ for all $a \in G$.
- **The set G contains an inverse for each of its elements**, meaning that for each $a \in G$, there exists an element $b \in G$ such that $a \cdot b = e = b \cdot a$.

Note that the operation in G need not be commutative. However, as we have seen, the operation is commutative in some instances. This gives rise to the next definition.

Definition 20.3. A group G is an **Abelian group** if $a \cdot b = b \cdot a$ for all $a, b \in G$ (in other words, if the operation is commutative in G).

An Abelian group is also called a *commutative* group.† Consistent with the usual convention, we will often omit the symbol for the group operation $(\cdot)$, writing ab instead of $a \cdot b$. In some cases, it is more natural to use the $+$ sign to denote the operation in a group; our choice will depend on the situation and should be clear from the context.

*The French mathematician Evariste Galois appears to be the first to use the word *group* in the expression *groupe de l'équation* in his paper *Mémoir sur les conditions de résolubilité des équations par redicaux* when referring to a subset of the collection of permutations of the roots of a polynomial. His idea of group was slightly different than the modern one.

†Abelian groups are named after the Norwegian mathematician Neils Henrik Abel (1802 – 1829), famous for (among other things) proving the impossibility of solving the general quintic polynomial equation by radicals.

Examples of Groups

We know of many sets on which we can define a binary operation, but not all of these sets are groups.

Activity 20.4. Which of the sets in the following list is a group under the indicated operation? If a set forms a group, is it an Abelian group? Explain.

- The set of symmetries of an object under composition
- $\mathbb{Z}$, the set of integers under addition
- $\mathbb{Z}$, the set of integers under multiplication
- any of the sets $\mathbb{Q}$, $\mathbb{R}$, $\mathbb{Z}_n$, $\mathcal{M}_{n\times n}(\mathbb{R})$ under their standard addition operation
- any of the sets $\mathbb{Q}$, $\mathbb{R}$, $\mathbb{Z}_n$, $\mathcal{M}_{n\times n}(\mathbb{R})$ under their standard multiplication operation
- the set $\text{GL}_n(\mathbb{R})$ of all invertible $n \times n$ matrices with real entries under the operation of matrix multiplication (in other words, the units in $\mathcal{M}_{n\times n}(\mathbb{R})$)
- $\mathbb{Q}^*$, the set of all nonzero rational numbers under multiplication
- $\mathbb{R}^+$, the set of all positive real numbers under multiplication
- $\{-1, 1\}$, the two element subset of the integers under integer multiplication (in other words, the units in $\mathbb{Z}$)
- $\mathcal{B}$, the set of all bijections mapping the reals to the reals under the operation of composition of functions.

As the next activity illustrates, there are also unfamiliar sets that turn out to be groups.

Activity 20.5. Let $S = \{a_1, a_2, a_3, a_4, a_5, a_6\}$ be the set on which an operation is defined by Table 20.1.

$\cdot$	a_1	a_2	a_3	a_4	a_5	a_6
a_1	a_1	a_2	a_3	a_4	a_5	a_6
a_2	a_2	a_1	a_5	a_6	a_3	a_4
a_3	a_3	a_6	a_1	a_5	a_4	a_2
a_4	a_4	a_5	a_6	a_1	a_2	a_3
a_5	a_5	a_4	a_2	a_3	a_6	a_1
a_6	a_6	a_3	a_4	a_2	a_1	a_5

Table 20.1
Operation table for the set S.

(a) How can we tell if the set S is closed under its operation?

(b) What is the identity element in S? How can we tell?

(c) What is the inverse of element a_5? How can we tell? What is the inverse of a_3?

(d) The operation in S is associative, although it is difficult to see this from the table. Verify the associative property in one case by computing $a_2(a_3a_4)$ and $(a_2a_3)a_4$.

The next activity compares various group properties for a few groups.

Activity 20.6. The sets listed in Table 20.2 under their standard operations (addition for $\mathbb{R}$, multiplication for $\mathbb{R}^+$, matrix multiplication for $\mathrm{GL}_2(\mathbb{R})$, function composition for $\mathcal{B}$, and the operation from Table 20.1 for S) can easily be seen to satisfy the four group axioms. (Verifying the associative law in each set can be a bit of work, but it is fairly straightforward in each case.) Complete Table 20.2, by determining: (i) the identity in each group; (ii) the inverse of the arbitrary element x; and (iii) whether the group's operation is commutative (answer (Y)es or (N)o). If the operation is not commutative, provide a counterexample. No proofs are required here, but you should be able to provide justification for your responses if asked.

	$\mathbb{R}$	$\mathbb{R}^+$	$\mathrm{GL}_2(\mathbb{R})$	$\mathcal{B}$	S
Identity					
Inverse of x					
Commutative?					

Table 20.2
Properties of groups.

Basic Properties of Groups

The group axioms should seem quite familiar since we have used them both implicitly and explicitly throughout previous investigations. But what about the other properties we have studied? You may have noticed that at least a few of these properties (including some that were satisfied by all of the number systems from Investigation 6) were not included in Definition 20.2. Part of this is due to the fact that many of the systems we encountered in Investigation 6 came with two binary operations and a group has only one binary operation. But even if we focus on only one binary operation at a time, there are certain properties that the sets in Investigation 6 satisfy that are not part of our definition of a group. Do these properties in fact hold for all groups, and if so, why are they not included in our list of group axioms?

To answer this question, we must think back to our discussion in Investigation 1 regarding the difference between axioms and theorems. There are certain properties, such as cancellation, that are in fact satisfied in all groups. These properties, however, can be proved from the group axioms. Thus, to include them would cause our axiom system to be redundant. The primary benefit of our definition of a group is that it provides a minimal set of axioms from which numerous other algebraic properties and theorems can be proved. Moreover, any property that we can prove using only the group axioms must necessarily hold in *every* group. Thus, the theory of groups gives us a way to study algebra more abstractly, instead of just within the context of specific number systems. In fact, the entire field of abstract algebra revolves around the study of general algebraic structures, such as groups, and their applications.

With that said, let's now formally state and prove the cancellation law we just mentioned. The integer version of this property was part of Investigation 1 (specifically, Exercise 2 on page 10). As it turns out, the argument we used there generalizes easily to the context of arbitrary groups. As you read the proof below, see if you can fill in the missing details and provide an explanation or justification wherever you see the ? symbol.

Theorem 20.7 (Group Cancellation Law). *Let G be a group, and let a, b, $c \in G$. If $ac = bc$, then $a = b$. Similarly, if $ab = ac$, then $b = c$.*

Proof. Let G be a group with identity e and let a, b, $c \in G$. Since G is a group, G contains an inverse d for c. Thus,

$$(ac)d = (bc)d,$$

which implies that

$$a(cd) = b(cd). ^{?}$$

Thus,

$$ae = be, ^{?}$$

and so $a = b$,$^{?}$ as desired. The proof that $ab = ac$ implies $b = c$ is left for you to complete. (Do you see why we need to prove both?) ■

Identities and Inverses in a Group

As we have mentioned, our list of group axioms leaves out a few important properties that we might expect to be satisfied in groups. In this section, we will verify some of these properties for arbitrary groups.

Activity 20.8.

(a) How many additive identities does $\mathbb{Z}$ have? Do you think your statement generalizes to arbitrary groups? In other words, how many identity elements does an arbitrary group have? Prove your answer. (Hint: Suppose a group G has two identities, e and e'. Evaluate ee' in two different ways, and compare.)

(b) How many additive inverses does each element in $\mathbb{Z}$ have? Does this property generalize to arbitrary groups? In other words, how many inverses does an element in an arbitrary group have? Prove your answer. (Hint: Begin, as in part (a), by assuming that an element $a \in G$ has two inverses, say b and c.)

If a is an element of a group G, then the inverse element b of a whose existence is guaranteed by the fourth group axiom is unique by Activity 20.8 so we can call this element *the* inverse of a. The notation we will use to denote the inverse of a group element a is either $-a$ (if we are using additive notation for our operation) or a^{-1} (if we are using multiplicative notation for our operation).

Activity 20.9. Let G be a group. Which of the following properties do you believe are satisfied in every group? Prove or disprove each property. (Hint: Use the definition of inverse from the group axioms.)

(a) $\left(a^{-1}\right)^{-1} = a$ for all $a \in G$,

(b) $(ab)^{-1} = a^{-1}b^{-1}$ for all $a, b \in G$,

(c) $(ab)^{-1} = b^{-1}a^{-1}$ for all $a, b \in G$,

The Order of a Group

In our examples, we have seen that some groups have an infinite number of elements ($\mathbb{Z}$, $\mathbb{Q}$, $\mathbb{R}$, $\mathrm{GL}_n(\mathbb{R})$) and some contain a finite number of elements ($\mathbb{Z}_n$, the symmetries of a square). This leads us to define the *order* of a group.

Definition 20.10. Let G be a group. If G contains only a finite number of elements, then G has **finite order** and we say G is a **finite group**. If G contains exactly m distinct elements, then the **order** of G, denoted $|G|$, is m. If G contains infinitely many elements, then G has **infinite order** and we say G is an **infinite group**.

For example, the groups $\mathbb{Z}$, $\mathbb{Q}$, $\mathbb{R}$, $\mathbb{Q}^*$, $\mathbb{R}^+$, $\mathrm{GL}_n(\mathbb{R})$ all have infinite order while $\mathbb{Z}_n$ has order n. In other words, $|\mathbb{Z}_n| = n$.

Activity 20.11.

(a) What is the order of the group of symmetries of an equilateral triangle?

(b) What is the order of the group of symmetries of square?

(c) What is the order of the group of symmetries of a regular pentagon?

(d) Do you see a pattern in the previous examples? If so, what do you expect the order of the group of symmetries of a regular n-gon to be?

We will see many other examples of finite groups as we proceed through our investigations.

Groups of Units

In earlier investigations we defined the units in the sets $\mathbb{Z}$, $\mathbb{Q}$, $\mathbb{C}$, $\mathbb{Z}_n$, and $\mathcal{M}_{n\times n}(\mathbb{R})$. Recall that a unit in a set that has both addition and multiplication operations is an element that has a multiplicative inverse. To define units in general, all we need is a set with an associative multiplication operation and an identity element.

Definition 20.12. Let S be a set on which an associative binary operation of multiplication is defined such that S contains an identity element 1_S. An element $u \in S$ is a **unit** in S if there is an element $v \in S$ such that

$$uv = 1_S = vu.$$

The element v for which $uv = vu = 1$ is unique (the argument is the same as the uniqueness of inverses in a group), so we call this element the *inverse* of u in S and denote it as u^{-1}. Note also that if $uv = vu = 1_S$, then $v = u^{-1}$ and $u = v^{-1}$.

Activity 20.13.

(a) Let n be a positive integer. In Investigation 5, we classified all of the units in $\mathbb{Z}_n$. How can we tell if a class $[a]$ is a unit in $\mathbb{Z}_n$? How can we find the inverse of a unit in $\mathbb{Z}_n$?

(b) Find or describe all of the units in each of the following sets:

(i) $\mathbb{Z}_5$

(ii) $\mathbb{Z}_8$

(iii) $\mathbb{Q}$

(iv) $\mathcal{M}_2(\mathbb{R})$

(c) Let U_n be the set of units in $\mathbb{Z}_n$. Construct the multiplication tables for the sets U_3, U_4, U_5, U_6, U_7 and U_8.

The examples above indicate that the set of units in $\mathbb{Z}_n$ forms a group under multiplication. In the next activity, we will decide if this is true in general.

Activity 20.14. Let S be a set on which an associative binary operation of multiplication is defined such that S contains an identity element 1_S. Let $U(S)$ be the set of units in S.

(a) Does $U(S)$ contain an identity element? If so, what is it? Prove your answer.

(b) Is $U(S)$ closed under multiplication? Explain. (Be careful not to assume that multiplication in S is commutative.)

(c) Why is multiplication associative in $U(S)$?

(d) Does $U(S)$ contain an inverse for each of its elements? Explain.

(e) Explain how we have just proved the following theorem.

Theorem 20.15. *Let S be a set on which an associative binary operation of multiplication is defined such that S contains an identity element 1_S. Let $U(S)$ be the set of units in S. Then $U(S)$ is a group under the operation of multiplication.*

We have seen that the number systems $\mathbb{Z}$, $\mathbb{Z}_n$, $\mathbb{Q}$, $\mathbb{R}$, $\mathbb{C}$, $\mathcal{M}_{n\times n}(\mathbb{R})$, and $\mathcal{P}_n$ are all groups under their additive operations. They are not, however, groups under their multiplicative operations. (Do you see why?) The result of Activity 20.14 is that the *sets of units* in the number systems $\mathbb{Z}$, $\mathbb{Z}_n$, $\mathbb{Q}$, $\mathbb{R}$, $\mathbb{C}$, $\mathcal{M}_{n\times n}(\mathbb{R})$, and $\mathcal{P}_n$ are groups under their multiplicative operations.

Concluding Activities

Activity 20.16. Let G be a group. Show that in the operation table for G, every element in G appears once and only once in each row and column.

Activity 20.17.

(a) In a group G with identity e, if $ab = e$ for some $a, b \in G$ must it follow that $b = a^{-1}$? Prove your answer.

(b) In a group G with identity e, if $ba = e$ for some $a, b \in G$ must it follow that $b = a^{-1}$? Prove your answer.

(c) Let f and g be functions from a set S to S. Let I be the identity function on S—that is $I(x) = x$ for all x in S. Show by example that it is possible to have $fg = I$, but $f \neq g^{-1}$. Does this violate part (a)? Explain.

Activity 20.18. Recall that a symmetry of an object O is a bijective, distance-preserving function f such that $f(O) = O$. In this activity, we will verify that the set S of symmetries of an object O forms a group under the operation of composition, called the **group of symmetries of** O. ‡

(a) Let f and g be bijective, distance-preserving functions with $f(O) = O$ and $g(O) = O$. To show that S is closed, we need to verify that $f \circ g$ is a bijective, distance preserving function with $(f \circ g)(O) = O$.

  (i) Prove that $|(f \circ g)(x) - (f \circ g)(y)| = |x - y|$ for all x, y in the domain of $f \circ g$.

  (ii) Use part (a) to deduce that $f \circ g$ is an injection.

  (iii) Show that $(f \circ g)(O) = O$. Deduce that $f \circ g$ is a surjection.

  (iv) Explain how parts (a) – (c) establish that S is closed under composition.

(b) Prove that composition of functions is an associative operation.

(c) What is the identity element of S? Verify your answer.

(d) If $f \in S$, what is the inverse of f in S? Verify your answer. (Please note that there is quite a bit to do to complete this problem.)

(e) Explain why S is a group. Is S an Abelian group? Explain.

Activity 20.19. Write a short summary that describes the important concepts, and the relationships between these concepts, that were introduced in this investigation. Explain how the material in this investigation is connected to the material in Investigation 19.

‡This activity assumes an understanding of injective, surjective, and bijective functions. For a review of these topics, see Appendix A.

Exercises

(1) Assume G is a group. Suppose that, due to a printer error, the operation table for G was printed with several entries missing, as shown below:

	a	b	c	d
a				
b			d	
c				b
d		c		

Complete the table using only the group axioms and consequent properties. Specifically explain how you determined each element.

(2) In a group, is it true that $AB = BC$ implies $A = C$? (Such a property is called a "cross cancellation" property.) If the answer is no, find a specific counterexample. If the answer is yes, then prove this property.

(3) Let $\mathbb{Z}^\star$ be the number system consisting of the set of all integers.

(a) Define an operation $\oplus$ on $\mathbb{Z}^\star$ by

$$a \oplus b = a + b - 1.$$

Note that $+$ denotes the normal operation of addition in $\mathbb{Z}$. Which of the group axioms are satisfied by $\mathbb{Z}^\star$ using the operation $\oplus$, and which are not? Is $\mathbb{Z}^\star$ a group with the operation $\oplus$? If so, is $\mathbb{Z}^\star$ an Abelian group? Prove your answers.

(b) Now define a different operation $\otimes$ on $\mathbb{Z}^\star$ by

$$a \otimes b = a + b - a \cdot b.$$

Here $+$ and $\cdot$ denote the normal operations of addition and multiplication in $\mathbb{Z}$. Which of the group axioms are satisfied by $\mathbb{Z}^\star$ using the operation $\otimes$, and which are not? Is $\mathbb{Z}^\star$ a group with the operation $\otimes$? If so, is $\mathbb{Z}^\star$ an Abelian group? Prove your answers.

(4) Determine if the set G is a group under the indicated operation. If G is a group, verify that each group property is satisfied. If G is not a group, provide examples that show which of the group properties are not satisfied.

(a) G is the set of odd integers under addition

(b) $G = \{[2], [4], [6], [8]\} \subset \mathbb{Z}_{10}$, with the operation of multiplication of congruence classes.

(c) $G = \{[0], [2], [4], [6], [8]\} \subset \mathbb{Z}_{10}$, with the operation of addition of congruence classes.

(d) $G = \{q \in \mathbb{Q} : q \neq 1\}$, with the operation $*$ defined by $a * b = a + b - ab$

(e) $G = \{[x] \in \mathbb{Z}_9 : x = 1, 2, 4, 5, 7, \text{ or } 8\}$, with the operation $[x] * [y] = [x][y]$

(5) Let $\mathbb{R}^+$ denote the set of positive real numbers.

(a) Is $\mathbb{R}^+$ a group using standard multiplication on $\mathbb{R}$? Prove your answer.

(b) Is $\mathbb{R}^+$ a group using standard division on $\mathbb{R}$? Prove your answer.

(6) Is $\mathbb{R}^-$, the set of negative real numbers, a group using the operation $x * y = -(xy)$, where xy is the standard product in $\mathbb{R}$? Prove your answer.

(7) Let k be an integer, and let $Z(k)$ be the set of integers on which an operation $\oplus_k$ is defined as follows:

$$a \oplus_k b = a + b - k,$$

where $a + b$ denotes the standard sum of a and b in $\mathbb{Z}$. Note that the set $Z(0)$ is the group of integers under the standard addition. For which values of k is $Z(k)$ a group under the operation $\oplus_k$? Prove your answer.

(8) Prove that a group G is Abelian if and only if $(ab)^2 = a^2b^2$ for all $a, b \in G$.

(9) Consider a strip of three equally spaced I's:

I I I

Describe the group of symmetries of this strip. Is the group of symmetries of the strip Abelian?

(10) Consider an infinitely long strip of equally spaced I's:

$\cdots$ I I I I $\cdots$

Describe the group of symmetries of this strip. Is the group of symmetries of the strip Abelian?

(11) (From a GRE Practice Exam) Let p and q be distinct primes. Suppose that H is a proper subset of the integers and H is a group under addition that contains exactly three elements of the set $\{p, p+q, pq, p^q, q^p\}$. Determine which of the following are the three elements in H.

(i) pq, p^q, q^p

(ii) $p+q, pq, p^q$

(iii) $p, p+q, pq$

(iv) p, p^q, q^p

(v) p, pq, p^q

(12) Prove that a group G is Abelian if and only if $(ab)^{-1} = a^{-1}b^{-1}$ for all $a, b \in G$.

(13) Let n be a nonnegative integer, and let $n\mathbb{Z} = \{nx : x \in \mathbb{Z}\}$, with addition defined as in $\mathbb{Z}$. Is $n\mathbb{Z}$ a group under this addition? If so, is $n\mathbb{Z}$ an Abelian group? Does your answer depend on the value of n? Prove your answers.

(14) Let n and k be natural numbers, both greater than 1. Let $\mathcal{M}_{n\times n}(\mathbb{Z}_k)$ be the set of all $n \times n$ matrices whose entries are in $\mathbb{Z}_k$.

(a) How many elements does $M_{n\times n}(\mathbb{Z}_k)$ have?

(b) Is $M_{n\times n}(\mathbb{Z}_k)$ a group under standard addition of matrices?

(c) Is $M_{n\times n}(\mathbb{Z}_k)$ a group under standard multiplication of matrices?

(d) Find all of the units in $M_{2\times 2}(\mathbb{Z}_2)$ using standard multiplication of matrices. Does this collection of units form an Abelian group? Explain.

(15) Let $\mathcal{F}(\mathbb{R})$ denote the set of all functions from $\mathbb{R}$ to $\mathbb{R}$. Define addition and multiplication on $\mathcal{F}(\mathbb{R})$ as follows:

- For all $f, g \in \mathcal{F}(\mathbb{R})$, $(f+g) : \mathbb{R} \to \mathbb{R}$ is the function defined by
$$(f+g)(x) = f(x) + g(x)$$
for all $x \in \mathbb{R}$.
- For all $f, g \in \mathcal{F}(\mathbb{R})$, $(fg) : \mathbb{R} \to \mathbb{R}$ is the function defined by
$$(fg)(x) = f(x)g(x)$$
for all $x \in \mathbb{R}$.

(a) Prove that $\mathcal{F}(\mathbb{R})$ is an Abelian group under addition.

(b) Does $\mathcal{F}(\mathbb{R})$ have an identity element for multiplication?

(c) Find an element in $\mathcal{F}(\mathbb{R})$ that does not have a multiplicative inverse in $\mathcal{F}(\mathbb{R})$. Explain how this shows $\mathcal{F}(\mathbb{R})$ is not a group under multiplication.

(d) Find necessary and sufficient conditions for an element in $\mathcal{F}(\mathbb{R})$ to be a unit in $\mathcal{F}(\mathbb{R})$. State your result in a lemma of the form "The function $f \in \mathcal{F}(\mathbb{R})$ is a unit in $\mathcal{F}(\mathbb{R})$ if and only if ...". Your lemma must say something more than just a rehash of the definition of a unit; rather, it must actually characterize the functions that are invertible under multiplication in $\mathcal{F}(\mathbb{R})$.

(16) Define a **blip** to be a pair of integers, denoted $\langle a, b\rangle$, and define two blips $\langle a, b\rangle$ and $\langle x, y\rangle$ to be equal whenever $a + b = x + y$ (so that, for instance, $\langle 3, 5\rangle$ and $\langle 10, -2\rangle$ would be considered equal since $3 + 5 = 8 = 10 + (-2)$). Define an operation $*$ on the set B of all blips as follows:
$$\langle a, b\rangle * \langle c, d\rangle = \langle a + c, b + d\rangle.$$
Is B a group under the operation $*$? If so, prove it. If not, determine which of the group axioms are satisfied and which are not.

(17) Let x and y be units in $\mathbb{Z}$. Prove or disprove: $x + y$ is a unit in $\mathbb{Z}$.

Connections

This investigation introduced the concept of a group. Groups are algebraic objects that share the same basic additive structure as the integers and the different number systems discussed in Investigation 6. There is a great deal of power to be found in recognizing the features these number systems have in common and then creating a larger category (groups) that encapsulates all of these features. Indeed, we can then learn about all of these number systems at one time by studying arbitrary groups.

If you studied ring theory before group theory, you should notice connections between the topics in this investigation and those in Investigation 7. In particular, groups and rings are both algebraic objects—that is, sets on which an operation or operations are defined, yielding an algebraic structure of some sort. The main difference between a group and a ring is that a group comes with one binary operation while a ring comes with two. In fact, every ring is a group under its addition operation, but not all groups can be made into rings in a natural way.

Since there is only one operation in a group, groups are simpler objects than rings. For that reason, a good argument can be made that the study of abstract algebra should begin with groups. From a structural standpoint, however, rings may be more familiar to us than groups in that many of the sets with which we have worked in our mathematical pasts (e.g., $\mathbb{Z}$, $\mathbb{Q}$, $\mathbb{R}$, and sets of polynomials) are all rings. For this reason, starting our exploration of abstract algebra with rings is also a reasonable choice. In either case, many of the concepts we will encounter in these investigations will apply to both rings and groups.

Investigation 21

Integer Powers of Elements in a Group

Focus Questions

By the end of this investigation, you should be able to give precise and thorough answers to the questions listed below. You may want to keep these questions in mind to focus your thoughts as you complete the investigation.

- How can integer exponentiation be defined in an arbitrary group?
- What properties are satisfied by integer exponentiation in groups?

Our study of groups began in Investigation 19, where we learned about the set of symmetries of an object. We then saw how many familiar sets, like $\mathbb{Z}$, $\mathbb{Z}_n$, and sets of invertible square matrices, all had a structure that was similar to that of a set of symmetries—namely, the structure of a group. One of the defining axioms of a group is that it is closed under its operation. In this investigation, we will define and study a familiar shorthand notation for repeatedly applying a group's operation.

Preview Activity 21.1. Use your intuition to calculate the quantities listed below. Throughout your calculations, you will be applying the definitions that we will formally develop in this investigation. Recall that G is the group of symmetries of a square (with the operation of composition), $\mathbb{Z}_6$ is the set of integers modulo 6 (with the operation of addition of congruence classes), and $\text{GL}_2(\mathbb{R})$ is the set of all invertible square matrices with real entries (with the operation of matrix multiplication).

In G:	In $\mathbb{Z}_6$:	In $\text{GL}_2(\mathbb{R})$:
$\begin{pmatrix} 1 & 2 & 3 & 4 \\ 2 & 3 & 4 & 1 \end{pmatrix}^0$	$0[4]$	$\begin{bmatrix} 1 & \pi \\ -\pi & 3 \end{bmatrix}^0$
$\begin{pmatrix} 1 & 2 & 3 & 4 \\ 4 & 3 & 2 & 1 \end{pmatrix}^2$	$2[5]$	$\begin{bmatrix} 1 & 1 \\ 1 & 1 \end{bmatrix}^2$
$\begin{pmatrix} 1 & 2 & 3 & 4 \\ 2 & 1 & 4 & 3 \end{pmatrix}^{-2}$	$(-2)[3]$	$\begin{bmatrix} 1 & 1 \\ 0 & 1 \end{bmatrix}^{-2}$

Introduction

In Preview Activity 21.1, we began to intuitively develop the notions of integer exponentiation for groups. You may have performed the requested calculations by simply thinking of exponentiation as repeatedly applying the group operation. For instance, in the group of symmetries of the square, we can calculate $\begin{pmatrix} 1 & 2 & 3 & 4 \\ 4 & 3 & 2 & 1 \end{pmatrix}^3$ as follows:

$$\begin{aligned} \begin{pmatrix} 1 & 2 & 3 & 4 \\ 4 & 3 & 2 & 1 \end{pmatrix}^3 &= \left[\begin{pmatrix} 1 & 2 & 3 & 4 \\ 2 & 3 & 4 & 1 \end{pmatrix}\begin{pmatrix} 1 & 2 & 3 & 4 \\ 2 & 3 & 4 & 1 \end{pmatrix}\right]\begin{pmatrix} 1 & 2 & 3 & 4 \\ 2 & 3 & 4 & 1 \end{pmatrix} \\ &= \begin{pmatrix} 1 & 2 & 3 & 4 \\ 3 & 4 & 1 & 2 \end{pmatrix}\begin{pmatrix} 1 & 2 & 3 & 4 \\ 2 & 3 & 4 & 1 \end{pmatrix} \\ &= \begin{pmatrix} 1 & 2 & 3 & 4 \\ 4 & 1 & 2 & 3 \end{pmatrix}. \end{aligned}$$

This intuitive formulation of integer exponentiation makes sense as long as we are exponentiating by a positive integer. For nonpositive integers, however, we will need to be a bit more careful. Furthermore, in order to prove that integer exponentiation works the way we expect it to, we will need to make use of a more formal definition.

We will develop such a definition in the next section, and we will use this definition to prove several fundamental properties of integer exponentiation in groups. As we move forward, it is important to note that in groups, which have only one operation, we will use the term *exponentiation* to refer to the intuitive idea of repeatedly applying a group's operation, regardless of whether that operation is multiplication (as in $\mathrm{GL}_n(\mathbb{R})$), addition (as in $\mathbb{Z}_n$), or something else (for example, composition in the group of symmetries of an object). When the group operation is addition, it is more natural to write mx instead of x^m. (In this case, integer exponentiation could alternatively be called integer multiplication.)

Powers of Elements in a Group

Let G be a group (with the operation written multiplicatively), and let $a \in G$. The elements $a, aa, aaa, \ldots, a^{-1}, a^{-1}a^{-1}, a^{-1}a^{-1}a^{-1}, \ldots$ are also in G. To represent these types of elements in a more convenient and natural way, we will let a^0 be the identity element. We will then define

$$a^1 = a, \quad a^2 = a^1a, \quad a^3 = a^2a, \quad a^4 = a^3a,$$

and so on. Similarly,

$$a^{-1} = a^{-1}, \quad a^{-2} = a^{-1}a^{-1} = \left(a^{-1}\right)^2, \quad a^{-3} = a^{-1}a^{-1}a^{-1} = \left(a^{-1}\right)^3,$$

and so on.

To be more formal, we can define powers of a recursively as follows:

Definition 21.2. Let G be a group with identity e, and let $a \in G$. Then for each integer m, we define a^m as follows:

- $a^0 = e$.
- $a^1 = a$.
- If m is a positive integer, then $a^m = a^{m-1}a$.

- If m is a positive integer, then $a^{-m} = \left(a^{-1}\right)^m$.

When we use additive notation for the operation in our group, a^m is written ma and these definitions can be written as $0a = e$, $1a = a$, $ma = (m-1)a + a$ and $(-m)a = m(-a)$.

Throughout your studies of mathematics, you have undoubtedly used exponentiation by an integer in many settings. This familiarity should raise a number of questions about Definition 21.2 . For example, is it true that $a^m = aa^{m-1}$ for positive integers m? Is it true that $a^m = a^{m-1}a$ if m is a negative integer? Is it true in a group that $a^m a^n = a^{m+n}$? If $b \in G$ is it true that $(ab)^m = a^m b^m$? We will address these questions in the remainder of this investigation and in the exercises.

Let's begin with the important question of whether $(ab)^m = a^m b^m$.

Activity 21.3. Let G be the symmetries of the square, and let $a = \begin{pmatrix} 1 & 2 & 3 & 4 \\ 3 & 4 & 1 & 2 \end{pmatrix}$, $b = \begin{pmatrix} 1 & 2 & 3 & 4 \\ 4 & 3 & 2 & 1 \end{pmatrix}$, and $c = \begin{pmatrix} 1 & 2 & 3 & 4 \\ 2 & 3 & 4 & 1 \end{pmatrix}$.

(a) Calculate $(bc)^2$ and b^2c^2. Are they the same?

(b) Calculate $(ab)^2$ and a^2b^2. Are they the same?

(c) In one of the preceding parts, we have $(xy)^2 = x^2y^2$, and in the other we don't. There is a significant difference in the relationship between x and y that accounts for this difference. How is the relationship between x and y different in the two parts, and why does this difference affect whether $(xy)^2 = x^2y^2$?

Activity 21.3 illustrates an important point when working in groups. In general we **<u>cannot</u>** assume that $(ab)^m = a^m b^m$. However, if we know an additional fact about a and b, then we can use this property.

Theorem 21.4. *Let G be a group with identity e, and let $a, b \in G$ such that $ab = ba$. Then $(ab)^m = a^m b^m$ for every integer m.*

When $ab = ba$ in a group G, we say that the elements a and b *commute* (or commute with each other). An outline of the proof of Theorem 21.4 is given below. As you read the proof, try to fill in the missing details and provide additional explanation or justification where indicated.

Proof of Theorem 21.4. Let G be a group with identity e, and let a, b elements in G such that $ab = ba$. First we will show that $(ab)^m = a^m b^m$ for every positive integer m. To do so, we will use induction on m. We know that

$$(ab)^1 \overset{?}{=} ab \overset{?}{=} a^1 b^1,$$

so our theorem is true for $m = 1$. Now assume that $(ab)^m = a^m b^m$ for some integer $m \geq 1$. We will show that $(ab)^{m+1} = a^{m+1}b^{m+1}$. Note that

$$(ab)^{m+1} \overset{?}{=} (ab)^m(ab) \overset{?}{=} (a^m b^m)(ab) \overset{?}{=} a^m(b^m a)b. \tag{21.1}$$

To complete this portion of the proof, we need to know that a commutes with b^m, or that $b^m a = ab^m$. This will also require an induction proof.

Claim. If n is a positive integer, then $b^n a = ab^n$.

Proof of Claim.. Notice that

$$b^1 a \overset{?}{=} ba \overset{?}{=} ab \overset{?}{=} ab^1,$$

so the claim is true for $n = 1$. For the induction step, we will assume that $b^n a = ab^n$ for some integer $n \geq 1$. We will show that $b^{n+1}a = ab^{n+1}$. Now

$$b^{n+1}a \overset{?}{=} (b^n b)a \overset{?}{=} b^n(ba) \overset{?}{=} b^n(ab) \overset{?}{=} (b^n a)b \overset{?}{=} (ab^n)b \overset{?}{=} a(b^n b) \overset{?}{=} ab^{n+1},$$

and so we have verified that $b^n a = ab^n$ for every positive integer n. ■

We will now return to our proof that $(ab)^{m+1} = a^{m+1}b^{m+1}$. Continuing with (21.1), we have

$$\begin{aligned}
(ab)^{m+1} &= (ab)^m(ab) \\
&= (a^m b^m)(ab) \\
&= a^m(b^m a)b \\
&\overset{?}{=} a^m(ab^m)b \\
&\overset{?}{=} (a^m a)(b^m b) \\
&\overset{?}{=} a^{m+1}b^{m+1}.
\end{aligned}$$

Thus we have that $(ab)^m = a^m b^m$ for all ___________ m.

To complete the proof of Theorem 21.4, we need to verify that $(ab)^m = a^m b^m$ for $m = 0$ and $m < 0$. Since

$$(ab)^0 \overset{?}{=} e \overset{?}{=} ee \overset{?}{=} a^0 b^0,$$

the theorem is true if $m = 0$. We now need to consider the case where m is negative.

Suppose $m = -1$. Since $ab = ba$, we have

$$a^{-1}b^{-1} \overset{?}{=} (ba)^{-1} \overset{?}{=} (ab)^{-1} \overset{?}{=} b^{-1}a^{-1},$$

so we see that $(ab)^{-1} = a^{-1}b^{-1}$. Moreover, we have shown that a^{-1} and b^{-1} commute with each other. Now let m be a negative integer and let $k = -m$. Then

$$\begin{aligned}
(ab)^m &\overset{?}{=} (ab)^{-k} \\
&\overset{?}{=} \left((ab)^{-1}\right)^k \\
&\overset{?}{=} \left((ba)^{-1}\right)^k \\
&\overset{?}{=} \left(a^{-1}b^{-1}\right)^k \\
&\overset{?}{=} \left(a^{-1}\right)^k \left(b^{-1}\right)^k \\
&\overset{?}{=} a^{-k}b^{-k} \\
&\overset{?}{=} a^m b^m.
\end{aligned}$$

Therefore, we can conclude that __________ = __________ for all __________ m. This, along with our previous cases, completes the proof. ■

Now that we have established that $(ab)^m = a^m b^m$ whenever a and b commute, let's consider some other familiar properties of exponentiation. For example, is it true that $(a^m)^{-1} = (a^{-1})^m$? To answer this question, note that since $aa^{-1} = e = a^{-1}a$, we see that a and a^{-1} commute. So

$$\left(a^{-1}\right)^m a^m = \left(a^{-1}a\right)^m = \left(aa^{-1}\right)^m = a^m \left(a^{-1}\right)^m.$$

It follows that $\left(a^{-1}\right)^m$ is an inverse of a^m. Since inverses are unique, we can conclude that

$$\left(a^{-1}\right)^m = (a^m)^{-1}$$

for any integer m.

Recall that, by Definition 21.2, $a^{-m} = \left(a^{-1}\right)^m$ whenever m is a positive integer. If m is nonpositive, is this also true?

We know that $\left(a^{-1}\right)^{-1} = a$ for every $a \in G$, so if m is a negative integer, then

$$\left(a^{-1}\right)^m = \left[\left(a^{-1}\right)^{-1}\right]^{-m} = a^{-m}.$$

Also,

$$a^0 = e = \left(a^{-1}\right)^0.$$

Putting all of this together, we arrive at part (i) of the next theorem.

Theorem 21.5. *Let G be a group. For every $a \in G$ and every $m, n \in \mathbb{Z}$:*

(i) $a^{-m} = \left(a^{-1}\right)^m = (a^m)^{-1}$ (or, if the group operation is written using additive notation, $(-m)a = m(-a) = -(ma)$).

(ii) $a^m a^n = a^{m+n}$ (or, if the group operation is written using additive notation, $ma + na = (m+n)a$).

(iii) $(a^m)^n = a^{mn}$ (or, if the group operation is written using additive notation, $n(ma) = (nm)a$).

The proofs of parts (ii) and (iii) of Theorem 21.5 are more complicated than part (i). Induction seems like a natural approach for these results, but there is a catch: both statements involve two arbitrary integers instead of just one. There are several different ways to proceed, and we will explore one approach to proving part (ii) ($a^m a^n = a^{m+n}$) in Activity 21.6. The proof of part (iii) is left as an exercise for you.

Activity 21.6. The statement in part (ii) of Theorem 21.5—namely, that $a^m a^n = a^{m+n}$ for all $a \in G$ and all $m, n \in Z$—involves three universal quantifiers. Our approach will be to choose an arbitrary $a \in G$ and $m \in \mathbb{Z}$, and then use induction on the integer n to resolve the cases where n is nonnegative. We will then use a separate argument to deal with the case where $n < 0$. So let $a \in G$, and let $m \in \mathbb{Z}$.

(a) To perform our inductive argument, we first need to establish a base case. Apply Definition 21.2 to prove that $a^m a^n = a^{m+n}$ when $n = 0$.

(b) Now let $n > 0$, and let $P(n)$ be the predicate, "$a^m a^n = a^{m+n}$." Below is an argument showing that $P(1)$ is true. Explain the rationale for each equality marked with the ? symbol.

If $n = 1$, then

$$a^m a^n = a^m a^1 \stackrel{?}{=} a^m a \stackrel{?}{=} a^{m+1} \stackrel{?}{=} a^{m+n}.$$

(c) To continue our induction proof, let n be a positive integer and assume that $P(n)$ (as defined in part (b)) is true. We then need to prove that $P(n+1)$ is true—that is,

$$a^m a^{n+1} = a^{m+(n+1)}.$$

Use the assumption that $P(n)$ is true, along with Definition 21.2 and whatever group axioms you need, to show that $P(n+1)$ is true.

(d) Deduce from parts (a) – (c) that the statement $a^m a^n = a^{m+n}$ is true for all $a \in G$, all integers m, and all nonnegative integers n.

(e) To complete our proof, we need to consider the case where $n < 0$. So assume that n is a negative integer. Give a justification for each step in the following argument:

$$\begin{aligned} \left(a^{-1}\right)^{-m} \left(a^{-1}\right)^{-n} &= \left(a^{-1}\right)^{(-m)+(-n)} \text{ (?)} \\ a^m a^n &= a^{-((-m)+(-n))} \text{ (?)} \\ a^m a^n &= a^{m+n}. \text{ (?)} \end{aligned}$$

(f) Use your work from the previous parts to write a complete, clear, and convincing proof that $a^m a^n = a^{m+n}$ for all $a \in G$ and all integers m and n.

Notice that our work in Activity 21.6 also answers (in the affirmative) the questions we asked earlier about whether $a^m = a^{m-1}a = aa^{m-1}$ for all $a \in G$ and all $m \in \mathbb{Z}$. (Do you see why?) Although these properties may seem obvious, they are not immediate consequences of Definition 21.2, which simply states that $a^m = a^{m-1}a$ for all $a \in G$ and all *positive* $m \in \mathbb{Z}$.

Concluding Activities

Activity 21.7. Let G be a group and a an element of G. Show that any two powers of a commute. That is, show that $a^m a^n = a^n a^m$ for any integers m and n.

Activity 21.8. Let G be a group and $a \in G$. Prove part (iii) of Theorem 21.5—that is, prove that $(a^m)^n = a^{mn}$ for all integers m and n.

Activity 21.9. Write a short summary that describes the important concepts, and the relationships between these concepts, that were introduced in this investigation. Explain how the material in this investigation is connected to the material in Investigation 20.

Exercises

(1) Let G be a group.

(a) Let $a, b, c \in G$. What element is $(abc)^{-1}$? Prove your answer.

(b) Let m be a positive integer, and let $a_1, a_2, \ldots, a_m$ be elements in G. What element is $(a_1 a_2 \cdots a_m)^{-1}$? Prove your answer.

$\star$ (2) Prove that if G is a group with identity e in which $a^2 = e$ for every $a \in G$, then G is an Abelian group. Is the converse true? Explain.

(3) Let G be a group and $a \in G$. A **conjugate** of a in G is any element of the form bab^{-1} for some $b \in G$. Show that if $c = bab^{-1}$ is a conjugate of a in G and n is any integer, then $c^n = ba^n b^{-1}$.

(4) We can generalize Theorem 21.4 as follows: Let G be a group and $a_1, a_2, \ldots, a_m$ elements in G for some positive integer $m \geq 2$ so that $a_i a_j = a_j a_i$ for all i and j. Prove that

$$(a_1 a_2 \cdots a_m)^n = a_1^n a_2^n \cdots a_m^n$$

for any $n \in \mathbb{Z}$.

(5) **Fibonacci sequences in groups.** The Fibonacci numbers F_n are defined recursively by $F_0 = 0$, $F_1 = 1$, and $F_n = F_{n-1} + F_{n-2}$ for $n \geq 2$. The definition of this sequence only depends on a binary operation. Since every group comes with a binary operation, we can define Fibonacci-type sequences in any group. Let G be a group, and define the sequence $\{f_n\}$ in G as follows: Let a_0, a_1 be elements of G, and define

$$f_0 = a_0, \quad f_1 = a_1, \quad \text{and} \quad f_n = a_{n-1} a_{n-2} \quad \text{for } n \geq 2.$$

In his paper "Fibonacci Series Modulo m" (*American Mathematical Monthly*, Vol. 67, 1960), D.D. Wall writes the following about his introduction to these sequences:

> "The problem arose in connection with a method for generating random numbers, but it turned out to be unexpectedly intricate, and so quickly became of interest in its own right."

In an interesting application of Fibonacci sequences in groups, Iannis Xenakis (in *Formalized Music*, Indiana University Press, 1971) uses Fibonacci sequences in groups to create "Fibonacci motions," which are sequences of musical properties such as pitch, volume, and timbre that give the composition its framework. We will see in this problem that these Fibonacci sequences become intricate quite quickly.

(a) Explain why $f_n = F_n$ if $G = \mathbb{Z}$, $a_0 = 0$, and $a_1 = 1$.

(b) Find the elements in $\{f_n\}$ if $G = \mathbb{Z}_3$, $a_0 = [1]$ and $a_1 = [2]$.

(c) Some of the sequences $\{f_n\}$ are periodic—that is, the same list of elements repeats in the same order. The number of elements in one cycle is called the **period** of the sequence.

(i) If a is a non-identity element of a group with identity e and $a^2 = e$, what is the period of the sequence that begins $a\ a\ \ldots$?

(ii) If a is a non-identity element of a group with identity e and $a^3 = e$, what is the period of the sequence that begins $a\ a\ \ldots$?

(iii) If a is a non-identity element of a group with identity e and $a^4 = e$ but $a^2 \neq e$, what is the period of the sequence that begins $a\ a\ \ldots$?

(iv) If a is a non-identity element of a group with identity e and $a^5 = e$, what is the period of the sequence that begins $a\ a\ \ldots$?

(v) In general, the pattern of periods of the sequences that begin $a\ a\ \ldots$ is not so easy to see. While we won't establish the pattern here (see Wall's paper for details), we can begin to see how to approach the problem. Explain how the general sequence of group elements that begins $a\ a\ \ldots$ is specifically related to the Fibonacci sequence of integers.

Connections

In this investigation, we studied integer powers of elements in groups (or integer multiples if we use additive notation). If you studied ring theory before group theory, you should notice connections between the topics in this investigation and those in Investigation 8. The major difference between groups and rings in this context is that there is only one operation in a group but two in a ring. As a result, we need to understand both integer multiples (under addition) and integer powers (under multiplication) of ring elements. With only one operation in a group, we only need one of these ideas. However, we still use integer multiples when we represent our group operation as addition and integer powers when we write our group operation multiplicatively. Consequently, we need to understand both notations, even though we will only use one of them in any given group.

Investigation 22

Subgroups

Focus Questions

By the end of this investigation, you should be able to give precise and thorough answers to the questions listed below. You may want to keep these questions in mind to focus your thoughts as you complete the investigation.

- What is a subgroup of a group? What conditions must be verified in order to show that a subset of a group is a subgroup?
- What is the center of a group, and what kind of subgroup is it?
- What is the subgroup generated by an element of a group? What is a cyclic group? How are the two related?
- How can cyclic groups be used to define the order of an element in a group?

Preview Activity 22.1. Throughout mathematics, the relationship between mathematical objects and their sub-objects is of central importance. For instance, in linear algebra, we study vector spaces and their subspaces. In discrete mathematics, many graph theory problems can be solved by finding a subgraph that is optimal in some sense. Furthermore, many other applied optimization problems involve minimizing or maximizing a certain function subject to certain constraints. These constraints define what is known as a *feasible region*, which is nothing more than a subset of the space of all possible solutions.

In light of these examples and our recent investigations of groups, it seems natural that we would be interested in defining and characterizing *subgroups*. To begin thinking along these lines, consider the set $\mathbb{E}$ of even integers as a subset of the set $\mathbb{Z}$ of integers. Since $\mathbb{E}$ is a subset of a group, it is natural to ask which (if any) of the group axioms $\mathbb{E}$ satisfies.

(a) Critique the following proof that addition is associative in $\mathbb{E}$. Is the proof correct? If so, could it be improved in any way? If not, what is the main error in the argument?

Proof. Let a, b, $c \in \mathbb{E}$. Since $\mathbb{E} \subseteq \mathbb{Z}$, it follows that a, b, $c \in \mathbb{Z}$ as well. Thus,

$$a + (b + c) = (a + b) + c$$

since addition is associative in $\mathbb{Z}$. This proves that addition is associative in $\mathbb{E}$. ■

(b) Critique the following proof that $\mathbb{E}$ is closed under addition. Is the proof correct? If so, could it be improved in any way? If not, what is the main error in the argument?

Proof. Let $a, b \in \mathbb{E}$. Since $\mathbb{E} \subseteq \mathbb{Z}$, it follows that $a, b \in \mathbb{Z}$ as well. But $\mathbb{Z}$ is closed under addition, so $a + b \in \mathbb{Z}$. This shows that $\mathbb{E}$ is closed under addition. ■

(c) Which of the group axioms (using the operation of addition) does the set $\mathbb{E}$ satisfy?

Introduction

In Preview Activity 22.1, we argued that the subset $\mathbb{E}$ of the set of integers was not only a subset of $\mathbb{Z}$, but also a group itself using the same operation as in $\mathbb{Z}$. The next definition formalizes this terminology.

Definition 22.2. A subset H of a group G is a **subgroup** of G if H is a group using the same operation as in G.

Definition 22.2 is not terribly surprising. Nevertheless, there is one important caveat to note—namely, the condition that both H and G use the same operation. The operation imposes a structure on the set, and different operations impose different structures. For example, the set $\mathbb{R}^*$ of nonzero real numbers is a subset of the group $\mathbb{R}$, but we would not want to consider the group $\mathbb{R}^*$ under multiplication as a subgroup of $\mathbb{R}$ under addition because the elements behave so differently with respect to the two operations. (As an example, $3 \cdot 3 = 9$ in $\mathbb{R}^*$, but $3 + 3 = 6$ in $\mathbb{R}$.)

Activity 22.3. For each part below, decide whether the set H is a subgroup of the group G. (You may assume that all of the groups listed use their standard addition operation.) Explain your answers.

(a) $H = \mathbb{Z}, G = \mathbb{R}$

(b) $H = \mathbb{Z}^+, G = \mathbb{R}$

(c) $H = \{3n : n \in \mathbb{Z}\}, G = \mathbb{Z}$

(d) $H = \{3n + 1 : n \in \mathbb{Z}\}, G = \mathbb{Z}$

(e) $H = \mathbb{Z}_5, G = \mathbb{Z}$

(f) $H = \mathbb{Z}_4, G = \mathbb{Z}_8$

The Subgroup Test

To prove that a subset H of a group G is a subgroup of G, it appears that we need to verify that H satisfies all four of the defining group properties described in Definition 20.2. (See page 284.) As it turns out, we can use the fact that H is a subset of G to simplify our work a bit.

As you may have observed in Preview Activity 22.1, the associativity of the operation in a group is a property of the operation itself and not of the underlying set. Thus, as long as we use the same operation on a subset of the group, the operation retains its associativity. A proof of this fact is essentially given in part (a) of Preview Activity 22.1.

What about the remaining three axioms: closure, the existence of an identity element, and the existence of inverses? What do they have in common, and why do they require more attention than the others?

The answer to this question lies in one small phrase: "there exists." Notice that two of the three axioms mentioned above contain this phrase, which, as you may remember from previous courses, is called an *existential quantifier*. Note also that the condition

$$ab \in G,$$

which appears in the closure axiom, is equivalent to the following:

$$\text{There exists } c \in G \text{ such that } ab = c.$$

Thus, the closure axiom can be rephrased in an equivalent form as follows:

$$\text{For each } a, b \in G, \text{ there exists } c \in G \text{ such that } c = ab.$$

In other words, these three axioms all assert that the set itself contains certain elements. Since these are properties of sets, they are not automatically inherited by subsets.

So, to summarize, there is one group axiom—namely, associativity—that asserts a property of the group operation and will therefore automatically be satisfied in any subset of a group that uses the same operation. The remaining properties all contain an existential quantifier and assert that the set itself contains certain elements. These existential axioms (as we might call them) are what need to be established to show that a subset of a group G is in fact a subgroup. The following theorem states these observations more formally:

Theorem 22.4 (The Subgroup Test). *A subset H of a group G is a subgroup of G if and only if*

(i) *H is closed under the operation from G;*

(ii) *H contains the identity element e from G; and*

(iii) *H contains the inverse of each of its elements—that is, if $h \in H$ and h^{-1} is the inverse of h in G, then $h^{-1} \in H$.*

Activity 22.5. Use the Subgroup Test to prove that the indicated subset H is a subgroup of the group G.

(a) $H = \{[0], [2], [4]\}$, $G = \mathbb{Z}_6$

(b) $H = \left\{ \begin{pmatrix} a & 0 \\ b & 0 \end{pmatrix} : a, b \in \mathbb{R} \right\}$, $G = M_{2\times 2}(\mathbb{R})$

While Theorem 22.4 may seem straightforward, there are a couple of details that need to be verified. For example, if we assume H is a subgroup of G, then H will contain an identity element. However, there is no reason we can assume that the identity element in H is the same as the identity element in G. A similar statement holds for inverses. We will deal with these issues in the proof of the Subgroup Test, which is given below. You may notice that, unlike previous proofs we have considered, we have not included any instances of the ? symbol. This is because you have now gained enough experience reading proofs to be able to decide for yourself where additional details or explanations are necessary. Although we will still occasionally use the ? symbol throughout the remainder of the text, we will do so less frequently than in past investigations. You should still try to fill in missing details and add clarifying information to the proofs we consider, even when you are not explicitly prompted to do so. You may even want to use the ? symbol as we have in the past to remind yourself where these additional details are necessary.

Proof of the Subgroup Test. Let G be a group, and let H be a subset of G. For the forward implication, we assume that H is a subgroup of G. So, by definition, H is closed. Now H must also contain an identity element, but we cannot assume that the identity element in H is the same as the identity element in G. Let e_H be the identity element in H and e_G the identity element in G. Then $e_H e_H = e_H e_G$ in G and so the cancellation law in G (Theorem 20.7 on page 287) shows that $e_H = e_G$. Thus, the identity element in H is the same as the identity element in G and H contains the identity element from G. So condition (ii) is satisfied. To verify condition (iii), let $h \in H$. Since H is a group we know that H contains an inverse h_H^{-1} for H. However, we cannot assume that h_H^{-1} is the same as h^{-1}, the inverse of h in G. But,

$$h_H^{-1} = h_H^{-1} e_G = h_H^{-1}\left(hh^{-1}\right) = \left(h_H^{-1}h\right)h^{-1} = e_H h^{-1} = e_G h^{-1} = h^{-1}.$$

So the inverse of h in H is the same as the inverse of H in G and condition (iii) is satisfied. For the converse, suppose that H is closed under the operation defined on G, that H contains the identity e_G from G, and that H contains an inverse for each of its elements. To show that H is a group, we only need verify associativity. But associativity is a property of the operation and is thus inherited by H (as in part (a) of Preview Activity 22.1). Thus, H is a subgroup of G. ■

The Center of a Group

Every group contains certain important subsets, one of which is its *center*.

Activity 22.6. In Activity 19.5 (see page 275), we determined the group of symmetries of a square. The operation table for this group (which we call D_4 for reasons we will see in Investigation 24) is reproduced in Table 22.1 for convenience.

$\circ$	I	$\overline{r}_1$	$\overline{r}_2$	r_1	r_2	R_1	R_2	R_3
I	I	$\overline{r}_1$	$\overline{r}_2$	r_1	r_2	R_1	R_2	R_3
$\overline{r}_1$	$\overline{r}_1$	I	R_2	R_1	R_3	r_1	$\overline{r}_2$	r_2
$\overline{r}_2$	$\overline{r}_2$	R_2	I	R_3	R_1	r_2	$\overline{r}_1$	r_1
r_1	r_1	R_3	R_1	I	R_2	$\overline{r}_2$	r_2	$\overline{r}_1$
r_2	r_2	R_1	R_3	R_2	I	$\overline{r}_1$	r_1	$\overline{r}_2$
R_1	R_1	r_2	r_1	$\overline{r}_1$	$\overline{r}_2$	R_2	R_3	I
R_2	R_2	$\overline{r}_2$	$\overline{r}_1$	r_2	r_1	R_3	I	R_1
R_3	R_3	r_1	r_2	$\overline{r}_2$	$\overline{r}_1$	I	R_1	R_2

Table 22.1
Symmetries of a square.

(a) The group D_4 is not an Abelian group, but there are some elements in D_4 that commute with all of the other elements. Find one such element. Are there any others?

(b) Let $Z(D_4)$ be the set of all elements in D_4 that commute with *every* element in D_4.

(i) Create an operation table for $Z(D_4)$ using the operation from D_4.

(ii) Is $Z(D_4)$ a group? Explain.

Activity 22.6 motivates the following definition:

Definition 22.7. Let G be a group. The **center** of G is the set

$$Z(G) = \{a \in G : ab = ba \text{ for all } b \in G\}.$$

In other words, the center of G is the set of all elements in G that commute with every element in G.

Activity 22.8. Find the center of each of the following groups:

(a) $\mathbb{Z}$

(b) The group G of symmetries of an equilateral triangle.

As we saw in Activity 22.6, in at least one case, the center of G is a subgroup of G. The next activity explores whether this relationship holds in general.

Activity 22.9. Let G be a group with identity element e.

(a) Is e in $Z(G)$? Explain.

(b) Is $Z(G)$ closed under the operation in G? Prove your answer.

(c) If $a \in Z(G)$, is $a^{-1} \in Z(G)$? Prove your answer.

(d) Is $Z(G)$ a subgroup of G? Explain. If $Z(G)$ is a subgroup of G, is $Z(G)$ an Abelian subgroup of G?

(e) Complete the statement of the following theorem as specifically and precisely as you can.

Theorem 22.10. *Let G be a group. The center of G is a(n) __________ __________ of G.*

So every group has at least three defined subgroups: the trivial subgroup consisting of just the identity, the entire group itself, and the center of the group. It is important to note that these subgroups need not be distinct. For example, the center of every Abelian group is the entire group. It is also possible for the center of a group to contain only the identity.

The Subgroup Generated by an Element

Given a group G and an element $a \in G$, it is natural to look for the smallest subgroup of G that contains a. The next activity demonstrates what this subgroup looks like for one particular example.

Activity 22.11. Let H be the smallest subgroup of $\mathbb{Z}$ containing 5.

(a) Find three other elements that must be in H.

(b) Explain why the set $H' = \{5m : m \in \mathbb{Z}\}$ is a subset of H.

(c) Explain why H' must equal H.

Let's now generalize what we saw in Activity 22.11. From this point on, we will assume the operation in G is written multiplicatively.

Activity 22.12. Denote by $\langle a \rangle$ the smallest subgroup of G containing the element a.

(a) Since $\langle a \rangle$ is a group, $\langle a \rangle$ must be closed under the group operation. Given $a \in \langle a \rangle$, list 5 other elements that must be in $\langle a \rangle$.

(b) Since $\langle a \rangle$ is a group, $\langle a \rangle$ must contain the inverse of each of its elements. Given $a \in \langle a \rangle$, list 5 more elements that must be in $\langle a \rangle$.

(c) Explain why $\langle a \rangle$ must contain the set $\{a^n \mid n \in \mathbb{Z}\}$. What is a^0?

(d) Let $H = \{a^n \mid n \in \mathbb{Z}\}$. If H is a group, then the previous results tell us that H must be the smallest group containing a. That is, $\langle a \rangle = H$. Prove that $H = \{a^n \mid n \in \mathbb{Z}\}$ is a subgroup of G.

The subgroup $\langle a \rangle$ of G is called the *subgroup generated by* a or the *cyclic subgroup* generated by a. The element a is called a *generator* of the group $\langle a \rangle$.

Definition 22.13. Let G be a group, and let $a \in G$. The **cyclic subgroup generated by** a, denoted $\langle a \rangle$, is defined by

$$\langle a \rangle = \{a^n \mid n \in \mathbb{Z}\}$$

if the group operation is written in multiplicative notation, or

$$\langle a \rangle = \{na \mid n \in \mathbb{Z}\}$$

if the group operation is written in additive notation.

As stated in the next definition, any group that is generated by a single element is said to be *cyclic*.

Definition 22.14. A group G is a **cyclic group** if $G = \langle a \rangle$ for some $a \in G$.

Activity 22.15. For the given group G and element a, determine the elements in the cyclic subgroup generated by a.

(a) $G = \mathbb{Z}_{10}$, $a = [3]$

(b) $G = U_{15}$, $a = [8]$ (Recall that U_n is the group of units in $\mathbb{Z}_n$.)

Activity 22.16. Which of the following groups is a cyclic group?

(a) $\mathbb{Z}$

(b) $\mathbb{Z}_n$ for a positive integer n

(c) $\mathbb{R}$

(d) U_{10}

Just as we defined the order of a group in Investigation 20, we can use cyclic groups to define the order of an *element* in a group.

Definition 22.17. Let G be a group and $a \in G$. If $\langle a \rangle$ is a finite group, then the element a has **finite order**. In this case, the **order** of a is equal to the order of the subgroup generated by a. If $\langle a \rangle$ is an infinite group, the element a has **infinite order**.

The notation we use for the order of an element a in a group G is $|a|$ (not to be confused with absolute value; the meaning of the symbol should be clear from the context). Definition 22.17 tells us $|a| = |\langle a \rangle|$. For example, in $\mathbb{Z}_{10}$, we have $|[4]| = |\langle [4] \rangle| = |\{[4], [8], [2], [6], [0]\}| = 5$.

Every group has a cyclic subgroup for each of its elements, although they may not all be different. Cyclic groups are very important in group theory, as we will learn throughout our investigations of group theory. As one example, we will see later that cyclic groups are the building blocks of all finite Abelian groups.

Concluding Activities

Activity 22.18. Let G be a group with identity e, and let H be a subset of G. To use the Subgroup Test to show that H is a subgroup of G, we need to verify three things: H is closed under the operation in G, $e \in H$, and $h^{-1} \in H$ whenever $h \in H$. In this activity, we will see that, with a little bit of thought, we can reduce the number of things we need to show from three down to two. Assume that G is a group and H is a nonempty subset of G.

(a) Show that if $hk^{-1} \in H$ for all $h, k \in H$, then H is closed.

(b) Show that if $hk^{-1} \in H$ for all $h, k \in H$, then H contains e.

(c) Show that if $hk^{-1} \in H$ for all $h, k \in H$, then H contains an inverse for each of its elements.

(d) Explain why if $hk^{-1} \in H$ for all $h, k \in H$, then H is a subgroup of G.

(e) Use the results from parts (a) – (d) to complete the following alternative form of the Subgroup Test. (Note that your statement should include two conditions.)

A subset H of a group G is a subgroup of G if ...

Activity 22.19. In this activity, we will explore a simple relationship between the subgroup generated by an element and the subgroup generated by its inverse.

(a) Determine the elements in the group $\langle [2] \rangle$ in $\mathbb{Z}_6$. What is the inverse of $[2]$ in $\mathbb{Z}_6$? Now determine the elements in the group $\langle -[2] \rangle$. What do you notice?

(b) Let $\alpha = \begin{pmatrix} 1 & 2 & 3 & 4 \\ 2 & 3 & 4 & 1 \end{pmatrix}$ in the group D_4 of symmetries of a square. Determine the elements in the group $\langle \alpha \rangle$. What is the inverse of α in D_4? Now determine the elements in the group $\langle \alpha^{-1} \rangle$. What do you notice?

(c) Let G be a group, and let a be an element of G. Based on your observations in parts (a) and (b), what relationship do you think exists between $\langle a \rangle$ and $\langle a^{-1} \rangle$? (Although it is dangerous to generalize from a small number of examples, in this case you should see a fairly clear relationship.) Prove this relationship.

(d) Let G be a group, and let a be an element of G. Using the relationship you found in part (c), what relationship do you think exists between the order of a and the order of a^{-1}? Explain. (Hint: Consider both the finite and infinite cases.)

Activity 22.20. Write a short summary that describes the important concepts, and the relationships between these concepts, that were introduced in this investigation. Explain how the material in this investigation is connected to the material in Investigations 20 and 21.

Exercises

(1) Determine the orders of each of the indicated elements.

(a) 2 in $\mathbb{Z}$

(b) $[10]$ in $\mathbb{Z}_{18}$

(c) $\begin{pmatrix} 1 & 2 & 3 & 4 \\ 2 & 3 & 4 & 1 \end{pmatrix}$ in D_4, the group of symmetries of the square

(2) Let H denote the set of all 2×2 matrices of the form

$$\begin{bmatrix} x & 0 \\ y & 0 \end{bmatrix},$$

where $x, y \in \mathbb{R}$. Is H a subgroup of $\mathcal{M}_{2\times 2}(\mathbb{R})$? Prove your answer.

(3) Let H denote the set of all 2×2 matrices of the form

$$\begin{bmatrix} x & y \\ -y & x \end{bmatrix},$$

where $x, y \in \mathbb{R}$. Is H a subgroup of $\mathcal{M}_{2\times 2}(\mathbb{R})$? Prove your answer.

(4) Let G be a group and H a subgroup of G. Which of the following conjectures do you think are true, and which do you think are false? Provide brief arguments or examples to justify your answers.

(a) If G is finite, then H is finite.

(b) If H is finite, then G is finite.

(c) If G is Abelian, then H is Abelian.

(d) If H is Abelian, then G is Abelian.

* (5) Let G be an Abelian group, and let H and K be subgroups of G. Let

$$HK = \{hk : h \in H \text{ and } k \in K\}.$$

(a) Let $H = \langle [6] \rangle$ and $K = \langle [8] \rangle$ in $G = \mathbb{Z}_{12}$. Find the elements of HK in this example.

(b) Prove that HK is a subgroup of G.

(c) Is HK a subgroup of G if G is non-Abelian? Verify your answer.

$\star$ (6) In this exercise, we will explore some special subsets of certain groups.

(a) Show that the set $H = \{n \in \mathbb{Z} : |n| = 1\}$ is a group under multiplication. Draw a picture to geometrically illustrate this group. Is H a subgroup of $\mathbb{Z}$?

(b) Show that the set $K = \{q \in \mathbb{Q} : |q| = 1\}$ is a group under multiplication. Draw a picture to geometrically illustrate this group. Is K a subgroup of $U(\mathbb{Q})$, the group of units in $\mathbb{Q}$?

(c) Show that the set $M = \{x \in \mathbb{R} : |x| = 1\}$ is a group under multiplication. Draw a picture to geometrically illustrate this group. Is M a subgroup of $U(\mathbb{R})$, the group of units in $\mathbb{R}$?

(d) Show that the set $S = \{z \in \mathbb{C} : |z| = 1\}$ is a group under multiplication. (Here $|z|$ is the norm or modulus of the complex number z—that is, if $z = a + bi$, then $|z| = \sqrt{a^2 + b^2}$.) Draw a picture to geometrically illustrate this group. Is S a subgroup of $U(\mathbb{C})$, the group of units in $\mathbb{C}$?

(e) Show that the set $\text{SL}_n(\mathbb{R}) = \{A \in \mathcal{M}_{n\times n}(\mathbb{R}) : |A| = 1\}$ is a group under multiplication. (Here $|A|$ is the determinant of the matrix A.) This group is called the **special linear group**. Is $\text{SL}_n(R)$ a subgroup of $\text{GL}_n(\mathbb{R}) = U(\mathcal{M}_{n\times n}(\mathbb{R}))$, the group of units in $\mathcal{M}_{n\times n}(\mathbb{R})$? (Note that the group $\text{GL}_n(\mathbb{R})$ is called the **general linear group**.)

(7) **Euler's formula**,

$$e^{ix} = \cos(x) + i\sin(x),$$

relates complex exponentials to the polar form of a complex number. Using Euler's formula we can see that if $z = e^{i(2\pi k)/n}$ for integers k between 0 and $n - 1$, then

$$z^n = \cos(2\pi k) + i\sin(2\pi k) = 1.$$

The numbers of the form $e^{i(2\pi k)/n}$ are called the **complex roots of unity**. Let n be a positive integer and let

$$H = \left\{e^{i(2\pi k)/n} : 0 \leq k < n, k \in \mathbb{Z}\right\}.$$

Show that H is a subgroup of the group of complex numbers with norm 1 (see Exercise 6). Draw a picture of H with $n = 6$ to illustrate.

$\star$ (8) **Intersections of subgroups.** Let G be a group with subgroups H and K.

(a) Is $H \cap K$ a subgroup of G? Prove your answer.

(b) Can we generalize? That is, if $\{H_\alpha\}$ is a collection of subgroups of G indexed by α in an indexing set I, is it the case that $\bigcap_{\alpha\in I} H_\alpha$ is a subgroup of G? Prove your answer.

(9) **Unions of subgroups.** Let G be a group with subgroups H and K.

(a) Is $H \cup K$ necessarily a subgroup of G? Prove your answer.

(b) Under what conditions is $H \cup K$ a subgroup of G? Prove your answer.

(10) Let G be a group, and let $a \in G$. The **centralizer** of a is defined to be the set of all $g \in G$ such that $ga = ag$. In other words, the centralizer of a is the set of all elements that commute with a.

(a) Find the centralizer of the 180° rotation R_2 in D_4, the group of symmetries of a square. (See Activity 22.6 for the notation and operation table for this group.)

(b) Find the centralizer of the reflection r_1 around the vertical axis in D_4.

(c) How is the centralizer of an element different than the center of the group? Explain.

(d) Prove that the centralizer of $a \in G$ is a subgroup of G.

(e) Is the centralizer of $a \in G$ necessarily an Abelian group? Give a proof or counterexample to justify your answer.

(f) Show that $Z(G)$, the center of G, is equal to the intersection of all of the centralizers of elements in G—that is,

$$Z(G) = \cap_{a \in G} C(a).$$

(11) If G is a group and H a subgroup of G, the **centralizer of the subgroup** H is the set

$$C(H) = \{a \in G : gh = hg \text{ for all } h \in H\}.$$

(a) We encountered the group of symmetries of a regular hexagon in Activity 19.5. (See page 275.) The operation table for this group, which we will label as D_6, is reproduced in Table 22.2. Find $C(H)$ if $H = \{I, R^2, R^4\}$.

$\circ$	I	r_1	r_2	r_3	$\overline{r}_1$	$\overline{r}_2$	$\overline{r}_3$	R_1	R_2	R_3	R_4	R_5
I	I	r_1	r_2	r_3	$\overline{r}_1$	$\overline{r}_2$	$\overline{r}_3$	R_1	R_2	R_3	R_4	R_5
r_1	r_1	I	R_4	R_2	R_5	R_3	R_1	$\overline{r}_3$	r_3	$\overline{r}_2$	r_2	$\overline{r}_1$
r_2	r_2	R_2	I	R_4	R_1	R_5	R_3	$\overline{r}_1$	r_1	$\overline{r}_3$	r_3	$\overline{r}_2$
r_3	r_3	R_4	R_2	I	R_3	R_1	R_5	$\overline{r}_2$	r_2	$\overline{r}_1$	r_1	$\overline{r}_3$
$\overline{r}_1$	$\overline{r}_1$	R_1	R_5	R_3	I	R_4	R_2	r_1	$\overline{r}_3$	r_3	$\overline{r}_2$	r_2
$\overline{r}_2$	$\overline{r}_2$	R_3	R_1	R_5	R_2	I	R_4	r_2	$\overline{r}_1$	r_1	$\overline{r}_3$	r_3
$\overline{r}_3$	$\overline{r}_3$	R_5	R_3	R_1	R_4	R_2	I	r_3	$\overline{r}_2$	r_2	$\overline{r}_1$	r_1
R_1	R_1	$\overline{r}_1$	$\overline{r}_2$	$\overline{r}_3$	r_2	r_3	r_1	R_2	R_3	R_4	R_5	I
R_2	R_2	r_2	r_3	r_1	$\overline{r}_2$	$\overline{r}_3$	$\overline{r}_1$	R_3	R_4	R_5	I	R_1
R_3	R_3	$\overline{r}_2$	$\overline{r}_3$	$\overline{r}_1$	r_3	r_1	r_2	R_4	R_5	I	R_1	R_2
R_4	R_4	r_3	r_1	r_2	$\overline{r}_3$	$\overline{r}_1$	$\overline{r}_2$	R_5	I	R_1	R_2	R_3
R_5	R_5	$\overline{r}_3$	$\overline{r}_1$	$\overline{r}_2$	r_1	r_2	r_3	I	R_1	R_2	R_3	R_4

Table 22.2
Symmetries of a regular hexagon.

(b) Let G be an arbitrary group and H a subgroup of G. Is $C(H)$ always a subgroup of G? Prove your answer.

(12) Determine whether H is a subgroup of G.

(a) $G = \mathbb{Z}_{20}$ under addition, $H = \{[0], [3], [6], [9], [12], [15], [18]\}$.

(b) $G = U_7$ under multiplication, $H = \{[1], [2], [4]\}$.

(c) $G = U_{16}$ and $H = \{[1], [7], [9], [15]\}$

⋆ (13) **Subgroups of $\mathbb{Z}$.** In this exercise, we will show that the only subgroups of $\mathbb{Z}$ are the subgroups of the form $n\mathbb{Z} = \{nk : k \in \mathbb{Z}\}$ for some $n \in \mathbb{Z}$.

(a) First show that $n\mathbb{Z}$ is a subgroup of $\mathbb{Z}$ for any $n \in \mathbb{Z}$.

(b) Now follow the steps below to show that any subgroup of $\mathbb{Z}$ is equal to $n\mathbb{Z}$ for some $n \in \mathbb{Z}$.

(i) Let H be a subgroup of $\mathbb{Z}$. Explain why there are two cases to consider: $H = \{0\}$ and $H \neq \{0\}$. Complete the case where $H = \{0\}$.

(ii) Assume $H \neq \{0\}$. Show that H must contain a positive integer.

(iii) Let $T = \{h \in H : h > 0\}$. Explain why T contains a smallest positive element. (Hint: Think of the Well-Ordering Principle.)

(iv) Let a be the smallest positive element in T. Prove that $H = a\mathbb{Z}$. (Hint: Use the Division Algorithm to show $a \mid h$ for each $h \in H$.)

(14) If m and n are positive integers, what can we say about the relationship (as groups) between $m\mathbb{Z}$ and $n\mathbb{Z}$? (See Exercise 13.) Be as specific as possible, and prove your answer.

* (15) Prove that any group of order 2 or 3 is cyclic.

(16) Find the order of each element of the group U_{11}. Also find all of the subgroups of U_{11}.

(17) (a) Find all of the elements that generate $\mathbb{Z}_6$, $\mathbb{Z}_8$, and $\mathbb{Z}_{15}$.

(b) Let $\langle a \rangle$ be a cyclic group of order 6, $\langle b \rangle$ a cyclic group of order 8, and $\langle c \rangle$ a cyclic group of order 15. Find all of the elements that generate $\langle a \rangle$, $\langle b \rangle$, and $\langle c \rangle$. Can you see a pattern here that characterizes the elements that generate a cyclic group?

(18) Is every cyclic group Abelian? Is every Abelian group cyclic? Justify your answers.

(19) Recall that U_{21} is the group of units in the ring $\mathbb{Z}_{21}$. List the elements in U_{21}. Then find the subgroup of U_{21} generated by the element $[5]$.

(20) Consider the two objects shown in Figure 22.1: a square S on the right and a square R with both solid and dashed sides on the left. We have already determined the group D_4 of symmetries of S (see Activity 22.6), but any symmetry of R will have to preserve the shading of the sides.

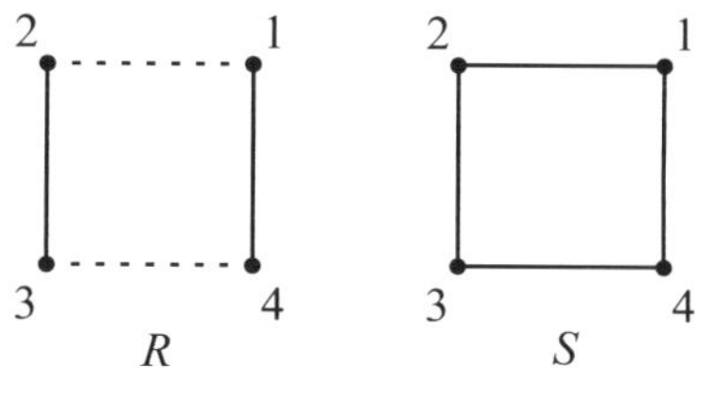

Figure 22.1
Rigid motions in the plane.

(a) Explain why any symmetry of R is also a symmetry of S.

(b) Determine the set H of symmetries of the object R.

(c) Which of the group axioms (using the operation of composition of symmetries) does the set of symmetries of R satisfy?

(d) Create the operation table for the set of symmetries of R.

(e) Explain why H is a subgroup of D_4.

(21) **The Subgroup Test for finite groups.** We can reduce the number of items we need to prove in the Subgroup Test if we are working in a finite group. Assume that G is a finite group with identity e and that the order of G is n.

(a) Let H be a nonempty subset of G that is closed under the operation in G, and let $h \in H$. Explain why the set of elements $S = \{h, h^2, h^3, \ldots\}$ must be finite.

(b) Use the fact that the elements in S repeat to show that $e \in H$.

(c) Again use the fact that the elements in S repeat to show that $h^{-1} \in H$.

(d) Write a formal proof of the following theorem:

Theorem. *Let G be a finite group. A nonempty subset H of G is a subgroup of G if and only if H is closed under the operation in G.*

* (22) **Subgroups of order 2.** In general, it is a difficult task to determine which subgroups a group has. In subsequent investigations, we will develop some powerful tools to help in this regard. For now, however, we will begin with a relatively straightforward argument. Show that any group of even order must contain an element (and hence a subgroup) or order 2. (Hint: Count elements and their inverses.)

(23) In this exercise, we will prove **Wilson's Theorem**:

Theorem (Wilson's Theorem). *A positive integer p is prime if and only if*

$$(p-1)! \equiv -1 \pmod{p}.$$

Wilson's Theorem* gives a test (although not a practical one) for determining if a positive integer is prime. Wilson's Theorem also shows that $n! + 1$ is composite for infinitely many different positive integers n. It is unknown if $n! + 1$ is prime for infinitely many different integers n.

(a) Let p be a prime. Show that $[p-1]$ is the only element in U_p of order 2.

(b) Prove that if p is prime, then $(p-1)! \equiv -1 \pmod{p}$. (Hint: If p is an odd prime, pair each element in U_p with its inverse.)

(c) Show that if m is composite, then $(m-1)! \not\equiv -1 \pmod{m}$.

(d) Explain why parts (b) and (c) prove Wilson's Theorem.

Connections

In this investigation, we studied subgroups of groups. If you studied ring theory before group theory, you should notice connections between the topics in this investigation and those in Investigation 9.

*This theorem was conjectured by John Wilson in the 18th century, but appears to have been known to Ibn al-Haythan (according to Oystein Ore in *Number Theory and its History*, Dover, 1988) as early as 1000 AD. Lagrange appears to have given the first proof in 1771.

The idea of a subring is the same is that of a subgroup. In particular, if R is a ring, then a subring of R is just a subset of R that is also a ring under the same operations as R. The only significant difference is that a group comes with one operation and a ring comes with two, so it is a bit more work to determine if a subset of a ring is a subring than if a subset of a group is a subgroup. Because of the simpler structure of groups, there are special types of subgroups that we identify (e.g., cyclic groups) that do not have a direct counterpart in rings. We will be able to exploit these special subgroups to classify some important families of groups—a task that is much more difficult to do with rings.

Investigation 23

Subgroups of Cyclic Groups

Focus Questions

By the end of this investigation, you should be able to give precise and thorough answers to the questions listed below. You may want to keep these questions in mind to focus your thoughts as you complete the investigation.

- What property must be satisfied by all subgroups of a cyclic group?
- What is the subgroup structure of a finite cyclic group?
- What is the subgroup structure of an infinite cyclic group?
- What important properties does the order of an element in a group have?

Preview Activity 23.1. The family of groups denoted by $\mathbb{Z}_n$ (where n is a positive integer) are the canonical examples of finite cyclic groups. In fact, it can be shown that, for each $n \in \mathbb{Z}^+$, $\mathbb{Z}_n$ is the only cyclic group of order n. Recall that every element in $\mathbb{Z}_n$ has the form $[k]_n = k[1]_n$ for some integer k, so $\mathbb{Z}_n$ is generated by $[1]_n$—that is, $\mathbb{Z}_n = \langle [1]_n \rangle$. (Note that, when the context is clear, we will typically omit subscripts and simply write $\mathbb{Z}_n = \langle [1] \rangle$.) One question we will address in this investigation is what the subgroup structure of $\mathbb{Z}_n$ looks like. In other words, if H is a subgroup of $\mathbb{Z}_n$, what kind of things can we say about H? In this activity, we will consider the specific example where H is a non-trivial subgroup of $\mathbb{Z}_{100}$.

(a) Suppose $[20] \in H$. List at least 3 other elements that are also in H. What is the smallest number of elements H can have? Explain.

(b) If $[20] \in H$, must it be true that H contains $\langle [20] \rangle$? Why or why not?

(c) If $[20] \in H$, is it possible that $[k] \in H$ for some integer k with $1 < k < 20$? If so, in what situations could that happen? If not, why is it impossible?

(d) Let $S = \{k \in \mathbb{Z}^+ : k[1] \in H\}$.

  (i) Why is S nonempty?

  (ii) What important conclusion can we draw about S as a nonempty subset of $\mathbb{Z}^+$? Explain.

  (iii) How do you think this important conclusion is related to H?

Introduction

Which subgroups a group contains can tell us a lot about the group. In general, it is difficult to explicitly describe all subgroups of a given group, but for some groups the subgroup structure is more accessible. The simplest type of group is a cyclic group, and the subgroup structure of a cyclic group is fairly simple, as we will see in this investigation.

Subgroups of Cyclic Groups

Throughout this section, we will let G be a cyclic group generated by an element $a \in G$. Recall that this means every element in G can be written as an integer power of a, or

$$G = \langle a \rangle = \{a^k : k \in \mathbb{Z}\}.$$

Our goal in this investigation is to determine all of the subgroups of G. Since every element in G is a power of a generator a, it follows that every element in any subgroup of G is also a power of a. This might lead us to conjecture the following theorem.

Theorem 23.2. *Every subgroup of a cyclic group is cyclic.*

The details of a proof of this theorem are presented in Activity 23.3, but a few words might be in order before we proceed. Let H be a subgroup of a cyclic group $G = \langle a \rangle$. Since $H \subseteq G$, every element of H is a power of a. The problem we encounter, though, is that it is not obvious exactly which powers of a are in H. This is the detail we need to address, and the main idea we will use to do so was introduced in Preview Activity 23.1.

Activity 23.3. Let $G = \langle a \rangle$ be a group.

(a) We want to prove that every subgroup of G is cyclic. How is this statement quantified? What should we assume to begin our proof?

(b) Let H be a subgroup of G. There is one subgroup of every group that is cyclic. What subgroup is that? If H is this subgroup, we are done. So we can assume H is not this subgroup. What additional assumption does that allow us to make?

(c) Let $h \in H$. Since $h \in G$, the element h must have a specific form. What form does h have?

(d) Let H be a non-trivial subgroup of G, and let $S = \{k \in \mathbb{Z}^+ : a^k \in H\}$. Why is S nonempty?

(e) What result tells us that S contains a smallest element?

(f) Let m be the smallest positive integer in S. That is, m is the smallest positive integer such that $a^m \in H$. This element a^m is a candidate for a generator for H. What must we do to show $\langle a^m \rangle = H$?

(g) Suppose $b \in H$. Why does b equal a^l for some integer l?

(h) Since m is positive, we can divide m into l and obtain a unique quotient q and remainder r. What result tells us we can do this? What is true about the remainder? What is the specific relationship between m and l?

(i) Why can we say $a^l = a^{qm+r}$? How can we use this equation to show $r = 0$? What conclusion can we then draw?

Theorem 23.2 tells us an important fact about every subgroup of a cyclic group, but there is much more that we can say. First we will need to know more about the orders of the elements in a group.

Properties of the Order of an Element

Preview Activity 23.4. The orders of the elements in $\mathbb{Z}_4$, $\mathbb{Z}_5$, $\mathbb{Z}_6$, and $\mathbb{Z}_7$ are given in Table 23.1.

(a) Complete Table 23.1 by calculating the orders of the elements in $\mathbb{Z}_8$ and $\mathbb{Z}_9$.

(b) For any positive integer n and any element $[k] \in \mathbb{Z}_n$, there is an explicit relationship between n and the order of $[k]$. Review the entries in Table 23.1, and make a conjecture about this relationship.

k	$\lvert[k]\rvert$ in $\mathbb{Z}_4$	$\lvert[k]\rvert$ in $\mathbb{Z}_5$	$\lvert[k]\rvert$ in $\mathbb{Z}_6$	$\lvert[k]\rvert$ in $\mathbb{Z}_7$	$\lvert[k]\rvert$ in $\mathbb{Z}_8$	$\lvert[k]\rvert$ in $\mathbb{Z}_9$
0	1	1	1	1		
1	4	5	6	7		
2	2	5	3	7		
3	4	5	2	7		
4	1	5	3	7		
5	4	1	6	7		
6	2	5	1	7		
7	4	5	6	1		
8	1	5	3	7		

Table 23.1
Orders of elements in $\mathbb{Z}_4$ through $\mathbb{Z}_9$.

Preview Activity 23.4 illustrates the relationship between the order of a power of a generator in a cyclic group (such as $k[1] = [k]$ in $\mathbb{Z}_n$) and the order of the generator itself ($[1]$ in $\mathbb{Z}_n$). In what follows, we will prove that relationship and use it to completely determine the subgroup structure of finite and infinite cyclic groups. We will first determine when a power of an element of finite order in G can equal the identity.

Theorem 23.5. *Let G be a group with identity e, and let a be an element of G of order n. Then:*

(i) $a^n = e$ *and, moreover, n is the smallest positive integer so that* $a^n = e$.

(ii) *If s is an integer so that $a^s = e$, then n divides s.*

Proof. Let G be a group with identity e, and let a be an element of G of order n.

To prove (i), note that since $a \in \langle a \rangle$ and $\langle a \rangle$ is closed under the group operation, the elements $a, a^2, a^3, \ldots, a^n \ldots$ are all in $\langle a \rangle$. Because the order of $\langle a \rangle$ is finite, these powers cannot all be distinct. So there must exist $0 < i < j$ so that

$$a^j = a^i.$$

Multiplying both sides of this equation by a^{-i} gives us

$$a^{j-i} = e.$$

So there is at least one positive power of a that is equal to the identity. The Well-Ordering Principle tells us that there is a smallest positive power k of a so that $a^k = e$. To complete the proof, we will show that $k = n$.

First, we will show that $k \leq n$. We proceed by contradiction and assume $k > n$. Consider the elements $a, a^2, a^3, \ldots, a^k$ in $\langle a \rangle$. Since $|\langle a \rangle| = n$, these powers cannot all be distinct. So there exist $0 < i < j \leq k$ so that $a^j = a^i$. This implies $a^{j-i} = e$. However, $0 < j - i < j \leq k$, which contradicts the fact that k is the smallest positive power of a that equals the identity. We can thus conclude that $k \leq n$.

Next, we will show that $k \geq n$. Again we proceed by contradiction and assume $k < n$. Let $t \in \mathbb{Z}$. By the Division Algorithm, there are integers q and r such that

$$t = qk + r \qquad \text{with } \ 0 \leq r < k.$$

Then

$$a^t = a^{qk+r} = (a^k)^q a^r = e^q a^r = a^r.$$

So any power of a is equal to a^r for some $0 \leq r < k$. This means that there are only $k < n$ distinct powers of a, and $|\langle a \rangle| \leq k < n$, a contradiction. Therefore, $k \geq n$.

Combining the inequalities $k \leq n$ and $k \geq n$ yields $k = n$ as desired.

To prove (ii), assume $a^s = e$ for some integer s. By the Division Algorithm, there are integers q and r such that

$$s = qn + r \qquad \text{with } \ 0 \leq r < n. \tag{23.1}$$

Then

$$e = a^s = a^{qn+r} = (a^n)^q a^r = a^r.$$

But n is the smallest positive integer such that $a^n = e$. Therefore, we must have $r = 0$, and (23.1) shows us that n divides s. ■

Now let's turn our attention to the result indicated in Preview Activity 23.4 about orders of elements in finite cyclic groups.

Theorem 23.6. *Let $G = \langle a \rangle$ be a cyclic group of order n, and let $k \in \mathbb{Z}$. Then*

(i) $\langle a^k \rangle = \langle a^{\gcd(k,n)} \rangle$

(ii) $|\langle a^k\rangle| = \frac{n}{\gcd(k,n)}$.

Proof. Let $G = \langle a\rangle$ be a cyclic group of order $n \in \mathbb{Z}^+$, and let $k \in \mathbb{Z}$. Let $d = \gcd(k, n)$. Then there exist integers u, v so that $du = k$ and $dv = n$.

To prove (i), we will show that $\langle a^k\rangle \subseteq \langle a^d\rangle$ and $\langle a^d\rangle \subseteq \langle a^k\rangle$. First note that $d \mid k$, and so $a^k = a^{du} = (a^d)^u$, which implies that $a^k \in \langle a^d\rangle$. By closure, it follows that $\langle a^k\rangle \subseteq \langle a^d\rangle$. To prove containment in the other direction, we will show that $a^d \in \langle a^k\rangle$.

By Bezout's Identity, there exist integers x, y so that $d = kx + ny$. Using part (i) of Theorem 23.5 we see that

$$a^d = a^{kx+ny} = (a^k)^x(a^n)^y = (a^k)^x \in \langle a^k\rangle.$$

Therefore, $\langle a^d\rangle \subseteq \langle a^k\rangle$ by closure. Combining the two containments gives us $\langle a^k\rangle = \langle a^d\rangle$.

To prove (ii), we must show that the order of a^d is $\frac{n}{d}$. Part (i) of Theorem 23.5 implies that

$$(a^d)^{\frac{n}{d}} = a^n = e.$$

Then part (ii) of Theorem 23.5 tells us that $|a^d| \leq \frac{n}{d}$. To show $|a^d| \geq \frac{n}{d}$, we need to show that no smaller positive power of a^d is equal to e. If $0 < i < \frac{n}{d}$, then $0 < di < n$. Since n is the smallest positive power of a that is equal to e (by part (ii) of Theorem 23.5), we see that $(a^d)^i \neq e$. Therefore, $|a^d| \geq \frac{n}{d}$. The two inequalities ($|a^d| \leq \frac{n}{d}$ and $|a^d| \geq \frac{n}{d}$) show that $|a^d| = \frac{n}{d}$, and part (i) allows us to conclude that $|a^k| = |a^d| = \frac{n}{d}$. ■

In the next section, we will use Theorem 23.6 to completely determine the subgroup structure of all finite cyclic groups.

Finite Cyclic Groups

Let $G = \langle a\rangle$ be a cyclic group of order n. Our previous theorems in this investigation allow us to completely classify the subgroup structure of G, as stated in the next theorem.

Theorem 23.7. *Let G be a finite cyclic group of order n. For each positive divisor m of n, there is exactly one subgroup of G of order m, and these are the only subgroups of G.*

An outline of a proof of Theorem 23.7 is contained in the next activity.

Activity 23.8. Let $G = \langle a\rangle$ be a finite cyclic group of order n, and let m be a positive divisor of n.

(a) Use Theorem 23.6 to find a subgroup of G of order m. This shows that G contains at least one subgroup of order m.

(b) Now suppose that G contains subgroups H and K of order m.

(i) How can we conclude that $H = \langle a^s\rangle$ and $K = \langle a^t\rangle$ for some integers s and t?

(ii) Explain how Theorem 23.6 tells us that $H = K$. What can we conclude about the number of subgroups of G of order m?

(c) Explain how we have completely classified the subgroup structure of finite cyclic groups.

As one final note, Theorem 23.6 also shows that a^k is a generator for G if and only if $\gcd(k, n) = 1$.

Activity 23.9.

(a) Find all the generators of $\mathbb{Z}_{30}$.

(b) Let a be an element of a group with $|a| = 15$. Find the orders of a^2, a^6, and a^{10}.

(c) Let G be a group, and let $a \in G$ be an element of order n. Explain why $\langle a^k \rangle = \langle a^{n-k} \rangle$. Use this result to prove that $\gcd(n, k) = \gcd(n - k, n)$.

Infinite Cyclic Groups

The subgroup structure of infinite cyclic groups is also known. If $G = \langle a \rangle$ is an infinite cyclic group, Theorem 23.2 tells us that every subgroup of G is also cyclic. Note that G will contain at least one finite cyclic subgroup. (Which subgroup must this be?) The structure of all other subgroups of G is described in the next theorem, whose proof is left as an exercise.

Theorem 23.10. *Let $G = \langle a \rangle$ be an infinite cyclic group with identity e, and let $b \neq e$ be an element in G. Then $\langle b \rangle$ is an infinite cyclic group.*

Concluding Activities

Activity 23.11. Use the steps outlined in Activity 23.3 to write a formal proof of Theorem 23.2.

Activity 23.12. Use the steps outlined in Activity 23.8 to write a formal proof of Theorem 23.7.

Activity 23.13. Let G be a group, and let $a \in G$ be an element of order n. If m is a positive integer and $|a^m| = k$, is it true that $|a^k| = m$? If yes, then prove your answer. If no, what can we say about $|a^k|$?

Activity 23.14. Write a short summary that describes the important concepts, and the relationships between these concepts, that were introduced in this investigation. Explain how the material in this investigation is connected to the material in Investigations 21 and 22.

Exercises

(1) Let $G = \mathbb{Z}_{24} = \langle [1] \rangle$.

(a) Explain why $\langle [2] \rangle = \langle [10] \rangle$ in G.

(b) Find all of the elements $[m] \in G$ so that $\langle [m] \rangle = \langle [2] \rangle$.

(c) Find $|[6]|$ in G.

(2) Explicitly verify Theorem 23.6 for the group U_{26}.

(3) What is the order of $[10]$ in $\mathbb{Z}_{18}$?

(4) (a) Show that U_{22} is cyclic.

(b) Find all of the generators of U_{22}. Explain how you know that each element is a generator.

(5) Let $A = \begin{bmatrix} 1 & 0 \\ 0 & -1 \end{bmatrix}$ and $B = \begin{bmatrix} 1 & 1 \\ 0 & -1 \end{bmatrix}$ in $\mathrm{GL}_2(\mathbb{R})$.

(a) Find $|A|$ and $|B|$.

(b) Determine $|AB|$. Does your answer surprise you? Explain.

(6) Is it possible for a group G to contain a non-identity element of finite order and also an element of infinite order? If yes, illustrate with an example. If no, give a convincing explanation for why it is not possible.

(7) Suppose $G = \langle a \rangle$ is a cyclic group of order 12.

(a) Find all the generators of G. Explain your reasoning.

(b) Find all of the proper subgroups of G, and list their elements. Find all the generators of each subgroup. Explain your reasoning.

(8) Let G be a group with identity element e, and let a be an element in G. Label each of the following statements as either true or false. Justify your answers.

(a) If n is a positive integer and $a^n = e$, then $|a| = n$.

(b) If $a^n \neq e$ for every nonzero integer n, then $G = \langle a \rangle$.

(9) Prove Theorem 23.10. (Hint: Let $G = \langle a \rangle$ be an infinite cyclic group with identity e, and let $b \neq e$ be an element in G. Show that the positive integer powers of b are all distinct.)

(10) Let G be a group with identity e, and let $a \in G$. If $a^6 = e$ but $a^2 \neq e$, what can we say about $|a|$? Prove your answer.

(11) Suppose that the only distinct subgroups a cyclic group G with identity e are G, $\{e\}$, and a subgroup of order 23. What is $|G|$?

(12) Let G be a group, and let a be an element of G of order n. Prove that $|a^k| = |a^{n-k}|$ for any integer k.

(13) Find two distinct subgroups of order 2 of the group D_3 of symmetries of an equilateral triangle. Explain why this fact alone shows that D_3 is not a cyclic group.

(14) Suppose G is a group of order n so that every proper subgroup of G is cyclic. Must G be cyclic? Prove your answer.

(15) Let $n \in \mathbb{Z}^+$, and let d be a positive divisor of n. Theorem 23.7 tells us that $\mathbb{Z}_n$ contains exactly one subgroup of order d, but not how many *elements* $\mathbb{Z}_n$ has of order d. We will determine that number in this exercise.

(a) Determine the number of elements in $\mathbb{Z}_{12}$ of each order d. Fill in the table below to compare your answers to the number of integers between 1 and d that are relatively prime to d.

Divisor d	Number of elements of order d	Number of integers between 1 and d relatively prime to d
1		
2		
3		
4		
6		
12		

(b) Part (a) appears to indicate that there is a relationship between the number of elements in $\mathbb{Z}_n$ of a given order and the number of integers relatively prime to that order. There is a useful function that describes this number. The *Euler phi function* φ is defined to give the number of positive integers less than or equal to and relatively prime to a given positive integer. For example, $\varphi(2) = 1$ since 1 is the only positive integer less than 2 and relatively prime to 2. Similarly, $\varphi(3) = 2$, $\varphi(4) = 2$, $\varphi(5) = 4$, etc. Show that the number of elements in $\mathbb{Z}_n$ of order d is $\varphi(d)$.

(16) Let G be a finite group.

(a) If G is cyclic, can G be the union of proper subgroups of G? Prove your answer.

(b) If G is not cyclic, is it always true that G can be written as a union of proper subgroups of G?

(c) Can an infinite cyclic group G be written as a union of proper subgroups of G?

(17) Let G be an Abelian group, let k be a positive integer, and let $O(k) = \{a \in G : |a| \text{ divides } k\}$.

(a) Find the elements in $O(6)$ if $G = \mathbb{Z}_{12}$. Construct an operation table for $O(6)$ in this case.

(b) Prove that $O(k)$ is a subgroup of G. (Hint: In G, why is $(ab)^n = a^n b^n$ for any $a, b \in G$ and any $n \in \mathbb{Z}^+$?)

⋆ (18) Let G be a group, and let $a, b \in G$ with $|a| = n$ and $|b| = m$.

(a) Is it necessarily true that $|ab| = mn$? Prove your answer.

(b) If $ab = ba$, is it necessarily true that $|ab| = mn$? Prove your answer.

(c) Prove that if $ab = ba$ and $\gcd(m, n) = 1$, then the order of ab is mn.

Investigation 24

The Dihedral Groups

Focus Questions

By the end of this investigation, you should be able to give precise and thorough answers to the questions listed below. You may want to keep these questions in mind to focus your thoughts as you complete the investigation.

- How can we describe the symmetries of a regular polygon? What similarities do all such sets of symmetries have?
- What are the dihedral groups, and what are some of their important properties?
- What are generators and relations, and how can they be used to describe a group's structure? What is a presentation of a group?

Preview Activity 24.1. In Investigation 19, we determined the group D_3 of symmetries of an equilateral triangle. For convenience, the operation table for D_3 is reproduced in Table 24.1.

$\circ$	I	r_1	r_2	r_3	R_1	R_2
I	I	r_1	r_2	r_3	R_1	R_2
r_1	r_1	I	R_1	R_2	r_2	r_3
r_2	r_2	R_2	I	R_1	r_3	r_1
r_3	r_3	R_1	R_2	I	r_1	r_2
R_1	R_1	r_3	r_1	r_2	R_2	I
R_2	R_2	r_2	r_3	r_1	I	R_1

Table 24.1
D_3, the group of symmetries of an equilateral triangle.

(a) Use Table 24.1 to determine if each of the following is a rotation or reflection in D_3:

(i) The composition of two rotations.

(ii) The composition of a rotation and a reflection.

(iii) The composition of two reflections.

(b) Which elements in D_3 can be written in the form $r_1^i R_1^j$, where i and j are integers?

(c) If t is an integer, can $R_1^t r_1$ be written in the form $r_1 R_1^j$ for some j? If yes, for which j?

(d) Let $r = r_1$ and $R = R_1$. Rewrite Table 24.1 so that every element is in the form $r^i R^j$ for some i and j.

Introduction

In Investigation 19, we saw that the symmetries of certain regular polygons (e.g., an equilateral triangle, a square, a regular pentagon, and a hexagon) consist of rotations around the center and reflections through various axes. Now we will consider the general case of a regular polygon with $n \geq 3$ sides, also called a regular n-gon (an n-sided figure whose sides all have the same length and in which the angles formed by adjacent sides are all congruent).

First, note that there can be at most $2n$ symmetries of a regular n-gon. To see why, pick a vertex of our regular n-gon and label it as 1. Label the vertex adjacent to 1 counterclockwise from it as 2. A symmetry will need to send vertices to vertices, so there are n vertices to which vertex 1 can be sent. Once the image of vertex 1 is determined, then the image of vertex 2 must be adjacent to the image of vertex 1. So there are two choices for the image of vertex 2. It is not difficult to see that the images of vertices 1 and 2 completely determine the images of all other points on the n-gon. Thus, there are at most $2n$ symmetries of a regular n-gon.

Could a regular n-gon have fewer than n symmetries? The answer is no; in fact, we can explicitly describe all of the $2n$ symmetries that a regular n-gon must have. To begin, let θ be the central angle of a regular n-gon—that is, the angle made by one side and the center. Then there are n rotational symmetries of angles $i\theta$ around the center of the n-gon for $0 \leq i \leq n-1$. If n is odd, then there are also n reflections around the lines passing through each vertex and the midpoint of the opposite side. If n is even, there are still n reflections: $\frac{n}{2}$ through lines connecting the midpoints of opposite sides and $\frac{n}{2}$ through lines connecting pairs of antipodal (opposite) vertices. In each case, the set of symmetries of a regular n-gon contains n rotations and n reflections, for a total of $2n$ elements. We will denote the set of symmetries of a regular n-gon as D_n and call this set the *dihedral group* of order $2n$.*

Definition 24.2. Let n be an integer with $n \geq 3$. The **dihedral group** of order $2n$ is the group of symmetries of a regular n-gon.

Before we proceed, a word of caution about notation is in order. Some texts denote the dihedral group of order $2n$ as D_{2n}, but we will use the notation D_n. In the remainder of this investigation, we will examine the structure of D_n in more detail.

Relationships between Elements in D_n

Let n be an integer with $n \geq 3$. Activity 24.1 provides a framework from which we can succinctly and efficiently represent all of the elements of D_n. Here we will establish the notation that we will use in this and subsequent investigations to represent elements in D_n.

*The term *dihedral* seems to have come from Felix Klein's study of symmetry groups as subgroups of motions in $\mathbb{R}^3$. Using a degenerate polyhedron, he invented and studied the *dihedron* (from the Greek "di," meaning two, and "hedron," meaning face or surface), which is a figure obtained by gluing together two congruent regular polygons of zero thickness.

Let $R_0, R_1, \ldots, R_{n-1}$ be the rotations in D_n, where R_i is a counterclockwise rotation of $i\theta$ around the center of the polygon. Note that $|R_1| = n$, so $\langle R_1 \rangle = \{R_0, R_1, \ldots, R_{n-1}\}$. For ease of notation. we will let $R = R_1$. Since we will only work with rotations in the plane, we know that R^j can be represented by the matrix transformation $\begin{bmatrix} \cos(j\theta) & -\sin(j\theta) \\ \sin(j\theta) & \cos(j\theta) \end{bmatrix}$. (See Exercise 7 on page 280 of Investigation 19.)

Now label one vertex of the regular n-gon with 1 and the others in order proceeding counterclockwise. Without loss of generality, we can rotate the n-gon so that vertex 1 is on the positive y-axis. (An example is illustrated by the pentagon to the right.) Let r be the reflection of the n-gon around the line through vertex 1 and the origin. In this case, r can be represented by the matrix transformation $\begin{bmatrix} -1 & 0 \\ 0 & 1 \end{bmatrix}$.

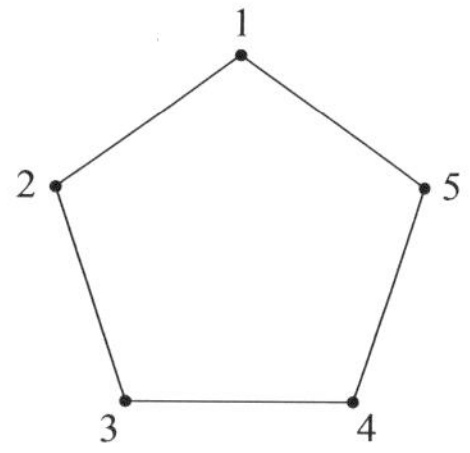

Activity 24.3. With the notation established above, answer the following.

(a) What is the order of r? Why?

(b) Explain why $R^i \neq r$ for any $0 \leq i \leq n-1$.

(c) Now explain why rR^i cannot equal rR^j for $0 \leq i, j \leq n-1$ unless $i = j$.

(d) Explain why

$$D_n = \{I, R, R^2, R^3, \ldots, R^{n-1}, r, rR, rR^2, \ldots, rR^{n-1}\}.$$

Activity 24.3 gives us a succinct way of representing all of the elements of the dihedral group D_n; in particular, every element of D_n can be written in the form r^iR^j, where $0 \leq i \leq 1$ and $0 \leq j \leq n-1$. This is very convenient and useful. Next we will see how this representation allows us to quickly generate the operation table for D_n.

Since we already know how to combine powers of the rotation R, to completely determine the structure of D_n we just need to know how to write R^ir in the form rR^j for some j. Let's begin with Rr. Notice that

$$Rr = \begin{bmatrix} \cos(\theta) & -\sin(\theta) \\ \sin(\theta) & \cos(\theta) \end{bmatrix} \begin{bmatrix} -1 & 0 \\ 0 & 1 \end{bmatrix} = \begin{bmatrix} -\cos(\theta) & -\sin(\theta) \\ -\sin(\theta) & \cos(\theta) \end{bmatrix}$$

is the reflection around the line through the origin making an angle of $\left(\frac{\theta}{2} + 90\right)^{\circ}$ with the positive x-axis. (See Exercise 7 on page 280 of Investigation 19.) Thus, Rr is a reflection and $(Rr)^{-1} = Rr$. This can also be seen by taking the square of the matrix form for Rr. Hence,

$$\begin{aligned} (Rr)^{-1} &= Rr \\ r^{-1}R^{-1} &= Rr \\ rR^{-1} &= Rr. \end{aligned}$$

Therefore, $Rr = rR^{-1} = rR^{n-1}$. Note that this last inequality shows that D_n is a non-Abelian group for $n \geq 3$.

Activity 24.4. Now that we can write the product Rr in the form rR^j, to complete the operation table for D_n we need to determine R^ir for $i > 1$. We will do so in this activity.

(a) Use the fact that $Rr = rR^{-1}$ to explain why $R^2r = rR^{-2}$.

(b) Assume $R^i r = rR^{-i}$ for some $i \geq 1$. Use this fact to show that $R^{i+1}r = rR^{-(i+1)}$.

(c) Explain how we have shown that $R^i r = rR^{-i}$ for all positive integers i.

Now that we know $R^i r = rR^{-i}$ for each i, we can compute every possible product in D_n in an efficient manner. For example, in D_4 we have

$$(rR^3)(rR^2) = r(R^3 r)R^2 = r(rR^{-3})R^2 = r^2 R^{-1} = IR^3 = R^3.$$

We can perform similar calculations to complete the operation table for D_4 (Table 24.2) in a format that is different from the one in Investigation 19 (Table 19.1 on page 275).

$\circ$	I	R	R^2	R^3	r	rR	rR^2	rR^3
I	I	R	R^2	R^3	r	rR	rR^2	rR^3
R	R	R^2	R^3	I	rR^3	r	rR	rR^2
R^2	R^2	R^3	I	R	rR^2	rR^3	r	rR
R^3	R^3	I	R	R^2	rR	rR^2	rR^3	r
r	r	rR	rR^2	rR^3	I	R	R^2	R^3
rR	rR	rR^2	rR^3	r	R^3	I	R	R^2
rR^2	rR^2	rR^3	r	rR	R^2	R^3	I	R
rR^3	rR^3	r	rR	rR^2	R	R^2	R^3	I

Table 24.2
Operation table for D_4.

Generators and Group Presentations

We have seen that in a cyclic group, every element can be written as a power of a single element. That is, if G is a cyclic group, then $G = \{a^n : n \in \mathbb{Z}\} = \langle a \rangle$ for some $a \in G$. In other words, we say that the single element a generates the group G. Something similar, but a bit more complicated, happens with the dihedral group D_n. In the previous section, we found that every element in D_n can be written in terms of a single rotation and a single reflection. In other words,

$$D_n = \{r^i R^j : i, j \in \mathbb{Z}\},$$

with the stipulations that $r^2 = I$, $R^n = I$, and $rR = R^{-1}r$. In this situation, we can say that r and R generate D_n, or that D_n is generated by r and R. The next definition generalizes this idea.

Definition 24.5. A subset S of a group G **generates** G if every element in G can be written as a finite product of the elements of S (or a finite sum if the group operation is written additively).

The elements of the set S are called *generators* of G. Unless G is a single-element group, we assume that the identity is not an element of S. We can produce all of the elements of the group G from generators for G, but the structure of G is also determined by how the generators interact with each other. For example, in D_4 the rotation R has order 4 ($R^4 = I$), the reflection r has order 2 ($r^2 = I$), and we have the relation that $rR = R^{-1}r$. These relations among the generators, along with the generators themselves, completely determine the group. We call the pair consisting of the generating subset S and the set of relations among these generators a *presentation* of the group G. We denote a group presentation by

$$\langle S \mid \textit{Relations} \rangle.$$

For example, D_n has the following presentation:

$$\langle r, R \mid r^2 = 1, R^n = 1, rR = R^{-1}r \rangle.$$

As another example, a cyclic group of order n has the presentation $\langle a \mid a^n = 1 \rangle$, where 1 denotes the identity element in the group.[†]

Activity 24.6.

(a) Create the operation table for the group $\mathcal{V}$ with presentation

$$\langle a, b \mid a^2 = b^2 = (ab)^2 = 1 \rangle.$$

(b) Let $a = \begin{bmatrix} -1 & 0 \\ 0 & 1 \end{bmatrix}$ and $b = \begin{bmatrix} 1 & 0 \\ 0 & -1 \end{bmatrix}$. Demonstrate that the group generated by a and b in $\mathrm{GL}_2(\mathbb{R})$ (using matrix multiplication) is an example of a group of order 4 with presentation

$$\langle a, b \mid a^2 = b^2 = (ab)^2 = 1 \rangle.$$

Although we won't show it, every group has a presentation. Moreover, a group can have more than one presentation. Presentations provide a concise and simple way to describe many groups, but there are subtle issues that can make working with group presentations more complicated. For instance, it can be difficult to determine the size of a group from a presentation, or to identify when two seemingly different products of generators are equal. It can also be a challenge to determine when two different presentations yield the same group.

Concluding Activities

Activity 24.7. Use the notation from this investigation to create the operation tables for D_3, D_5, and D_6.

Activity 24.8. We defined the groups D_n for $n \geq 3$. You may wonder what we can do if $n = 1$ or $n = 2$.

(a) We can consider the group D_1 as the group of symmetries of a line segment. Find the elements of D_1, and write an operation table for D_1.

(b) We can define D_2 as the group of symmetries of a non-square rectangle. Find the elements of D_2, and write an operation table for D_2.

(c) The two groups D_1 and D_2 differ from the groups D_n for $n \geq 3$ in some important ways. Describe at least two of these differences.

[†]Although we often denote the identity in a group by e, it is standard notation for group presentations to use 1 instead.

Activity 24.9. Write a short summary that describes the important concepts, and the relationships between these concepts, that were introduced in this investigation. Explain how the material in this investigation is connected to the material in Investigations 19 and 20.

Exercises

(1) Determine if each of the following is a rotation or reflection in D_n:

(a) The composition of two rotations

(b) The composition of a rotation and a reflection

(c) The composition of a reflection and a rotation

(d) The composition of two reflections

(2) Let n be an integer with $n \geq 3$.

(a) If n is even, show that the center of D_n is not trivial. Then find all of the elements in $Z(D_n)$.

(b) If n is odd, find all elements in $Z(D_n)$.

(3) The Connected Mathematics Project (CMP, a collection of instructional materials for school mathematics) has, as part of its materials, various investigations in geometry for middle school students. One particular investigation deals with symmetries. As part of this investigation, students are asked to construct an operation table for the symmetries of various figures.

(a) A figure similar to one presented to students in the CMP investigation is shown in Figure 24.1. Find all of the symmetries of this figure, and construct an operation table for it. Clearly describe and label your symmetries. Place this group of symmetries in the context of group theory. Where have we seen something like this before?

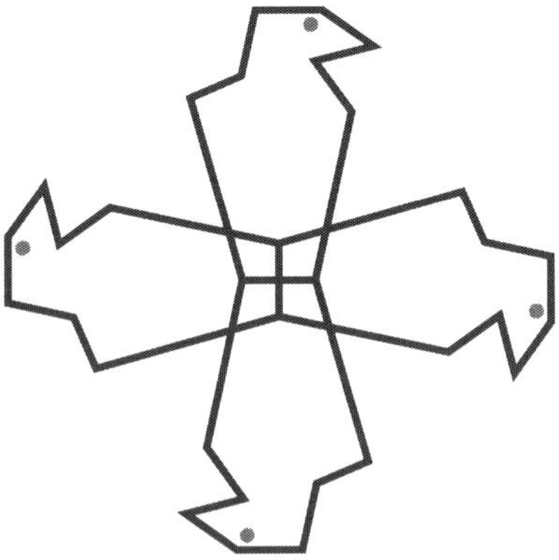

Figure 24.1
A Connected Mathematics Project figure.

At one point in the CMP exercise, students are asked to make a table of the symmetries of an equilateral triangle. They are then asked to, "Make an operation table for multiplication of the whole numbers 1, 2, 3, 4, 5, and 6. Compare the patterns in your multiplication table with the patterns in your table of transformation combinations (symmetries). Describe any interesting similarities and differences you discover." Parts (b) and (c) refer to this task.

(b) What possible responses could students give to this question? Use appropriate terminology from group theory. Your response should include the terms "closed," "identity," and a discussion of operations.

(c) Provide a more complete response to this CMP problem from the point of view of group theory, using appropriate terminology and notation from this course. Your discussion here should include the words "group," "modulus," "units," and "operations."

(4) **Symmetries of a tetrahedron.** We can also define symmetries in three-dimensions. For example, there are 12 rotational symmetries of the regular tetrahedron. (A regular tetrahedron has 4 vertices and 4 faces, all of which are equilateral triangles, as shown in Figure 24.2.) Describe these symmetries, and then represent each in permutation notation. Finally, construct the operation table for this group of symmetries. (We will call this group *Tetra*. As a hint, note that nine symmetries are fairly easy to find, and the other three are more difficult to see. One of these more difficult symmetries is shown in Figure 24.2 (on the right) where the dashed line connects the midpoints of two non-adjacent sides.)

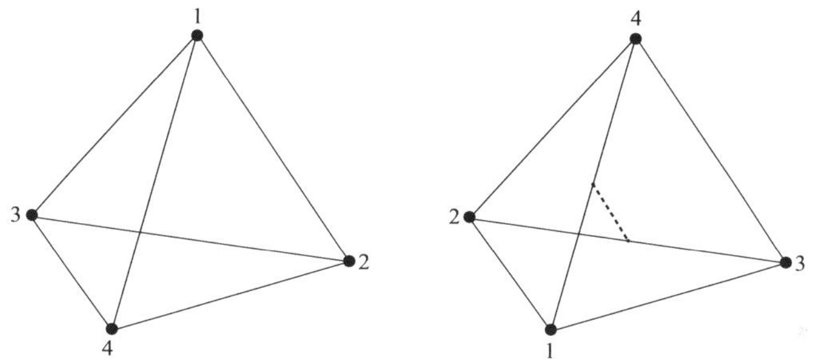

Figure 24.2
A tetrahedron and one of its symmetries.

(5) Find a presentation for the group $\mathbb{Z}$.

* (6) Let $\mathbf{Q}$ be the group with presentation $\langle a, b \mid a^4 = 1, a^2 = b^2, ab = b^{-1}a \rangle$. (The group $\mathbf{Q}$ is called the *quaternion* group or the group of *quaternions*.)

(a) Show that the elements 1, a, a^2, a^3, b, ab, a^2b, and a^3b are all distinct.

(b) Create the operation table for the set $\{1, a, a^2, a^3, b, ab, a^2b, a^3b\}$ to show that this set is a group. This is the group $\mathbf{Q}$.

(c) Let $a = \begin{bmatrix} 0 & 1 \\ -1 & 0 \end{bmatrix}$, and $b = \begin{bmatrix} 0 & i \\ i & 0 \end{bmatrix}$, where $i^2 = -1$. Demonstrate that the group generated by a and b in $\mathrm{GL}_2(\mathbb{C})$ (using matrix multiplication) is an example of a group of order 8 with presentation $\langle a, b \mid a^4 = 1, a^2 = b^2, ab = b^{-1}a \rangle$.

(7) (a) Show that the quaternion group $\mathbf{Q}$ (see Exercise 6 in this investigation) contains exactly one subgroup of order 2.

(b) Show that the quaternion group $\mathbf{Q}$ contains more than one cyclic subgroup of order 4. How many distinct subgroups of order 4 does $\mathbf{Q}$ have? Verify your answer.

(8) As we have seen, the group D_n has the presentation

$$\langle r, R \mid r^2 = 1, R^n = 1, rR = R^{-1}r \rangle.$$

Let $a = r$ and $b = rR$. Show that D_n also has the presentation

$$\langle a, b \mid a^2 = b^2 = (ab)^n = 1 \rangle.$$

This will show that D_n can be generated by elements of order 2. (You could construct the group with presentation $\langle a, b \mid a^2 = b^2 = (ab)^n = 1 \rangle$ and then compare to D_n, but it might be easier to show that the generators and relations from each presentation can be obtained from the other.)

$\star$ (9) Let G be a group with identity e. If $a, b \in G$, the element $[a, b] = a^{-1}b^{-1}ab$ is called the **commutator** of the pair a, b.

(a) Calculate $[a, b]$ if $G = D_4$, $a = rR$, and $b = R^3$.

(b) Let $a, b \in G$. Prove that $[a, b] = e$ if and only if a and b commute.

(c) Let G' be the subgroup (called the **commutator** subgroup) of G generated by all the commutators in G, so that every element of G' can be written as a finite product of elements of the form $[a_i, b_i]$, where $a_i, b_i \in G$. Prove that $G' = \{e\}$ if and only if G is Abelian.

(d) Determine G' if $G = D_n$. How does your answer depend on n? Explain.

Connections

The dihedral groups are special examples of the groups of symmetries that we introduced in Investigation 19. The dihedral groups have a well-defined structure that can be described in a concise manner, and so these groups are a good source of examples in mathematics. As symmetry groups of fundamental geometric objects, the dihedral groups play an important role in geometry. The dihedral groups also have applications outside of mathematics. For example, we will see how dihedral groups are used to create a check digit scheme in Investigation 35. They also have applications in chemistry where the structure of molecules can often be modeled with regular polygons, and in other areas where an idea depends on the geometric properties of some underlying structure—for example, in the design of vision filters.

Investigation 25

The Symmetric Groups

Focus Questions

By the end of this investigation, you should be able to give precise and thorough answers to the questions listed below. You may want to keep these questions in mind to focus your thoughts as you complete the investigation.

- What is a permutation of a set? Under what operation does the collection of permutations of a set form a group?
- What are the symmetric groups? What notation can be used to represent elements in symmetric groups?
- What is the cycle decomposition of a permutation, and how can we find it?
- What is a transposition, and why are transpositions important?
- What is the alternating group, and what important properties does it satisfy?

Preview Activity 25.1. Functions are important throughout mathematics, and permutations are special types of functions* that will allow us to form an important family of groups. A permutation of a set is simply a rearrangement of the elements of the set, defined formally as follows:

Definition 25.2. A **permutation** of a set S is a bijection $f : S \to S$.

(a) Let S be a set. If we compose two permutations from S to S, is the resulting composite function a permutation? Why or why not?

(b) Let $f : S \to S$ be a permutation.

(i) How can we define the inverse function f^{-1} of f? (Hint: Define a function $g : S \to S$ by completing the following statement: $g(s) = t$ if $f(\_\_\_) = \_\_\_$. This function g will be our candidate for the inverse of f.)

(ii) Is the function you defined in part (a) a permutation of S? Verify your answer.

(iii) Show that the function you defined in part (a) is actually the inverse of f.

*This investigation assumes an understanding of injective, surjective, and bijective functions. For a review of these topics, see Appendix A.

Introduction

We have seen that the set of symmetries of an equilateral triangle, a square, a regular pentagon, and a regular hexagon form groups under the operation of composition. These groups are actually subgroups of much larger groups called the *symmetric groups*, which we will study in this investigation. The symmetric groups are groups of permutations. As we will see later, symmetric groups are important in mathematics the sense that every group can be viewed as a subgroup of a symmetric group. Symmetric groups also have applications to molecular chemistry, quantum mechanics, games, and many other areas.

The Symmetric Group of a Set

Preview Activity 25.1 showed that the collection of permutations of a set satisfies some of the group axioms using the operation of composition of functions. To show that the collection of permutations of a given set actually does form a group under composition, there are just a few more details we need to verify.

Activity 25.3. Let S be a set.

(a) Why is the identity function from S to S a permutation of S?

(b) Why is the composition of permutations on S an associative operation?

(c) Explain how the results in this activity and Preview Activity 25.1 establish the following theorem:

Theorem 25.4. *Let S be a set, and let $P(S)$ denote the collection of permutations of S. Then $P(S)$ is a group under the operation of composition of functions.*

The group $P(S)$ is called the *permutation group of S*. It is not difficult to see that if S is a set with n elements, then $P(S)$ is basically the same as $P(\{1, 2, \ldots, n\})$. For this reason, and to make the notation easier, we typically focus on studying the permutations of the set $\{1, 2, \ldots, n\}$. This group of permutations is given a special name:

Definition 25.5. The **symmetric group** of degree n is the group

$$S_n = P(\{1, 2, \ldots, n\}).$$

In the next sections, we will investigate ways to represent elements of S_n.

Permutation Notation and Cycles

Let n be a positive integer. As we did with the symmetries of regular polygons we examined in Investigation 19, we can represent elements of S_n using permutation notation. If σ is a permutation in S_n such that $\sigma(i) = a_i$ for each i, then the permutation notation for σ is

$$\begin{pmatrix} 1 & 2 & 3 & \cdots & n \\ a_1 & a_2 & a_3 & \cdots & a_n \end{pmatrix}.$$

There is an alternate, more elegant, way to represent a permutation in S_n using *cycle notation*. To illustrate, note that the permutation $\sigma = \begin{pmatrix} 1 & 2 & 3 & 4 & 5 \\ 2 & 3 & 5 & 4 & 1 \end{pmatrix}$ has a special form. This permutation fixes 4 but sends 1 to 2, then 2 to 3, 3 to 5, and then *cycles* 5 back to 1. We could visualize the action of this permutation as follows:

Such a permutation is called a *cycle*. More generally, a permutation $\sigma \in S_n$ is a *cycle* if for some distinct integers $a_1, a_2, \ldots, a_k$ between 1 and n we have

$$\begin{aligned} \sigma(a_1) &= a_2 \\ \sigma(a_2) &= a_3 \\ \sigma(a_3) &= a_4 \\ &\vdots \\ \sigma(a_{k-1}) &= a_k \\ \sigma(a_k) &= a_1 \end{aligned}$$

and $\sigma(i) = i$ for all other i. In other words, a cycle has the form

$$\begin{pmatrix} a_1 & a_2 & a_3 & \cdots & a_k & b_1 & b_2 & \cdots & b_{n-k} \\ a_2 & a_3 & a_4 & \cdots & a_k & b_1 & b_2 & \cdots & b_{n-k} \end{pmatrix} \tag{25.1}$$

where $b_1, b_2, \ldots, b_{n-k}$ are the remaining integers between 1 and n that are not in $\{a_1, a_2, \ldots, a_k\}$. For example, the permutation

$$\begin{pmatrix} 1 & 2 & 3 & 4 & 5 & 6 \\ 1 & 4 & 3 & 6 & 2 & 5 \end{pmatrix}$$

can be rewritten as

$$\begin{pmatrix} 2 & 4 & 6 & 5 & 1 & 3 \\ 4 & 6 & 5 & 2 & 1 & 3 \end{pmatrix}$$

and has the form of a cycle. We can represent this cycle more concisely using the following *cycle notation*:

$$(2\,4\,6\,5).$$

We can represent the general cycle from (25.1) in cycle notation as

$$(a_1\, a_2\, a_3\, \ldots\, a_k).$$

The next definition formalizes these ideas.

Definition 25.6. Let k be a positive integer. The k-cycle

$$\alpha = (a_1\ a_2\ a_3\ \cdots\ a_{k-1}\ a_k)$$

is the permutation satisfying $\alpha(a_i) = a_{i+1}$ for $1 \leq i \leq k-1$, $\alpha(a_k) = a_1$, and $\alpha(j) = j$ for all $j \notin \{a_1, a_2, \ldots, a_{k-1}, a_k\}$.

Activity 25.7. Which of the following permutations is a cycle? Write each cycle in cycle notation.

(a) $\begin{pmatrix} 1 & 2 & 3 & 4 & 5 & 6 & 7 \\ 6 & 2 & 5 & 3 & 1 & 4 & 7 \end{pmatrix}$

(b) $\begin{pmatrix} 1 & 2 & 3 & 4 & 5 & 6 & 7 \\ 1 & 2 & 4 & 6 & 5 & 7 & 3 \end{pmatrix}$

(c) $\begin{pmatrix} 1 & 2 & 3 & 4 & 5 & 6 & 7 \\ 2 & 1 & 4 & 3 & 6 & 7 & 5 \end{pmatrix}$

The Cycle Decomposition of a Permutation

Not every permutation in S_n is a cycle, but it is not difficult to see that every permutation can be written as a product of cycles. Moreover, it turns out that each permutation can be written in some unique way as a product of disjoint cycles. To see how this works, consider the following example.

Let

$$\sigma = \begin{pmatrix} 1 & 2 & 3 & 4 & 5 & 6 & 7 & 8 \\ 2 & 3 & 1 & 7 & 5 & 6 & 4 & 8 \end{pmatrix}$$

in S_8. We can rewrite σ as

$$\sigma = \begin{pmatrix} 1 & 2 & 3 & 4 & 7 & 5 & 6 & 8 \\ 2 & 3 & 1 & 7 & 4 & 5 & 6 & 8 \end{pmatrix}.$$

Notice that σ does not have the form indicated in (25.1) and is therefore not a cycle. However, σ is made up of two cycles: $\sigma_1 = (1\ 2\ 3)$ and $\sigma_2 = (4\ 7)$. If we consider the composition of σ_1 and σ_2, we find the following:

$$\begin{aligned}
\sigma_1\sigma_2(1) &= \sigma_1(\sigma_2(1)) = \sigma_1(1) = 2 \\
\sigma_1\sigma_2(2) &= \sigma_1(\sigma_2(2)) = \sigma_1(2) = 3 \\
\sigma_1\sigma_2(3) &= \sigma_1(\sigma_2(3)) = \sigma_1(3) = 1 \\
\sigma_1\sigma_2(4) &= \sigma_1(\sigma_2(4)) = \sigma_1(7) = 7 \\
\sigma_1\sigma_2(5) &= \sigma_1(\sigma_2(5)) = \sigma_1(5) = 5 \\
\sigma_1\sigma_2(6) &= \sigma_1(\sigma_2(6)) = \sigma_1(6) = 6 \\
\sigma_1\sigma_2(7) &= \sigma_1(\sigma_2(7)) = \sigma_1(4) = 4 \\
\sigma_1\sigma_2(8) &= \sigma_1(\sigma_2(8)) = \sigma_1(8) = 8
\end{aligned}$$

So $\sigma_1\sigma_2 = \sigma$, and we call $\sigma_1\sigma_2$ the *cycle decomposition* of σ.

There is an easy way to find the cycle decomposition of an arbitrary permutation $\sigma \in S_n$.

We can begin by finding the image of 1 under σ (any starting place will do, but it is nice to have some consistency) and then follow the permutation along until we return to 1. This forms the first cycle. We can then repeat this process beginning with the smallest integer not in the first cycle, and continue until we have included each integer between 1 and n in some cycle. To illustrate, consider the permutation

$$\sigma = \begin{pmatrix} 1 & 2 & 3 & 4 & 5 & 6 & 7 & 8 & 9 \\ 3 & 6 & 4 & 2 & 7 & 1 & 9 & 8 & 5 \end{pmatrix}.$$

Note that

$$\sigma(1) = 3, \sigma(3) = 4, \sigma(4) = 2, \sigma(2) = 6, \text{ and } \sigma(6) = 1,$$

so one cycle in the cycle decomposition of σ is $(1\ 3\ 4\ 2\ 6)$. Next we see that

$$\sigma(5) = 7, \sigma(7) = 9, \text{ and } \sigma(9) = 5,$$

so another cycle in the cycle decomposition of σ is $(5\ 7\ 9)$. Since $\sigma(8) = 8$, the last cycle in the cycle decomposition of σ is simply (8). However, the cycle (8) is just the identity cycle, so we will omit it from the cycle decomposition. Therefore, the cycle decomposition of σ is

$$\sigma = (1\ 3\ 4\ 2\ 6)(5\ 7\ 9).$$

Activity 25.8. Find the cycle decomposition of each of the indicated permutations.

(a) $\begin{pmatrix} 1 & 2 & 3 & 4 & 5 & 6 & 7 \\ 2 & 4 & 3 & 1 & 7 & 6 & 5 \end{pmatrix}$

(b) $\begin{pmatrix} 1 & 2 & 3 & 4 & 5 & 6 & 7 & 8 \\ 1 & 2 & 5 & 3 & 4 & 6 & 8 & 7 \end{pmatrix}$

(c) $\begin{pmatrix} 1 & 2 & 3 & 4 & 5 & 6 & 7 & 8 & 9 & 10 \\ 6 & 10 & 9 & 4 & 7 & 1 & 5 & 8 & 2 & 3 \end{pmatrix}$

Activity 25.9. Write each product of cycles in permutation notation.

(a) $\sigma = (1\ 3\ 5\ 6)(2\ 4)$ in S_6

(b) $\sigma = (2\ 5\ 3)(7\ 4)(6\ 9)$ in S_{10}

One thing to note in each of our cycle decompositions is that every permutation was written as a product of non-overlapping (that is, *disjoint*) cycles. This observation motivates our next definition.

Definition 25.10. Two cycles $\sigma = (a_1\ a_2\ \ldots\ a_k)$ and $\tau = (b_1\ b_2\ \ldots\ b_m)$ are **disjoint** if $a_i \neq b_j$ for all $1 \leq i \leq k$ and $1 \leq j \leq m$.

That we can always decompose a permutation as a product of disjoint cycles is the subject of the following theorem.

Theorem 25.11. *Let n be a positive integer. Every permutation in S_n is either a cycle or can be written as a product of disjoint cycles.*

Proof. We will show via strong induction that any permutation of a finite collection of numbers is either a cycle or can be written as a product of disjoint cycles. Let α be a permutation of m numbers. If $m = 1$, then α must be the identity permutation, and $\alpha = (1)$ is a cycle. For our induction hypothesis, we assume, for some integer $m \geq 1$, that any permutation of m or fewer numbers is a cycle or a product of disjoint cycles. Now suppose α permutes $m + 1$ numbers. Since

α is not the identity, there is some number that is not fixed by α. For the sake of convenience, label this number as a_1. For each positive integer s, consider the sequence $\alpha(a_1), \alpha^2(a_1), \alpha^3(a_1), \cdots$, where

$$\alpha^s(a_1) = \alpha(\alpha(\alpha(\cdots(\alpha(a_1))))),$$

with α appearing s times in this iterated composition. Since $m + 1$ is finite, this sequence must repeat at some point. Let i and j be positive integers with $i < j$ such that $\alpha^j(a_1) = \alpha^i(a_1)$. Then $\alpha^{j-i}(a_1) = a_1$, and there is a positive power of α that fixes a_1. By the Well-Ordering Principle, there is a smallest positive integer k such that $\alpha^k(a_1) = a_1$. Since a_1 is not fixed by α, it must be that $k \geq 2$. For each integer t with $2 \leq t \leq k$, let $a_t = \alpha^{t-1}(a_1)$, and let $\sigma = (a_1\ a_2\ a_3\ \cdots\ a_k)$. Note that $\sigma(a_i) = \alpha(a_i)$ for each i between 1 and k.

Now let α' be the permutation in S_n defined by

$$\alpha'(s) = \begin{cases} a_i, & \text{if } s = a_i \text{ for some } i \\ \alpha(s), & \text{otherwise.} \end{cases}$$

We will show that $\alpha = \sigma\alpha'$. To do so, choose $q \in \{1, 2, \ldots, m+1\}$. We will show that $(\sigma\alpha')(q) = \alpha(q)$ by considering cases.

For the first case, assume $q = a_i$ for some $1 \leq i \leq k$. In this case, we have

$$(\sigma\alpha')(q) = \sigma(\alpha'(q)) = \sigma(a_i) = \alpha(a_i).$$

For the remaining case, suppose $q \neq a_i$ for all i. To apply σ to $\alpha'(q)$, we need to know if $\alpha'(q)$ can be one of the a_i for i between 1 and k. The following claim answers this question.

Claim. $\alpha'(q) \neq a_i$ for all i.

Proof of Claim. We will prove the claim by contradiction. Suppose $\alpha'(q) = a_i$ for some $1 \leq i \leq k$. By definition, $\alpha'(q) = \alpha(q) = a_i$. Since α is an injection, this then implies $q = a_{i-1 (\text{mod } k)}$, a contradiction to the fact that q is not equal to one of the a_i. ■

Since our claim establishes that $q \neq a_i$ for all i, it follows that

$$(\sigma\alpha')(q) = \sigma(\alpha'(q)) = \alpha'(q) = \alpha(q).$$

We can therefore conclude that $(\sigma\alpha')(q) = \alpha(q)$ for all q. Hence, $\sigma\alpha' = \alpha$.

Now α' is a permutation of $(m + 1) - k$ integers, so by our induction hypothesis, we know α' is either a cycle or a product of disjoint cycles. Moreover, by our claim, these cycles are all disjoint from σ. Therefore, $\alpha = \sigma\alpha'$ is a cycle or a product of disjoint cycles. ■

Activity 25.12.

(a) List the elements in S_3 as cycles or products of disjoint cycles.

(b) Construct the operation table for S_3.

(c) Is S_3 an Abelian group? Explain.

(d) List all the elements of S_4 as cycles or products of disjoint cycles.

(e) Is S_4 an Abelian group?

As you may suspect, S_n (for $n \geq 2$) is never an Abelian group. The proof of this fact is left as an exercise (see Exercise 6), although it is not much more difficult than what you were asked to show in parts (c) and (e) of Activity 25.12.

Transpositions

Just like integers, permutations can be classified as being either even or odd. This classification is a useful tool in group theory and will ultimately allow us to define a special class of groups called the *alternating groups*. The parity of a permutation—that is, whether it is even or odd—is determined by first writing the permutation as a product of 2-cycles or *transpositions*. A straightforward computation shows that any cycle can be written as a product of transpositions as follows:

$$\sigma = (a_1\ a_2\ \dots\ a_k) = (a_1\ a_2)(a_2\ a_3)(a_3\ a_4)\cdots(a_{k-2}\ a_{k-1})(a_{k-1}\ a_k). \tag{25.2}$$

The price we pay for this 2-cycle decomposition is that our 2-cycles are not disjoint and do NOT commute. However, since every permutation can be written as a product of cycles and every cycle can be written as a product of 2-cycles, then every permutation can be written as a product of 2-cycles. We should also note that the decomposition in (25.2) is not the only way to decompose a cycle into a product of transpositions. For example, you should convince yourself that

$$(a_1\ a_2\ \dots\ a_k) = (a_1\ a_k)(a_1\ a_{k-1})(a_1\ a_{k-2})\cdots(a_1\ a_2)$$

is another 2-cycle decomposition of $(a_1\ a_2\ \dots\ a_k)$.

Activity 25.13. Write each permutation as a product of 2-cycles.

(a) $\begin{pmatrix} 1 & 2 & 3 & 4 & 5 & 6 & 7 \\ 2 & 4 & 3 & 1 & 7 & 6 & 5 \end{pmatrix}$

(b) $\begin{pmatrix} 1 & 2 & 3 & 4 & 5 & 6 & 7 & 8 \\ 1 & 2 & 5 & 3 & 4 & 6 & 8 & 7 \end{pmatrix}$

(c) $\begin{pmatrix} 1 & 2 & 3 & 4 & 5 & 6 & 7 & 8 & 9 & 10 \\ 6 & 10 & 9 & 4 & 7 & 1 & 5 & 8 & 2 & 3 \end{pmatrix}$

Why should we care about writing a permutation as a product of transpositions? Since the transpositions in the resulting product are not usually disjoint, this doesn't seem to be an improvement over the disjoint cycle decomposition we developed in Theorem 25.11. In fact, as we noted earlier, there are many different ways we can write a permutation as a product of transpositions. For example,

$$(1\ 2)(1\ 3)(1\ 4) = (1\ 4\ 3\ 2) = (1\ 4)(4\ 3)(3\ 2) = (1\ 4)(4\ 3)(3\ 2)(2\ 4)(4\ 2).$$

Notice that all of these decompositions are made up of an odd number of transpositions. This important observation holds in general, as stated in the next theorem.

Theorem 25.14. *Let $n \in \mathbb{Z}^+$, and let $\sigma \in S_n$. No matter how σ is written as a product of transpositions, the number of transpositions in the product will always have the same parity.*

In order to prove Theorem 25.14, we will first need to make some conclusions about the parity of the identity permutation, I.

Lemma 25.15. *Let $n \in \mathbb{Z}^+$, and let I be the identity permutation in S_n. If*

$$I = \tau_1 \tau_2 \cdots \tau_k,$$

where $\tau_1, \tau_2, \dots, \tau_k$ are transpositions, then k is even.

Proof. Let $n \in \mathbb{Z}^+$, and let I be the identity permutation in S_n. Suppose

$$I = \tau_1\tau_2 \cdots \tau_k$$

for some transpositions $\tau_1, \tau_2, \ldots, \tau_k$. First note that the identity permutation is not itself a transposition, so $k > 1$. We will proceed by induction on k. For the base case, note that if $k = 2$, then k is even and we are done. For the inductive step, let $k \geq 2$, and assume that if I is written as a product of m transpositions with $m \leq k$, then m is even. Now suppose

$$I = \tau_1\tau_2 \cdots \tau_k\tau_{k+1},$$

where $\tau_1, \tau_2, \ldots, \tau_k, \tau_{k+1}$ are transpositions. Let $\tau_{k+1} = (a\ b)$ and $\tau_k = (r\ s)$. There are four different possibilities for the product of these two transpositions:

(1) If $(r\ s) = (a\ b)$, then $\tau_k\tau_{k+1} = (a\ b)(a\ b) = I$.

(2) If $r = a$ and $s \neq b$, then

$$\tau_k\tau_{k+1} = (r\ s)(a\ b) = (a\ s)(a\ b) = (a\ b\ s) = (a\ b)(s\ b).$$

(3) If $s = b$ and $r \neq a$, then

$$\tau_k\tau_{k+1} = (r\ s)(a\ b) = (r\ b)(a\ b) = (a\ r\ b) = (a\ r)(r\ b).$$

(4) If τ_k and τ_{k+1} are disjoint cycles, then $(r\ s)(a\ b) = (a\ b)(r\ s)$.

In cases (2) – (4), we can rewrite $\tau_k\tau_{k+1}$ in the form $\alpha_1\tau_k'$ for some transpositions τ_k' and α_1, where $\tau_k'(a) = a$ and $\alpha_1(a) \neq a$. Then

$$I = \tau_1\tau_2 \cdots \tau_{k-1}\alpha_1\tau_k'.$$

As long as $\alpha_1 \neq \tau_{k-1}$, we can repeat this process, interchanging α_1 with τ_{k-1}. In this case, $\tau_{k-1}\alpha_1 = \alpha_2\tau_{k-1}'$, where $\tau_{k-1}'(a) = a$ and $\alpha_2(a) \neq a$, and

$$I = \tau_1\tau_2 \cdots \tau_{k-2}\alpha_2\tau_{k-1}'\tau_k'.$$

Continuing in this same manner, if we never encounter case (1)—that is, if we never find i and j so that $\tau_j = \alpha_i$—then we can interchange each α_i with each τ_j, leaving τ_j' and α_{i+1} so that $\tau_j'(a) = a$ and $\alpha_{i+1}(a) \neq a$, as shown in cases (2), (3), and (4). In this situation, we will ultimately have

$$I = \alpha_k\tau_1'\tau_2' \cdots \tau_{k-1}'\tau_k',$$

where $(\tau_1'\tau_2' \cdots \tau_{k-1}'\tau_k')(a) = a$ and $\alpha_k(a) \neq a$. Then

$$I(a) = (\alpha_k\tau_1'\tau_2' \cdots \tau_{k-1}'\tau_k')(a) = \alpha_k(\tau_1'\tau_2' \cdots \tau_{k-1}'\tau_k'(a)) = \alpha_k(a) \neq a,$$

which is impossible.

Therefore, as we interchange the α_i with the τ_j, we must encounter a j so that $\tau_j = \alpha_i$. As in case (1), the product $\tau_j\alpha_i$ will equal the identity and will drop out of the product. Then

$$I = \tau_1\tau_2 \cdots \tau_{j-1}\tau_{j+1}'\tau_{j+2}' \cdots \tau_k',$$

and I is the product of $k - 1$ transpositions. By our induction hypothesis, it then follows that $k - 1$ is even, and therefore $k + 1$ is also even, as desired. ■

We are now ready to prove Theorem 25.14, which establishes that the parity of the number of transpositions in the 2-cycle decomposition of a permutation is an invariant.

Proof of Theorem 25.14. Let $n \in \mathbb{Z}^+$, and let $\sigma \in S_n$. Assume $\tau_1, \tau_2, \ldots, \tau_k$ and $\alpha_1, \alpha_2, \ldots, \alpha_m$ are transpositions such that

$$\sigma = \tau_1\tau_2\cdots\tau_k = \alpha_1\alpha_2\cdots\alpha_m. \tag{25.3}$$

We need to show that $k \equiv m \pmod 2$ or, equivalently, that $k + m$ is even. Multiplying the two decompositions in (25.3) on the right by $\alpha_m\alpha_{m-1}\cdots\alpha_2\alpha_1$, keeping in mind that a transposition is its own inverse, we obtain

$$\tau_1\tau_2\cdots\tau_k\alpha_m\alpha_{m-1}\cdots\alpha_2\alpha_1 = I.$$

We have thus written the identity permutation I as a product of $k+m$ transpositions. Lemma 25.15 now implies that $k + m$ must be even, as desired. ■

Even and Odd Permutations and the Alternating Group

Theorem 25.14 shows that the number of transpositions in the 2-cycle decomposition of a permutation is always either even or odd. We can use this result to define the parity of a permutation.

Definition 25.16. Let $n \in \mathbb{Z}^+$. A permutation $\sigma \in S_n$ is **even** if σ can be written as a product of an even number of transpositions. A permutation $\sigma \in S_n$ is **odd** if σ is not even.

So, for example, the permutation $(1\ 2)(2\ 4)(3\ 5)(4\ 1)$ in S_5 is an even permutation.

Activity 25.17. Determine whether the indicated permutation is even or odd.

(a) $\begin{pmatrix} 1 & 2 & 3 & 4 & 5 & 6 & 7 \\ 2 & 4 & 3 & 1 & 7 & 6 & 5 \end{pmatrix}$

(b) $\begin{pmatrix} 1 & 2 & 3 & 4 & 5 & 6 & 7 & 8 \\ 1 & 2 & 5 & 3 & 4 & 6 & 8 & 7 \end{pmatrix}$

(c) $\begin{pmatrix} 1 & 2 & 3 & 4 & 5 & 6 & 7 & 8 & 9 & 10 \\ 6 & 10 & 9 & 4 & 7 & 1 & 5 & 8 & 2 & 3 \end{pmatrix}$

An interesting question is the following: If we collect all of even (or odd) permutations together, does the resulting set form a subgroup of S_n?

Activity 25.18.

(a) Explain why the collection of odd permutations is not a subgroup of S_n.

(b) Is the product of two even permutations always an even permutation? Explain.

(c) Is the inverse of an even permutation always an even permutation? Explain.

(d) Let A_n be the set of all even permutations in S_n. Explain why A_n is a subgroup of S_n.

The subgroup A_n defined in part (d) of Activity 25.18 is an important one and is given a special name.

Definition 25.19. For $n \geq 2$, the **alternating group** A_n is the subgroup of S_n consisting of the even permutations in S_n.

Activity 25.20.

(a) Find all the elements in A_3.

(b) Find all the elements in A_4.

You will show in Exercise 11 that the order of S_n is $n!$. But what is the order of A_n? Activity 25.20 seems to indicate that the order of A_n is half that of the order of S_n. To see that this is true in general, let $\alpha, \beta \in S_n$. If α is odd, then $(1\ 2)\alpha$ is even. Also, if $(1\ 2)\alpha = (1\ 2)\beta$, then $\alpha = \beta$. So the number of even permutations in S_n is greater than or equal to the number of odd permutations in S_n. If α is even, then $(1\ 2)\alpha$ is odd. Also, if $(1\ 2)\alpha = (1\ 2)\beta$, then $\alpha = \beta$. So the number of odd permutations in S_n is greater than or equal to the number of even permutations in S_n. Thus, the permutations in S_n are evenly divided between even and odd permutations, and $|A_n| = \frac{|S_n|}{2} = \frac{n!}{2}$.

Concluding Activities

Activity 25.21. Let n be an integer with $n \geq 3$. What is the connection between D_n and S_n? Explain.

Activity 25.22. One advantage of decomposing a permutation as a product of disjoint cycles is that disjoint cycles commute. To verify this property, let $\alpha = (a_1\ a_2\ \ldots\ a_k)$ and $\beta = (b_1\ b_2\ \ldots\ b_m)$ be disjoint cycles in S_n.

(a) What can we say about the sets $A = \{a_1, a_2, \ldots, a_k\}$ and $B = \{b_1, b_2, \ldots, b_m\}$?

(b) What is $\alpha(b_j)$ for any $1 \leq j \leq m$? What is $\beta(a_i)$ for any $1 \leq i \leq k$? Explain.

(c) If $r \in \{1, 2, \ldots, n\} \setminus (A \cup B)$, what are $\alpha(r)$ and $\beta(r)$? Explain.

(d) Show that $(\alpha\beta)(t) = (\beta\alpha)(t)$ for every $t \in \{1, 2, \ldots, n\}$ by considering cases.

Activity 25.23. Write a short summary that describes the important concepts, and the relationships between these concepts, that were introduced in this investigation. Explain how the material in this investigation is connected to the material in Investigations 19, 20, and 24.

Exercises

(1) Write $(3\ 1)(4\ 2)$ in S_5 in permutation notation.

(2) Write the permutation $\begin{pmatrix} 1 & 2 & 3 & 4 & 5 & 6 \\ 4 & 5 & 6 & 1 & 3 & 2 \end{pmatrix}$ as a product of disjoint cycles.

(3) Write the cycle $(1\ 2\ 4\ 3)$ as a product of transpositions.

(4) Write the given cycle as a product of transpositions. Identify if the permutation is even or odd.

(a) $(1\,2\,4\,3)$

(b) $(2\,4\,8\,9\,3)$

(5) A friend of yours gives you the permutation $\begin{pmatrix} 1 & 2 & 3 & 4 & 5 & 6 & 7 & 8 & 9 \\ 7 & 5 & 3 & & 9 & & 8 & 1 & 2 \end{pmatrix}$ in S_9 and tells you that the images of 4 and 6 are lost. Your friend recalls, however, that this permutation is even. What are the images of 4 and 6? Explain.

⋆ (6) Prove that for $n \geq 2$, S_n is non-Abelian.

(7) Find a subgroup of S_4 of order 8, or show that one does not exist.

(8) Find a subgroup of S_4 of order 6, or show that one does not exist.

(9) Let $n \in \mathbb{Z}^+$, and let k be a positive integer with $k \leq n$. What is the order of a k-cycle in S_n? Prove your result.

(10) Let n be an integer greater than 2. Prove that the center of S_n is $\{I\}$, where I is the identity permutation in S_n. (Hint: If α is a non-identity element in S_n, then there exist distinct integers i and j between 1 and n (inclusive) such that $\alpha(i) = j$.)

⋆ (11) Prove that the number of permutations of a set with n elements is $n!$. Based on this fact, what is the order of S_n?

(12) When is the cycle $(a_1\ a_2\ \cdots\ a_k)$ in S_n even, and when is it odd? Prove your answer.

(13) A fixed point of a function f is an input p so that $f(p) = p$. As functions, permutations can have fixed points.

(a) Determine all of the permutations in S_4 that have 2 as a fixed point. Let G_2 be this set of permutations. Is G_2 a subgroup of S_4?

(b) Let $n \in \mathbb{Z}^+$, and let k be a fixed integer between 1 and n. Define

$$G_k = \{\sigma \in S_n : \sigma(k) = k\}.$$

Prove that G_k is a subgroup of S_n. (This group G_k is called a **stabilizer subgroup** of S_n.)

(c) Can we generalize the result of the previous part to the following?

Let T be a nonempty subset of $\{1, 2, \ldots, n\}$ and define G_T as

$$G_T = \{\sigma \in S_n : \sigma(k) = k \text{ for all } k \in T\}.$$

Then G_T is a subgroup of S_n.

Give a proof or counterexample to justify your answer.

(d) Suppose $T \subset \{1, 2, \ldots, n\}$ contains m numbers. What do you expect the order of G_T to be? Explain.

(14) Let $n \geq 2$.

(a) Let α and β be disjoint cycles in S_n. Determine $|\alpha\beta|$ in terms of $|\alpha|$ and $|\beta|$.

(b) Let $\alpha_1, \alpha_2, \ldots, \alpha_m$ be disjoint cycles in S_n for some integer $m \geq 2$. Find and prove a formula similar to that from part (a) for $|\alpha_1\alpha_2 \cdots \alpha_m|$. (Hint: Exercise 4 on page 301 of Investigation 21 might be useful.)

(15) Let G be an Abelian group, and let $H = \{x^2 : x \in G\}$.

(a) Is H a subgroup of G? Prove your answer.

(b) In your proof in part (a), did you use the assumption that G is Abelian? Do you think it would be possible to prove this result without the assumption that G is Abelian? If not, then give a counterexample where G is non-Abelian and H is not a subgroup of G.

⋆ (16) The permutation α in S_n is **conjugate** to $\beta \in S_n$ if $\alpha = \sigma\beta\sigma^{-1}$ for some $\sigma \in S_n$.

(a) Find all of the permutations in S_3 that are conjugate to $(1\ 2\ 3)$.

(b) Let $\beta = (b_1\ b_2\ \ldots\ b_k)$ be a k-cycle in S_n, and let $\sigma \in S_n$. Show that

$$\sigma\beta\sigma^{-1} = (\sigma(b_1)\ \sigma(b_2)\ \cdots\ \sigma(b_k)).$$

Conclude that the conjugate of a k-cycle is a k-cycle.

(c) Let $\pi \in S_n$ and suppose π is written as the product

$$\pi = \pi_1\pi_2 \cdots \pi_m$$

of disjoint cycles. Let $\sigma \in S_n$. Show that

$$\sigma\pi\sigma^{-1} = \tau_1\tau_2 \cdots \tau_m,$$

where the τ_i are disjoint cycles and, for each i, τ_i is a cycle of the same length as π_i. Conclude that conjugation preserves the cycle structure of a permutation.

(17) Write the operation table for A_4. (See Activity 25.20.)

⋆ (18) Let $T = \langle \alpha, \beta \rangle$, where

$$\alpha = (1\ 2\ 3\ 4\ 5\ 6)(7\ 8\ 9\ 10\ 11\ 12) \quad \text{and} \quad \beta = (1\ 7\ 4\ 10)(2\ 12\ 5\ 9)(3\ 11\ 6\ 8)$$

in S_{12}. This group T has presentation $\langle s, t \mid s^6 = 1, s^3 = t^2, sts = t \rangle$.

(a) Find all of the elements of T, and construct the operation table for T.

(b) Show that T also has the presentation

$$\langle x, y \mid x^4 = y^3 = 1, yxy = x \rangle.$$

(19) Theorem 25.11 shows that every permutation in S_n is either a cycle or can be written as a product of disjoint cycles. Equation (25.2) then demonstrates how every cycle can be written as a product of transpositions. Thus, the transpositions generate S_n. However, it turns out that we do not need all of the transpositions to generate S_n. To see why this is the case, prove that the transpositions $(1\ 2)$, $(1\ 3)$, $(1\ 4)$, $\ldots$, $(1\ n)$ generate S_n.

(20) **Card shuffling.** Many magicians excel at card tricks. One important skill in card tricks is the perfect shuffle. A *shuffle* of a deck of $2n$ cards is obtained when the deck is split into two piles and then the cards from each pile are rearranged into one pile. A *perfect shuffle* occurs when the deck is split into two piles A and B of equal size and the cards are rearranged into

one pile by alternating cards from piles A and B. There are two types of perfect shuffles: the in-shuffle and the out-shuffle. The difference between the two shuffles is their effect on the top card; after an out-shuffle, the top card of the pre-shuffled deck remains on top, but an in-shuffle moves the top card of the pre-shuffled deck to the second position.

Label the top position in a deck as $\#1$, the second card from the top as $\#2$, and so on. A perfect in-shuffle can then be represented by the permutation

$$\begin{pmatrix} 1 & 2 & 3 & \cdots & n & n+1 & n+2 & \cdots & 2n \\ 2 & 4 & 6 & \cdots & 2n & 1 & 3 & \cdots & 2n-1 \end{pmatrix}.$$

(a) Determine the fewest number of perfect shuffles required to return a deck of 14 cards to its original position using only in-shuffles. Clearly explain your reasoning.

(b) Determine the fewest number of perfect shuffles required to return a deck of 52 cards to its original position using only in-shuffles. Explain.

(c) Represent a perfect out-shuffle in permutation notation.

(d) Determine the fewest number of perfect shuffles required to return a deck of 52 cards to its original position using only out-shuffles. Explain.

(21) **The Futurama Theorem.** In the Season 6 episode "The Prisoner of Benda" of the animated TV series *Futurama*, a machine is created that allows beings to swap minds. However, due to "cerebral immune response," the machine does not allow them to reverse the process. To solve the problem, the professor states "I'm afraid we need to use math!". The characters in the episode spend their time figuring out a way to get all minds back with their original bodies.

The solution to the problem faced in this episode is contained in the following theorem (due to *Futurama* writer Ken Keeler, who holds a Ph.D. in applied mathematics), which states that all mind switches can be undone by introducing two more characters into the situation.

Theorem (The Futurama Theorem). *Let A be a finite set, and let x and y be distinct objects that do not belong to A. Any permutation of A can be reduced to the identity permutation by applying a sequence of distinct transpositions of $A \cup \{x, y\}$, each of which includes just one of x, y.*

(a) Since the group of permutations of any set with n elements can be identified with the permutations of the set $\{1, 2, \ldots, n\}$, we can assume that our set A is the set $\{1, 2, \ldots, n\}$. We will first prove the Futurama Theorem for cycles.

(i) Assume we have a k-cycle σ that indicates mind swaps between k individuals (that is, individual i's mind is in individual $\sigma(i)$'s body). We can relabel to assume, without loss of generality, that

$$\sigma = (1\ 2\ 3\ \cdots\ k).$$

Explain how the Futurama Theorem states that we can find a permutation $\gamma \in S_{n+2}$ that is a product of transpositions, each of which permutes just one of $n+1$ or $n+2$, so that $\sigma\gamma = I$.

(ii) Suppose we introduce two new characters x and y. Without loss of generality, we can let $x = n+1$ and $y = n+2$. For each i from 1 to k, let

$$\begin{aligned} \tau_i = &(n+1\ \ 1)(n+1\ \ 2)\cdots(n+1\ \ i) \\ &(n+2\ \ i+1)(n+2\ \ i+2)\cdots(n+2\ \ k) \\ &(n+1\, i+1)(n+2\, 1). \end{aligned}$$

Consider σ as a permutation in S_{n+2}. What permutation is $\sigma\tau_i$? Explain.

(iii) Explain how we can reverse the mind swaps defined by the k-cycle σ so that all minds are back in their original bodies.

(b) Now prove the Futurama Theorem for any permutation.

Connections

A symmetry of a set is a bijection from the set to itself. Every symmetry of an object (as described in Investigation 19) is also a bijection from the object to itself. If we think of an object as a set of points, then every symmetry of the object corresponds to a symmetry of the set of points that represents it. In this sense, we can identify the collection of symmetries of an object with a subgroup of the symmetries of the set of points of the object. In particular, we can identify the dihedral group D_n, the set of symmetries of a regular n-gon, with a subset of S_n corresponding to permutations of the vertices of the n-gon. (We will make this identification more specific after Investigation 29 on group isomorphisms.) It is important to note, however, that not every symmetry of a set that defines an object is a symmetry of the object itself. By noting the difference between the orders of D_n and S_n, you should be able to come up with an example to convince yourself of this.

Investigation 26

Cosets and Lagrange's Theorem

Focus Questions

By the end of this investigation, you should be able to give precise and thorough answers to the questions listed below. You may want to keep these questions in mind to focus your thoughts as you complete the investigation.

- What are left and right cosets?
- What are some important properties of cosets? How do cosets partition the groups on which they are defined?
- What important fact does Lagrange's Theorem tell us about subgroups of finite groups?
- What are some of the consequences and corollaries of Lagrange's Theorem?

Preview Activity 26.1. In Investigation 23, we classified all of the subgroups of finite cyclic groups. In particular, we learned in Theorem 23.7 (see page 321) that in a finite cyclic group of order n, there is exactly one subgroup of order m for each positive divisor m of n, and these are the only subgroups. It is natural to ask if Theorem 23.7 generalizes to *all* finite groups. As we will see, part of it does and part of it does not.

(a) Find all of the non-trivial, proper subgroups of D_3. How are the orders of these subgroups related to the order of D_3?

(b) For each divisor m of $|D_6|$, does D_6 contain a subgroup of order m? If yes, exhibit a subgroup for each divisor. If no, explain why.

Introduction

The subgroup structure of a group tells us a lot about the group. Theorem 23.7 completely classified the subgroup structure of finite cyclic groups, but Activity 26.1 shows us that this theorem does not generalize to arbitrary finite groups. For one thing, the group D_3 contains more than one subgroup of order 2. On the other hand, the order of every subgroup of D_3 does divide the order of D_3, and both D_3 and D_4 contain subgroups of each order that divides the order of the larger group. But what can we say in general? In other words, exactly what we can say about the subgroups of an arbitrary finite group? This is an important question, and one that we will begin to answer throughout this investigation.

To uncover the underlying subgroup structure of a group, we often use counting arguments. Such arguments usually involve arranging the elements of a group into disjoint subsets so that we can count the elements without overlap. Throughout this and subsequent investigations, we will use counting arguments to prove important theorems (such as Lagrange's Theorem and the Sylow theorems) and to derive other critical tools (such as the Class Equation). In this investigation, we will introduce one of the more basic counting tools: the coset decomposition of a group. We will use cosets to prove Lagrange's Theorem, which is one of the most important theorems in finite group theory. In the next investigation (Investigation 27), we will also see how different types of subgroups lead to different coset structures, an observation that will motivate our study of normal subgroups and quotient groups.

A Relation in Groups

In Investigation 5, we constructed the set $\mathbb{Z}_n$, which we defined in terms of the congruence modulo n relation on $\mathbb{Z}$. Recall that if n is a positive integer and $a, b \in \mathbb{Z}$, we say that a is congruent to b modulo n if $b - a$ is a multiple of n.* We can rephrase this idea in terms of the subgroup $n\mathbb{Z} = \{nk : k \in \mathbb{Z}\}$ and say that $a \equiv b \pmod{n}$ if $b - a \in n\mathbb{Z}$. The advantage of this perspective is that we can apply it to other groups, as the next activity suggests.

Activity 26.2. Let H be the subset of $\mathbb{Z}_8$ defined by $H = \langle [4] \rangle = \{[0], [4]\}$. Define a relation $\sim_H$ on $\mathbb{Z}_8$ by

$$a \sim_H b \text{ if and only if } b - a \in H.$$

(a) Find all of the elements in $\mathbb{Z}_8$ that are related to $[2]$.

(b) For each $g \in \mathbb{Z}_8$, find all of the elements in $\mathbb{Z}_8$ related to g using the relation $\sim_H$.

(c) Let $[g]$ denote the set of elements in $\mathbb{Z}_8$ related to g. List three different things you notice about the collection of sets $[g]$ for $g \in \mathbb{Z}_8$.

The relations we defined on $\mathbb{Z}$ in Investigation 5, and on $\mathbb{Z}_8$ in Activity 26.2, can be extended to arbitrary groups and subgroups. The result of Activity 26.2 suggests that we might expect to be able to partition any group into a disjoint union of classes using a similar relation, in much the same way we partitioned $\mathbb{Z}$ into a disjoint union of congruence classes. Partitioning groups into classes will turn out to be a very important tool for us. Before we proceed to the general definition, however, we need to stop for a moment to talk about notation. Note that since the group operation for both $\mathbb{Z}$ and $\mathbb{Z}_8$ is addition, we used additive notation to define a congruence relation on these groups. We usually assume that an additive operation is commutative, so we could write $b - a$ as $(-a) + b$, which will be more convenient in what follows. When we use multiplicative notation, we replace $(-a) + b$ with $a^{-1}b$. This brings us to the next definition.

*In Investigation 5, we defined a to be congruent to b modulo n if $a - b$ (rather than $b - a$) was a multiple of n. Because of the symmetry of the congruence relation, these two definitions are equivalent. However, the version that uses $b - a$ will be more convenient for the purposes of working with cosets.

Definition 26.3. Let G be a group and H a subgroup of G. Let $\sim_H$ be the relation on G such that for all $a, b \in G$,

$$a \sim_H b \quad \text{if and only if} \quad a^{-1}b \in H.$$

For example, let $G = S_3$ and $H = \langle(1\ 2)\rangle$. Since $(1\ 3)^{-1}(1\ 2\ 3) = (1\ 2)$, the definition shows us that $(1\ 3) \sim_H (1\ 2\ 3)$. The groups G and H will usually be clear from the context, so we will normally suppress the subscript and just write $\sim$ instead of $\sim_H$.

Activity 26.4. Let $G = S_3$ and $H = \langle(1\,2)\rangle$. Find all of the elements α in S_3 satisfying $\alpha \sim (1\,2\,3)$.

Whenever we have a relation, it is natural to ask what properties the relation satisfies.

Activity 26.5. Let G be a group with identity e and H a subgroup of G.

(a) Prove that $\sim$ is a reflexive relation on G.

(b) Prove that $\sim$ is a symmetric relation on G.

(c) Prove that $\sim$ is a transitive relation on G.

Recall that any relation that is reflexive, symmetric, and transitive is an *equivalence relation*. In Investigation 5, we learned that every equivalence relation partitions the underlying set into disjoint equivalence classes. (See Theorem 5.6 on page 49.) We will exploit this important property throughout the next sections.

Cosets

Let G be a group, let H be a subgroup of G, and let $\sim$ be the relation defined in the preceding section. It turns out that the equivalence classes corresponding to $\sim$ have a special form and can be easily described. Let $a \in G$, and choose $b \in [a]$ (where $[a]$ denotes the equivalence class of a under $\sim$). Then $a \sim b$. So $a^{-1}b \in H$ and $a^{-1}b = h$ for some $h \in H$. Thus, $b = ah$. If we let $aH = \{ah : h \in H\}$, then it follows that $[a] \subseteq aH$. Conversely, let $b \in aH$. Then $b = ah'$ for some $h' \in H$. So $a^{-1}b = h' \in H$ and $a \sim b$, which implies $b \in [a]$. Thus, $aH \subseteq [a]$, and so we have shown that $[a] = aH$. We can therefore recognize the equivalence classes under the relation $\sim$ as the sets of the form aH. These sets play an important role in counting techniques in finite group theory and are given a special name.

Definition 26.6. Let G be a group and H a subgroup of G. Let $g \in G$. The **left coset**[†] of H in G containing g is the set

$$gH = \{gh : h \in H\}.$$

Similarly, the **right coset** of H in G containing g is the set

$$Hg = \{hg : h \in H\}.$$

The element g is called the coset representative of gH (or Hg).

[†]The idea of a coset seems to have been first used by Evariste Galois in his paper *Mémoir sur les conditions de résolubilité des équations par redicaux* in 1830. The term *coset* was first applied by G.A. Miller in the Quarterly Journal of Mathematics in 1910. The word coset appears to have literally meant "co-set" to Miller (in the same vein that a word like co-pilot is used) and replaced the previous term *Nebengruppen* for this idea.

The sets in Definition 26.6 are written using multiplicative notation. If we were using additive notation instead, we would write $a + H$ for a left coset rather than aH.

Since the equivalence classes under $\sim$ are exactly the left cosets of H in G, we have the following result.

Theorem 26.7. *Let G be a finite group and H a subgroup of G.*

(i) *If a and b are in G, then $aH = bH$ or $aH \cap bH = \emptyset$.*

(ii) *The group G can be written as a disjoint union of left cosets of H.*

Theorem 26.7 is important in that once we have partitioned a finite group into a disjoint union of left cosets, we can count the elements in the group by simply summing the elements in the cosets, without having to worry about repeated elements. We will see several applications of this type of partition in what follows. It should be noted at this point that everything we have done with left cosets can be replicated for right cosets as well.

Some additional examples are in order.

Activity 26.8.

(a) Find all the left and right cosets of $H = \langle(1\ 2)\rangle$ in $G = S_3$.

(b) Find all the left and right cosets of $H = \langle(1\ 2\ 3)\rangle$ in $G = S_3$.

(c) If G is any group and H is a subgroup of G, must the left coset of H containing $g \in G$ be the same as the right coset of H in G containing g? Explain.

Note that, in general, a left coset of G is NOT a subgroup of G.

Activity 26.9. In each of the cases below, write the group G as a disjoint union of left cosets of the subgroup H.

(a) $G = \mathbb{Z}_{16}$, $H = \langle[4]\rangle$

(b) $G = S_3$, $H = \langle(1\ 2)\rangle$

Lagrange's Theorem

Lagrange's Theorem, which was hinted at in Preview Activity 26.1, is one of the most important theorems in all of finite group theory. In this section, we will discover what Lagrange's Theorem tells us about the subgroups of a finite group. Throughout this section, we will let G be a group of finite order and H a subgroup of G. For any finite set X, we will denote by $|X|$ the number of elements in X.

Preview Activity 26.10. To prove Lagrange's Theorem, we need to formalize an observation we have made about the number of elements in each left coset. Let $a \in G$. There is a natural function $\varphi : aH \to H$ defined by $\varphi(ah) = h$.

(a) Is φ an injection? Prove your answer.

(b) Is φ a surjection? Prove your answer.

(c) Is φ a bijection?

(d) What do your answers to parts (a) – (c) tell us about the number of elements in aH and in H? Explain.

You have probably noticed by now that in all of the examples we have seen, when G has been a finite group and H a subgroup of G, the order of H has been a divisor of the order of G. Lagrange's Theorem says that this is always true.

Theorem 26.11 (Lagrange's Theorem). *If G is a finite group and H is a subgroup of G, then the order of H divides the order of G.*

To understand why Lagrange's Theorem holds, we need to connect the result of Preview Activity 26.10 to the partition of G determined by the relation $\sim$. Recall that the relation $\sim$ on G defined by $a \sim b$ if and only if $a^{-1}b \in H$ is an equivalence relation on G that partitions G into a disjoint union of left cosets.

Activity 26.12.

(a) Suppose G is partitioned into m distinct equivalence classes with representatives $a_1, a_2, \ldots, a_{m-1}, a_m$. Fill in the blanks to write G as a disjoint union of left cosets.

$$G = \_\_\_ \cup \_\_\_ \cup \_\_\_ \cup \_\_\_ \quad \cdots \quad \cup \_\_\_ \tag{26.1}$$

(b) What does Preview Activity 26.10 tell us about how the number of elements in each left coset is related to $|H|$?

(c) You may have learned at one point that if X and Y are disjoint finite sets, then $|X \cup Y| = |X| + |Y|$. Use this fact and your response to parts (a) and (b) to write an equation that relates $|G|$ to $|H|$.

(d) Explain how your response to part (c) proves Lagrange's Theorem.

Lagrange's Theorem provides important information about the subgroup structure of a group and tells us that a finite group of order n cannot have a subgroup whose order does not divide n. We must be careful though, since **Lagrange's Theorem does NOT say that if m is a divisor of $|G|$, then G contains a subgroup of order m.** This conclusion is FALSE. (See Activity 26.20.) Lagrange's Theorem tells us only about the *possible* orders of subgroups of a finite group, but it does not tell us that subgroups of these possible orders must exist. For all we know at this point, it may be possible for a group of order 10 to have no subgroups of order 2 or 5. The first corollary to Lagrange's Theorem shows that that cannot happen. There are also other theorems that we will encounter in later investigations that will tell us something more about the existence of certain types of subgroups of any group.

Corollary 26.13. *Let G be a group of order n with $n > 1$. Then there is a prime integer p such that G contains a subgroup of order p.*

Proof. Let G be a group of order $n > 1$ with identity e. Choose an element $a \neq e$ in G. Then $|a| > 1$. Since $|G|$ is finite, Lagrange's Theorem tells us $|a| = |\langle a \rangle|$ divides n, and $|a|$ is finite. Let $m = |a|$. Let p be a prime factor of m and let $k \in \mathbb{Z}$ so that $pk = m$. Theorem 23.6 (see page 320) then shows that

$$|a^k| = \frac{m}{\gcd(m, k)} = \frac{m}{k} = p,$$

and so G contains an element a of order p. Therefore, $\langle a^k \rangle$ is a subgroup of G of order p. ■

We should be cautious about Corollary 26.13. This corollary does not say that if a prime divides the order of a finite group G, then G contains a subgroup of order p—only that there is a subgroup of G of *some* prime order. For example, we can conclude that a group of order 10 contains either a subgroup of order 2 or a subgroup of order 5, but we can't say which one. Furthermore, such a group may contain subgroups of both orders (as in $\mathbb{Z}_{10}$), but Corollary 26.13 does not guarantee this.

Another simple, yet important corollary of Lagrange's Theorem is the following:

Corollary 26.14. *Let G be a finite group with identity element e. Then $a^{|G|} = e$ for every $a \in G$.*

Proof. Let G be a finite group of order n with identity element e and let $a \in G$. Lagrange's Theorem shows that $|a| = |\langle a \rangle|$ divides n, so there exists an integer m such that $|a|m = n$. Then

$$a^n = a^{m|a|} = \left(a^{|a|}\right)^m = e^m = e.$$

■

The exercises will provide several additional examples of the usefulness of Corollary 26.14

Activity 26.12 shows us that if m is the number of left cosets of a subgroup H in a finite group G, then $m|H| = |G|$. This number of left cosets of H in G is given a name:

Definition 26.15. Let G be a group and H a subgroup of G. The **index** of H in G is the number of distinct left cosets of H in G.

We denote the index of H in G as $[G : H]$. If G is a finite group and H a subgroup of G, Activity 26.12 tells us that $[G : H] = \frac{|G|}{|H|}$. It should be noted though that Definition 26.15 also applies to infinite groups. For more details, see Exercise 17.

Activity 26.16. In each of the cases below, find $[G : H]$. Refer to Activity 26.9.

(a) $G = \mathbb{Z}_{16}$, $H = \langle [4] \rangle$

(b) $G = S_3$, $H = \langle (1\ 2) \rangle$

We will conclude this investigation with an application of Lagrange's Theorem to groups of prime order. In previous investigations, we have seen that any group of order 2 or 3 is cyclic (see Exercise 15 in Investigation 22), but that this is not true of groups of order 4. (Can you think of an example of a non-cyclic group of order 4?) A major goal in finite group theory is to classify all groups of a given type. While this problem is far from completed, much progress has been made. Lagrange's Theorem is a key tool in the classification process, and the next activity illustrates a small portion of its power in this regard.

Activity 26.17.

(a) Let G be a group of order p, where p is a prime, and let H be a subgroup of G. What does Lagrange's Theorem tell us must be true of H? Explain. How many subgroups does G have?

(b) Let G be a group of order p, where p is a prime, and let $a \in G$. What must be true of $|a|$? Explain. Must G be Abelian? Must G be cyclic? Explain. What can we conclude about every group of prime order?

Concluding Activities

Activity 26.18. Write a formal proof of Lagrange's Theorem (Theorem 26.11).

Activity 26.19. You might be wondering why we focus on the relation $\sim_H$ instead of some other relation. In general, a relation just partitions the underlying set without concern for any structure the set possesses. With groups, however, there is a structure that we want to have preserved. An equivalence relation on a group that preserves the group structure as well is called a *congruence relation* (a generalization of the relation called congruence). In other words, an equivalence relation $\sim$ on a group G is a **congruence relation** if whenever $a, b, c, d \in G$ so that $a \sim b$ and $c \sim d$, then

(i) $ac \sim bd$ and

(ii) $a^{-1} \sim b^{-1}$.

As we will see in this activity, there is only one type of congruence relation on a group, and that is why we focus on the relation $\sim_H$. Let G be a group with identity e, and assume that $\sim$ is a congruence relation on G.

(a) Let $H = \{x \in G : x \sim e\}$. Show that H is a subgroup of G.

(b) Let $a, b \in G$. Prove that $a \sim b$ if and only if $a^{-1}b \in H$.

(c) Explain why $\sim_H$ is the only possible congruence relation on a group G. (Note that this does not say that $\sim_H$ is always a congruence relation on G. We will address that issue in Investigation 27.)

Activity 26.20. The Converse of Lagrange's Theorem. Lagrange's Theorem tells us that the order of any subgroup H of finite group G must divide the order of G. We did not prove the converse of Lagrange's Theorem for the simple reason that it is not true. In this activity, we'll see why this is the case.

(a) State the converse of Lagrange's Theorem. What do we need to do to show that the converse of Lagrange's Theorem is not true?

(b) Consider the group $G = A_4$. List the elements of A_4 in cycle notation, and determine the order of A_4.

(c) Assume that H is a subgroup of A_4 of order 6.

  (i) Explain why the non-identity elements of H must have order 2 or 3.

  (ii) Explain why there must be an element α of A_4 of order 3 that is not in H.

  (iii) Explain why the left cosets H, αH and $\alpha^2 H$ cannot all be distinct.

  (iv) Show that it is not possible for any two of H, αH and $\alpha^2 H$ to be equal.

(d) Explain why the converse of Lagrange's Theorem is not true.

Activity 26.21. Write a short summary that describes the important concepts, and the relationships between these concepts, that were introduced in this investigation. Explain how the material in this investigation is connected to the material in Investigation 20 and 22.

Exercises

(1) Suppose a classmate told you that there is a subgroup of Z_{15} of order 10. What would your response be? Explain as clearly and completely as possible, citing all relevant theorems.

(2) Let G be a group and H a subgroup of G.

(a) If $a \in G$ and $aH = H$, what must be true about a? Prove your answer.

(b) If $a \in H$, what coset is aH? Prove your answer.

(c) Correctly complete the statement of the following theorem

Theorem. *Let G be a group and H a subgroup of G. If $a \in G$, then $aH = H$ if and only if ________.*

(3) Let G be a group of order n. If H is a subgroup of G and $|H| \geq \frac{n}{2} + 1$, what group must H be? Prove your answer.

(4) A group G contains elements of every order from 1 to 10. What is the smallest order G could have? Find a group G of that order that contains elements of every order from 1 through 10.

(5) For each of the examples given below, list the distinct cosets of H in G.

(a) $G = \mathbb{Z}_{12}$, $H = \langle [4] \rangle$

(b) $G = U_{15}$, $H = \langle [11] \rangle$

(6) Let $H = \{I, r\}$ in D_4.

(a) Determine all of the distinct left cosets of H in D_4.

(b) Determine all of the distinct right cosets of H in D_4.

(7) Let G be a group, and let $a \in G$ be an element of order 21. Let $H = \langle a \rangle$ and $K = \langle a^7 \rangle$.

(a) How many left cosets are there of K in H?

(b) Find all of the left cosets of K in H.

(8) Let G be a group of order 840, and suppose K is a subgroup of G of order 42. If H is a subgroup of G that contains K as a subgroup, what could the order of H be? Explain.

(9) Let $G = \langle a \rangle$ be a cyclic group of order n. If k is an integer between 1 and n, what is the index of $\langle a^k \rangle$ in G? Prove your answer.

(10) The subgroup structure of a group tells us much about the group. For example, there are groups that have no non-trivial proper subgroups (subgroups of order larger than 1 that are not the entire group), and these groups have a very simple structure.

(a) Suppose G is a finite group that contains no non-trivial proper subgroups. What can you say about $|G|$? Prove your answer.

(b) What conditions are there on the order of a finite group G that ensure that G contains a non-trivial proper subgroup? Prove your answer.

(c) Correctly complete the statement of the following theorem

Theorem. *A group G of order n has a non-trivial proper subgroup if and only if n is* ________.

(11) **Subgroups of p-groups.** A group G is called a p-group if $|G| = p^k$ for some prime p and some integer $k \geq 1$. For example, the groups D_4, $\mathbb{Z}_8$, U_{15}, U_{24}, and $\mathbf{Q}$ from Exercise 6 in Investigation 24 (see page 331) are all 2-groups of order 8. In this exercise, we will examine subgroups of p-groups in more detail.

(a) Exercise 22 of Investigation 22 (see page 314) shows that any 2-group must contain an element of order 2. Must a p-group of order p^k contain elements of order p^t for all $1 \leq t \leq k$? Explain.

(b) Can we generalize Exercise 22 of Investigation 22 to show that a p-group must contain an element of order p? Prove your answer.

(12) Let G be a finite group with identity e, and let H and K be subgroups of G.

(a) If $|H| \neq |K|$, must $H \cap K = \{e\}$? Prove your answer.

(b) Is there any condition on $|H|$ and $|K|$ that ensures that $H \cap K = \{e\}$? Prove your conjecture. (Hint: If $a \in H \cap K$, what does Corollary 26.14 tell us?)

(13) Over the real numbers, we are able to solve all equations of the form $x^m = b$ for x if m is a positive integer. A solution to the equation $x^m = b$ is called an mth *root* of b.

(a) If G is a finite group, $b \in G$, and $m \in \mathbb{Z}^+$, can we always solve the equation $x^m = b$ in G? That is, does every group contain all m^{th} roots of each of its elements? Prove your answer.

(b) There are certain situations in which a finite group must contain some of the mth roots of its elements. We will explore that situation in this part of the exercise. Let G be a group of order n with identity e.

(i) If $g \in G$ and $g^m = e$ for some $m \in \mathbb{Z}^+$ with $\gcd(m, n) = 1$, show that $g = e$. (Hint: Use Corollary 26.14 on page 352.)

(ii) Let $m \in \mathbb{Z}^+$ so that $1 = rn + sm$ for some integers r and s. If $g \in G$, show that g^s is an mth root of g. Conclude that if $\gcd(m, n) = 1$, then G contains an mth root for each of its elements.

(iii) Illustrate the process in the previous part to find a 3rd root of the element $[4]$ in U_{11}.

(c) A non-trivial Abelian group G is said to be **divisible** if G contains an mth root for each of its elements. That is, G is divisible if for each $a \in G$ and each nonzero $m \in \mathbb{Z}$, there exists $x \in G$ so that $x^m = a$.

(i) Is the group $\mathbb{Q}$ of rational numbers divisible? Prove your answer.

(ii) Is the group $U(\mathbb{Q})$ (the group of units in $\mathbb{Q}$, which consists of all nonzero rational numbers) divisible? Prove your answer.

(iii) Can a finite Abelian group be divisible? Prove your answer.

(14) (a) Let G be a group of order 4 with identity e. Show that G is either cyclic or $a^2 = e$ for all $a \in G$.

(b) Does the result of part (a) generalize to groups of order p^2 for any positive integer p? In other words, is it the case that if G is a group of order p^2 with identity e, then G is either cyclic or $a^p = e$ for every $a \in G$. Prove your answer.

(c) Is there any condition on p that will make the answer to the question in part (b) yes? If so, state and prove a conjecture. If no, explain why.

(d) Does the result of part (a) generalize to groups of order 2^k for any positive integer $k > 2$? That is, if G is a group of order 2^k for some $k \in \mathbb{Z}$ with $k > 2$, must it be the case that either G is cyclic or $a^2 = e$ for every $a \in G$? Prove your answer.

⋆ (15) Show that every group of order 4 is Abelian. Must every group of order 4 be cyclic? Explain. (Hint: Use part (a) of Exercise 14 on page 355.)

(16) Let n be a positive integer. Let $H = n\mathbb{Z} = \{nk : k \in \mathbb{Z}\}$ (that is, H is the set of all integer multiples of n). Find all left and right cosets of H in $\mathbb{Z}$. What is $[\mathbb{Z} : n\mathbb{Z}]$?

⋆ (17) If G is a finite group and H a subgroup of G, then Lagrange's Theorem tells us about $[G : H]$. But Definition 26.15 also applies to infinite groups.

(a) Is it possible to find an infinite group G and subgroup H of G so that $[G : H]$ is finite? If yes, find and explain such an example. If no, explain why.

(b) Is it possible to find an infinite group G and subgroup H of G so that there are infinitely many different left cosets of H in G? (In this case we would say that $[G : H]$ is infinite.) If yes, find and explain such an example. If no, explain why.

(18) Let G be a group and H a subgroup of G. We have indicated that the number of left cosets of H in G is equal to the number of right cosets of H in G. Prove that statement. (Hint: Consider the function f that maps the left coset aH of H in G to the right coset Ha^{-1}.)

(19) Let $G = D_n$, the group of symmetries of a regular n-gon.

(a) Determine the distinct left cosets of $H = \langle R \rangle$ in G.

(b) Use the result of (a) to explain why $D_n = \{r^i R^j : 0 \leq i \leq 1, 0 \leq j \leq n-1\}$.

⋆ (20) A famous theorem in number theory is Fermat's Little Theorem. (We will use this theorem in Investigation 34 to prove the validity of RSA encryption.)

Theorem (Fermat's Little Theorem). *Let p be a prime number. If $a \in \mathbb{Z}$, then*

$$a^p \equiv a \bmod p.$$

Prove Fermat's Little Theorem. (Hint: Use the result of Corollary 26.14 in the group U_p.)

(21) The Euler phi function φ is defined to give the number of positive integers less than or equal to and relatively prime to a given positive integer. For example, $\varphi(2) = 1$ since 1 is the only positive integer less than 2 and relatively prime to 2. Similarly, $\varphi(3) = 2$, $\varphi(4) = 2$, $\varphi(5) = 4$, and so on. There is a generalization of Fermat's Little Theorem (from Exercise 20) due to Euler:

Theorem. *Let n be a positive integer and $a \in \mathbb{Z}$ with $\gcd(a, n) = 1$. Then*

$$a^{\varphi(n)} \equiv 1 \pmod{n}.$$

Prove this theorem. Then explain why this theorem is a generalization of Fermat's Little Theorem. (Hint: Use the result of Corollary 26.14 in the group U_n.)

(22) If G is a finite group and H and K are subgroups of G with K a subgroup of H, then Lagrange's Theorem tells us that $[G : H] = \frac{|G|}{|H|}$, $[H : K] = \frac{|H|}{|K|}$, and $[G : K] = \frac{|G|}{|K|}$. So in this case we have $[G : K] = [G : H][H : K]$. If G is an infinite group, we cannot apply Lagrange's Theorem to obtain this index property. In this exercise, we will determine if $[G : K] = [G : H][H : K]$ when G is infinite and H and K are subgroups of a group G so that $K \subset H$ and $[G : H]$ and $[H : K]$ are finite.

(a) Let $G = \mathbb{Z}$, $H = 4\mathbb{Z}$ and $K = 12\mathbb{Z}$.

(i) Let $a_1 + H, a_2 + H, \ldots, a_n + H$ be the collection of distinct left cosets of H in G. Find n and a representative a_i for each coset. What is $[G : H]$?

(ii) Let $b_1 + K, b_2 + K, \ldots, b_m + K$ be the collection of distinct left cosets of K in H. Find m and a representative b_j for each coset. What is $[H : K]$?

(iii) Find all of the left cosets of K in G in terms of the a_i and b_j. What is $[G : K]$? Is $[G : K] = [G : H][H : K]$?

(b) Let G be an infinite group, and let H and K be subgroups of G such that $K \subset H$ and both $[G : H]$ and $[H : K]$ are finite. Is it true that $[G : K]$ is finite and $[G : K] = [G : H][H : K]$? Prove your answer.

Connections

Given a subgroup H of a group G, we showed in this investigation that congruence modulo H is an equivalence relation. As we will see in subsequent investigations, this relation will lead us to the useful construction of quotient structures. We are familiar with other notions of congruence in groups; in particular, congruence modulo an integer n led us to the group $\mathbb{Z}_n$ in Investigation 5. If you studied ring theory before group theory, you should notice connections between the topics in this investigation and those in Investigations 15 and 16. In particular, congruence modulo a subgroup is the same idea as congruence modulo a polynomial in a polynomial ring $F[x]$, or congruence modulo an ideal I in a ring R.

Investigation 27

Normal Subgroups and Quotient Groups

Focus Questions

By the end of this investigation, you should be able to give precise and thorough answers to the questions listed below. You may want to keep these questions in mind to focus your thoughts as you complete the investigation.

- What is a normal subgroup of a group? Why are normal subgroups important?
- Under what conditions is the quotient of a group by a subgroup a group itself?
- What does Cauchy's Theorem say about the subgroups that a finite Abelian group must have?
- What is a simple group, and what are some examples of simple groups?

Preview Activity 27.1. Recall that if G is a group and H is a subgroup of G, the cosets of H partition G into disjoint equivalence classes. In this activity, we will investigate what these cosets might look like.

(a) Table 27.1 presents the operation table for a group G of order 6. (Note that we have used the letters a through f to denote the elements of this group, but we are not saying that e is the identity of the group.) We are going to rearrange the elements of G so that the operation table for G allows us to observe more easily what some of the cosets in G look like. The table on the right will eventually be the rearranged table for G, but we will replace each of the labels g_1, g_2, g_3, g_4, g_5, and g_6 with the elements a, b, c, d, e, and f from the group G.

	a	b	c	d	e	f
a	b	d	f	a	c	e
b	d	a	e	b	f	c
c	f	e	a	c	d	b
d	a	b	c	d	e	f
e	c	f	d	e	b	a
f	e	c	b	f	a	d

	g_1	g_2	g_3	g_4	g_5	g_6
g_1						
g_2						
g_3						
g_4						
g_5						
g_6						

Table 27.1
Operation table for a group G of order 6 (left) and a rearrangement of the table (right).

To begin, notice that f is an element of order 2 in this group. Let the element g_1 in the table on the right be f and let $H = \langle f \rangle$ be the subgroup of G generated by f. Let the element g_2 in the table on the right be the other element in $\langle f \rangle$ (which will be d). The table on the right now appears as shown in Table 27.2.

	a	b	c	d	e	f
a	b	d	f	a	c	e
b	d	a	e	b	f	c
c	f	e	a	c	d	b
d	a	b	c	d	e	f
e	c	f	d	e	b	a
f	e	c	b	f	a	d

	f	d				
f	d	f				
d	f	d				

Table 27.2
Operation table for a group G of order 6 (left) and a rearrangement of the table (right).

(i) We know that H is one left coset of H in G, and now we will find the others. Choose any element $g_3 \in G$ so that $g_3 \notin H$. (That is, choose g_3 to be either a, b, c, or e.) Then replace g_3 in the table on the right with the element you chose. List the elements in g_3H, and replace g_4 with the remaining element in g_3H.

(ii) There is an element $g_5 \in G$ with $g_5 \notin H \cup g_3H$. Replace g_5 and g_6 in the table on the right with the elements in g_5H. What is the relationship between H, g_3H, and g_5H?

(iii) Now fill in the operation table for G writing the elements in the order you used when you re-labeled the table on the right in parts (i) and (ii). Color the cells in this table that correspond to elements in H with the color blue (or any other color you choose). Then color the cells that correspond to elements in g_3H with a different color. Finally, select a third color, and color the cells that correspond to elements in g_5H with that third color. What do you notice about this table if you ignore the labels and just focus on the colors?

(b) Repeat part (a) with the group G whose operation table is shown in Table 27.3. That is, let $H = \langle g_1 \rangle$ be a subgroup of G of order 2, and then: (i) find the distinct left cosets H, g_3H, and g_5H of H in G; (ii) complete Table 27.3 at right; and (iii) color-code the cells of the table corresponding to distinct cosets of H. What do you notice? Compare your observations to those from part (a).

Introduction

Let G be a group and H a subgroup of G. Preview Activity 27.1 seems to indicate that in some cases the collection of distinct cosets of H in G has a group structure, and in other cases it does not.

	a	b	c	d	e	f
a	a	b	c	d	e	f
b	b	a	e	f	c	d
c	c	f	a	e	d	b
d	d	e	f	a	b	c
e	e	d	b	c	f	a
f	f	c	d	b	a	e

	g_1	g_2	g_3	g_4	g_5	g_6
g_1						
g_2						
g_3						
g_4						
g_5						
g_6						

Table 27.3
Operation table for a second group G of order 6 (left) and a rearrangement of the table (right).

In this investigation, we will explore the structure of the set of distinct left cosets of a subgroup of a group.

An Operation on Cosets

Let G be a group (with the operation written multiplicatively), H a subgroup of G, and $g \in G$. Recall that the left coset of H in G containing g is the set

$$gH = \{gh : h \in H\}.$$

We will denote by G/H the collection of all distinct left cosets of the subgroup H in the group G. We want to understand when the set G/H can be made into a group. To identify a group structure on G/H, we will first need to define an operation on G/H. There is a natural way to try to define such an operation. In particular, for any elements aH, bH in G/H, we could define the product $(aH)(bH)$ as

$$(aH)(bH) = (ab)H. \tag{27.1}$$

Activity 27.2. One question we must address with the operation on G/N proposed in (27.1) is whether or not the operation is well-defined.

(a) What would it mean for the operation from (27.1) to be well-defined on G/H? Be specific. In general, why is it important for operations to be well-defined?

(b) Let G be the 12 element group in Table 27.4. (You may assume that G is a group.) Let $H = \{I, a_9, a_{10}, a_{11}\}$ and $K = \{I, a_3, a_4\}$. It can be shown that H and K are subgroups of G.

(i) Find the distinct left cosets of H in G. What is G/H?

(ii) Find the right cosets of H in G. How are the left and right cosets of H in G related?

(iii) Find the distinct left cosets of K in G. What is G/K?

(iv) Find the right cosets of K in G. What do you notice about the left and right cosets of K in G?

	I	a_1	a_2	a_3	a_4	a_5	a_6	a_7	a_8	a_9	a_{10}	a_{11}
I	I	a_1	a_2	a_3	a_4	a_5	a_6	a_7	a_8	a_9	a_{10}	a_{11}
a_1	a_1	a_2	I	a_{10}	a_6	a_7	a_{11}	a_9	a_3	a_5	a_8	a_4
a_2	a_2	I	a_1	a_8	a_{11}	a_9	a_4	a_5	a_{10}	a_7	a_3	a_6
a_3	a_3	a_{11}	a_5	a_4	I	a_{10}	a_8	a_1	a_9	a_6	a_2	a_7
a_4	a_4	a_7	a_{10}	I	a_3	a_2	a_9	a_{11}	a_6	a_8	a_5	a_1
a_5	a_5	a_3	a_{11}	a_9	a_7	a_6	I	a_{10}	a_2	a_1	a_4	a_8
a_6	a_6	a_9	a_8	a_1	a_{10}	I	a_5	a_4	a_{11}	a_3	a_7	a_2
a_7	a_7	a_{10}	a_4	a_5	a_9	a_{11}	a_1	a_8	I	a_2	a_6	a_3
a_8	a_8	a_6	a_9	a_{11}	a_2	a_3	a_{10}	I	a_7	a_4	a_1	a_5
a_9	a_9	a_8	a_6	a_7	a_5	a_4	a_2	a_3	a_1	I	a_{11}	a_{10}
a_{10}	a_{10}	a_4	a_7	a_6	a_1	a_8	a_3	a_2	a_5	a_{11}	I	a_9
a_{11}	a_{11}	a_5	a_3	a_2	a_8	a_1	a_7	a_6	a_4	a_{10}	a_9	I

Table 27.4
Operation table for a group G of order 12.

(v) Give an example to show that the operation proposed in (27.1) is not well-defined on G/K.

There is a substantive difference between the cosets of H in G and the cosets of K in G (from Activity 27.2) that makes the operation from (27.1) well-defined on G/H but not on G/K. As we saw, for each $a \in G$, we had $aH = Ha$, but this is not true for K. Note that $aH = Ha$ does not mean $ah = ha$ for all $h \in H$; rather, for each $h \in H$ there is an element $h' \in H$ so that $ah = h'a$. Under this condition, we can show that the operation from (27.1) is well-defined on G/H.

Activity 27.3. Let G be a group, and let H be a subgroup of G such that $gH = Hg$ for all $g \in G$. Let a, a', b, b' be elements of G with $aH = a'H$ and $bH = b'H$.

(a) Explain why there must be elements $h_a, h_b \in H$ so that

$$a = a'h_a \quad \text{and} \quad b = b'h_b.$$

(b) Explain why there is an element $h' \in H$ so that $h_a b' = b'h'$. (Hint: Every left coset of H is a right coset of H.)

(c) Use the previous two results to show that ab is an element of $a'b'H$. What does this tell us about the relationship between abH and $a'b'H$? Explain.

(d) Explain why the operation from (27.1) is well-defined on G/H.

The condition on H that makes the operation from (27.1) well-defined is an important one, and we will study it in more detail in the next section.

Normal Subgroups

In the previous section we saw that if G is a group and H is a subgroup of G so that $aH = Ha$ for all $a \in G$, then we can construct a well-defined operation on the collection of left cosets of H in G. The subgroups that allow us to define such an operation are called *normal* subgroups.

Definition 27.4. Let G be a group. A subgroup N of G is **normal** in G (or is a **normal subgroup** of G) if $aN = Na$ for all $a \in G$.*

The normal subgroups of G are exactly the subgroups for which we have a well-defined operation on G/N, the collection of all distinct left cosets of N in G. When N is a normal subgroup of G, we write $N \triangleleft G$.

To show that a subgroup N is normal in a group G, we can use Definition 27.4, which involves proving two different containment statements—specifically, that $aN \subseteq Na$ and $Na \subseteq aN$ for all $a \in G$. However, we can simplify this process a bit. Informally, if $aN \subseteq Na$, then we might think that we could "multiply" both sides of this containment on the right by a^{-1} to obtain the equivalent statement that $aNa^{-1} \subseteq N$. Similarly, we might think that the statement $Na \subseteq aN$ is equivalent to $a^{-1}Na \subseteq N$. The next theorem formalizes these intuitive ideas.

Theorem 27.5. *Let G be a group and N a subgroup of G. Then N is normal in G if and only if $aNa^{-1} \subseteq N$ for all $a \in G$, where*

$$aNa^{-1} = \{ana^{-1} : n \in N\}.$$

Proof. The usefulness of this theorem is in the reverse implication, so we leave the proof of the forward implication to the exercises. (See Exercise 6.) Let G be a group and N a subgroup of G. Assume $aNa^{-1} \subseteq N$ for all $a \in G$. To show $N \triangleleft G$, we must show $aN = Na$ for all $a \in G$. Choose $a \in G$. To prove $aN = Na$, we will demonstrate containment in both directions. Let $x \in aN$. Then there exists $n \in N$ such that $x = an$. By hypothesis we know that $aNa^{-1} \subseteq N$, so $ana^{-1} \in N$. Let $n' = ana^{-1}$. Then

$$x = an = n'a \in Na.$$

Therefore, $aN \subseteq Na$. To prove containment in the opposite direction, let $x \in Na$. So there exists $n \in N$ such that $x = na$. By hypothesis, we know that

$$a^{-1}Na = (a)^{-1}N\left(a^{-1}\right)^{-1} \subseteq N,$$

so $a^{-1}na \in N$. Let $n' = a^{-1}na$. Then

$$x = na = an' \in aN.$$

The two containments we have proved show that $Na = aN$, and so N is a normal subgroup of G. ■

A few examples are in order.

Activity 27.6. For each part below, decide if the given subgroup N is a normal subgroup of the group G. You may want to use Theorem 27.5.

(a) $G = \mathbb{Z}_8$, $N = \langle [2] \rangle$

(b) $G = \mathbb{Z}$, $N = \langle 5 \rangle$

(c) $G = S_3$, $N = \langle (1\ 2\ 3) \rangle$.

*The term *normal* seems to be used here because in Galois theory (not discussed in these investigations), the Galois correspondence between subgroups and subfields has normal subgroups corresponding to subfields that are normal extensions of the base field. Normal subgroups are also called *invariant subgroups* because they are invariant under inner automorphisms.

Quotient Groups

Let G be a group and $N \lhd G$. Recall that the operation on G/N defined as

$$(aN)(bN) = (ab)N \tag{27.2}$$

is well-defined. It is natural to ask what kind of structure this operation imposes on G/N.

Activity 27.7. By definition, G/N is closed under the operation from (27.2). Here we will investigate some other properties of G/N under this same operation.

(a) There is an identity element in G/N. What is it? Verify your answer.

(b) Prove that the operation from (27.2) is associative on G/N.

(c) Prove that G/N contains an inverse for each of its elements.

The responses to Activity 27.7 show that G/N is a group under the operation from (27.2). This group is called the *quotient group* (or *factor group*) of G by N.

Definition 27.8. Let G be a group and N a normal subgroup of G. The **quotient group** (or **factor group**) of G by N is the group

$$G/N = \{aN : a \in G\}$$

with the operation $(aN)(bN) = (ab)N$ for all $a, b \in G$.

At this point, it will be helpful for us to explore a few examples of quotient groups and their structures.

Activity 27.9. For each choice of G and N below, construct an operation table for the quotient group G/N.

(a) $G = \mathbb{Z}_8$, $N = \langle [2] \rangle$

(b) $G = \mathbb{Z}$, $N = \langle 5 \rangle$

(c) $G = D_4$, $N = \langle R^2 \rangle$

You may have noticed that if G is a finite group, then the order of G/N is related to the orders of G and N. In fact, Lagrange's Theorem (see page 351) tells us that $|G/N| = \frac{|G|}{|N|}$ when G is finite. Notice that G/N can be a finite group even if G is an infinite group.

Quotient groups are very useful, in part because if G is a finite group and N a non-trivial normal subgroup of G, then G/N has smaller order than G. Consequently, it may be easier to prove properties of G/N than it is for G. Often, information about G/N tells us something important about G, as the next theorem shows. (Recall that for any group G, $Z(G)$ denotes the center of G—that is, the set of elements that commute with every element of G.)

Theorem 27.10. *If G is a group and $G/Z(G)$ is cyclic, then G is Abelian.*

Proof. Let G be a group such that $G/Z(G)$ is cyclic. So that the proof is easier to follow, let $N = Z(G)$. Since G/N is cyclic, we know there is an element $gN \in G/N$ that generates G/N.

That is, $G/N = \langle gN \rangle$. To show G is Abelian, we need to show $ab = ba$ for all $a, b \in G$. Let $a, b \in G$. Then $aN, bN \in G/N$. Moreover, $aN = (gN)^k = g^k N$ and $bN = (gN)^m = g^m N$ for some integers k, m. Since $a \in aN$, we have $a = g^k z_a$ for some $z_a \in N$. Since $b \in bN$, we also have $b = g^m z_b$ for some $z_b \in N$. Since z_a and z_b are in the center of G, we know z_a and z_b commute with all elements of G. Therefore,

$$\begin{aligned} ab &= (g^k z_a)(g^m z_b) \\ &= z_b(g^k g^m) z_a \\ &= z_b g^{k+m} z_a \\ &= z_b g^{m+k} z_a \\ &= z_b g^m g^k z_a \\ &= (z_b g^m)(z_a g^k) \\ &= ba. \end{aligned}$$

It follows that G is an Abelian group. ■

Cauchy's Theorem for Finite Abelian Groups

In our study of groups so far, we have seen that it is important to understand the subgroup structure of a group. Lagrange's Theorem (see page 351) tells us what orders are possible for subgroups of a finite group, but it does not tell us anything about the existence of subgroups of any particular order. One result we have so far in that direction is Corollary 26.13 (see page 351), which shows that every finite group must contain a subgroup of prime order. However, the drawback of Corollary 26.13 is that it doesn't tell us *which* prime order subgroup must exist. A more general result, Cauchy's Theorem, does just that. We will focus here on Cauchy's Theorem for Finite Abelian Groups, and we will later extend the theorem to all finite groups.

Theorem 27.11 (Cauchy's Theorem for Finite Abelian Groups)**.** *Let G be an Abelian group of finite order n. If p is a prime divisor of n, then G contains an element of order p.*

Preview Activity 27.12. In the proof of Cauchy's Theorem for Finite Abelian Groups, we will see that the construction of a quotient group allows us to reduce the order of a group and then apply the principle of mathematical induction. This is a powerful idea, and in this activity we will explore an example to see how it works. Let $G = U_{21} = \{[1], [2], [4], [5], [8], [10], [11], [13], [16], [17], [19], [20]\}$. Once proved, Cauchy's Theorem will tell us that if a prime p divides the order of G, then G must contain an element of order p. Thus, U_{21} must contain elements of order 2 and 3. Since U_{21} is a finite group of relatively small order, we could easily determine the orders of each of its elements, but our goal in this activity is to create a general process that will allow us to find such an element in any finite Abelian group. In this example, we will find an element of order 3 in U_{21}. To find such an element in U_{21}, we will start with Corollary 26.13, which tells us that G contains an element of some prime order q that divides $|G|$. Recall, however, that we don't know exactly what prime q is.

(a) Explain why we are done with this activity if $q = 3$.

(b) Now assume $q \neq 3$. For the purposes of this illustration, we will consider the case that $q = 2$. Recall that we are still looking for an element of order 3 in G. We are assuming there is an element of order 2 in G. In fact, there are three such elements, and [8] happens to be one of them. Let $N = \langle [8] \rangle$. Since G is Abelian, we know that N is a normal subgroup of G, and so the quotient $\overline{G} = G/N$ is a group.

(i) Explain why $|\overline{G}| < |G|$. Since $|\overline{G}| < |G|$, we can apply the idea of induction. We will be more specific about the induction hypothesis a bit later, but the basic concept is that since $|\overline{G}| < |G|$, we can assume the result of Theorem 27.11 for $\overline{G}$. Explain why doing so allows us to assume that $\overline{G}$ contains an element of order 3.

(ii) Note that $\overline{G} = \{N, [2]N, [4]N, [5]N, [10]N, [13]N\}$. The operation table for $\overline{G}$ is given in Table 27.5. Show that $[2]N$ has order 3 in $\overline{G}$.

	N	$[2]N$	$[4]N$	$[5]N$	$[10]N$	$[13]N$
N	N	$[2]N$	$[4]N$	$[5]N$	$[10]N$	$[13]N$
$[2]N$	$[2]N$	$[4]N$	N	$[10]N$	$[13]N$	$[5]N$
$[4]N$	$[4]N$	N	$[2]N$	$[13]N$	$[5]N$	$[10]N$
$[5]N$	$[5]N$	$[10]N$	$[13]N$	$[4]N$	N	$[2]N$
$[10]N$	$[10]N$	$[13]N$	$[5]N$	N	$[2]N$	$[4]N$
$[13]N$	$[13]N$	$[5]N$	$[10]N$	$[2]N$	$[4]N$	N

Table 27.5
Operation table for $U_{21}/\langle [8] \rangle$.

(iii) If $\overline{G} = G/N$ contains an element of order 3, it might be natural to expect that there is a corresponding element of order 3 in G. (See Exercise 16.) This, however, is not always the case. For instance, the element $[2]N$ has order 3 in $\overline{G}$, but $|[2]| \neq 3$ in G. Find $|[2]|$ in G. Then, using only what we know about finite cyclic groups and $|[2]|$ in G, find an element of order 3 in G. (This shows that G contains an element of order 3, as desired.)

If G is a finite Abelian group of order n and p is a prime that divides n, Theorem 27.11 tells us that G must contain an element of order p. Preview Activity 27.12 illustrates the method we will use to prove Theorem 27.11. First, we will find an element $a \in G$ of some prime order q that divides n. If $q \neq p$, then the quotient group $\overline{G} = G/\langle a \rangle$ has smaller order than G. If p divides $|\overline{G}|$, then we can apply induction and find an element $\overline{b} \in \overline{G}$ so that $|\overline{b}| = p$. We can associate the element $\overline{b}$ in $\overline{G}$ with an element $b \in G$ so that $\overline{b} = bN$. If $|b|$ is a multiple of p, then we can find an element of order p in G. The proof of Cauchy's Theorem presented below completes these steps in detail.

Proof of Cauchy's Theorem for Finite Abelian Groups. This proof is a bit more complicated than some others we have done, so you should be sure that you can provide the rationale for any steps that are omitted or labeled with the ? symbol. To begin, let G be an Abelian group of finite order n with identity element e, and let p be a prime divisor of n. We will show that G contains an element of order p. Our proof will be by induction on the order of the group.

The base case will be when $n = 2$.? When $|G| = 2$, the only prime p that divides $|G|$ is $p = 2$. Let $a \in G$ be the non-identity element. By Lagrange's Theorem, $|a| = |\langle a \rangle| = 2$, so a is an element of order 2. Thus, our theorem is true when $n = 2$.

For the inductive step, assume that for any Abelian group G' of order $k < n$ and for any prime divisor q of k, G' contains an element of order q.(?) By Corollary 26.13 (see page 351), we know that G contains an element a of some prime order q, where q divides n. If $q = p$, then a has order p and we are done because we have found an element in G of order p. So assume that $q \neq p$. Let $N = \langle a \rangle$. Since $N \triangleleft G$,(?) the quotient $\overline{G} = G/N$ is a group. Since a has prime order, we know $|a| = |N|$ is greater than 1. Thus, $|\overline{G}| = |G/N| = \frac{|G|}{|N|} < |G|$, and so $\overline{G}$ has order smaller than $|G|$—that is, $|\overline{G}| < |G| = n$. Furthermore, since G is an Abelian group, so is $\overline{G}$.(?) Thus, we can apply the induction hypothesis to $\overline{G}$ and any prime divisor of $|\overline{G}|$. Now p divides n and p does not divide q, so p divides $|\overline{G}|$.(?) We can then apply our induction hypothesis to conclude there is an element $\overline{b}$ of order p in $\overline{G}$. Now all we need to do is translate this conclusion about $\overline{G}$ to one about G. To do so, note that since $\overline{b} \in \overline{G}$, there exists $b \in G$ such that $\overline{b} = bN$. Recall that $|\overline{b}| = p$, so bN is not the identity in $\overline{G}$, and $(bN)^p = N$.(?) Since $(bN)^p = b^pN$, it follows that $b^p \in N$.(?) Note that $b \neq e$, since that would imply $bN = N$ is the identity element in $\overline{G}$. Thus, b is a non-identity element in G such that $b^p \in N$. We will consider two possibilities: $b^p = e$ or $b^p \neq e$.

If $b^p = e$ and $b \neq e$, then b must have order p,(?) and we have found an element of order p in G, as desired. So assume $b^p \neq e$. Recall that $|N| = |\langle a \rangle| = q$, and so the order of b^p must be q.(?) Thus, $b^{pq} = e$, and the order of b divides pq. So the possible orders of b are 1, p, q, and pq. We have already argued that $b \neq e$, and we have considered the case where $|b| = p$. Therefore, we are left to consider the cases where $|b| = q$ or $|b| = pq$. Now if $|b| = q$, then $\overline{b}^q = (bN)^q = b^qN = eN = N$, and $|\overline{b}|$ divides q. But p and q are distinct primes and $|\overline{b}| = p$, so this is impossible.(?) Thus, the only case left to consider is when $|b| = pq$. If $|b| = pq$, then Theorem 23.6 (see page 320) shows that $|b^q| = \frac{pq}{\gcd(pq,q)} = p$. Therefore, b^q is an element in G of order p. In each case, we have demonstrated that G contains an element of order p, as desired. ■

One elementary consequence of Cauchy's Theorem for Finite Abelian Groups is a determination of the structure of all Abelian groups of order pq, where p and q are distinct primes.

Corollary 27.13. *Any Abelian group of order pq, where p and q are distinct primes, is cyclic.*

Proof. Let p and q be distinct primes, and let G be an Abelian group of order pq. Cauchy's Theorem for Finite Abelian Groups tells us that G contains an element a of order p and an element b of order q. Since $\gcd(|a|, |b|) = 1$ and G is Abelian, it follows (see Exercise 18 on page 324 of Investigation 23) that $|ab| = pq$, and so $G = \langle ab \rangle$ is a cyclic group. Hence, every Abelian group of order pq (where p and q are distinct primes) is a cyclic group. ■

Corollary 27.13 tells us, for example, that the Abelian groups of order 6, 10, 15, 21, etc., are all cyclic.

Simple Groups and the Simplicity of A_n

While mathematicians have not been able to classify all of the finite groups (that is, determine all of the groups of any given order), it is a remarkable accomplishment that the classification of all of the finite *simple* groups has been completed. Simple groups are important in group theory because they form the building blocks of all groups. (We'll explain this statement in more detail a little later.)

Definition 27.14. A group G is **simple** if G has no non-trivial proper normal subgroups.

Note that by *non-trivial proper normal subgroups*, we mean normal subgroups of more than one element that are not the entire group.

There is one straightforward piece of the classification of simple groups: the finite Abelian simple groups. (See Exercise 34.) The best known examples of non-Abelian simple groups are the alternating groups A_n. In fact, it is the simplicity of these groups that can be used to show that there is no general method for finding roots of polynomials with complex coefficients of degree 5 or higher. After proving a couple of preliminary results, we will be able to prove that A_n is simple for $n \geq 5$.

To begin, recall that A_n is the set of even permutations in S_n—that is, the permutations that can be written as a product of an even number of transpositions. The first fact we will prove is that every element in A_n can also be written as a product of 3-cycles. The general idea behind the proof is contained in the next activity.

Activity 27.15. Let $n \geq 3$ be an integer and let $a, b, c \in \{1, 2, 3, \ldots, n\}$.

(a) Write the product $(a\ b)(a\ c)$ as a cycle or a product of disjoint cycles. Why does this show that A_n contains every 3-cycle?

(b) Write each of the following products as a product of two transpositions.

(i) $(1\ 3\ 2)(1\ 3\ 4)$

(ii) $(1\ 2\ 3)$

(iii) $(1\ 2\ 3)(1\ 3\ 2)$

Activity 27.15 illustrates how A_n contains all of the 3-cycles (for $n \geq 3$) and also how products of 3-cycles are related to products of even numbers of transpositions. This leads us to the next result.

Lemma 27.16. *Let n be an integer with $n \geq 3$. Any permutation in A_n can be written as a product of 3-cycles.*

Proof. Let n be an integer with $n \geq 3$. Note that each 3-cycle $(a\ b\ c) = (a\ c)(a\ b)$ is an even permutation. So A_n contains the 3-cycles.

Let $\sigma \in A_n$. Then there exist transpositions $\tau_1, \tau_2, \ldots, \tau_m$ so that

$$\sigma = \tau_m \cdots \tau_3\tau_2\tau_1.$$

Since $\sigma \in A_n$, we know that m is even. So there is an integer k such that $m = 2k$. We can then collect the τ_i into groups of 2 so that

$$\sigma = (\tau_{2k}\tau_{2k-1}) \cdots (\tau_4\tau_3)(\tau_2\tau_1).$$

If in any pair we have $\tau_{2i} = \tau_{2i-1}$, then $\tau_{2i}\tau_{2i-1}$ is the identity and can be removed from the product for σ. Therefore, we can assume that $\tau_{2i} \neq \tau_{2i-1}$ for each pair of transpositions $(\tau_{2i}\ \tau_{2i-1})$. It follows that each pair has the form $(a\ b)(c\ d)$ with $a \neq b$, $c \neq d$, and $(a\ b) \neq (c\ d)$. There are two cases that we must consider.

(1) The first case is when a, b, c, and d are all distinct. In this situation, we have $(a\ b)(c\ d) = (a\ c\ b)(a\ c\ d)$.

(2) The other case is when $b = c$ and $a \neq d$. (This case is equivalent to those in which $a = d$ and $b \neq c$, or $a = c$ and $b \neq d$. Therefore, we will assume, without loss of generality, that $b = c$ and $a \neq d$.) Then $(a\ b)(c\ d) = (a\ b)(b\ d) = (a\ b\ d)$.

In each case we see that we can write a product of two transpositions in terms of 3-cycles. Therefore, every even permutation—that is, every element of A_n—can be written as a product of 3-cycles. ■

The fact that A_n is generated by the 3-cycles also tells us something about the normal subgroups of A_n.

Activity 27.17. Let $n \geq 3$ and suppose N is a normal subgroup of A_n that contains the 3-cycle $(1\ 2\ 3)$.

(a) Why must N contain the 3-cycle $(1\ 3\ 2)$?

(b) Let $r, s, t \in \{1, 2, 3, \ldots, n\}$. Write each of the following products in A_n as a cycle or a product of disjoint cycles. Why must each of these elements be in N?

(i) $(1\ t\ 3)(1\ 2\ 3)(1\ 3\ t)$,

(ii) $(r\ 1\ t)(1\ t\ 2)(r\ t\ 1)$,

(iii) $(1\ 2\ s)(r\ 2\ t)(1\ s\ 2)$

(c) How does the result of part (b) show that $N = A_n$?

Activity 27.17 seems to provide a method for generating all 3-cycles in a normal subgroup of A_n that contains a 3-cycle. The next lemma shows this in general.

Lemma 27.18. *Let n be an integer with $n \geq 3$. If N is a normal subgroup of A_n and N contains a 3-cycle, then $N = A_n$.*

Proof. Let n be an integer with $n \geq 3$. Let N be a normal subgroup of A_n, and let $(a\ b\ c)$ be a 3-cycle in N. We will show that $N = A_n$ by showing that N contains all possible 3-cycles. Then Lemma 27.16 will establish that $N = A_n$. Let r, s, t be integers between 1 and n. Since N is normal in A_n, we know $\alpha(a\ b\ c)\alpha^{-1}$ is in N for every $\alpha \in A_n$. From Lemma 27.16, we know that A_n contains all 3-cycles. Thus,

- $(a\ t\ c)(a\ b\ c)(a\ c\ t) = (a\ t\ b) \in N$, which implies that
- $(r\ a\ t)(a\ t\ b)(r\ t\ a) = (r\ b\ t) \in N$, which implies that
- $(b\ s\ a)(r\ b\ t)(b\ a\ s) = (r\ s\ t) \in N$.

Since r, s, and t were chosen to be arbitrary integers between 1 and n, this shows that N contains every possible 3-cycle, and so $N = A_n$. ■

We will now show that A_n is a simple group for all $n \geq 5$.

Theorem 27.19. *Let n be an integer with $n \geq 5$. Then A_n contains no non-trivial normal subgroups.*

Proof. Let n be an integer with $n \geq 5$. Suppose N is a non-trivial normal subgroup of A_n. Let σ be a non-identity element in N that permutes the fewest number k of integers. If $k = 3$, then σ must be a 3-cycle and Lemma 27.18 shows $N = A_n$, a contradiction. Assume $k \geq 4$, and decompose σ into a product of disjoint cycles:

$$\sigma = \sigma_1\sigma_2\sigma_3 \cdots \sigma_m.$$

We will consider two cases:

- Case 1: For some i, σ_i is a q-cycle with $q \geq 3$.

  In this case, we can assume, without loss of generality, that $\sigma = (1\ 2\ 3\ \ldots)\tau$, where τ is disjoint from $(1\ 2\ 3\ \ldots)$. If σ permutes exactly 4 integers, then $\sigma = (1\ 2\ 3\ a)$ for some integer $a \neq 1, 2, 3$, and σ is odd. Thus, σ is not in N. Therefore, σ must permute at least 5 integers. Without loss of generality, we assume $\sigma(4) = a_4$ and $\sigma(5) = a_5$. Let $\beta = (3\ 4\ 5)$, and let

  $$\gamma = \sigma^{-1}\beta\sigma\beta^{-1}.$$

  Since N is normal in A_n, we know $\beta\sigma\beta^{-1} \in N$. Therefore, $\gamma \in N$. Now

  $$\gamma(1) = \sigma^{-1}\beta\sigma\beta^{-1}(1) = \sigma^{-1}\beta\sigma(1) = \sigma^{-1}\beta(2) = \sigma^{-1}(2) = 1,$$

  so $\gamma \neq \sigma$. In addition,

  $$\gamma(2) = \sigma^{-1}\beta\sigma\beta^{-1}(2) = \sigma^{-1}\beta\sigma(2) = \sigma^{-1}\beta(3) = \sigma^{-1}(4) \neq 2.$$

  Thus, γ is not the identity permutation. Also, if $t > 5$ and $\sigma(t) = t$, then $\gamma(t) = t$ as well. Since $\gamma(1) = 1$, γ permutes fewer elements than σ, a contradiction to our assumption that σ is the element of N that permutes the fewest number of integers.

- Case 2: For each i, σ_i is a transposition.

  In this case, we can assume, without loss of generality, that $\sigma = (1\ 2)(3\ 4)\cdots$. We again let $\beta = (3\ 4\ 5)$ and let

  $$\gamma = \sigma^{-1}\beta\sigma\beta^{-1}.$$

  Then $\gamma(1) = 1$ and $\gamma(2) = 2$. If $\sigma(t) = t$ for any $t > 5$, then $\gamma(t) = t$. Also,

  $$\gamma(3) = \sigma^{-1}\beta\sigma\beta^{-1}(4) = \sigma^{-1}\beta\sigma(3) = \sigma^{-1}\beta(4) = \sigma^{-1}(5) \neq 3,$$

  so γ is not the identity. Once again, γ permutes fewer elements than σ, a contradiction.

Since both cases lead to a contradiction, we can conclude that A_n contains no non-trivial normal subgroup group when $n \geq 5$. ■

The simplicity of A_n will allow us to explicitly determine the normal subgroups of S_n. (See Exercise 37.)

We will close this section with a few comments about the importance of simple groups. Simply put (no pun intended!), simple groups form the building blocks of all groups. While we won't provide a rigorous formulation or proof of this statement, we can understand some of the ideas behind it.

Suppose G is a finite group with identity e. Let G_1 be a normal subgroup of G (other than G) of largest order. (Note that if G is simple, then $G_1 = \{e\}$.) Now G/G_1 must be a simple group. To see why, suppose M is a normal subgroup of G/G_1. Then there is a normal subgroup N of G with $G_1 \subset N \subset G$ so that $M = N/G_1$. This contradicts the maximal order of G_1.

If $G_1 \neq \{e\}$, then we can repeat this process on G_1 to obtain a second group $G_2 \subset G_1$, normal in G_1 (but not necessarily in G), so that G_1/G_2 is simple. Continuing this process, we obtain a sequence of groups

$$G \supseteq G_1 \supseteq G_2 \supseteq G_3 \cdots \supseteq G_n,$$

with $G_n = \{e\}$, such that for each i, G_{i+1} is a normal subgroup of G_i and G_i/G_{i+1} is simple. The quotient groups G_i/G_{i+1} are called the *composition factors* of G and turn out to be independent of the choices of the groups G_i. (In other words, we may have different choices for the normal subgroups G_i, but the quotient groups are always the same in some sense that we will make clear later.) A consequence of this result is that if we can classify all simple groups, then understand how the composition groups of an arbitrary group determine the group, we will be able to classify all finite groups.

In a remarkable collection of work, mathematicians have been able to classify all of the finite simple groups. The final result is a combined effort of hundreds of researchers producing over 500 articles of more than 14,000 journal pages. In essence, the proof states that finite simple groups fall into two categories: several infinite families of groups for which there is an established pattern, and 26 other groups known as the sporadic groups:

(1) Cyclic groups of prime group order,

(2) Alternating groups of degree at least five,

(3) Lie-type Chevalley groups,

(4) Lie-type twisted Chevalley groups or the Tits group, and

(5) Sporadic groups:

- Mathieu groups $M_{11}, M_{12}, M_{22}, M_{23}, M_{24}$
- Janko groups J_1, J_2 (also known as the Hall-Janko group HJ), J_3, J_4
- Conway groups Co_1, Co_2, Co_3
- Fischer groups F_{22}, F_{23}, F_{24}
- Higman-Sims group HS
- McLaughlin group McL
- Held group He
- Rudvalis group Ru
- Suzuki sporadic group Suz
- O'Nan group $O'N$
- Harada-Norton group HN
- Lyons group Ly
- Thompson group Th
- Baby Monster group B
- Monster group M

Five of these sporadic groups were found in the 1860s and the other 21 were found between 1965 and 1975. The appropriately named Monster group is the largest of the sporadic groups and has order

$$808,017,424,794,512,875,886,459,904,961,710,757,005,754,368,000,000,000.$$

Quite a monster, indeed!

Concluding Activities

Activity 27.20. Let G be a group and H a subgroup of G. Theorem 27.5 allows us to tell if H is normal in G by determining if

$$aHa^{-1} = \{aha^{-1} : h \in H\}$$

is equal to H for every $a \in G$. As we will demonstrate in this activity, the set aHa^{-1} is a subgroup of G for any $a \in G$ and is called a **conjugate** of the subgroup H (or the conjugate of H by a).

(a) Let $H = \langle R^2 \rangle$ in the group D_6 of symmetries of a regular hexagon. Determine the elements in rHr^{-1}.

(b) Let $K = \langle r \rangle$ in D_6. Determine the elements in $(rR)K(rR)^{-1}$.

(c) Let $a \in G$. Prove that aHa^{-1} a subgroup of G.

Activity 27.21. In Activity 26.19 (see page 353) we showed that the only congruence relations on a group G with subgroup H are the relations $\sim_H$, where $a \sim_H b$ if $a^{-1}b \in H$. However, we did not show that $\sim_H$ is always a congruence relation. In this activity, we will show that the congruence relations on a group G are exactly the relations $\sim_N$, where N is a normal subgroup of G. This explains why we study only the relations $\sim_N$ and not others. Recall that an equivalence relation $\sim$ on a group G is a congruence relation if whenever $a, b, c, d \in G$ with $a \sim b$ and $c \sim d$ we also have

(i) $ac \sim bd$ and

(ii) $a^{-1} \sim b^{-1}$.

In other words, a congruence relation preserves the group structure.

(a) Let $\sim$ be a relation on a group G with identity e and let $N = \{x \in G : x \sim e\}$. Activity 26.19 shows that N is a subgroup of G. Prove that N is a normal subgroup of G. This establishes that every congruence relation on a group G corresponds to a normal subgroup of G.

(b) Now show that if N is a normal subgroup of a group G, then the relation $\sim_N$ is a congruence relation on G.

Activity 27.22. Let H and K be subgroups of the group G, with K normal in G. Let

$$KH = \{kh : k \in K \text{ and } h \in H\}.$$

(See Exercise 5 on page 310 in Investigation 22.)

(a) Let $G = D_4$, $K = \langle R^2 \rangle$, and $H = \langle r \rangle$. Assume K is a normal subgroup of G. Find the elements of KH.

(b) Let G be an arbitrary group, and let H and K be subgroups of G with K normal in G. Prove that KH is a subgroup of G.

Activity 27.23. Write a short summary that describes the important concepts, and the relationships between these concepts, that were introduced in this investigation. Explain how the material in this investigation is connected to the material in Investigations 20, 22, and 26.

Exercises

(1) Certain subgroups of every group are normal. Let G be any group, and let e be the identity in G.

(a) Prove that $G \triangleleft G$.

(b) Prove that $\{e\} \triangleleft G$.

(c) Prove that if H is a subgroup of $Z(G)$, then $H \triangleleft G$. In particular, conclude that $Z(G) \triangleleft G$.

(2) Determine if the indicated subgroup H is a normal subgroup of G.

(a) $G = A_4$ and $H = \{e, (1\,2)(3\,4), (1\,3)(2\,4), (1\,4)(2\,3)\}$.

(b) $G = \mathrm{GL}_2(\mathbb{R})$ and $H = \left\{ \begin{bmatrix} a & 0 \\ 0 & a \end{bmatrix} : a \neq 0 \right\}$.

(c) $G = \mathrm{GL}_2(\mathbb{R})$ and $H = \left\{ \begin{bmatrix} a & b \\ 0 & d \end{bmatrix} : a, b, d \in \mathbb{R}, ad \neq 0 \right\}$.

(d) $G = D_4$ and H is the subgroup consisting of the rotations in D_4.

(e) $G = \mathrm{GL}_n(\mathbb{R})$ and $H = \mathrm{SL}_n(\mathbb{R})$.

(3) If N is a normal subgroup of a group G and H is a subgroup of G that contains N, must N be normal in H? Prove your answer.

(4) (a) Give an example of a non-trivial normal subgroup of S_3, or explain why one doesn't exist.

(b) Give an example of a non-trivial subgroup of S_3 that is not normal, or explain why one doesn't exist.

(5) Let G be a group and H a subgroup of G.

(a) If $a, b \in G$ so that $aH = bH$, must it follow that $Ha = Hb$? Prove your answer.

(b) For any $a, b \in G$, suppose that $Ha = Hb$ whenever $aH = bH$. Prove that $H = gHg^{-1}$ for any $g \in G$.

* (6) (a) Prove the forward implication of Theorem 27.5. That is, show that if N is a normal subgroup of a group G, then $aNa^{-1} \subseteq N$ for all $a \in G$.

(b) Let G be a group and N a subgroup of G. To use Theorem 27.5 to show that N is a normal subgroup of G, we show that $aNa^{-1} \subseteq N$ for every $a \in G$. Is this the same as showing $a^{-1}Na \subseteq N$ for all $a \in G$? In other words, is the statement $aNa^{-1} \subseteq N$ for all $a \in G$ equivalent to $a^{-1}Na \subseteq N$ for all $a \in G$? Prove your answer.

(c) Let N be a normal subgroup of a group G. Is the reverse containment in Theorem 27.5 true? That is, must $N \subseteq aNa^{-1}$ for every $a \in G$? Prove your answer.

(7) The result of Exercise 6 is that if N is a normal subgroup of a group G, then $aNa^{-1} = N = a^{-1}Na$ for every $a \in G$. Is the converse true—that is, if H is a subgroup of a group G and $aHa^{-1} = a^{-1}Ha$ for all $a \in G$, is $H \triangleleft G$? Prove your answer.

(8) Write out an operation table for G/N for the given group G and subgroup N. Then explain how you know that N is normal in G.

(a) $G = U_{13}$ and $N = \{[1], [3], [9]\}$.

(b) $G = S_3$ and $N = \langle(1\ 2\ 3)\rangle$.

(9) (a) Find the order of the element $[10] + \langle[6]\rangle$ in the quotient group $\mathbb{Z}_{24}/\langle[6]\rangle$.

(b) Find all of the elements in $\mathbb{Z}_{24}/\langle[6]\rangle$, and determine the order of each.

(10) Let $G = U(\mathbb{R})$ (the group of units in $\mathbb{R}$) and let $N = \mathbb{R}^+$ (the subgroup of G consisting of all positive real numbers). Find the elements of G/N, and create the operation table for this quotient group. What kind of a group is G/N?

⋆ (11) Let H and K be subgroups of the group G, with K normal in G. In Activity 27.22 we showed that

$$KH = \{kh : k \in K \text{ and } h \in H\}$$

is a subgroup of G.

(a) Is the set

$$HK = \{hk : h \in H \text{ and } k \in K\}$$

also a subgroup of G? Prove your answer.

(b) Does $KH = HK$? Prove your answer.

(12) Where in the proof of Theorem 27.11 did we use the fact that G is an Abelian group? Why doesn't our proof apply to non-Abelian groups?

(13) The operation table for D_6, the dihedral group of order 12, is given in Table 27.6.

	I	R	R^2	R^3	R^4	R^5	r	rR	rR^2	rR^3	rR^4	rR^5
I	I	R	R^2	R^3	R^4	R^5	r	rR	rR^2	rR^3	rR^4	rR^5
R	R	R^2	R^3	R^4	R^5	I	rR^5	r	rR	rR^2	rR^3	rR^4
R^2	R^2	R^3	R^4	R^5	I	R	rR^4	rR^5	r	rR	rR^2	rR^3
R^3	R^3	R^4	R^5	I	R	R^2	rR^3	rR^4	rR^5	r	rR	rR^2
R^4	R^4	R^5	I	R	R^2	R^3	rR^2	rR^3	rR^4	rR^5	r	rR
R^5	R^5	I	R	R^2	R^3	R^4	rR	rR^2	rR^3	rR^4	rR^5	r
r	r	rR	rR^2	rR^3	rR^4	rR^5	I	R	R^2	R^3	R^4	R^5
rR	rR	rR^2	rR^3	rR^4	rR^5	r	R^5	I	R	R^2	R^3	R^4
rR^2	rR^2	rR^3	rR^4	rR^5	r	rR	R^4	R^5	I	R	R^2	R^3
rR^3	rR^3	rR^4	rR^5	r	rR	rR^2	R^3	R^4	R^5	I	R	R^2
rR^4	rR^4	rR^5	r	rR	rR^2	rR^3	R^2	R^3	R^4	R^5	I	R
rR^5	rR^5	r	rR	rR^2	rR^3	rR^4	R	R^2	R^3	R^4	R^5	I

Table 27.6
Operation table for D_6

(a) Find the elements of the set $D_6/Z(D_6)$.

(b) Write the operation table for the group $D_6/Z(D_6)$.

(c) The examples of quotient groups we have seen so far have all been Abelian groups. Is it true that every quotient group is Abelian? Explain.

(d) Give a necessary condition on a group G if G/N is a non-Abelian group. Is this necessary condition sufficient as well? Explain.

(14) Prove the following theorem:

Theorem 27.24. *Let G be a group and N a normal subgroup of G. If $a \in G$ and $n \in \mathbb{Z}$, then $(aN)^n = a^n N$.*

(15) (a) Can the order of a non-identity element aN in a quotient group G/N be smaller than the order of a in G? If yes, provide an example to illustrate. If no, prove it.

(b) Can the order of an element aN in a quotient group G/N be greater than the order of a in G? If yes, provide an example to illustrate. If no, prove it.

(c) Can the order of a non-identity element aN in a quotient group G/N be equal to the order of a in G? If yes, provide an example to illustrate. If no, prove it.

⋆ (16) (a) If N is a normal subgroup of a finite group G and G contains an element of order n, must G/N contain an element of order n? If yes, provide an example to illustrate. If no, prove it.

(b) If N is a finite normal subgroup of a group G and G/N contains an element of order n, must G contain an element of order n? If yes, provide an example to illustrate. If no, prove it.

(17) (a) Is every quotient group of an Abelian group Abelian? Prove your answer.

(b) Is every quotient group of a cyclic group cyclic? Prove your answer.

(c) Is every quotient group of a non-Abelian group non-Abelian? Prove your answer.

⋆ (18) Let G be a group and H a subgroup of G. Even though H may not be normal in G, we can consider in some sense how close H is to being normal by determining how many elements $g \in G$ have the property that $gHg^{-1} = H$. This leads us to define the **normalizer** of H in G as the set

$$N(H) = \{g \in G : gHg^{-1} = H\}.$$

(a) Find $N(\langle r \rangle)$ in D_4.

(b) Prove that $N(H)$ is always a subgroup of G. Must $N(H)$ be a normal subgroup of G? Prove your answer.

(c) Show that $H \subseteq N(H)$. Is H a normal subgroup of $N(H)$?

⋆ (19) Determining when a subgroup is normal can be a challenge. In this exercise, we will consider a specific situation in which we can tell that a subgroup is normal.

(a) Prove that if G is a finite group and N is a subgroup of G so that $|G| = 2|N|$, then N is a normal subgroup of G.

(b) If H is a subgroup of G with $|G| = 2|H|$, then $[G : H] = 2$. Prove that, under this condition, if $x, y \in G$ and $x, y \notin H$, then $xy \in H$.

(c) Explain how the result of part (b) shows each of the following.

- The sum of two odd integers is an even integer.
- The product of two odd permutations is an even permutation.
- The product of two reflections is a rotation.

(20) Similar to Exercise 19, prove that if G is a group and has exactly one subgroup H of order n, then H is a normal subgroup of G. (Hint: Refer to Activity 27.20.)

(21) Let $\mathbf{Q}$ be the group of quaternions as introduced in Exercise 6 of Investigation 24. (See page 331.) If we let $i = a$, $j = b$, and $k = ab$, (with a and b as in Exercise 6 of Investigation 24), then the operation table for $\mathbf{Q}$ can be written as shown in Table 27.7. The structure of this

	1	i	j	k	$-i$	$-j$	$-k$	-1
1	1	i	j	k	$-i$	$-j$	$-k$	-1
i	i	-1	k	$-j$	1	$-k$	j	$-i$
j	j	$-k$	-1	i	k	1	$-i$	$-j$
k	k	j	$-i$	-1	$-j$	i	1	$-k$
$-i$	$-i$	1	$-k$	j	-1	k	$-j$	i
$-j$	$-j$	k	1	$-i$	$-k$	-1	i	j
$-k$	$-k$	$-j$	i	1	j	$-i$	-1	k
-1	-1	$-i$	$-j$	$-k$	i	j	k	1

Table 27.7
Operation table for the quaternions $\mathbf{Q}$.

group can be easily remembered if we think of i, j, and k as the standard unit vectors in $\mathbb{R}^3$ with the cross product as the operation and 1 as the identity.

(a) Find all of the normal subgroups of $\mathbf{Q}$. Make sure to verify that each subgroup you find is normal.

(b) From Exercise 17, we know that every subgroup of an Abelian group is normal. Is it possible to have a non-Abelian group in which every subgroup is normal? Prove your answer.

(22) An operation table for the group $\mathbf{Q}$ of quaternions is given in Table 27.7. Determine the elements of $\mathbf{Q}/\langle -1 \rangle$, and create the operation table for this group.

(23) Let G be a group and N a normal subgroup of G. If a and b are elements of G with $|aN| = |bN|$ in G/N, must it follow that $|a| = |b|$ in G? Prove your answer.

(24) Let N be a normal subgroup of G, and let $a \in G$. Suppose that the order of aN in G/N is 5 and the order of N is 12. What are the possible orders of a in G? Prove your answer, and provide examples that illustrate each possibility.

(25) (a) Let $G = U_{21}$ and $N = \langle [4] \rangle$. Since G is Abelian, we know that $N \lhd G$. Let $H = \langle [5] \rangle$ and let $\overline{H} = \{hN : h \in H\}$. Find the elements of $\overline{H}$. Is $\overline{H}$ a subgroup of G/N? If yes, write the operation table for $\overline{H}$. If no, explain why.

(b) Let G be an arbitrary group and N a normal subgroup of G. If H is a subgroup of G, let $\overline{H} = \{hN : h \in H\}$. Prove that $\overline{H}$ a subgroup of G/N.

(26) Let p and q be primes (not necessarily distinct). Let G be a group of order pq. Prove that $|Z(G)| = 1$ or $|Z(G)| = pq$. Provide examples to illustrate each.

(27) It is possible to have a group that has elements of any order. Let $G = \mathbb{R}/\mathbb{Z}$.

(a) Find the order of the element $\frac{1}{2} + \mathbb{Z}$ in G.

(b) Let n be a positive integer. Does G contain an element of order n? Prove your answer.

(c) Does every element of G have finite order? Prove your answer.

(d) Determine the conditions on $x \in \mathbb{R}$ so that $x + \mathbb{Z}$ has finite order in $\mathbb{R}/\mathbb{Z}$. In other words, correctly complete and prove the following theorem:

Theorem. *An element $x+\mathbb{Z}$ in $\mathbb{R}/\mathbb{Z}$ has finite order if and only if* ____________________.

(e) Is it possible to find a group G so that every element in G has finite order, but we can find elements with any finite order? Explain.

(28) Let G be a group generated by a set S. Let N be a normal subgroup of G.

(a) Find a set of generators for G/N. Prove your answer.

(b) Recall that the transpositions $(1\ 2)$, $(1\ 3)$, and $(1\ 4)$ generate S_4 (see Exercise 19 in Investigation 25). Use this fact to find a set of generators for the group S_4/N, where $N = \langle I, (1\ 2)(3\ 4), (1\ 3)(2\ 4), (1\ 4)(2\ 3)\rangle$. (See Exercise 2.) Write the operation table for S_4/N.

(29) Let H and K be subgroups of a group G, with K normal in G. Let

$$KH = \{kh : k \in K \text{ and } h \in H\}$$

as in Activity 27.22

(a) Can KH be a normal subgroup of G? Is it possible that KH is not a normal subgroup of G? Explain.

(b) Prove that if K and H are normal subgroups of a group G, then KH is a normal subgroup of G. Is the converse true?

(30) Let N be a subgroup of a group G. We have seen that if every left coset aN of N in G is equal to the corresponding right coset Na, then N is normal in G. What if we change the assumption to something more general? Suppose that, for each $a \in G$, there exists $b \in G$ such that $aN = Nb$. In other words, every left coset of N in G is equal to *some* right coset of N in G. Must N then be a normal subgroup of G? Prove your answer.

(31) Let $N = \{e, n\}$ be a normal subgroup of order 2 in a group G, where e is the identity in G. Prove that n commutes with every element in G. Conclude that $N \subseteq Z(G)$.

(32) **Intersections of normal subgroups.** Is the intersection of two normal subgroups of a group G always a normal subgroup of G? Prove your answer.

(33) (a) Can we extend Corollary 27.13 to Abelian groups whose orders are the products of three distinct primes? That is, if G is an Abelian group of order pqr where p, q, and r are distinct primes, must G be a cyclic group? Prove your answer.

(b) Can we extend Corollary 27.13 even more to Abelian groups whose orders are the products of any number of distinct primes? That is, if n is a positive integer and G is an Abelian group of order $p_1p_2\cdots p_n$ where $p_1, p_2, \ldots, p_n$ are distinct primes, must G be a cyclic group? Prove your answer.

$\star$ (34) **Finite Abelian simple groups.** Find all of the finite Abelian simple groups. Prove your answer.

(35) Theorem 27.19 shows that A_n is a simple group if $n \geq 5$. What if $n < 5$? Determine if A_2, A_3, and A_4 are simple.

⋆ (36) **Normal subgroups of D_n.** Determining which subgroups of a given group are normal can be a difficult task. However, there are some groups whose normal subgroups can be completely classified. In this exercise, we will classify the normal subgroups of the dihedral groups. Let n be a positive integer.

(a) If i and j are integers, show that

$$(rR^j)(rR^i)(rR^j)^{-1} = rR^{2j-i} \quad \text{and} \quad (rR^j)(R^i)(rR^j)^{-1} = R^{-i}.$$

(b) Suppose N is a normal subgroup of D_n. Show that if N contains a reflection, then N contains R^2 and either r or rR.

(c) When are the groups $\langle r, R^2\rangle$ and $\langle rR, R^2\rangle$ proper subgroups of D_n? Your answers should depend on n. When $\langle r, R^2\rangle$ and $\langle rR, R^2\rangle$ are proper subgroups of D_n, what are their indices in D_n? Prove your answer.

(d) Prove the following theorem:

Theorem. *A non-trivial proper subgroup N of D_n is normal in D_n if and only if N is a subgroup of $\langle R\rangle$ or n is even and N is one of $\langle r, R^2\rangle$ or $\langle rR, R^2\rangle$.*

⋆ (37) **Normal subgroups of S_n.** As with the dihedral groups (see Exercise 36), the normal subgroups of S_n are also well known. We will investigate them in this exercise.

(a) Explain why $A_n \lhd S_n$ for $n \geq 2$.

(b) Let $n \geq 2$, and let N be a normal subgroup of S_n. Use Exercise 16 of Investigation 25 (see page 344) to show that if $\pi \in N$ can be written as the product $\pi_1\pi_2\cdots\pi_k$ of disjoint cycles, then N contains all elements of the form $\tau_1\tau_2\cdots\tau_k$, where the τ_i are disjoint cycles and τ_i is a cycle of the same length as π_i for each i.

(c) Determine the non-trivial proper normal subgroups of S_2 and S_3.

(d) Determine the normal subgroups of S_4.

(e) In the remainder of this exercise, we will prove the following theorem:

Theorem. *Let $n \geq 5$. Then A_n is the only non-trivial normal subgroup of S_n.*

Let $n \geq 5$, and let N be a non-trivial normal subgroup of S_n.

(i) Explain why $N \cap A_n$ is a normal subgroup of A_n. Then explain why $N \cap A_n = A_n$ or $N \cap A_n = \{I\}$.

(ii) What can be said about N if $N \cap A_n = A_n$?

(iii) If $N \cap A_n = \{I\}$, why can N contain only odd permutations (in addition to I)? Show that N cannot contain more than one odd permutation. How does this show that $|N| = 2$?

(iv) To complete our proof, show that if $N = \{I, \alpha\}$, where α is an odd permutation of order 2, then N is not normal in S_n.

(38) Let G be a group and N, K subgroups of G with K a subset of N. Prove or disprove: If N is a normal subgroup of G and K is a normal subgroup of N, then K is a normal subgroup of G. In other words, is normality transitive?

(39) Let G be a group, N a normal subgroup of G, and H any subgroup of G.

(a) Must $N \cap H$ be a normal subgroup of G? Prove your answer.

(b) Must $N \cap H$ be a normal subgroup of N? Prove your answer.

(c) Must $N \cap H$ be a normal subgroup of H? Prove your answer.

(40) The result of Exercise 38 shows that, in general, normality is not transitive. However, there are some special instances in which normality is transitive. In this exercise, we will investigate one such situation. Let G be a group and N a normal subgroup of G. Assume that N is cyclic with generator a. Let H be a subgroup of N.

(a) Explain why we can write $H = \langle a^m \rangle$ for some integer m.

(b) Let $y \in H$ and $x \in G$. Write y in terms of a. Then use the fact from Exercise 3 in Investigation 21 (see page 301) that

$$xy^k x^{-1} = (xyx^{-1})^k$$

for all $k \in \mathbb{Z}$ to show that H is a normal subgroup of G. Conclude that if N is a cyclic subgroup of G, $N \triangleleft G$, and H is a subgroup of N (and so H is normal in N because N is cyclic), then $H \triangleleft G$.

Connections

Given a normal subgroup N of a group G, we can define congruence modulo N and form the quotient group G/N. Quotient structures are useful in mathematics in that they often have a simpler structure than the original set and yet can provide important information about the original set. In Investigation 5, we studied congruence and quotient structures in the integers, constructing the set $\mathbb{Z}_n$. If you studied ring theory before group theory, you should also notice connections between the topics in this investigation and those in Investigations 15 and 16. In particular, the set G/N represents the set of distinct congruence classes modulo N in the same way that $F[x]/(f(x))$ represents the set of distinct congruence classes modulo $f(x)$, and R/I represents the distinct congruence classes modulo an ideal I. Since a ring R is an Abelian group under its addition operation, any ideal I of R is always a normal subgroup of R under addition. So, in the case of rings, we do not have to worry about normality as we do when constructing quotient groups.

Investigation 28

Products of Groups

Focus Questions

By the end of this investigation, you should be able to give precise and thorough answers to the questions listed below. You may want to keep these questions in mind to focus your thoughts as you complete the investigation.

- What is a direct product, and how can direct products be used to construct larger groups from smaller ones?
- What is the external direct product of two groups, and how is the operation in an external direct product defined?
- How can we find the order of an element in a direct product of groups?
- What is an internal direct product of groups? Under what conditions is the internal direct product defined?
- How are the external and internal direct products of groups similar, and how are they different?

Preview Activity 28.1. If we have two integers, say k and m, we can combine these integers in different ways (e.g., using addition or multiplication) to obtain another integer. In a similar manner, if we have two groups G and H, we can combine them together to make another group, called the *direct product* of G and H, that contains copies of both G and H as subgroups. We will soon define direct products formally, but before doing so, let's take a look at an example. An operation table for the direct product of $\mathbb{Z}_2$ and $\mathbb{Z}_3$, denoted $\mathbb{Z}_2 \oplus \mathbb{Z}_3$, is shown in Table 28.1. We use the notation $[a]_n$ to indicate the congruence class of a in $\mathbb{Z}_n$.

(a) Describe precisely how the elements of $\mathbb{Z}_2 \oplus \mathbb{Z}_3$ are related to the elements of $\mathbb{Z}_2$ and $\mathbb{Z}_3$.

(b) How does the operation in $\mathbb{Z}_2 \oplus \mathbb{Z}_3$ seem to be related to the operations in $\mathbb{Z}_2$ and $\mathbb{Z}_3$?

(c) Does $\mathbb{Z}_2 \oplus \mathbb{Z}_3$ seem to be a group under the operation given in Table 28.1? If so, what is the identity element of $\mathbb{Z}_2 \oplus \mathbb{Z}_3$? How could we find the inverse of an element $([a]_2, [b]_3)$ in $\mathbb{Z}_2 \oplus \mathbb{Z}_3$?

(d) Consider the set S defined by

$$S = \{([0]_2, [x]_3) : [x]_3 \in \mathbb{Z}_3\}.$$

Is S a subgroup of $\mathbb{Z}_2 \oplus \mathbb{Z}_3$? Why or why not?

$+$	$([0]_2,[0]_3)$	$([0]_2,[1]_3)$	$([0]_2,[2]_3)$	$([1]_2,[0]_3)$	$([1]_2,[1]_3)$	$([1]_2,[2]_3)$
$([0]_2,[0]_3)$	$([0]_2,[0]_3)$	$([0]_2,[1]_3)$	$([0]_2,[2]_3)$	$([1]_2,[0]_3)$	$([1]_2,[1]_3)$	$([1]_2,[2]_3)$
$([0]_2,[1]_3)$	$([0]_2,[1]_3)$	$([0]_2,[2]_3)$	$([0]_2,[0]_3)$	$([1]_2,[1]_3)$	$([1]_2,[2]_3)$	$([1]_2,[0]_3)$
$([0]_2,[2]_3)$	$([0]_2,[2]_3)$	$([0]_2,[0]_3)$	$([0]_2,[1]_3)$	$([1]_2,[2]_3)$	$([1]_2,[0]_3)$	$([1]_2,[1]_3)$
$([1]_2,[0]_3)$	$([1]_2,[0]_3)$	$([1]_2,[1]_3)$	$([1]_2,[2]_3)$	$([0]_2,[0]_3)$	$([0]_2,[1]_3)$	$([0]_2,[2]_3)$
$([1]_2,[1]_3)$	$([1]_2,[1]_3)$	$([1]_2,[2]_3)$	$([1]_2,[0]_3)$	$([0]_2,[1]_3)$	$([0]_2,[2]_3)$	$([0]_2,[0]_3)$
$([1]_2,[2]_3)$	$([1]_2,[2]_3)$	$([1]_2,[0]_3)$	$([1]_2,[1]_3)$	$([0]_2,[2]_3)$	$([0]_2,[0]_3)$	$([0]_2,[1]_3)$

Table 28.1
Operation table for $\mathbb{Z}_2 \oplus \mathbb{Z}_3$.

(e) What is the relationship between $\mathbb{Z}_3$ and the set S defined in part (d)? Explain.

External Direct Products of Groups

Preview Activity 28.1 introduces an operation on pairs of groups and indicates a way to construct new groups from given ones. The formal definition uses the Cartesian product of two sets. If R and S are sets, then the *Cartesian product* $R \times S$ of R and S is the set of ordered pairs of elements, where the first coordinate comes from R and the second coordinate comes from S. In other words,

$$R \times S = \{(r, s) : r \in R, s \in S\}.$$

Definition 28.2. Let G and H be groups with operations $\cdot_G$ and $\cdot_H$, respectively. The **external direct product** $G \oplus H$ of G and H is the Cartesian product $G \times H$, together with an operation defined componentwise—that is,

$$(g_1, h_1)(g_2, h_2) = (g_1 \cdot_G g_2, h_1 \cdot_H h_2).$$

It is important to note in Definition 28.2 that there are three distinct operations involved: the operation in G, the operation in H, and the operation in $G \oplus H$. In general, we will suppress any symbolism that indicates which operation is which and leave the context in which the notation is used to alleviate any potential ambiguities. For instance, when we write $(g_1, h_1)(g_2, h_2)$ with $(g_1, h_1), (g_2, h_2) \in G \oplus H$, it is understood that we are using the operation in $G \oplus H$. Likewise, when we write $g_1 g_2$ for a pair of elements $g_1, g_2 \in G$, then we know that we are using the operation from G.

A word about notation and terminology is in order here. Some authors write the external direct product as $G \times H$, and some just call it a direct product. We will use the word *external* to distinguish this product from the *internal* direct product that we will define later. The word external is used because there may be no relationship at all between the groups G and H, and the construction of $G \oplus H$ takes place outside of both groups. Some authors only use the notation $G \oplus H$ for the external direct product (and call it a direct sum) when both of the groups G and H are Abelian under additive operations. The definition of the internal direct product, and the connection between the external and internal direct products, will ultimately remove any conflict in the notation. We have adopted the $\oplus$ notation to distinguish between the external direct product and the internal direct product, and to be consistent with the notation we use for the direct sum in ring theory. (See Investigation 9.)

Activity 28.3.

(a) List all of the elements in $\mathbb{Z}_2 \oplus \mathbb{Z}_2$.

(b) Construct the operation table for $\mathbb{Z}_2 \oplus \mathbb{Z}_2$ using the operation defined on the direct product.

You may have observed in Preview Activity 28.1 and Activity 28.3 that direct products appear to satisfy a number of important properties. We will explore some of these properties in the next activity.

Activity 28.4. Let G and H be groups.

(a) Why is $G \oplus H$ closed with respect to the operation defined on the direct product?

(b) Is the operation defined on $G \oplus H$ an associative operation? Prove your answer.

(c) Is there an identity element for the operation defined on $G \oplus H$? Explain.

(d) Does the element $(g, h) \in G \oplus H$ have an inverse in $G \oplus H$? Explain.

(e) Is $G \oplus H$ a group under the operation defined on $G \oplus H$? Explain.

Activity 28.4 provides the rationale for the following theorem.

Theorem 28.5. *Let G and H be groups. Then $G \oplus H$ is also a group.*

Finally, it is worth noting that $G \oplus H$ contains copies of both G and H as subgroups, as formalized in the next theorem. Its proof is left as an exercise. (See Exercise 3.)

Theorem 28.6. *Let G and H be groups with identities e_G and e_H, respectively. Then*

$$G \oplus \{e_H\} = \{(g, e_H) : g \in G\}$$

and

$$\{e_G\} \oplus H = \{(e_G, h) : h \in H\}$$

are both subgroups of $G \oplus H$.

A word of caution about Theorem 28.6: the group $G \oplus \{e_H\}$ is not equal to G; it just looks like G. The same is true for the groups $\{e_G\} \oplus H$ and H. However, you may have noticed that we can identify each element (g, e_H) of $G \oplus \{e_H\}$ with the element g in G. So while $G \oplus \{e_H\}$ is not equal to G, there is a natural identification of the elements of $G \oplus \{e_H\}$ with the elements of G in such a way that the two groups are virtually identical, both in their makeup and in the way they behave with respect to their operations. In that sense, the groups G and $G \oplus \{e_H\}$ are essentially the same, a notion that we will make more precise in Investigation 29 when we study group isomorphisms. Likewise, H and $\{e_G\} \oplus H$ can also be considered to be essentially the same group, which implies that $G \oplus H$ in some sense contains a copy of both G and H. Thus, direct products provide a way to construct a larger group that contains each of two smaller groups. This construction is useful in numerous examples, and it will play an essential role in the classification of finite Abelian groups, as we will see in Investigation 31.

Activity 28.7. If G is a group of order m and H a group of order n, what is $|G \oplus H|$? Explain.

Orders of Elements in Direct Products

As we have discussed, the subgroup structure of a group tells us much about the group. Each element in a group determines a subgroup—namely, the cyclic subgroup generated by that element—and we can learn a lot about a group by understanding its cyclic subgroups. In this section, we will examine the orders of elements in direct products, which will provide important insights about cyclic subgroups of direct products.

Let G and H be groups. If $g \in G$ has order m and $h \in H$ has order n, what can we say about the order of (g, h) in $G \oplus H$? It is natural at first to think that $|(g, h)| = |g|\,|h|$. In the next activity, we will see if this is actually the case.

Activity 28.8.

(a) Determine the orders of the elements $[4]_8 \in \mathbb{Z}_8$, $[3]_{12} \in \mathbb{Z}_{12}$, and $([4]_8, [3]_{12}) \in \mathbb{Z}_8 \oplus \mathbb{Z}_{12}$. Does $|([4]_8, [3]_{12})|$ equal $|[4]_8|\,|[3]_{12}|$?

(b) Determine the orders of the elements $[4]_8 \in \mathbb{Z}_8$, $[4]_{12} \in \mathbb{Z}_{12}$, and $([4]_8, [4]_{12}) \in \mathbb{Z}_8 \oplus \mathbb{Z}_{12}$. Does $|([4]_8, [4]_{12})|$ equal $|[4]_8|\,|[4]_{12}|$?

Activity 28.8 shows that there are times when $|(g, h)| = |g|\,|h|$ and times when $|(g, h)| \neq |g|\,|h|$. If we were to do enough examples, we would undoubtedly conjecture the formula for determining the order of an element (g, h) in $G \oplus H$ that is given in the next theorem. You may want to do enough other examples on your own to convince yourself that the theorem makes sense. In the proof of the theorem, one result we will use that we have not yet verified (although you will be asked to do so in Activity 28.18) is that for any element $(g, h) \in G \oplus H$ and any integer k, $(g, h)^k = (g^k, h^k)$.

Theorem 28.9. *Let G and H be groups with $g \in G$ and $h \in H$, both of finite order. Then*

$$|(g, h)| = \operatorname{lcm}(|g|, |h|).$$

Proof. Let e_G be the identity in G and e_H the identity in H. Let $m = |g|$ and $n = |h|$. First, we will show that $(g, h)^{\operatorname{lcm}(m,n)} = (e_G, e_H)$. Since $\operatorname{lcm}(m, n)$ is a common multiple of m and n, we know that $mr = \operatorname{lcm}(m, n) = ns$ for some integers r and s. Therefore,

$$(g, h)^{\operatorname{lcm}(m,n)} = \left(g^{\operatorname{lcm}(m,n)}, h^{\operatorname{lcm}(m,n)}\right) = ((g^m)^r, (h^n)^s) = (e_G, e_H).$$

So $|(g, h)| \leq \operatorname{lcm}(m, n)$.

Now suppose that $(g, h)^k = (e_G, e_H)$ for some positive integer k. Then

$$(e_G, e_H) = (g, h)^k = (g^k, h^k),$$

and so $g^k = e_G$ and $h^k = e_H$. Theorem 23.5 (see page 320) shows that m divides k and n divides k. By the definition of the least common multiple, we can therefore conclude that $\operatorname{lcm}(m, n) \leq k$. This, however, implies that $|(g, h)| \geq \operatorname{lcm}(m, n)$. Since we showed earlier that $|(g, h)| \leq \operatorname{lcm}(m, n)$, it follows that $|(g, h)| = \operatorname{lcm}(|g|, |h|)$. ∎

As the next activity demonstrates, Theorem 28.9 can be applied to yield important conclusions about the subgroup structure of direct products.

Activity 28.10.

(a) Is $\mathbb{Z}_9 \oplus \mathbb{Z}_{10}$ a cyclic group? Answer using Theorem 28.9.

(b) Does $\mathbb{Z}_9 \oplus \mathbb{Z}_{10}$ contain an element of order 6? If yes, find such an element. If no, explain why not.

Although we defined the external direct product for only two groups, we can extend the definition to any finite number of groups. The next activity illustrates how this can be done.

Activity 28.11. We can generalize the construction of the external direct product of two groups to any finite number of groups. Let $n \geq 2$ be an integer, and let G_1, G_2, ..., G_n be groups. We define the elements in $G_1 \oplus G_2 \oplus \cdots \oplus G_n$ to be all ordered n-tuples of the form $(g_1, g_2, \ldots, g_n)$, where $g_i \in G_i$ for each i. We can then define an operation on $G_1 \oplus G_2 \oplus \cdots \oplus G_n$ by

$$(a_1, a_2, \ldots, a_n)(g_1, g_2, \ldots, g_n) = (a_1g_1, a_2g_2, \ldots, a_ng_n)$$

for any $(a_1, a_2, \ldots, a_n), (g_1, g_2, \ldots, g_n) \in G_1 \oplus G_2 \oplus \cdots \oplus G_n$.

(a) Show that $G_1 \oplus G_2 \oplus \cdots \oplus G_n$ is closed under its operation.

(b) For each i, let e_i be the identity in G_i. What is the identity element in $G_1 \oplus G_2 \oplus \cdots \oplus G_n$? Prove your answer.

(c) Show that the operation defined above is associative in $G_1 \oplus G_2 \oplus \cdots \oplus G_n$.

(d) What conclusion can we draw about the set $G_1 \oplus G_2 \oplus \cdots \oplus G_n$?

Internal Direct Products in Groups

We have seen that if G and H are groups, we can form the direct product $G \oplus H$ of G and H, and this direct product is a group under the obvious componentwise operation. The group $G \oplus H$ is an *external* direct product because it is constructed outside of both G and H.

We can take a different approach to the group $G \oplus H$ and view it as a product of subgroups—that is, an *internal* direct product. Recall that Theorem 28.6 shows that the groups $G' = G \oplus \{e_H\} = \{(g, e_H) : g \in G\}$ and $H' = \{e_G\} \oplus H = \{(e_G, h) : h \in H\}$ are subgroups of $G \oplus H$. Furthermore, both G' and H' are normal subgroups of $G \oplus H$. (See Exercise 3.) These two subgroups G' and H' generate all of the elements of $G \oplus H$ in a natural way. We can write any element $(g, h) \in G \oplus H$ as $(g, h) = (g, e_H)(e_G, h)$, and so every element in $G \oplus H$ is of the form $g'h'$ with $g' \in G'$ and $h' \in H'$. In addition, this decomposition of an element $(g, h) \in G \oplus H$ into a product $g'h'$ with $g' \in G'$ and $h' \in H'$ is unique. As we will see, the uniqueness of the decomposition is due to the fact that $G' \cap H' = \{(e_G, e_H)\}$. So, to summarize, we have two normal subgroups G' and H' of a group $G \oplus H$, the intersection of G' and H' is trivial, and every element in $G \oplus H$ can be written uniquely as a product of an element in G' and an element in H'. Since G' and H' are subgroups of $G \oplus H$, we call this decomposition of $G \oplus H$ into products of elements from its subgroups G' and H' an *internal direct product*.

As it turns out, we can generalize this type of construction. Suppose that K and N are both normal subgroups of a group G and $K \cap N = \{e\}$ (where e is the identity in G). Then we can make the set

$$KN = \{kn : k \in K, n \in N\}$$

into a subgroup of G using the operation from G. (See Activity 27.22 on page 372.) The resulting group, denoted $K \times N$, is contained within G and is called an *internal* direct product of K and N.

Definition 28.12. Let G be a group with identity e, and let K and N be normal subgroups of G such that $K \cap N = \{e\}$. The **internal direct product** of K and N is the subgroup

$$K \times N = \{kn : k \in K, n \in N\}$$

of G.

When a subgroup H of a group G is decomposed as the product $H = K \times N$ for some normal subgroups K and N of G, these normal subgroups are called *factors* of H. The product $K \times N$ is called an *internal* product because all of the products of group elements in $K \times N$ are performed within the group G.

Before we move on, a word of caution about notation is in order. Note that we use the $\times$ symbol for several different purposes, and you should be careful not to confuse the internal direct product $K \times N$ with the Cartesian product $K \times N$. The internal direct product $K \times N$ is a subgroup of the group G that contains K and N as normal subgroups, while the Cartesian product $K \times N$ is just a set of ordered pairs. The context should make clear which product is meant when the $\times$ symbol is used.

Activity 28.13. The operation table for

$$U_{28} = \{[1], [3], [5], [9], [11], [13], [15], [17], [19], [23], [25], [27]\}$$

is given in Table 28.2. Let $K = \langle [9] \rangle$ and $N = \langle [13] \rangle$. Note that $K \cap N = \{[1]\}$. Find the elements

	[1]	[3]	[5]	[9]	[11]	[13]	[15]	[17]	[19]	[23]	[25]	[27]
[1]	[1]	[3]	[5]	[9]	[11]	[13]	[15]	[17]	[19]	[23]	[25]	[27]
[3]	[3]	[9]	[15]	[27]	[5]	[11]	[17]	[23]	[1]	[13]	[19]	[25]
[5]	[5]	[15]	[25]	[17]	[27]	[9]	[19]	[1]	[11]	[3]	[13]	[23]
[9]	[9]	[27]	[17]	[25]	[15]	[5]	[23]	[13]	[3]	[11]	[1]	[19]
[11]	[11]	[5]	[27]	[15]	[9]	[3]	[25]	[19]	[13]	[1]	[23]	[17]
[13]	[13]	[11]	[9]	[5]	[3]	[1]	[27]	[25]	[23]	[19]	[17]	[15]
[15]	[15]	[17]	[19]	[23]	[25]	[27]	[1]	[3]	[5]	[9]	[11]	[13]
[17]	[17]	[23]	[1]	[13]	[19]	[25]	[3]	[9]	[15]	[27]	[5]	[11]
[19]	[19]	[1]	[11]	[3]	[13]	[23]	[5]	[15]	[25]	[17]	[27]	[9]
[23]	[23]	[13]	[3]	[11]	[1]	[19]	[9]	[27]	[17]	[25]	[15]	[5]
[25]	[25]	[19]	[13]	[1]	[23]	[17]	[11]	[5]	[27]	[15]	[9]	[3]
[27]	[27]	[25]	[23]	[19]	[17]	[15]	[13]	[11]	[9]	[5]	[3]	[1]

Table 28.2
Operation table for U_{28}.

of the internal direct product $K \times N$, and then write the operation table for $K \times N$.

In some cases, we can write an entire group G as an internal direct product.

Activity 28.14. Let $G = U_{28}$, $K = \langle [3] \rangle$, and $N = \langle [13] \rangle$.

(a) Determine the elements in K and N. What is $K \cap N$?

	[1]	[13]
[1]	[1]	[13]
[3]	[3]	
[9]	[9]	
[19]	[19]	
[25]	[25]	
[27]	[27]	

Table 28.3
Products in $\langle[3]\rangle \times \langle[13]\rangle$ in U_{28}.

(b) By completing the calculation of all of the products in Table 28.3, show that every element in U_{28} can be written uniquely in the form kn for some $k \in K$ and $n \in N$.

These last activities serve as a warm-up for a later investigation (Investigation 31) in which we will prove the Fundamental Theorem of Finite Abelian Groups, which will tell us that every finite Abelian group can be decomposed into an internal (and, in a sense, external) direct product of cyclic groups.

There are two important properties of internal direct products that deserve mention. First, if a group G is non-Abelian, we know that we cannot assume that elements in G commute. However, with an internal direct product, we can show that certain elements do commute. Theorem 28.15 formalizes this result. Second, each element in an internal direct product $K \times N$ can be written as a product kn where $k \in K$ and $n \in N$, but we don't know if such a representation is unique. Theorem 28.15 resolves this issue as well.

Theorem 28.15. *Let G be a group with identity e, and let K and N be normal subgroups of G such that $K \cap N = \{e\}$. Then*

(i) *$kn = nk$ for all $k \in K$ and $n \in N$; and*

(ii) *the representation of an element in $K \times N$ in the form kn, where $k \in K$ and $n \in N$, is unique.*

Proof. Let G be a group with identity e, and let K and N be normal subgroups of G such that $K \cap N = \{e\}$. To prove part (i), let $k \in K$ and $n \in N$. We know that $knk^{-1} \in N$ and $nk^{-1}n^{-1} \in K$, so $(knk^{-1})n^{-1} \in N$ and $k(nk^{-1}n^{-1}) \in K$. Therefore, $knk^{-1}n^{-1} \in K \cap N$ and so $knk^{-1}n^{-1} = e$. Now $e = knk^{-1}n^{-1} = (kn)(nk)^{-1}$, so right multiplication by nk yields $kn = nk$.

For part (ii), let $x \in K \times N$, and suppose that $x = k_1n_1 = k_2n_2$ for some $k_1, k_2 \in K$ and $n_1, n_2 \in N$. Then $k_2^{-1}k_1 = n_2n_1^{-1} \in K \cap N$ and so $k_2^{-1}k_1 = n_2n_1^{-1} = e$. Therefore, $k_1 = k_2$ and $n_1 = n_2$, which shows that the representation of the element x in $K \times N$ is unique. ∎

Note that part (i) of Theorem 28.15 shows that the elements of K and N commute with each other, but K and N themselves may not be Abelian groups.

Activity 28.16. We can generalize the construction of the internal direct product of two groups to any finite number of groups. Let G be a group with identity e, $m \geq 2$ an integer, and let N_1, N_2, ..., N_m be normal subgroups of G with

$$N_i \cap (N_1 \times N_2 \times \cdots N_{i-1} \times N_{i+1} \times \cdots \times N_m) = \{e\}$$

for each i. We define the elements in $N_1 \times N_2 \times \cdots \times N_m$ to be elements in G of the form $n_1 n_2 \cdots n_m$ where $n_i \in N_i$ for each i. The operation is the same as the operation in G.

(a) Explain why $N_1 \times N_2 \times \cdots \times N_m$ is closed under the operation from G.

(b) Explain why $N_1 \times N_2 \times \cdots \times N_m$ contains the identity from G.

(c) Explain why $N_1 \times N_2 \times \cdots \times N_m$ contains an inverse for each of its elements.

(d) What conclusion can we draw from parts (a) – (c)?

(e) You may be wondering why, in the above construction, we required that

$$N_i \cap (N_1 \times N_2 \times \cdots N_{i-1} \times N_{i+1} \times \cdots \times N_m) = \{e\}$$

for each i. This condition seems to be more complicated than in the two subgroup case, but it is what we need in order to guarantee the same uniqueness of the representation of an element in $N_1 \times N_2 \times \cdots \times N_m$ that part (ii) of Theorem 28.15 gives in the $m = 2$ case. Prove that, with this condition, the representation of an element in $N_1 \times N_2 \times \cdots \times N_m$ in the form $n_1 n_2 \cdots n_m$, with $n_i \in N_i$ for each i, is in fact unique.

Concluding Activities

Activity 28.17. Write a formal proof of Theorem 28.5.

Activity 28.18. In this activity, we will verify a result that we used in the proof of Theorem 28.9.

(a) Let G and H be groups, and let $g \in G$ and $h \in H$. Prove that $(g, h)^n = (g^n, h^n)$ for each integer n.

(b) Let $n \geq 2$ be an integer, let G_1, G_2, ..., G_n be groups, and let $(g_1, g_2, \ldots, g_n) \in G_1 \oplus G_2 \oplus \cdots \oplus G_n$. What element is $(g_1, g_2, \ldots, g_n)^k$, where k is an integer? Prove your answer.

Activity 28.19. Let G and H be groups.

(a) Under what conditions is $G \oplus H$ Abelian? State your answer in the form of a biconditional statement, and then prove it.

(b) Is $G \oplus H$ cyclic when G and H are both cyclic? If $G \oplus H$ is cyclic, must G and H both be cyclic? Prove your answers.

Activity 28.20. Write a short summary that describes the important concepts, and the relationships between these concepts, that were introduced in this investigation. Explain how the material in this investigation is connected to the material in Investigations 20 and 22.

Exercises

(1) The operation table for the Klein 4-group $\mathcal{V}$ (see Activity 24.6 on page 329 of Investigation 24) is as follows:

	1	a	b	ab
1	1	a	b	ab
a	a	a	ab	b
b	b	ab	1	a
ab	ab	b	a	1

(a) Construct the operation table for the group $\mathbb{Z}_2 \oplus \mathbb{Z}_2$. (See Activity 28.3.)

(b) Compare the operation tables for $\mathbb{Z}_2 \oplus \mathbb{Z}_2$ and for $\mathcal{V}$. How are these groups similar? How are they different?

(2) Create the operation table for the group $(\mathbb{Z}_4 \oplus \mathbb{Z}_2)/\langle([2],[1])\rangle$.

* (3) Let G and H be groups with identities e_G and e_H, respectively.

(a) Prove Theorem 28.6, which states that if G and H are groups with identities e_G and e_H, respectively, then

$$G \oplus \{e_H\} = \{(g, e_H) : g \in G\}$$

and

$$\{e_G\} \oplus H = \{(e_G, h) : h \in H\}$$

are both subgroups of $G \oplus H$.

(b) Show that G' and H' are normal subgroups of $G \oplus H$.

(4) Let $G = \mathbb{Z}_4 \oplus \mathbb{Z}_6$.

(a) Does G contain an element of order 12? If so, find one. If not, prove that no such element exists.

(b) Calculate the orders of all of the elements in G.

(5) Let $n \geq 2$ be an integer, and let G_1, G_2, ..., G_n be groups. For each i, let N_i be a subgroup of G_i.

(a) Prove that $N_1 \oplus N_2 \oplus \cdots \oplus N_n$ is a subgroup of $G_1 \oplus G_2 \oplus \cdots \oplus G_n$.

(b) Find necessary and sufficient conditions on the groups N_i so that $N_1 \oplus N_2 \oplus \cdots \oplus N_n$ is a normal subgroup of $G_1 \oplus G_2 \oplus \cdots \oplus G_n$.

(c) Is it true that if M is a subgroup of $G_1 \oplus G_2 \oplus \cdots \oplus G_n$, then for each i there is a subgroup N_i of G_i so that $M = N_1 \oplus N_2 \oplus \cdots \oplus N_n$? Prove your answer.

(6) (a) Let G be a cyclic group of order m and H a cyclic group of order n. Find necessary and sufficient conditions m and n so that $G \oplus H$ is cyclic. Prove your answer.

(b) Now extend part (a) to any finite number of cyclic groups. That is, let $k \geq 2$ be a positive integer, and let $n_1, n_2, \ldots, n_k$ be positive integers. For each i between 1 and k, let G_i be a cyclic group of order n_i. Find and prove necessary and sufficient conditions so that the group

$$G = G_1 \oplus G_2 \oplus \cdots \oplus G_k$$

is cyclic.

(7) Generalize Theorem 28.9. That is, let $n \geq 2$ be an integer, let $G_1, G_2, \ldots, G_n$ be groups, and, for each i, let $g_i \in G_i$ be an element of finite order in G_i. Determine $|(g_1, g_2, g_3, \ldots, g_n)|$ in $G_1 \oplus G_2 \oplus \cdots \oplus G_n$. Prove your answer.

(8) (a) Construct the operation table for the group $D_3 \oplus \mathbb{Z}_2$.

(b) Find the orders of all of the elements in $D_3 \oplus \mathbb{Z}_2$.

(c) Exercise 3 shows that the groups $D_3 \oplus \{[0]\}$ and $\{I\} \oplus \mathbb{Z}_2$ are both normal subgroups of $D_3 \oplus \mathbb{Z}_2$.

(i) Find the elements of $(D_3 \oplus \mathbb{Z}_2)/(D_2 \oplus \{[0]\})$, and construct the operation table for this quotient group. What familiar group does $(D_3 \oplus \mathbb{Z}_2)/(D_3 \oplus \{[0]\})$ look like?

(ii) Find the elements of $(D_3 \oplus \mathbb{Z}_2)/(\{I\} \oplus \mathbb{Z}_2)$, and construct the operation table for this quotient group. What familiar group does $(D_3 \oplus \mathbb{Z}_2)/(\{I\} \oplus \mathbb{Z}_2)$ look like?

(9) Let G be a finite group with identity e.

(a) If K and N are normal subgroups of G with $K \cap N = \{e\}$ and $|K|\,|N| = |G|$, does $G = K \times N$? Prove your answer.

(b) With the same conditions as in (a), does $G = N \times K$? Explain.

(c) Does part (a) generalize to any number of normal subgroups? State your answer in the form of a conjecture, and then prove it.

(10) (a) Write the group U_{15} as an internal direct product of subgroups.

(b) Is the representation of U_{15} as an internal direct product of groups (from part (a)) unique? That is, can you find normal subgroups K' and N' of U_{15} so that $U_{15} = K' \times N'$ but either $K' \neq K$ or $N' \neq N$?

(11) In this problem, we will determine when D_n can be written as an internal direct product of normal subgroups. Two previous results might be helpful. First, Exercise 36 of Investigation 27 (see page 378) tells us that a non-trivial proper subgroup N of D_n is normal in D_n if and only if N is a subgroup of $\langle R \rangle$ or n is even and N is one of $\langle r, R^2 \rangle$ or $\langle rR, R^2 \rangle$. Also, Exercise 2 of Investigation 24 (see page 330) shows that $Z(D_n) = \{I, R^{n/2}\}$ if n is even and $Z(D_n) = \{I\}$ if n is odd.

(a) Explain why D_n cannot be written as an internal direct product of two normal subgroups if n is odd.

(b) Can D_4 be written as an internal direct product of normal subgroups? Explain your answer.

(c) Show that D_6 can be written as an internal direct product of normal subgroups.

(d) State and prove a complete characterization of the groups D_n that can be written as an internal direct product of normal subgroups.

(12) In this investigation, we introduced the external direct product of a finite number of groups. We can also define an external direct product of an infinite number of groups. For each natural number i, let G_i be a group with identity e_i. Define the **infinite direct product**

$$\bigoplus_{i=1}^{\infty} G_i = G_1 \oplus G_2 \oplus G_3 \oplus \cdots$$

to be the set of all infinite tuples of the form

$$a = (a_1, a_2, a_3, \ldots),$$

where $a_i \in G_i$ for all i, with an operation defined componentwise—that is,

$$(a_1, a_2, a_3, \ldots)(b_1, b_2, b_3, \ldots) = (a_1 b_1, a_2 b_2, a_3 b_3, \ldots).$$

(a) Prove that the infinite direct product $\bigoplus_{i=1}^{\infty} G_i$ is a group.

(b) The **restricted direct product** (or **direct sum**) of the groups G_i, for $i \in \mathbb{Z}^+$, is the subset of $\bigoplus_{i=1}^{\infty} G_i$ consisting of the infinite tuples of the form $a = (a_1, a_2, a_3, \ldots)$, where $a_i \in G_i$ for all i, and $a_i = e_i$ for all but finitely many values of i. To illustrate, let $G_i = \mathbb{Z}_{i+1}$ for every natural number i, and define G to be the restricted direct product of the group

$$\bigoplus_{i=1}^{\infty} G_i = \bigoplus_{i=1}^{\infty} \mathbb{Z}_{i+1} = \mathbb{Z}_2 \oplus \mathbb{Z}_3 \oplus \mathbb{Z}_4 \oplus \cdots$$

Then

$$a = ([1]_2, [0]_3, [0]_4, [4]_5, [3]_6, [0]_7, [0]_8, [0]_9, \ldots) \in G,$$

since a_i is the zero element in $\mathbb{Z}_{i+1}$ for all but 3 values of i. In contrast,

$$b = ([1]_2, [1]_3, [1]_4, [1]_5, \ldots) \notin G,$$

since b_i is nonzero for infinitely many values of i.

(i) Prove that the restricted direct product is a subgroup of the direct product.

(ii) Is the restricted direct product a normal subgroup of the direct product? Prove your answer.

(iii) Let $i \in \mathbb{Z}^+$, and let $G_i = \mathbb{Z}_{i+1}$ for each i. Show that $\bigoplus_{i=1}^{\infty} G_i$ has elements of infinite order, but every element of the corresponding restricted direct product has finite order.

Connections

In this investigation, we studied external and internal products of groups. If you studied ring theory before group theory, you should notice similarities between the external direct products of groups from this investigation and the direct sums of rings we considered in Investigation 9. Because of the simpler structure of groups, we can define more than one type of product of groups. Since there are two operations defined on a ring, rings are more complicated objects than groups and so we only defined one type of "product" (which we called a direct sum). The constructions of direct sums of rings and external direct products of groups are analogous. To create a direct sum/product of a pair of objects (either rings or groups), we make the Cartesian product of the two sets into the appropriate algebraic structure (either a ring or a group) using operations defined componentwise.

Investigation 29

Group Isomorphisms and Invariants

Focus Questions

By the end of this investigation, you should be able to give precise and thorough answers to the questions listed below. You may want to keep these questions in mind to focus your thoughts as you complete the investigation.

- Intuitively, what does it mean for two groups to be "essentially the same"?
- What does it mean for two groups to be isomorphic? How does the definition of isomorphism reflect the informal definition of "essentially the same"?
- What strategies can be used to prove that two groups are isomorphic? How are these strategies motivated by the definition of isomorphism?
- What is an invariant, and how does one prove that a property is an invariant?
- How can invariants be used to prove that two groups are not isomorphic? Can invariants be used to prove that two groups are isomorphic?
- What are isomorphism classes, and what is their role in the classification of finite groups?
- What does Cayley's Theorem say, and why is it important?

Preview Activity 29.1. The notion of "sameness" is very important in mathematics, for it allows us to identify when two objects should be considered indistinguishable, and thus treated identically. Identifying sameness also makes our analysis more efficient, since it allows us to consider entire classes of objects at the same time, instead of dealing with each object individually.

In this investigation, we will define precisely what it means for two groups to be essentially the same, or *isomorphic*. Before we do so, however, let's apply our intuitive ideas about sameness to a few examples.

The operation tables for four groups are shown below. Which of these groups would you consider to be essentially the same, and which would you consider to be different? Consider each possible pair of groups, and give a convincing argument to justify your answer for each.

G_1 :

	a	b	c	d	e	f	g	h
a	a	b	c	d	e	f	g	h
b	b	c	d	a	h	e	f	g
c	c	d	a	b	g	h	e	f
d	d	a	b	c	f	g	h	e
e	e	f	g	h	a	b	c	d
f	f	g	h	e	d	a	b	c
g	g	h	e	f	c	d	a	b
h	h	e	f	g	b	c	d	a

G_2 :

	α	β	γ	δ	ϵ	ζ	η	θ
α	α	β	γ	δ	ϵ	ζ	η	θ
β	β	γ	δ	ϵ	ζ	η	θ	α
γ	γ	δ	ϵ	ζ	η	θ	α	β
δ	δ	ϵ	ζ	η	θ	α	β	γ
ϵ	ϵ	ζ	η	θ	α	β	γ	δ
ζ	ζ	η	θ	α	β	γ	δ	ϵ
η	η	θ	α	β	γ	δ	ϵ	ζ
θ	θ	α	β	γ	δ	ϵ	ζ	η

G_3 :

	s	t	u	v	w	x	y	z
s	s	t	u	v	w	x	y	z
t	t	u	v	s	x	y	z	w
u	u	v	s	t	y	z	w	x
v	v	s	t	u	z	w	x	y
w	w	x	y	z	t	u	v	s
x	x	y	z	w	u	v	s	t
y	y	z	w	x	v	s	t	u
z	z	w	x	y	s	t	u	v

G_4 :

	♠	♡	♢	♣	✠	♯	†	℧
♠	♠	♡	♢	♣	✠	♯	†	℧
♡	♡	♢	♣	♠	♯	†	℧	✠
♢	♢	♣	♠	♡	†	℧	✠	♯
♣	♣	♠	♡	♢	℧	✠	♯	†
✠	✠	♯	†	℧	♠	♡	♢	♣
♯	♯	†	℧	✠	♡	♢	♣	♠
†	†	℧	✠	♯	♢	♣	♠	♡
℧	℧	✠	♯	†	♣	♠	♡	♢

Introduction

In Preview Activity 29.1, you were asked to decide which of the four groups shown were essentially the same, and which were not. At first glance, it would be easy to think that all of the groups are different. After all, their elements are certainly different. But is this enough to conclude that the groups themselves are different?

To answer this question, let's consider the group $\mathbb{Z}_8$:

$\mathbb{Z}_8$:

	[0]	[1]	[2]	[3]	[4]	[5]	[6]	[7]
[0]	[0]	[1]	[2]	[3]	[4]	[5]	[6]	[7]
[1]	[1]	[2]	[3]	[4]	[5]	[6]	[7]	[0]
[2]	[2]	[3]	[4]	[5]	[6]	[7]	[0]	[1]
[3]	[3]	[4]	[5]	[6]	[7]	[0]	[1]	[2]
[4]	[4]	[5]	[6]	[7]	[0]	[1]	[2]	[3]
[5]	[5]	[6]	[7]	[0]	[1]	[2]	[3]	[4]
[6]	[6]	[7]	[0]	[1]	[2]	[3]	[4]	[5]
[7]	[7]	[0]	[1]	[2]	[3]	[4]	[5]	[6]

Let's suppose also that we decided to abbreviate the names of the equivalence classes in $\mathbb{Z}_8$ by assigning a variable to each one. In particular, we'll let $\alpha = [0]$, $\beta = [1]$, $\gamma = [2]$, $\delta = [3]$, $\epsilon = [4]$, $\zeta = [5]$, $\eta = [6]$, and $\theta = [7]$.

Activity 29.2. Substitute α, β, γ, δ, ϵ, ζ, η, and θ for [0], [1], [2], [3], [4], [5], [6], and [7] in the operation table for $\mathbb{Z}_8$. That is, each time [0] appears in the tables, replace it with α. Do the same for the other classes as well, replacing [1] with β, [2] with γ, [3] with δ, [4] with ϵ, [5] with ζ, [6] with η, and [7] with θ. What do you notice?

If you completed Activity 29.2 correctly, you probably observed that the operation table for $\mathbb{Z}_8$ can be made to look exactly like the operation table of G_2 in Preview Activity 29.1, simply by renaming the elements. In other words, the only differences between $\mathbb{Z}_8$ and G_2 are the names of the elements. As far as their structure as groups is concerned, $\mathbb{Z}_8$ and G_2 are basically the same group. It is the operation—the way the elements interact with each other—rather than the names of the elements themselves that determines a group.

As it turns out, we can carry out a similar renaming to show that $\mathbb{Z}_8$ and G_3 are also essentially the same. Here we have to be a bit more careful, however, since the elements of G_3 seem to be arranged in a different order than those of $\mathbb{Z}_8$ and G_2. To illustrate, let's see what would happen if we replaced the names of the elements of G_3 with the names of the elements of $\mathbb{Z}_8$, keeping the elements in the same order as they are listed in the tables. We would replace s with [0], t with [1], u with [2], v with [3], w with [4], x with [5], y with [6], and z with [7], which would yield the following operation table:

	[0]	[1]	[2]	[3]	[4]	[5]	[6]	[7]
[0]	[0]	[1]	[2]	[3]	[4]	[5]	[6]	[7]
[1]	[1]	[2]	[3]	[0]	[5]	[6]	[7]	[4]
[2]	[2]	[6]	[0]	[1]	[6]	[7]	[4]	[5]
[3]	[3]	[0]	[1]	[2]	[7]	[4]	[5]	[6]
[4]	[4]	[5]	[6]	[7]	[1]	[2]	[3]	[0]
[5]	[5]	[6]	[7]	[4]	[2]	[3]	[0]	[1]
[6]	[6]	[7]	[4]	[5]	[3]	[0]	[1]	[2]
[7]	[7]	[4]	[5]	[6]	[0]	[1]	[2]	[3]

This table does not look like the operation table for $\mathbb{Z}_8$; the renaming of the elements of G_3 that we used did not yield a group that seemed to be the same as $\mathbb{Z}_8$. So what can we conclude from this? Does our seemingly failed attempt at renaming the elements of G_3 necessarily imply that G_3 and $\mathbb{Z}_8$ are different groups? In fact, it does not. All that we know at this point is that the particular renaming that we used yields a group that doesn't look like $\mathbb{Z}_8$. But what if we used a different renaming? For instance, notice that $w^2 = t$, $w^3 = x$, $w^4 = u$, $w^5 = y$, $w^6 = v$, $w^7 = z$, and $w^8 = s$, so w has order 8 and the group G_3 is cyclic. Because of this, if we want to identify elements of G_3 with the elements in $\mathbb{Z}_8$ so that the resulting group structures look identical, then we will need to identify w with an element in $\mathbb{Z}_8$ of order 8. This might lead us to try replacing w with $[1]$, $w^2 = t$ with $[2]$, $w^3 = x$ with $[3]$, $w^4 = u$ with $[4]$, $w^5 = y$ with $[5]$, $w^6 = v$ with $[6]$, $w^7 = z$ with $[7]$, and $w^8 = s$ with $[0]$. Doing so would yield the following operation table:

	[0]	[2]	[4]	[6]	[1]	[3]	[5]	[7]
[0]	[0]	[2]	[4]	[6]	[1]	[3]	[5]	[7]
[2]	[2]	[4]	[6]	[0]	[3]	[5]	[7]	[1]
[4]	[4]	[5]	[0]	[2]	[5]	[7]	[1]	[3]
[6]	[6]	[0]	[2]	[4]	[7]	[1]	[3]	[5]
[1]	[1]	[3]	[5]	[7]	[2]	[4]	[6]	[0]
[3]	[3]	[5]	[7]	[1]	[4]	[6]	[0]	[2]
[5]	[5]	[7]	[1]	[3]	[6]	[0]	[2]	[4]
[7]	[7]	[1]	[3]	[5]	[0]	[2]	[4]	[6]

Notice that, except for the order in which the elements are listed, this last operation table is identical to the operation table for $\mathbb{Z}_8$. Note also that the ordering of the elements affects only the way the table is displayed, and not the information that it contains. A simple rearrangement of the rows and columns puts the table in its more standard form. Thus, we can see that by renaming the elements of G_3, and possibly reordering the rows and columns of the resulting operation table, we are able to produce the operation table for $\mathbb{Z}_8$. Because of this, we say that G_3 and $\mathbb{Z}_8$ are basically the same group.

Now that we have seen a few examples, we are ready to state an informal definition, which we will use to formally define the notion of isomorphism in the next section.

Informal Definition 29.3. Let G and H be finite groups. The group G is said to be **essentially the same** as H if the operation table for G can be transformed into the operation table for H by doing nothing more than renaming the elements of G and/or reordering the rows and columns of G's operation table.

Note that this informal definition can easily be used to show that two groups are *not* essentially the same. For example, we saw that the group G_3 was cyclic, but neither G_1 nor G_4 contain an element of order 8. From these observations, it is clear that G_3 is a different group than either G_1 or G_3, since no matter how we rename and/or reorder the elements of G_3, we will still end up with a group that contains an element of order 8. Also, group G_1 is non-Abelian, so no matter how we relabel the elements of G_1, the resulting operation table will be that of a non-Abelian group. Since G_4 is an Abelian group, we can conclude that G_1 and G_4 are different groups.

To summarize, note that in order to show that two groups are the same, we must find a way to rename and/or reorder the elements of one group so that its operation table is identical to that of the other group. In order to show that two groups are different, however, it often suffices to identify a property that is different between the two groups—in particular, a property that could not possibly be different if one group had been obtained from the other by simply renaming and reordering elements.

Activity 29.4. We have already argued that groups G_1 and G_4 from Preview Activity 29.1 are different because G_1 is a non-Abelian group while G_4 is an Abelian group. Setting that difference aside, list at least two other properties that are different between G_1 and G_4 and would therefore show (by Informal Definition 29.3) that G_1 and G_4 are different groups.

Isomorphisms of Groups

Informal Definition 29.3 provides a helpful and intuitive way of thinking about what it means for two groups to be the same. This informal definition, however, has some significant limitations. First, it only works for finite groups. This is because it would be impossible to actually create the operation table for a group with infinitely many elements. Second, even for finite groups, the definition can be extremely cumbersome to work with, especially if the groups in question have more than a few elements. Can you imagine trying to create an operation table for a group with 50 or 1000 or even 50000 elements? The task would be daunting at best, and practically impossible at worst.

To deal with these difficulties, we will adopt a formal definition that captures the idea behind Informal Definition 29.3, but does so in a more precise manner. In order to motivate this definition, let's consider again the two main parts of Informal Definition 29.3.

Renaming Elements

When we argued that $\mathbb{Z}_8$ and G_2 were essentially the same group, we found a way to rename the elements of $\mathbb{Z}_8$ using the same names as the elements of G_2. This renaming was really just a bijective function (that is, a function that is both one-to-one and onto)* from $\mathbb{Z}_8$ to G_2. Denoting this function by $\varphi : \mathbb{Z}_8 \to G_2$, we could write:

$$\begin{array}{llll} \varphi([0]) = \alpha, & \varphi([1]) = \beta, & \varphi([2]) = \gamma, & \varphi([3]) = \delta, \\ \varphi([4]) = \epsilon, & \varphi([5]) = \zeta, & \varphi([6]) = \eta, & \varphi([7]) = \theta. \end{array}$$

Note that any function that actually corresponds to a valid renaming would have to be a bijection. This is because it wouldn't make sense to give two group elements the same name, or to leave a

*The remainder of this investigation assumes an understanding of injective, surjective, and bijective functions. For a review of these topics, see Appendix A.

group element out. If two groups are truly the same, then the elements of one group should be able to be matched in a one-to-one correspondence with the elements of the other. As such, our formal definition of "sameness" will begin with a bijective function.

Preserving Operations

As we saw in our earlier example, just having a bijective function from one group to another is not enough to say that the two groups are the same. Indeed, this bijective function must also transform the operation table of the first group into the operation table of the second. To see exactly what this means, let's look at an example.

Consider two groups G and H, each having three elements. Suppose also that we have defined a bijective "renaming" function $\varphi : G \to H$. Let's consider the operation table for G, which we can write generically as follows (using a, b, and c to denote the elements of G):

	a	b	c
a	aa	ab	ac
b	ba	bb	bc
c	ca	cb	cc

If we simply replace each entry in this table with its new name (as given by φ), we obtain the following table:

	$\varphi(a)$	$\varphi(b)$	$\varphi(c)$
$\varphi(a)$	$\varphi(aa)$	$\varphi(ab)$	$\varphi(ac)$
$\varphi(b)$	$\varphi(ba)$	$\varphi(bb)$	$\varphi(bc)$
$\varphi(c)$	$\varphi(ca)$	$\varphi(cb)$	$\varphi(cc)$

Is this table the multiplication table for H? Its entries are certainly elements of H, since φ maps from G to H. But the actual multiplication table for H would be defined as follows:

$\cdot$	$\varphi(a)$	$\varphi(b)$	$\varphi(c)$
$\varphi(a)$	$\varphi(a)\varphi(a)$	$\varphi(a)\varphi(b)$	$\varphi(a)\varphi(c)$
$\varphi(b)$	$\varphi(b)\varphi(a)$	$\varphi(b)\varphi(b)$	$\varphi(b)\varphi(c)$
$\varphi(c)$	$\varphi(c)\varphi(a)$	$\varphi(c)\varphi(b)$	$\varphi(c)\varphi(c)$

If the notation seems confusing here, just keep in mind that $\varphi(a)$, $\varphi(b)$, and $\varphi(c)$ are the elements of H, and we have formed the multiplication table for H in the usual way. In particular, each entry in the table is the product of the corresponding row and column headers (so, for instance, the entry in the $\varphi(a)$ row and $\varphi(b)$ column is just $\varphi(a)\varphi(b)$).

So what can we conclude? Recall that we wanted the renamed G table to be equal to the H table. In order for this to happen, each entry of the renamed G table must be equal to the corresponding entry of the H table. Thus, it must be the case that $\varphi(aa) = \varphi(a)\varphi(a)$, $\varphi(ab) = \varphi(a)\varphi(b)$, $\varphi(ac) = \varphi(a)\varphi(c)$, $\varphi(ba) = \varphi(b)\varphi(a)$, and so on.

In other words, we want φ to *preserve the operation*, which means:

For all x, $y \in G$, $\varphi(xy) = \varphi(x)\varphi(y)$.

This condition helps us to state in a more precise way exactly what it means for two groups to have the same operation table—or, in other words, the same algebraic structure. Any bijective function φ that is both a bijection and that preserves the operation is called an *isomorphism*, defined formally below.

Definition 29.5. Let G and H be groups. An **isomorphism** is a bijective function $\varphi : G \to H$ such that for all x, $y \in G$,

$$\varphi(xy) = \varphi(x)\varphi(y).$$

If there is an isomorphism from the group G to the group H we say that G is **isomorphic** to H.

The word *isomorphic* comes from two Greek words: *isos*, which means *equal* or *same*, and *morphe*, which means *form* or *structure*. Thus, when we say that two groups are isomorphic, we mean that they have the same structure. Likewise, we can think of an isomorphism as being an *operation-preserving bijection* or a *structure-preserving bijection*.

It is worth noting that the isomorphism relation is symmetric. (In fact, the isomorphism relation is an equivalence relation, as we will show later.) In particular, if $\varphi : G \to H$ is an isomorphism, then $\varphi^{-1} : H \to G$ is also an isomorphism. For this reason, we will often simply say that two groups G and H are isomorphic, rather than saying that G is isomorphic to H, or H is isomorphic to G. When G and H are isomorphic, we denote this relationship by writing $G \cong H$. So $G \cong H$ means that there is an isomorphism from G to H (or, equivalently, an isomorphism from H to G).

It's also important to note that the *function* in Definition 29.5 is called an *isomorphism*, while the *groups* G and H are said to be *isomorphic*. If the operations in G and H are written additively, then the operation-preserving condition for φ is written as

$$\varphi(x + y) = \varphi(x) + \varphi(y)$$

for all $x, y \in G$. If, on the other hand, the operations in G and H are written using different notations—for instance, suppose the operation in G is written multiplicatively and the operation in H is written additively—then the operation-preserving condition for φ must reflect this difference. In this case, we would write the condition as

$$\varphi(xy) = \varphi(x) + \varphi(y)$$

for all $x, y \in G$. Which notation is most appropriate should be clear from the context.

Activity 29.6.

(a) Use Definition 29.5 to explain why the function $\varphi : G_3 \to \mathbb{Z}_8$ (from Preview Activity 29.1) defined by

$$\varphi(w^k) = [k]$$

is an isomorphism.

(b) Is G_2 isomorphic to G_3? Use Definition 29.5 to justify your answer.

(c) Use Definition 29.5 to explain why the function $\varphi : G_2 \to G_4$ defined by

$$\begin{array}{llll} \varphi(\alpha) = \spadesuit, & \varphi(\beta) = \heartsuit, & \varphi(\gamma) = \diamondsuit, & \varphi(\delta) = \clubsuit, \\ \varphi(\epsilon) = \maltese, & \varphi(\zeta) = \sharp, & \varphi(\eta) = \dagger, & \varphi(\theta) = \mho \end{array}$$

is **not** an isomorphism.

(d) What does your answer to part (c) allow you to conclude about whether G_2 is isomorphic to G_4?

Proving Isomorphism

Now that we have precisely defined what it means for one group to be isomorphic to another, let's consider how we might use this definition in the context of groups that have more than just a few elements. Let $G = \left\{ \begin{bmatrix} 1 & x \\ 0 & 1 \end{bmatrix} : x \in \mathbb{R} \right\}$.

Activity 29.7.

(a) Notice that G (as defined above) is a subset of $\mathrm{GL}_2(\mathbb{R})$ (the group of invertible 2×2 matrices under multiplication). Show that G is actually a *subgroup* of $\mathrm{GL}_2(\mathbb{R})$.

(b) To which familiar group do you think G is isomorphic? You don't have to prove your answer now, but you should make a reasonable conjecture with some solid reasoning to back it up.

Looking again at the definition of G, it appears that each element of G corresponds to a unique real number (and vice versa). Thus, it seems reasonable that we would try to prove that G is isomorphic to the group $\mathbb{R}$ under addition. Since both G and $\mathbb{R}$ have infinitely many elements, we will not be able to simply work with their operation tables. Instead, we must use Definition 29.5, which suggests the following steps:

(1) We must define an appropriate function $\varphi : G \to \mathbb{R}$.

(2) We must show that φ is bijective; that is, we must show that φ is both injective and surjective.

(3) We must show that φ preserves the operation in G.

Activity 29.8. Carefully read the following proof that G is isomorphic to $\mathbb{R}$, filling in all of the missing details and providing additional explanations where appropriate. As you read the proof, try to identify where each of the three steps outlined above are taking place.

Theorem. *Let* $G = \left\{ \begin{bmatrix} 1 & x \\ 0 & 1 \end{bmatrix} : x \in \mathbb{R} \right\}$. *Then G is isomorphic to $\mathbb{R}$.*

Proof. Let $\varphi : G \to \mathbb{R}$ be defined by

$$\varphi\left(\begin{bmatrix} 1 & x \\ 0 & 1 \end{bmatrix}\right) = x.$$

Suppose that

$$\varphi\left(\begin{bmatrix} 1 & x \\ 0 & 1 \end{bmatrix}\right) = \varphi\left(\begin{bmatrix} 1 & y \\ 0 & 1 \end{bmatrix}\right).$$

Then $x = y$, which implies that

$$\begin{bmatrix} 1 & x \\ 0 & 1 \end{bmatrix} = \begin{bmatrix} 1 & y \\ 0 & 1 \end{bmatrix}.$$

Also observe that for all $x \in \mathbb{R}$,

$$\varphi\left(\begin{bmatrix} 1 & x \\ 0 & 1 \end{bmatrix}\right) = x.$$

Finally,

$$\begin{aligned}
\varphi\left(\begin{bmatrix}1 & x\\ 0 & 1\end{bmatrix}\begin{bmatrix}1 & y\\ 0 & 1\end{bmatrix}\right) &= \varphi\left(\begin{bmatrix}1 & x+y\\ 0 & 1\end{bmatrix}\right)\\
&= x+y\\
&= \varphi\left(\begin{bmatrix}1 & x\\ 0 & 1\end{bmatrix}\right)+\varphi\left(\begin{bmatrix}1 & y\\ 0 & 1\end{bmatrix}\right).
\end{aligned}$$

Therefore, G is isomorphic to $\mathbb{R}$. ■

Some Basic Properties of Isomorphisms

Since an isomorphism from a group G to a group H preserves the structure of G, there are certain things we should expect all isomorphisms to do. As an example, each group has exactly one identity element, so we should expect an isomorphism to map the identity to the identity. That property and a few others are given in the next activity.

Activity 29.9. Let φ be an isomorphism from a group G to a group H. Let e_G be the identity in G and e_H the identity in H.

(a) Prove that $\varphi(e_G) = e_H$. (Hint: It suffices to show that $h\varphi(e_G) = h$ for some $h \in H$. You may need to use the fact that φ is a surjection.)

(b) Prove that $\varphi\left(a^{-1}\right) = (\varphi(a))^{-1}$ for all $a \in G$. (Hint: It suffices to show that $\varphi(a)\varphi\left(a^{-1}\right) = e_H$.)

(c) Let $a \in G$. Prove that $\varphi(a^k) = (\varphi(a))^k$ for all positive integers k.

(d) Let $a \in G$. Use parts (b) and (c) to prove that $\varphi(a^k) = (\varphi(a))^k$ for all negative integers k. This will complete the proof of the following theorem.

Theorem. *Let φ be an isomorphism from a group G to a group H. Then $\varphi(a^k) = (\varphi(a))^k$ for all $a \in G$ and $k \in \mathbb{Z}$.*

We will discuss other properties of isomorphisms later.

Well-Defined Functions

Preview Activity 29.10. Let f assign to each element $[a]_3$ in $\mathbb{Z}_3$ the element $[a]_6$ in $\mathbb{Z}_6$.

(a) Does f preserve the group operation in $\mathbb{Z}_3$? Explain.

(b) Consider the following proof that f is an injection:

> Let $[a]_3$ and $[b]_3$ be in $\mathbb{Z}_3$, and assume $f([a]_3) = f([b]_3)$. Then $[a]_6 = [b]_6$, and so 6 divides $b - a$. Thus, 3 divides $b - a$, which implies that $[a]_3 = [b]_3$.

This proof might seem to imply that $\mathbb{Z}_3$ is isomorphic to the set $f(\mathbb{Z}_3) = \{f([a]_3) : [a]_3 \in \mathbb{Z}_3\} = \{[0]_6, [1]_6, [2]_6\}$. What do you think about this conclusion? Explain your answer in detail.

There is one additional consideration we need to keep in mind when proving isomorphism. Activity 29.10 shows that we can define a map that preserves a group's structure and seems to behave like an isomorphism, but if the map treats equal elements with different representations in different ways, then whatever conclusions we might draw will make little sense. We saw this same idea in Investigation 5 when we discussed well-defined operations. To emphasize the point, any time we have multiple ways to represent the elements in a set (like in $\mathbb{Z}_n$ or $\mathbb{Q}$), we need to be sure that anything that acts on the elements of that set (like an operation or a function) is well-defined. The next definition formalizes this idea.

Definition 29.11. Let S and T be sets. A mapping $f : S \to T$ is **well-defined** if $f(a) = f(b)$ whenever $a = b$ in S.

When we use the word *function*, we always mean a well-defined mapping. Well-defined mappings or functions are also called *single-valued*. In many cases, we do not need to worry about a function begin well-defined; in particular, if there is only one way to represent an element in the domain, then there is nothing to show. If, however, there are multiple ways to represent elements in the domain (like in $\mathbb{Z}_n$ or $\mathbb{Q}$), then we need to verify that any mapping we consider is well-defined before we worry about any other properties the mapping might possess.

Activity 29.12. Let f be the mapping from $\mathbb{Q}$ to $\mathbb{Z}$ defined by $f\left(\frac{a}{b}\right) = a + b$. Is f well-defined? Why or why not?

Disproving Isomorphism

In Activity 29.8, we saw an example of how Definition 29.5 can be used to show that two groups are isomorphic. Although other examples may require more sophisticated arguments, the basic structure will often be the same: we first define a particular function, and then we show that this function is bijective and operation-preserving.

What should we do, however, if we want to prove that a group G is *not* isomorphic to a group H? For instance, consider the groups $\mathbb{Z}$ and $\mathbb{Q}$ under addition. These sets are quite different under multiplication (we can divide by nonzero elements in $\mathbb{Q}$, but not in $\mathbb{Z}$), but is the additive structure of $\mathbb{Z}$ different from the additive structure of $\mathbb{Q}$?

If we wanted to use the definition of isomorphism to prove that $\mathbb{Z}$ is not isomorphic to $\mathbb{Q}$, we would have to show that there does not exist an isomorphism $\varphi : \mathbb{Z} \to \mathbb{Q}$. In other words, we would have to show that every function that we could possibly define from $\mathbb{Z}$ to $\mathbb{Q}$ would violate at least one of the conditions that define isomorphisms. To show this directly seems daunting, if not impossible. The next activity provides one doable example.

Activity 29.13. To show that $\mathbb{Z}$ is not isomorphic to $\mathbb{Q}$, suppose, to the contrary, that there does exist a function $\varphi : \mathbb{Z} \to \mathbb{Q}$ that is both bijective and operation-preserving. Since 1 generates $\mathbb{Z}$, we will consider the rational number $q = \varphi(1)$.

(a) Explain why $\varphi(n) = nq$ for every integer n.

(b) Is φ an injection? Prove your answer.

(c) Is φ a surjection? Prove your answer.

(d) Deduce that $\mathbb{Z}$ cannot be isomorphic to $\mathbb{Q}$. (While it may seem difficult to believe, there are bijections from $\mathbb{Z}$ to $\mathbb{Q}$, but they cannot preserve the structure of $\mathbb{Z}$.)

Invariants

In Activity 29.9, we identified a few properties (e.g., mapping inverses to inverses) that isomorphisms must satisfy. We then used those properties in Activity 29.13 to show that the group $\mathbb{Z}$ is not isomorphic to the group $\mathbb{Q}$.

Any property that isomorphic groups must share is known as an *invariant*, or technically, an invariant of group isomorphisms. Invariants are properties that must be preserved by any isomorphism. Thus, if P is an invariant and a group G satisfies P, then every group to which G is isomorphic must also satisfy P. Consequently, if two groups differ with respect to an established invariant, then they cannot be isomorphic.

As an example, being cyclic is an invariant of a group isomorphism. That is, any group isomorphic to a cyclic group must also be cyclic. This was the basic idea in Activity 29.13, where we showed that $\mathbb{Z}$ is not isomorphic to $\mathbb{Q}$. Showing that two groups are different with respect to a particular invariant (and *any* invariant will do) is sufficient to establish that the two groups are not isomorphic. However, even if two groups agree with respect to every invariant we can think of, this does not prove that they are isomorphic. **Although invariants can be used to prove that two groups are different, they cannot be used to prove that two groups are the same. In order to prove that two groups are isomorphic, we must find an appropriate function from one group to the other, and then prove that this function is in fact an isomorphism.**

Table 29.1 lists some of the more common and useful invariants. This list, however, is far from complete, and we will add to it as needed throughout the remainder of our study of groups.

Isomorphism Classes

When we introduced the notation $G \cong H$ to denote the isomorphism relation between groups G and H, we claimed that this relation was in fact an equivalence relation on the set of all groups. The next theorem formalizes this result.

Theorem 29.14. *The relation of group isomorphism ($\cong$) is an equivalence relation on the set of all groups.*

Proof. To show that $\cong$ is an equivalence relation on the set of all groups, we must show that $\cong$ is reflexive, symmetric, and transitive.

To show that $\cong$ is a reflexive relation, we need to show that every group G is isomorphic to

A Partial List of Invariants of Group Isomorphism

- Number of elements
- Commutativity
- Being cyclic
- Number of elements of each order
- Number of subgroups of each order
- Being simple

Table 29.1
Some common group invariants.

itself. Let G be a group. The natural map to use in this case is the identity map, which is always an isomorphism. (See Exercise 1.) So $G \cong G$, and $\cong$ is a reflexive relation.

To show that $\cong$ is a symmetric relation, let G and H be groups, and suppose that G and H are isomorphic—that is, suppose there is an isomorphism $\varphi : G \to H$. We need to show that there is also a corresponding isomorphism from H to G. Define $\varphi^{-1} : H \to G$ by $\varphi^{-1}(y) = x$ whenever $\varphi(x) = y$. To show that φ^{-1} is a group isomorphism, we will first show that φ^{-1} is well-defined. Suppose there is an element $y \in H$ so that $\varphi^{-1}(y) = x$ and $\varphi^{-1}(y) = x'$ for some $x, x' \in G$. Then $y = \varphi(x)$ and $y = \varphi(x')$. Since φ is an injection, it must be the case that $x = x'$. Therefore, φ^{-1} is well-defined.

Next we will show that φ^{-1} is an injection. Suppose $\varphi^{-1}(y) = \varphi^{-1}(y')$ for some $y, y' \in H$. Let $x = \varphi^{-1}(y) = \varphi^{-1}(y')$. Then $\varphi(x) = y$ and $\varphi(x) = y'$. Since φ is a function, we know $y = y'$. Therefore, φ^{-1} is an injection.

To show φ^{-1} is a surjection, let $x \in G$. Then let $y = \varphi(x)$. By definition, we have $\varphi^{-1}(y) = x$, so φ^{-1} is surjective.

To complete the proof that φ^{-1} is an isomorphism, we must show that φ^{-1} preserves the group operation. Let $y, y' \in H$, $x = \varphi^{-1}(y)$, and $x' = \varphi^{-1}(y')$. Then $\varphi(x) = y$ and $\varphi(x') = y'$. So $\varphi(xx') = \varphi(x)\varphi(x') = yy'$ and, by definition, $\varphi^{-1}(yy') = xx'$. Therefore,

$$\varphi^{-1}(yy') = xx' = \varphi^{-1}(y)\varphi^{-1}(y'),$$

and so φ^{-1} is operation-preserving. Therefore, φ^{-1} is an isomorphism,which verifies that $\cong$ is a symmetric relation.

To show that $\cong$ is a transitive relation, let G, H, and K be groups, and let $\varphi : G \to H$ and $\theta : H \to K$ be group isomorphisms. We must show that there is an isomorphism from G to K. The natural candidate is $\theta \circ \varphi$. We know that the composite of two bijective functions is also bijective. (See Appendix A.) Therefore, $\theta \circ \varphi$ is a bijection, and we only need show that $\theta \circ \varphi$ preserves the group operation. This argument is left as an exercise for the reader. (See Exercise 2.) Once it is completed, we will have established that $\theta \circ \varphi$ is a group isomorphism from G to K, and therefore $\cong$ is a transitive relation. Since $\cong$ is reflexive, symmetric, and transitive, it follows that $\cong$ is an equivalence relation. ■

As an equivalence relation, isomorphism partitions the set of all groups into disjoint equivalence classes—or, as we call them in this context, *isomorphism classes*. (See Theorem 5.6 on page 49.) In particular, the set of all groups isomorphic to a group G is called the isomorphism class of G. An important problem in group theory is to determine which isomorphism class a given group fits into. We have already learned (or will soon learn) a few things about isomorphism classes of groups of certain orders.

- Activity 26.17 (see page 352) shows that every group of prime order is cyclic, and Theorem 29.18 will show that every cyclic group of order n is isomorphic to $\mathbb{Z}_n$. Consequently, every group of prime order p is isomorphic to $\mathbb{Z}_p$, and there is exactly one isomorphism class of groups of each prime order. Thus, we have classified up to isomorphism all groups of order 2, 3, 5, 7, 11, and so on.

- Corollary 27.13 (see page 367) tells us that if p and q are distinct primes, then any Abelian group G of order pq is cyclic. Again, Theorem 29.18 will show that G is isomorphic to $\mathbb{Z}_{pq}$. This tells us that, for distinct primes p and q, there is only one isomorphism class of Abelian groups of order pq. Consequently, any Abelian group of order 6 is isomorphic to $\mathbb{Z}_6$, any Abelian group of order 10 is isomorphic to $\mathbb{Z}_{10}$, any Abelian group of order 14 is isomorphic to $\mathbb{Z}_{14}$, and so on. Note, however, that Corollary 27.13 tells us nothing about non-Abelian groups of order pq.

Informally, when we say that there are k groups of a given order n, we mean that there are k isomorphism classes of groups of order n. So we would say that there is only one group of each prime order. Another way of saying this is that there are k groups of order n, *up to isomorphism*. Thus, we have classified up to isomorphism all groups of prime order and all Abelian groups of order pq, where p and q are distinct primes.

These results only begin to solve the problem of classifying finite groups. For example, neither of them tells us anything about groups of order 4, where things are just a bit more complicated. We know of one group of order 4 that is not cyclic—namely, $\mathbb{Z}_2 \oplus \mathbb{Z}_2$. We do know, however, that every group of order 4 is Abelian. (See Exercise 15 on page 356 of Investigation 26.) In Exercise 17, we will show that, up to isomorphism, there are only two groups of order 4—namely, $\mathbb{Z}_4$ and $\mathbb{Z}_2 \oplus \mathbb{Z}_2$.

In a later investigation, we will actually classify all finite Abelian groups. However, the general classification of finite non-Abelian groups is a very difficult (and unsolved) problem. In spite of its difficulty, significant progress has been made on this problem. For instance, Table 29.2 provides the number of isomorphism classes of groups of order 1 through 100, which is only a small portion of the information we know about the classification of finite groups. We will study some specific cases of the classification problem in the exercises and in later investigations.

For now, however, there is one classification of finite non-Abelian groups that we can complete.

Activity 29.15. Let p be an odd prime, and let G be a non-Abelian group of order $2p$ with identity e. In this activity, we will determine the isomorphism class of G.

(a) We know of one non-Abelian group of order $2p$. What is it?

(b) Exercise 2 of Investigation 21 (see page 301) shows that if $a^2 = e$ for all $a \in G$, then G is Abelian. Consequently, we can assume that there is an element $b \in G$ such that $b^2 \neq e$. Explain why $|b| = p$.

(c) Let $N = \langle b \rangle$. Since $|G|$ is even, Exercise 22 in Investigation 22 (see page 314) shows that every group of even order contains an element of order 2, so G contains an element a of order 2. Explain why $a \notin N$.

n	$N(n)$	n	$N(n)$	n	$N(n)$	n	$N(n)$
1	1	26	2	51	1	76	4
2	1	27	5	52	5	77	1
3	1	28	4	53	1	78	6
4	2	29	1	54	15	79	1
5	1	30	4	55	2	80	52
6	2	31	1	56	13	81	15
7	1	32	51	57	2	82	2
8	5	33	1	58	2	83	1
9	2	34	2	59	1	84	15
10	2	35	1	60	13	85	1
11	1	36	14	61	1	86	2
12	5	37	1	62	2	87	1
13	1	38	2	63	4	88	12
14	2	39	2	64	267	89	1
15	1	40	14	65	1	90	10
16	14	41	1	66	4	91	1
17	1	42	6	67	1	92	4
18	5	43	1	68	5	93	2
19	1	44	4	69	1	94	2
20	5	45	2	70	4	95	1
21	2	46	2	71	1	96	230
22	2	47	1	72	50	97	1
23	1	48	52	73	1	98	5
24	15	49	2	74	2	99	2
25	2	50	5	75	3	100	16

Table 29.2
The number $N(n)$ of isomorphism classes for groups of order n. (From `http://mathworld.wolfram.com/FiniteGroup.html`)

(d) Now explain why $G = N \cup aN$, and then explain why $G = \{e, b, b^2, \ldots, b^{p-1}, a, ab, ab^2, \ldots, ab^{p-1}\}$. What familiar group does G look like?

(e) Exercise 19 of Investigation 27 (see page 375) shows that every subgroup of index 2 in a group is a normal subgroup. So $N \triangleleft G$. Explain why $aba \in N$ and why $aba = b^i$ for some integer $2 \leq i \leq p-1$.

(f) Explain why $a = b^i ab^{-1}$, and then use the fact that $a^2 = e$ to show that $b^{i-1}a = ab^{-i+1}$.

(g) Recall that D_p has presentation $\langle r, R \mid r^2 = 1, R^p = 1, rR = R^{-1}r \rangle$. (See Investigation 24.) In Exercise 31, we will show that two groups with the same presentation are isomorphic. Explain why G also has the same presentation as D_p. What elements play the role of r and R? (Hint: What is $|b^{-i+1}|$ in G?)

(h) Explain how we have proved the following theorem:

Theorem 29.16. *Let p be an odd prime and G a non-Abelian group of order $2p$. Then $G \cong D_p$.*

Using Theorem 29.16, we can complete our classification of the groups of order 6, 10, and 14. Earlier we argued that there is exactly one Abelian group of each of these orders. Theorem 29.16 now shows that there is exactly one non-Abelian group of each of these orders, namely D_3, D_{10}, and D_{14}, which verifies those entries in Table 29.2. We now have completely classified all of the groups of orders 1 through 7, 10, 11, 13, and 14, as shown in Table 29.3. Exercise 27 shows that there are at least 5 different isomorphism classes of groups of order 8, while Exercise 28 tells us that there are at least 5 different isomorphism classes of groups of order 12. We will complete the classification of groups of order 8, 12, and 15 in later investigations. Since there are 14 groups of order 16, we will stop at the groups of order 15.

Isomorphisms and Cyclic Groups

When we study groups, we are interested in the structure that the group operation imposes on the set. We don't really care what the elements of the group look like—only how they interact with each other. When two groups are isomorphic, corresponding elements behave the same way, so for all intents and purposes, the isomorphic groups are the same (except for how the elements are represented). We often talk about classifying a group G, which means determining to which set of groups G is isomorphic. In this section, we will show that any cyclic group of finite order n is isomorphic to $\mathbb{Z}_n$. As a result, there is only one finite cyclic group of each order, up to isomorphism. The infinite case is left for the exercises.

Activity 29.17. Let $G = \langle a \rangle$ be a cyclic group of order $n \in \mathbb{Z}^+$. To show G is isomorphic to $\mathbb{Z}_n$, we need to construct a function $\varphi : G \to \mathbb{Z}_n$ that is an isomorphism.

(a) What is the form of an arbitrary element in G?

(b) There is a natural mapping φ from G to $\mathbb{Z}_n$. Based on your answer to part (a), write down an equation to precisely define this mapping.

(c) Show that the mapping φ you defined in part (b) is a well-defined function. Why is this step necessary (and important)?

n	Distinct groups of order n
1	$\{1\}$
2	$\mathbb{Z}_2$
3	$\mathbb{Z}_3$
4	$\mathbb{Z}_4$ and $\mathbb{Z}_2 \oplus \mathbb{Z}_2$
5	$\mathbb{Z}_5$
6	$\mathbb{Z}_6$ and D_3
7	$\mathbb{Z}_7$
8	$\mathbb{Z}_8$, $\mathbb{Z}_4 \oplus \mathbb{Z}_2$, $\mathbb{Z}_2 \oplus \mathbb{Z}_2 \oplus \mathbb{Z}_2$, D_4, $\mathbf{Q}$, and ?
9	$\mathbb{Z}_9$, $\mathbb{Z}_3 \oplus \mathbb{Z}_3$, and ?
10	$\mathbb{Z}_{10}$ and D_5
11	$\mathbb{Z}_{11}$
12	$\mathbb{Z}_{12}$, $\mathbb{Z}_6 \oplus \mathbb{Z}_2$, D_6, A_4, T and ?
13	$\mathbb{Z}_{13}$
14	$\mathbb{Z}_{14}$ and D_7
15	$\mathbb{Z}_{15}$ and ?

Table 29.3
Known (to us) isomorphism classes of groups of orders 1 through 15.

(d) Prove that φ is a surjection.

(e) Prove that φ is an injection. (Hint: First review the definition of an injection. Then consider what $[k] = [l]$ in $\mathbb{Z}_n$ tells us about the relationship between k and l.)

(f) Explain how parts (b) – (e) establish that φ is an isomorphism, and thus G is isomorphic to $\mathbb{Z}_n$.

The result of Activity 29.17 is the following:

Theorem 29.18. *Any finite cyclic group of order n is isomorphic to $\mathbb{Z}_n$.*

Just as in the finite case, it is also true that, up to isomorphism, there is only one infinite cyclic group—namely, $\mathbb{Z}$. The next theorem states this result formally.

Theorem 29.19. *If G is an infinite cyclic group, then $G \cong \mathbb{Z}$.*

The proof of Theorem 29.19 is left for the exercises. (See Exercise 18.)

Cayley's Theorem

Group theory originated as the study of permutations, but we did not begin *our* study of group theory that way. Although the permutations of a set form a group, the abstract definition of a group may seem somewhat removed from what we know about permutations. In this section, we will study Cayley's Theorem, which establishes that all of group theory can in fact be considered as the study of permutations. Consequently, if we could completely understand permutation groups, then we would completely understand all groups. (Of course, we are currently far from such an understanding!)

Preview Activity 29.20. To understand Cayley's Theorem, we need to understand how a group element can act as a permutation on a group. Let G be a group with identity e, and let $a \in G$. Define

$$T_a : G \to G \text{ by } T_a(g) = ag.$$

(a) Illustrate the definition of T_a by determining $T_a(x)$ for each $a, x \in U_8$.

(b) What properties does the function T_a seem to possess? For example, is T_a an injection? Does T_a preserve the structure of U_8? List and verify as many properties as you can.

Cayley's Theorem is named after the mathematician Arthur Cayley and provides the surprising result that *every* group can be viewed as a group of permutations. Preview Activity 29.20 illustrates a method by which we can identify each element of a group with a permutation of the elements of that group. We will now investigate this construction in general.

Activity 29.21. Let G be a group with identity e, and let $a \in G$. Define

$$T_a : G \to G \text{ by } T_a(g) = ag. \tag{29.1}$$

(a) Is T_a is an injection? Verify your answer.

(b) Is T_a a surjection? Verify your answer.

(c) What specific function is T_e? Explain.

(d) If $b \in G$, the composite T_aT_b has the form T_c for some $c \in G$. For which c is this true? Prove your answer.

(e) What is the relationship between T_a and $T_{a^{-1}}$? Prove your conjecture.

Activity 29.21 shows us that for each $a \in G$, the function T_a is a permutation of G. Now we can see how G itself can be viewed as a subgroup of a group of permutations.

Activity 29.22. Let $P(G)$ be the collection of all permutations of G. Let $\Pi(G) = \{T_a : a \in G\}$. Explain why $\Pi(G)$ is a subgroup of $P(G)$.

Finally, let $\Theta : G \to \Pi(G)$ be defined by $\Theta(a) = T_a$.

Activity 29.23. Illustrate the definition of Θ by describing $\Theta : U_8 \to \Pi(U_8)$. (Hint: One way to do this is to specify what $\Theta(g)$ is for each $g \in U_8$.)

The major result of this section is the following:

Theorem 29.24 (Cayley's Theorem). *Every group is a subgroup of a group of permutations.*

Proof. Let G be a group with identity e, and define $\Theta : G \to \Pi(G)$ as above. We will show that Θ is an isomorphism.

First we will show that Θ is an injection. Suppose $\Theta(a) = \Theta(b)$ for some $a, b \in G$. Then $T_a = T_b$. This means $T_a(e) = T_b(e)$ or $ae = be$. Thus, $a = b$, and so Θ is injective. Next we will demonstrate that Θ is a surjection. If $T_a \in \Pi(G)$, then $\Theta(a) = T_a$. Therefore, the range of Θ is $\Pi(G)$, and so Θ is surjective. To complete the proof, we must show that Θ preserves the operation in G. Let $a, b \in G$. In part (d) of Activity 29.21, we showed that $T_aT_b = T_{ab}$. Therefore,

$$\Theta(a)\Theta(b) = T_aT_b = T_{ab} = \Theta(ab).$$

It follows that Θ is an isomorphism of groups, and so $G \cong \Pi(G)$. Thus, we have shown that G is isomorphic to a group of permutations. ∎

One corollary of Cayley's Theorem is the following:

Corollary 29.25. *If G is a finite group of order n, then G is isomorphic to a subgroup of the symmetric group S_n.*

Proof. The proof of this corollary is really just a matter of demonstrating that if T is a set with n elements, then $P(T) \cong S_n$. (See Exercise 35.) Then G is isomorphic to a subgroup of $P(G)$, which is then isomorphic to a subgroup of S_n. The transitivity of the isomorphism relation then shows that G is isomorphic to a subgroup of S_n. ∎

Corollary 29.25 can be useful in that it allows us to concretely represent any abstract finite group as a subgroup of S_n. The next activity illustrates how this representation works.

Activity 29.26. Let G be the group of order 4 with presentation $\langle a, b \mid a^2 = b^2 = (ab)^2 = 1\rangle$. (See Activity 24.6 on page 329.) This group has the operation table

	1	a	b	ab
1	1	a	b	ab
a	a	1	ab	b
b	b	ab	1	a
ab	ab	b	a	1

Identify the elements in G with the set $\{1, 2, 3, 4\}$ by $1 \leftrightarrow 1$, $a \leftrightarrow 2$, $b \leftrightarrow 3$, and $ab \leftrightarrow 4$.

(a) Recall that each element $g \in G$ is identified with the permutation $T_g \in P(G)$, where T_g is defined as in (29.1). Complete the table below to determine the actions of T_g for each $g \in G$.

x	1	a	b	ab
$T_1(x)$	1	a	b	ab
$T_a(x)$				
$T_b(x)$				
$T_{ab}(x)$				

(b) Use the identifications $1 \leftrightarrow 1$, $a \leftrightarrow 2$, $b \leftrightarrow 3$, and $ab \leftrightarrow 4$ to identify each T_g with an element in S_4. Then find a subgroup of S_4 isomorphic to G.

(c) Let S be the subgroup you found in part (b). Find a specific isomorphism from G to S.

Concluding Activities

Activity 29.27. Let G be a group with identity element e, and let K and N be normal subgroups of G with $K \cap N = \{e\}$. As we will demonstrate in this activity, $(K \times N) \cong (K \oplus N)$. Because of this isomorphism, we can identify an internal direct product with the corresponding external direct product, and so it doesn't matter which product we use.

To show that $(K \times N) \cong (K \oplus N)$, we need to exhibit an isomorphism between the two groups. The natural mapping to try is $\varphi : (K \times N) \to (K \oplus N)$ defined by

$$\varphi(kn) = (k, n).$$

To complete this activity, you may want to recall that the representation of an element in $K \times N$ as kn for some $k \in K$ and $n \in N$ is unique, and that if $k \in K$ and $n \in N$ then $kn = nk$. (See Theorem 28.15 on page 387.)

(a) Explain why φ is well-defined.

(b) Prove that φ is an isomorphism.

Activity 29.28. In Activity 29.13, we showed that if there were an isomorphism $\varphi : \mathbb{Z} \to \mathbb{Q}$, then φ would be completely determined by its action on 1, a generator of $\mathbb{Z}$. In this activity, we will show that this property is true for arbitrary cyclic groups. Let $G = \langle a \rangle$ be a cyclic group, and let $\varphi : G \to H$ be a group isomorphism. Show that the element $\varphi(a)$ completely determines the isomorphism φ.

Activity 29.29. Write a short summary that describes the important concepts, and the relationships between these concepts, that were introduced in this investigation. Explain how the material in this investigation is connected to the material in Investigations 20 and 25.

Exercises

* (1) If G is a group, prove that the identity map $I : G \to G$ defined by $I(a) = a$ for all $a \in G$ is a group isomorphism.

* (2) Let G, H, and K be groups, and let $\varphi : G \to H$ and $\theta : H \to K$ be group isomorphisms. Prove that $\theta \circ \varphi$ preserves structure.

(3) Let G be a group. In Exercise 1, we showed that the identity map $I : G \to G$ is an isomorphism. In this exercise, we will determine when the function $\varphi : G \to G$ defined by $\varphi(a) = a^{-1}$ is an isomorphism.

   (a) Is φ an isomorphism when $G = \mathbb{Z}_3$?

   (b) Is φ an isomorphism when $G = D_3$?

   (c) Determine necessary and sufficient conditions on G so that φ is an isomorphism. Prove your answer.

(4) Let G be a group. In this exercise, we will consider when the mapping $\varphi : G \to G$ defined by $\varphi(a) = a^2$ is an isomorphism.

   (a) Is φ an isomorphism if $G = \mathbb{Z}_3$? Explain.

   (b) Is φ an isomorphism if $G = \mathbb{Z}_4$? Explain.

   (c) Suppose G is a finite Abelian group and G has no element of order 2. Show that φ is an isomorphism. Then show by example that if G is infinite, then φ need not be an isomorphism.

   (d) If G is a finite Abelian group and φ is an isomorphism, must G contain no element of order 2? Prove your answer.

(5) Let G be a group. In this exercise, we will investigate isomorphisms $\varphi : G \to G$ of the form $\varphi(a) = a^n$ for some integer n.

   (a) Show that the function $\varphi : U_{22} \to U_{22}$ defined by $\varphi(a) = a^3$ is an isomorphism of groups.

   (b) Show that the function $\varphi : U_{22} \to U_{22}$ defined by $\varphi(a) = a^4$ is not an isomorphism of groups.

   (c) Let G be a finite Abelian group and n a positive integer that is relatively prime to $|G|$. Show that the mapping $\varphi : G \to G$ defined by $\varphi(a) = a^n$ is an isomorphism.

   (d) Let G be a finite Abelian group, and suppose $\varphi : G \to G$ defined by $\varphi(a) = a^n$ for some integer n is an isomorphism. Must $\gcd(n, |G|) = 1$? Prove your answer.

(6) To what familiar group is the group G_1 from Preview Activity 29.1 isomorphic? Prove your answer.

(7) Prove that commutativity is an invariant. That is, prove that if G and H are isomorphic groups and G is Abelian, then H must also be Abelian.

(8) Prove that the order of each element in a group is an invariant. That is, prove that if $\varphi : G \to H$ is an isomorphism between the groups G and H and $a \in G$, then the order of $\varphi(a)$ in H is the same as the order of a in G.

* (9) In Exercise 8, we showed that the order of each element in a group is an invariant. In this exercise, we will show that the number of elements of a given order in a group is also an invariant. We will then explore the converse; that is, we will consider whether two groups can have the same number of elements of each order and yet not be isomorphic.

  (a) Let G be a finite group, k a positive integer, and $\nu_G(k)$ the number of elements of order k in G. Prove that if G and G' are isomorphic groups, then $\nu_G(k) = \nu_{G'}(k)$ for every positive integer k.

  (b) Now we want to determine if the converse is true. That is, if G and G' are groups of the same order and $\nu_G(k) = \nu_{G'}(k)$ for every positive integer k, must it be true that $G \equiv G'$? Let p be an odd prime, and let $G = \left\{ \begin{bmatrix} [1] & [a] & [b] \\ [0] & [1] & [c] \\ [0] & [0] & [1] \end{bmatrix} : [a], [b], [c] \in \mathbb{Z}_p \right\}$.[†]

    (i) Show that G is a group of order p^3. You may assume that $\mathrm{GL}_3(\mathbb{Z}_p) = \{A \in M_{3\times 3}(\mathbb{Z}_p) : \det(A) \neq 0\}$ is a group under standard matrix multiplication.

    (ii) Show that the order of every non-identity element in G is p. (Hint: Show that

$$\begin{bmatrix} [1] & [a] & [b] \\ [0] & [1] & [c] \\ [0] & [0] & [1] \end{bmatrix}^n = \begin{bmatrix} [1] & n[a] & \frac{n(n-1)}{2}[ac] + n[b] \\ [0] & [1] & n[c] \\ [0] & [0] & [1] \end{bmatrix}$$

    for every positive integer n.)

    (iii) Is it true that if G and G' are groups of the same order and $\nu_G(k) = \nu_{G'}(k)$ for every positive integer k, then $G \cong G'$? Prove your answer.

(10) Prove that being cyclic is an invariant. That is, prove that if G and H are isomorphic groups and G is cyclic, then H must also be cyclic.

(11) **Direct sums of $\mathbb{Z}_m$ and $\mathbb{Z}_n$.** Under what circumstances is $\mathbb{Z}_n \oplus \mathbb{Z}_m$ isomorphic to $\mathbb{Z}_{mn}$? State (and prove) your answer in the form of a biconditional (if and only if) statement.

(12) Explain why the indicated groups are *not* isomorphic.

  (a) $\mathbb{Z}_6$ and S_3

  (b) $\mathbb{Z}_4 \oplus \mathbb{Z}_2$ and D_4

  (c) $Z_4 \oplus Z_2$ and $Z_2 \oplus Z_2 \oplus Z_2$

  (d) $\mathbb{Z}$ and $\mathbb{R}$

  (e) $\mathbb{R}^* = U(\mathbb{R})$ and $\mathbb{R}$

  (f) $\mathbb{Q}$ and $\mathbb{R}$ (Hint: If s and t are nonzero rational numbers, show that there are nonzero integers m, n such that $ms = nt$.)

[†]This group G is called the Heisenberg group modulo p (named after the Nobel prize winning physicist Werner Heisenberg). The Heisenberg group (with real entries) is related to the Heisenberg Uncertainty Principle in quantum physics.

(13) One of the following statements is true, and the other is false. Which is which? Prove your answer.

(a) $\mathbb{Q}$ is isomorphic to the multiplicative group $\mathbb{Q}^+$ of positive rational numbers

(b) $\mathbb{R}$ is isomorphic to the multiplicative group $\mathbb{R}^+$ of positive real numbers

(14) Let $\varphi : U_7 \to \mathbb{Z}_6$ be an isomorphism such that $\varphi([3]_7) = [5]_6$. Find $\varphi([a]_7)$ for every $[a]_7 \in U_7$.

(15) Let $H = \left\{ \begin{bmatrix} 1+n & -n \\ n & 1-n \end{bmatrix} : n \in \mathbb{Z} \right\}$.

(a) Prove that H is a group under matrix multiplication.

(b) To what familiar group is H isomorphic? Prove your answer.

(16) Determine if the given groups are isomorphic. Prove your answers.

(a) U_5 and U_{10}

(b) U_{20} and U_{24}

⋆ (17) **Groups of order 4.** In this exercise, we will classify all groups of order 4. Let G be a group of order 4 with identity e, and let a be a non-identity element in G.

(a) What must be true about G if $|a| = 4$?

(b) If $|a| \neq 4$, what are the possible values for $|a|$? Why?

(c) If $a^2 = e$, then there must be another non-identity element b in G. What, then, is the fourth element in G?

(d) What can we say about the relationship between ab and ba? Explain. What kind of group is G in this case?

(e) Explain why there are (up to isomorphism) exactly two groups of order 4 ($\mathbb{Z}_4$ and $\mathbb{Z}_2 \oplus \mathbb{Z}_2$).

⋆ (18) Prove Theorem 29.19.

(19) Explain why any two cyclic groups of the same order are isomorphic to each other.

(20) Is it possible for a group to be isomorphic to one of its proper subgroups? Prove your answer.

(21) Let k and n be positive integers with $n > 1$. Recall that

$$k\mathbb{Z}_n = \{k[x] : [x] \in \mathbb{Z}_n\}.$$

(a) Find (and prove) a necessary and sufficient condition for $k\mathbb{Z}_n$ to be isomorphic to $\mathbb{Z}_n$.

(b) In general, to what familiar group is $k\mathbb{Z}_n$ isomorphic? Prove your answer.

(22) (a) Let G_1 and G_2 be groups. Prove that $(G_1 \oplus G_2) \cong (G_2 \oplus G_1)$.

(b) Generalize part (a) to any number of factors to show that we can rearrange the factors in a direct product in any order and still have a group isomorphic to the original. A rearrangement of the order of the factors is just a permutation of the factors, so we can restate this problem more formally as follows. Let G_1, G_2, ..., and G_n be groups for some integer $n \geq 2$. For any $\sigma \in S_n$, prove that

$$(G_1 \oplus G_2 \oplus \cdots G_n) \cong \left(G_{\sigma(1)} \oplus G_{\sigma(2)} \oplus \cdots G_{\sigma(n)}\right).$$

(23) In this exercise, we will generalize the result of Activity 29.27 to direct products with any finite number of factors. Let G be a group with identity element e, and let N_1, N_2, ..., N_m be normal subgroups of G for some integer $m \geq 2$ such that

$$N_i \cap (N_1 \times N_2 \times \cdots \times N_{i-1} \times N_{i+1} \times \cdots \times N_m) = \{e\}$$

for each i. Prove that

$$(N_1 \times N_2 \times \cdots \times N_m) \cong (N_1 \oplus N_2 \oplus \cdots \oplus N_m).$$

(24) Let φ be the Euler phi function, as defined in Exercise 15 of Investigation 23 and Exercise 21 of Investigation 26. (See pages 323 and 356, respectively. Recall that $\varphi(n)$ is equal to the number of positive integers less than or equal to n that are relatively prime to n.)

(a) Show by example that $\varphi(mn)$ is not always equal to $\varphi(m)\varphi(n)$.

(b) Let s and t be positive integers with $\gcd(s,t) = 1$. Prove that $U_{st} \cong (U_s \oplus U_t)$.

(c) Show that if m and n are positive integers with $\gcd(m,n) = 1$, then $\varphi(mn) = \varphi(m)\varphi(n)$. (Functions that have this property are called *multiplicative functions*.)

(25) Prove, by comparing orders of elements, that $\mathbb{Z}_8 \oplus \mathbb{Z}_2$ is not isomorphic to $\mathbb{Z}_4 \oplus \mathbb{Z}_4$.

(26) Let $n = 2k$ be a positive even integer. In Exercise 2 of Investigation 24 (see page 330), we showed that $Z(D_n) = \langle R^k \rangle$. To what familiar group is $D_n/Z(D_n)$ isomorphic? Prove your answer.

⋆ (27) We have seen several groups of order 8: $\mathbb{Z}_8$, $\mathbb{Z}_4 \oplus \mathbb{Z}_2$, $\mathbb{Z}_2 \oplus \mathbb{Z}_2 \oplus \mathbb{Z}_2$, D_4, and the quaternions $\mathbf{Q}$. (See Exercise 21 on page 376 of Investigation 27 and Exercise 6 on page 359 of Investigation 24.)

(a) Determine the number of elements of each order in each of these groups.

(b) Based on your work in part (a), separate these groups into their potential isomorphism classes. Can you be completely sure of your identifications based only on the work from part (a)? If so, explain why. If not, why not? Which of your identifications can you be sure of, and why?

⋆ (28) We have seen several groups of order 12: A_4, D_6, $Tetra$ (see Exercise 4 in Investigation 24), T (see Exercise 18 of Investigation 24), $D_3 \oplus \mathbb{Z}_2$, $\mathbb{Z}_{12}$, and $\mathbb{Z}_6 \oplus \mathbb{Z}_2$.

(a) Determine the number of elements of each order in each of these groups.

(b) Based on your work in part (a), separate these groups into their potential isomorphism classes. Can you be completely sure of your identifications based only on the work from part (a)? If so, explain why. If not, why not? Which of your identifications can you be sure of, and why?

(29) (a) Let G_1, G_2, H_1, and H_2 be groups with $G_1 \cong H_1$ and $G_2 \cong H_2$. Prove that $(G_1 \oplus G_2) \cong (H_1 \oplus H_2)$.

(b) Does part (a) generalize to any finite number of groups? If so, formally state a claim and then prove it. If not, give a counterexample.

(30) Let $H = \langle 3 \rangle$ and $K = \langle 12 \rangle$ in $\mathbb{Z}$.

(a) To what familiar group is $\langle 3 \rangle / \langle 12 \rangle$ isomorphic? Prove your answer.

(b) To what familiar group is $\langle 8 \rangle / \langle 48 \rangle$ isomorphic? Prove your answer.

(c) Generalize your work in parts (a) and (b) to arbitrary integers k and m. To what familiar group is $\langle k\rangle/\langle m\rangle$ isomorphic if k divides m? Prove your answer.

$\star$ (31) Let G and H be two groups with presentations of the form

$$\langle a_1, a_2, \ldots, a_n : r_1 = r_2 = \cdots = r_m = 1\rangle.$$

In other words, G is generated by elements g_1, g_2, ..., g_n satisfying the relations $r_1 = r_2 = \cdots = r_m = 1$ (replacing each a_i in these relations with g_i), and H is generated by elements h_1, h_2, ..., h_n satisfying the relations $r_1 = r_2 = \cdots = r_m = 1$ (replacing each a_i in these relations with h_i). Prove that $G \cong H$.

(32) Prove that a group G of order 4 with presentation $\langle a, b \mid a^2 = b^2 = (ab)^2 = 1\rangle$ (see Activity 24.6 on page 329 and Activity 29.26) is isomorphic to the group $\mathbb{Z}_2 \oplus \mathbb{Z}_2$. (This group $\mathcal{V} = \mathbb{Z}_2 \oplus \mathbb{Z}_2$—and any group isomorphic to it—is called the Klein 4-group. The name comes from the German "Viergruppe," found in 1884 in the paper *Vorlesungen uber das Ikosaeder und die Aufloesung der Gleichungen vom funften Grade* by Felix Klein.)

(33) Let G be a group. An isomorphism from G to G is called an **automorphism** and the set of automorphisms of G is denoted $\text{Aut}(G)$. That is,

$$\text{Aut}(G) = \{\varphi : G \to G : \varphi \text{ is an isomorphism}\}.$$

(a) Prove that there are exactly two automorphisms of $\mathbb{Z}$. (Hint: Activity 29.9 should be helpful.)

(b) Prove that $\text{Aut}(G)$ is a group under the operation of composition of functions.

(c) To what familiar group is $\text{Aut}(\mathbb{Z})$ isomorphic? Explain.

$\star$ (34) Exercise 33 shows that $\text{Aut}(G)$ is a group whenever G is a group. In this exercise, we will determine $\text{Aut}(\mathbb{Z}_n)$. Let n be a positive integer.

(a) Suppose that $\varphi \in \text{Aut}(\mathbb{Z}_n)$. What is $|\varphi([1])|$? What, specifically, does this result tell us about which elements $\varphi([1])$ could be in $\mathbb{Z}_n$? Explain.

(b) For each of the possibilities you found for $\varphi([1])$ in part (a), which define an automorphism of $\mathbb{Z}_n$? Prove your answer.

$\star$ (35) Prove that if T is a set with n elements, then the permutation group $P(T)$ of T is isomorphic to the symmetric group S_n. (Hint: Label the elements in T in some order as t_1, t_2, ..., t_n. Then connect a permutation of T to a permutation of the subscripts.)

(36) The corollary to Cayley's Theorem tells us that if G is a group of order 8, then G is isomorphic to a subgroup of S_8. For each of the following groups, find a subgroup of S_8 to which the group is isomorphic. (Recall that $\mathbf{Q}$ denotes the group of quaternions.)

$$D_4 \qquad \mathbb{Z}_8 \qquad \mathbb{Z}_4 \oplus \mathbb{Z}_2 \qquad \mathbb{Z}_2 \oplus \mathbb{Z}_2 \oplus \mathbb{Z}_2 \qquad \mathbf{Q}$$

Connections

In this investigation, we studied isomorphisms of groups. If you studied ring theory before group theory, you should notice connections between isomorphisms of groups in this investigation and isomorphisms of rings in Investigation 10. The idea is the same in both contexts; isomorphic groups (or rings) are essentially the same, and an isomorphism is a bijection that preserves the underlying algebraic structure. Since there is only one operation defined in a group but two operations in a ring, the major difference between isomorphisms of groups and isomorphisms of rings is that an isomorphism of groups must preserve one operation, whereas an isomorphism of rings must preserve two operations. The process of verifying an isomorphism is the same in both contexts, but there is an extra step required for isomorphisms of rings.

Investigation 30

Homomorphisms and Isomorphism Theorems

Focus Questions

By the end of this investigation, you should be able to give precise and thorough answers to the questions listed below. You may want to keep these questions in mind to focus your thoughts as you complete the investigation.

- What is a homomorphism of groups, and how is a homomorphism different than an isomorphism?
- What are monomorphisms and epimorphisms of groups?
- What are the kernel and image of a group homomorphism, and what properties do they satisfy?
- What are the isomorphism theorems for groups, and how do they use homomorphisms to establish relationships between groups?

Preview Activity 30.1. As we saw in Investigation 29, the notion of isomorphism formalizes what it means for two groups to be essentially the same. Recall that an isomorphism of groups is a bijective, structure-preserving function. In group theory, structure-preserving maps are important even if they are not bijections. In this activity, we will explore three different kinds of structure-preserving functions. (Throughout the activity, recall that we use the notation $[k]_n$ to denote the congruence class of k in $\mathbb{Z}_n$.)

(a) Is the function $\varphi : \mathbb{Z}_3 \to \mathbb{Z}_6$ defined by $\varphi([k]_3) = [4k]_6$ structure-preserving? Is φ an injection? Is φ a surjection? Verify your answers. (You may assume that φ is well-defined.)

(b) Is the function $\varphi : \mathbb{Z}_6 \to \mathbb{Z}_3$ defined by $\varphi([k]_6) = [k]_3$ structure-preserving? Is φ an injection? Is φ a surjection? Verify your answers. (You may assume that φ is well-defined.)

(c) Is the function $\varphi : \mathbb{Z}_6 \to \mathbb{Z}_4$ defined by $\varphi([k]_6) = [2k]_4$ structure-preserving? Is φ an injection? Is φ a surjection? Verify your answers. (You may assume that φ is well-defined.)

Homomorphisms

Preview Activity 30.1 illustrates that it is possible to have structure-preserving maps that are injections but not surjections, surjections but not injections, or neither surjections nor injections. When we study groups, we are mostly interested in maps that preserve the group structure or operation. Such maps—whether they are injective, surjective, neither, or both—are called *homomorphisms*, defined formally as follows:

Definition 30.2. Let G and H be groups. A function φ from G to H is a **homomorphism** of groups if

$$\varphi(ab) = \varphi(a)\varphi(b)$$

for all $a, b \in G$.

Just like isomorphism, the word *homomorphism* comes from two Greek words: *homos*, which means *similar* or *like*, and *morphe*, which means *form* or *structure*. Thus, when there is a homomorphism from one group to another, it means that there is some similarity of structure between the two groups. Just like an isomorphism, a homomorphism is an *operation-preserving* or a *structure-preserving* function, but not necessarily a bijection.

Although it is not a requirement, some homomorphisms are also injections, surjections, or bijections (as seen in Preview Activity 30.1). Homomorphisms that satisfy these additional properties are given special names. In particular,

- a **monomorphism** is an injective homomorphism;
- an **epimorphism** is a surjective homomorphism; and
- an **isomorphism** is a bijective homomorphism.

If $\varphi : G \to G'$ is an epimorphism, we call G' a **homomorphic image** of G.

Activity 30.3. Determine whether each of the following functions is a homomorphism from G to H. If a function is a homomorphism, decide whether it is a monomorphism, an epimorphism, an isomorphism, or none of these.

(a) $G = \mathbb{Z}$, $H = \mathbb{Z}_5$, and $\varphi(k) = [k]_5$

(b) $G = \mathbb{Z}_3$, $H = \mathbb{Z}_{18}$, and $\varphi([k]_3) = [6k]_{18}$

(c) $G = \mathbb{Z}$, $H = \mathbb{Z}_2 \oplus \mathbb{Z}_4$, and $\varphi(k) = ([k]_2, [k]_4)$

(d) $G = \mathbb{R}^+$, $H = \mathbb{R}^+$, and $\varphi(k) = \sqrt{k}$

(e) $G = U_{12}$, $H = \mathbb{Z}_6$, and $\varphi([k]_{12}) = [k]_6$

The Kernel of a Homomorphism

In Investigation 29, we showed that every isomorphism between two groups G and H maps the identity in G to the identity in H. We also saw that isomorphisms map inverses to inverses. The same proofs show that these properties are also true for homomorphisms in general.

Theorem 30.4. *Let G and H be groups with identities e_G and e_H, respectively, and let $\varphi : G \to H$ be a homomorphism. Then:*

(i) $\varphi(e_G) = e_H$.

(ii) *If $a \in G$, then $\varphi(a^{-1}) = (\varphi(a))^{-1}$.*

Proof. Let G and H be groups with identities e_G and e_H, respectively, and let $\varphi : G \to H$ be a homomorphism. Then

$$\varphi(e_G)\varphi(e_G) = \varphi(e_G e_G) = \varphi(e_G) = e_H\varphi(e_G),$$

and cancellation shows that $\varphi(e_G) = e_H$. So φ must map the identity element of G to the identity element of H.

Now let $a \in G$. To show that $\varphi(a^{-1}) = (\varphi(a))^{-1}$, note that

$$e_H = \varphi(e_G) = \varphi\left(aa^{-1}\right) = \varphi(a)\varphi\left(a^{-1}\right).$$

Therefore, $(\varphi(a))^{-1} = \varphi(a^{-1})$. ■

Let $\varphi : G \to H$ be a homomorphism of groups. By part (i) of Theorem 30.4, we know that φ maps the identity in G to the identity in H. If φ is a monomorphism—that is, if φ is also injective—then φ maps *only* the identity to the identity. If φ is not a monomorphism, we can measure (in a sense) how close φ is to being a monomorphism by determining the number of elements φ maps to the identity. This important idea leads us to the next definition.

Definition 30.5. Let $\varphi : G \to H$ be a homomorphism of groups, and let e_H be the identity element in H. The **kernel** of φ is the set

$$\text{Ker}(\varphi) = \{a \in G : \varphi(a) = e_H\}.$$

You may have seen an idea related to the kernel of a homomorphism in linear algebra, where the set of objects that are sent to the zero vector under a matrix transformation is called the *null space* of the matrix. In this sense, the notion of the kernel of a homomorphism is not entirely new.

Activity 30.6. Find Ker(φ) for each of the homomorphisms in Activity 30.3.

Since Ker(φ) is a subset of G, is is natural to ask if Ker(φ) is a subgroup of G.

Activity 30.7. Let G and H be groups with identities e_G and e_H, respectively, and let $\varphi : G \to H$ be a homomorphism.

(a) Is e_G in Ker(φ)? Explain.

(b) Is Ker(φ) closed under the operation in G? Prove your answer.

(c) If $a \in \text{Ker}(\varphi)$, is $a^{-1} \in \text{Ker}(\varphi)$? Prove your answer.

(d) Is $\text{Ker}(\varphi)$ a subgroup of G? Explain.

The next theorem tells us an important fact about the kernel of a group homomorphism.

Theorem 30.8. *Let G and H be groups, and let $\varphi : G \to H$ be a group homomorphism. Then Ker(φ) is a normal subgroup of G.*

Proof. Let G and H be groups, and let $\varphi : G \to H$ be a group homomorphism. Let e_G and e_H be the identities for G and H, respectively. Let $K = \text{Ker}(\varphi)$. To that show K is a normal subgroup of G, we will show that $aKa^{-1} \subseteq K$ for all $a \in G$. Let $a \in G$. We then need to show $aka^{-1} \in K$ for every $k \in K$. Now let $k \in K$. Then $\varphi(k) = e_H$. So

$$\varphi(aka^{-1}) = \varphi(a)\varphi(k)\varphi(a^{-1}) = \varphi(a)e_H(\varphi(a))^{-1} = \varphi(a)(\varphi(a))^{-1} = e_H,$$

and $aka^{-1} \in K$. Thus, $aKa^{-1} \subseteq K$ for every $a \in G$, and so $K = \text{Ker}(\varphi)$ is a normal subgroup of G. ∎

The Image of a Homomorphism

Let $\varphi : G \to H$ be a homomorphism of groups G and H with identities e_G and e_H, respectively. As we have seen, $\varphi(e_G) = e_H$, and so e_H is always the image of the identity element under any homomorphism. If φ is an epimorphism, then every element in H is the image under φ of some element from G. If φ is not an epimorphism, then some elements in H are "missed" by φ—that is, they are not the images of any elements from G. The size of the set of images of elements in G under φ measures (in some sense) how close φ is to being an epimorphism. This leads us to the next definition.

Definition 30.9. Let $\varphi : G \to H$ be a homomorphism of groups. The **image** of φ is the set

$$\text{Im}(\varphi) = \{\varphi(a) : a \in G\}.$$

Activity 30.10. Find $\text{Im}(\varphi)$ for each of the homomorphisms in Activity 30.3.

Since $\text{Im}(\varphi)$ is a subset of H, it is natural to ask if $\text{Im}(\varphi)$ is a subgroup of H.

Activity 30.11. Let G and H be groups with identities e_G and e_H, respectively, and let $\varphi : G \to H$ be a homomorphism.

(a) Is e_H in $\text{Im}(\varphi)$? Explain.

(b) Is $\text{Im}(\varphi)$ closed under the operation in H? Prove your answer.

(c) If $y \in \text{Im}(\varphi)$, is $y^{-1} \in \text{Im}(\varphi)$? Prove your answer.

(d) Is $\text{Im}(\varphi)$ a subgroup of H? Explain.

The Isomorphism Theorems for Groups

Sometimes homomorphisms can be used to recognize isomorphic groups, even if the homomorphisms themselves are not isomorphisms. Four theorems, called the *isomorphism theorems* for groups, formalize this idea. The First Isomorphism Theorem connects the kernel and image of a homomorphism. The Second Isomorphism Theorem connects intersections and products of subgroups. The Third Isomorphism Theorem provides a sort of "cancellation" rule for quotient groups. Finally, the Fourth Isomorphism Theorem establishes a correspondence between certain subgroups of a group and one of its quotients.

The First Isomorphism Theorem for Groups

Preview Activity 30.12. Let $G = \mathbb{Z}_{24}$, $H = \mathbb{Z}_8$, and let $\varphi : G \to H$ be defined by $\varphi([m]_{24}) = [6m]_8$.

(a) Show that φ is well-defined. Why is this necessary?

(b) Show that φ is a homomorphism of groups. Is φ a monomorphism? Is φ an epimorphism? Is φ an isomorphism? Explain.

(c) Let $K = \text{Ker}(\varphi)$. Find all the elements of K.

(d) Determine the elements of the group G/K. Is G/K Abelian? Is G/K cyclic? Explain.

(e) Let $R = \text{Im}(\varphi)$. Find all the elements of R.

(f) What specific relationship is there between the groups G/K and R? Explain.

Preview Activity 30.12 shows that in one example where $\varphi : G \to H$ is a group homomorphism, we have $G/\text{Ker}(\varphi) \cong \text{Im}(\varphi)$. The First Isomorphism Theorem tells us that this is always true.

Theorem 30.13 (The First Isomorphism Theorem)**.** *Let G and H be groups, and let $\varphi : G \to H$ be a group homomorphism. Then $G/Ker(\varphi) \cong Im(\varphi)$.*

The proof of the First Isomorphism Theorem is outlined in the next activity.

Activity 30.14. Let G and H be groups, and let $\varphi : G \to H$ be a group homomorphism. For the sake of convenience, let $K = \text{Ker}(\varphi)$. We have already seen that $K = \text{Ker}(\varphi)$ is a normal subgroup of G, so G/K is a group. To prove Theorem 30.13, we first need to define a function from G/K to H. A natural choice for such a function is

$$\Phi : G/K \to H \quad \text{defined by} \quad \Phi(aK) = \varphi(a)$$

for all $aK \in G/K$.

(a) Prove that Φ is well-defined. Why do we need to do this?

(b) Prove that Φ is a homomorphism of groups.

(c) Prove that Φ is a monomorphism.

(d) Prove that $\text{Im}(\Phi) = \text{Im}(\varphi)$.

(e) Explain how the previous parts of this activity prove the First Isomorphism Theorem.

As an example of the use of the First Isomorphism Theorem, let n be a positive integer n. We know that the set $n\mathbb{Z} = \{nk : k \in \mathbb{Z}\}$ is a subgroup of $\mathbb{Z}$. It can be shown that the canonical map $\varphi : \mathbb{Z} \to \mathbb{Z}_n$ defined by $\varphi(k) = [k]$ is an epimorphism, and $\text{Ker}(\varphi) = n\mathbb{Z}$. (See Exercise 9.) Therefore, $\mathbb{Z}/n\mathbb{Z} \cong \mathbb{Z}_n$, and $\mathbb{Z}_n$ is really a quotient group of the group of integers. We will explore additional applications of the First Isomorphism Theorem in the exercises.

The Second Isomorphism Theorem for Groups

Before stating the Second Isomorphism Theorem, we need to recall two prior results. Let G be a group, and let K and N be subgroups of G. The result of Exercise 8 of Investigation 22 (see page 311) shows that $K \cap N$ is a subgroup of G. Furthermore, Exercise 11 of Investigation 27 (see page 374) establishes that if N is a normal subgroup of G, then the set

$$KN = \{kn : k \in K, n \in N\}$$

is a subgroup of G.

The next activity illustrates the Second Isomorphism Theorem.

Activity 30.15. Let $G = D_6$, $K = \langle r, R^3 \rangle = \{I, R^3, r, rR^3\}$, and $N = \langle r, R^2 \rangle = \{I, R^2, R^4, r, rR^2, rR^4\}$. Exercise 19 or 36 of Investigation 27 (see page 375 or 378) shows that N is a normal subgroup of G.

(a) Find the elements of $K \cap N$.

(b) Find the elements of KN.

(c) What are the elements in $K/(K \cap N)$? To what familiar group is $K/(K \cap N)$ isomorphic?

(d) What are the elements in KN/N? To what familiar group is KN/N isomorphic?

(e) What is the relationship between $K/(K \cap N)$ and KN/N? Explain.

The result of Activity 30.15 is that, for the given groups, we have $K/(K \cap N) \cong KN/N$. The Second Isomorphism states that this is always true.

Theorem 30.16 (The Second Isomorphism Theorem)**.** *Let G be a group, K a subgroup of G, and $N \triangleleft G$. Then $K/(K \cap N) \cong KN/N$.*

Proof. Let G be a group with identity e, K a subgroup of G, and $N \triangleleft G$. For each $k \in K$, we have $keN = kN \in KN/N$. Define a function $\varphi : K \to KN/N$ by

$$\varphi(k) = kN.$$

We will show that φ is an epimorphism with kernel $K \cap N$. The First Isomorphism Theorem will then allow us to conclude $K/(K \cap N) \cong KN/N$.

First we will show that φ is a homomorphism. Let $k_1, k_1 \in K$. Then

$$\varphi(k_1k_2) = (k_1k_2)N = (k_1N)(k_2N) = \varphi(k_1)\varphi(k_2).$$

Thus, φ is a homomorphism.

Next we will show that φ is an epimorphism. Let $(kn)N \in KN/N$ for some $k \in K$, $n \in N$. Then $(kn)N = (kN)(nN) = (kN)N = kN$, since N is the identity element in KN/N. So $\varphi(k) = (kn)N$, and φ is an epimorphism.

Finally, we will show that $\text{Ker}(\varphi) = K \cap N$. Let $k \in \text{Ker}(\varphi)$. Then $\varphi(k) = kN = N$. Thus, $k \in N$. Since $\text{Ker}(\varphi) \subseteq K$, we also know $k \in K$. So $k \in K \cap N$. This shows $\text{Ker}(\varphi) \subseteq K \cap N$. Now let $a \in K \cap N$. Then $a \in N$, and so $\varphi(a) = aN = N$, and $a \in \text{Ker}(\varphi)$. Therefore, $K \cap N \subseteq \text{Ker}(\varphi)$. Since $\text{Ker}(\varphi) \subseteq K \cap N$ also, it follows that $K \cap N = \text{Ker}(\varphi)$. The First Isomorphism Theorem then allows us to conclude that $K/(K \cap N) \cong KN/N$. ■

The Third Isomorphism Theorem for Groups

The Third Isomorphism Theorem establishes a cancellation law of sorts for quotient groups, as the next activity illustrates.

Activity 30.17. Let $G = D_6$, $N = \langle R \rangle = \{I, R, R^2, R^3, R^4, R^5\}$, and $K = \langle R^2 \rangle = \{I, R^2, R^4\}$. Exercise 36 of Investigation 27 (see page 378) shows that both N and K are normal subgroups of G.

(a) Find the elements of G/N. To what familiar group is G/N isomorphic?

(b) Find the elements of G/K and N/K. Why is N/K a normal subgroup of G/K?

(c) What are the elements of $(G/K)/(N/K)$? To what familiar group is $(G/K)/(N/K)$ isomorphic?

(d) What is the relationship between $(G/K)/(N/K)$ and G/N? Explain.

The result of Activity 30.17 is that, for these particular groups,

$$(G/K)/(N/K) \cong G/N.$$

In other words, we can "cancel" the Ks in this particular example. The Third Isomorphism Theorem tells us that this type of cancellation works in general.

Theorem 30.18 (The Third Isomorphism Theorem)**.** *Let G be a group, and let K and N be normal subgroups of G with $K \subseteq N$. Then $(G/K)/(N/K) \cong G/N$.*

The proof of the Third Isomorphism Theorem is left as an exercise. (See Exercise 14.)

The Fourth Isomorphism Theorem for Groups

Let G be a group and K a normal subgroup of G. Among other things, the Fourth Isomorphism Theorem provides a correspondence between the subgroups of G that contain K and the subgroups of G/K. The next activity addresses this correspondence for a particular example.

Activity 30.19. Let $G = D_4$, the group of symmetries of a square. We know all of the subgroups of G, and we can list them and indicate their containments in a *subgroup lattice*, as shown in Figure 30.1. The single lines indicate containment, and the meaning of the double lines should become clear shortly. Let $K = \langle R^2 \rangle$. Exercise 36 in Investigation 27 (see page 378) shows that K is a normal subgroup of G.

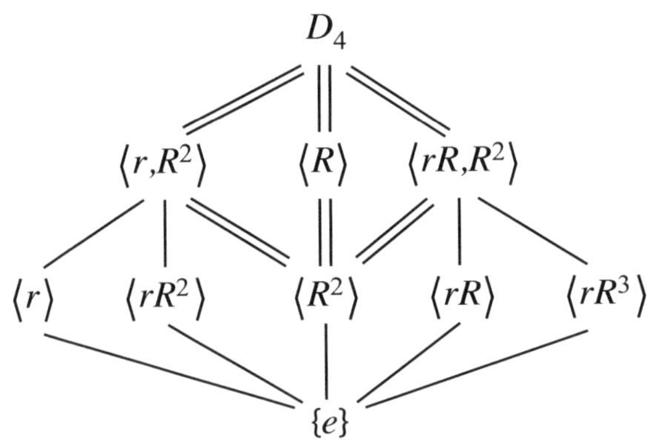

Figure 30.1
The subgroup lattice for D_4.

(a) Create the subgroup lattice for the quotient group $\overline{G} = G/K$.

(b) Compare the subgroups of $\overline{G}$ to the subgroups of G. List three things you notice about the two sets of subgroups. Include in your discussion how the normal subgroups of $\overline{G}$ are related to subgroups of G. In particular, compare the subgroups of $\overline{G}$ to the subgroups of G that contain K.

Activity 30.19 shows that, in the case of $G = D_4$ and $K = \langle R^2 \rangle$, there is a one-to-one correspondence between the subgroups of G/K and the subgroups of G that contain K. Moreover, the normal subgroups of G/K correspond to normal subgroups of G that contain K. Although it is stated in terms of group epimorphisms (rather than normal subgroups), the Fourth Isomorphism Theorem shows that this is true in general. In particular, since every normal subgroup K of G can be viewed as the kernel of a group epimorphism—namely, the mapping $G \to G/K$ defined by $a \mapsto aK$ (see Exercise 10)—the Fourth Isomorphism Theorem, when applied to this map, yields the conclusions about normal subgroups that we noted above.

Theorem 30.20 (The Fourth Isomorphism Theorem). *Let $\varphi : G \to G'$ be a group epimorphism, and let $K = Ker(\varphi)$.*

(i) *If A is a subgroup of G, then $\varphi(A) = \{\varphi(a) : a \in A\}$ is a subgroup of G'.*

(ii) *If A' is a subgroup of G', then $A' = \varphi(H)$, where $H = \{a \in G : \varphi(a) \in A'\}$ is a subgroup of G containing K.*

(iii) *Let Φ be the function that assigns to each subgroup A of G that contains K the subgroup $\varphi(A)$. Then Φ is a bijection.*

(iv) *The function Φ preserves inclusion—that is $K \subseteq A \subseteq B$ if and only if $\varphi(A) \subseteq \varphi(B)$.*

(v) *The function Φ preserves normality—that is, N is a normal subgroup of G containing K if and only if $\varphi(N)$ is a normal subgroup of G'.*

(vi) *If N is a normal subgroup of G containing K, then the function $(G/N) \to (G'/\varphi(N))$ given by $aN \mapsto \varphi(a)\varphi(N)$ for all $a \in G$ is an isomorphism.*

Because the Fourth Isomorphism Theorem gives us a correspondence between the subgroups in the subgroup lattices for G and G', this theorem is also referred to as the **lattice isomorphism theorem** or the **correspondence theorem**. The proof of the Fourth Isomorphism Theorem is left as an exercise. (See Exercise 17.)

Concluding Activities

Activity 30.21. Let G and H be arbitrary groups, and let $\varphi : G \to H$ be a homomorphism of groups.

(a) If φ is a monomorphism, what is $\text{Ker}(\varphi)$? Explain.

(b) Is the converse of part (a) true? That is, if $\text{Ker}(\varphi) = \{e_G\}$, must φ be a monomorphism? Explain.

(c) Write a formal proof of the following theorem:

Theorem 30.22. *Let $\varphi : G \to H$ be a homomorphism of groups, and let e_G be the identity element in G. Then φ is a monomorphism if and only if $\text{Ker}(\varphi) = \{e_G\}$.*

Activity 30.23. Write a formal proof of the First Isomorphism Theorem.

Activity 30.24. Write a short summary that describes the important concepts, and the relationships between these concepts, that were introduced in this investigation. Explain how the material in this investigation is connected to the material in Investigations 20, 22, and 29.

Exercises

(1) Let $\varphi : G \to H$ be a group homomorphism, and let $a \in G$. If φ is an isomorphism, we know that $|\varphi(a)| = |a|$. Is the same result true if φ is not an isomorphism? If yes, prove it. If no, provide a counterexample and then determine and prove what is true about the relationship between $|\varphi(a)|$ and $|a|$.

(2) Let m and n be positive integers and define $\varphi : \mathbb{Z}_m \to \mathbb{Z}_n$ by $\varphi([a]_m) = [a]_n$.

(a) Under what conditions on m and n is φ well-defined?

(b) When φ is well-defined, is φ a homomorphism? If yes, under what conditions is φ a monomorphism or an epimorphism? Prove your answers.

(3) We can take Exercise 2 a step further. Let m and n be positive integers, and let k be any integer. Define $\varphi : \mathbb{Z}_m \to \mathbb{Z}_n$ by $\varphi([a]_m) = [ka]_n$.

(a) Under what conditions on m, n, and k is φ well-defined?

(b) When φ is well-defined, is φ a homomorphism? If yes, under what conditions is φ a monomorphism or an epimorphism? Prove your answers.

(4) Let $n \in \mathbb{Z}^+$, and let $\varphi : \text{GL}_n(\mathbb{R}) \to \mathbb{R}^*$ be defined by $\varphi(A) = \det(A)$. (Here $\mathbb{R}^* = U(\mathbb{R})$ is the group of nonzero real numbers.)

(a) Explain why φ is a homomorphism.

(b) Show that φ is an epimorphism.

(c) Find Ker(φ). Use the First Isomorphism Theorem to find a quotient of $\text{GL}_n(\mathbb{R})$ that is isomorphic to $\mathbb{R}^*$.

(5) Let n be a positive integer, and let G_1, G_2, ..., G_n be groups with identities e_1, e_2, ..., e_n, respectively. For each $1 \leq i \leq n$, define

$$\varphi_i : G_i \to (G_1 \oplus G_2 \oplus \cdots \oplus G_n)$$

by

$$\varphi_i(g) = (e_1, e_2, \ldots, e_{i-1}, g, e_{i+1}, \ldots, e_n).$$

Show that each φ_i is a monomorphism and that

$$\text{Im}(\varphi_i) = \{e_1\} \oplus \{e_2\} \oplus \cdots \oplus \{e_{i-1}\} \oplus G_i \oplus \{e_{i+1}\} \oplus \cdots \oplus \{e_n\}.$$

The map φ_i shows that $G_1 \oplus G_2 \oplus \cdots \oplus G_n$ contains an isomorphic copy of G_i.

(6) Table 29.1 (see page 404) lists a number of invariants of group isomorphisms. Activity 30.4 shows that group homomorphisms preserve the identity and inverses, while the result of Exercise 1 demonstrates that group homomorphisms do not necessarily preserve the orders of elements. In this exercise, we will examine a few more invariants of isomorphisms to see if they are also invariants of group homomorphisms. Let $\varphi : G \to H$ be a group homomorphism.

(a) Prove or disprove: $|G| = |H|$. If the statement is false, are there any conditions on φ that are weaker than requiring φ to be an isomorphism but that still make the statement true? Prove your answer.

(b) Prove or disprove: if G is Abelian, then H is Abelian. If the statement is false, are there any conditions on φ that are weaker than requiring φ to be an isomorphism but that still make the statement true? Prove your answer.

(c) Prove or disprove: if G is cyclic, then H is cyclic. If the statement is false, are there any conditions on φ that are weaker than requiring φ to be an isomorphism but that still make the statement true? Prove your answer.

(7) (a) Determine the number of homomorphisms from $\mathbb{Z}_{12}$ to $\mathbb{Z}_{42}$.

(b) Let n and m be positive integers. Determine as best you can the number of homomorphisms from $\mathbb{Z}_m$ to $\mathbb{Z}_n$. (Note that Exercise 15 on page 323 of Investigation 23 might be useful.) Does your result agree with the specific example in part (a)?

(c) Let G and G' be finite groups. If $\gcd(|G|, |G'|) = 1$, explain why there is only one homomorphism from G to G'.

(8) Find a homomorphism $\varphi : \mathbb{Z}_8 \to U_{10}$, if possible. Either prove that your choice of φ is a homomorphism or that no such homomorphism exists. If you can find a homomorphism φ, compute Ker(φ).

* (9) Let n be a positive integer. Define $\varphi : \mathbb{Z} \to \mathbb{Z}_n$ by $\varphi(k) = [k]$.

(a) Show that φ is a homomorphism.

(b) Find Ker(φ) and Im(φ).

(c) To what familiar group is the quotient group $\mathbb{Z}/n\mathbb{Z}$ isomorphic? Explain.

⋆ (10) Let G and H be groups, and let $\varphi : G \to H$ be a homomorphism. Theorem 30.8 shows that $\text{Ker}(\varphi)$ is a normal subgroup of G. The converse of this statement is also true, as we will show in this exercise. Prove that if G is a group and N a normal subgroup of G, then the function $\varphi : G \to G/N$ defined by $\varphi(g) = gN$ is a homomorphism. Then show that $\text{Ker}(\varphi) = N$.

(11) As we will see in later investigations, the number of normal subgroups of a given order in a group tells us something about the subgroup structure of the group.

(a) Is it possible for a finite group G to have more than one normal subgroup of a given order? If yes, provide an example. If no, prove your answer.

(b) Let G be a group. In Exercise 20 of Investigation 27 (see page 376), we showed that if G has exactly one subgroup N of a given order, then N is a normal subgroup of G. This problem does not, however, show that if N is normal in G, then there is only one subgroup of order $|N|$ in G. However, there are certain situations in which we can be sure that there is only one normal subgroup of a given order in a finite group. Show that if G is a finite group and K a normal subgroup of G with $\gcd(|K|, [G : K]) = 1$, then K is the only subgroup of G of order $|K|$. (Hint: Use Exercise 25 from Investigation 27.)

(c) Let G be a group of order 15, and suppose G has a subgroup K of order 3 and a subgroup N of order 5. Explain why $G = K \times N$. Explain how this shows that G is cyclic.

(12) Let $G = \mathbb{Z}$ and $H = \mathbb{Z}_2 \oplus \mathbb{Z}_4$. Define $\varphi : \mathbb{Z} \to (\mathbb{Z}_2 \oplus \mathbb{Z}_4)$ by $\varphi(k) = ([k]_2, [k]_4)$.

(a) Prove that φ is a homomorphism.

(b) Find $\text{Ker}(\varphi)$.

(c) Find $\text{Im}(\varphi)$.

(d) Find the elements of $G/\text{Ker}(\varphi)$. How is $G/\text{Ker}(\varphi)$ related to $\text{Im}(\varphi)$?

(13) Let G_1 and G_2 be groups with identities e_1 and g_2, respectively, and let $G = G_1 \oplus G_2$. Show that $G/(G_1 \oplus \{e_2\}) \cong G_2$ and that $G/(\{e_1\} \oplus G_2) \cong G_1$.

⋆ (14) Prove the Third Isomorphism Theorem.

(15) Let a, b, and c be integers such that a divides b and b divides c. Then $c\mathbb{Z} \subseteq b\mathbb{Z} \subseteq a\mathbb{Z}$. Let $\varphi : a\mathbb{Z}/c\mathbb{Z} \to a\mathbb{Z}/b\mathbb{Z}$ be defined by $\varphi(at + c\mathbb{Z}) = at + b\mathbb{Z}$.

(a) Show that φ is a well-defined epimorphism.

(b) What is $\text{Ker}(\varphi)$? (Hint: Use the Third Isomorphism Theorem.)

(16) If N and K are normal subgroups of a group G with K a subgroup of N, the Third Isomorphism Theorem tells us that $(G/K)/(N/K) \cong G/N$. A consequence of this is that

$$[G/K : N/K] = [G : N]. \tag{30.1}$$

Recall that we can define $[G : N]$ as the number of distinct left cosets of N in G even if N is not normal in G. Is equation (30.1) true even if N and K are not normal subgroups of G? In other words, if G is any group, N is any subgroup of G (not necessarily normal), and K is any subgroup of N (not necessarily normal), must $[G/K : N/K] = [G : N]$? Prove your answer. (Hint: One way to show that two sets X and Y have the same number of elements is to find a bijection $F : X \to Y$.)

* (17) Prove the Fourth Isomorphism Theorem.

(18) Find all of the groups G' so that there is an epimorphism $\varphi : D_4 \to G'$. (Hint #1: The normal subgroups of D_4 are $\{I\}$, $\langle R^2 \rangle$, $\{I, r, rR^2, R^2\}$, $\langle R \rangle$, and D_4; see Exercise 36 on page 378 of Investigation 27. Hint #2: Use Theorem 30.8 and the First Isomorphism Theorem.)

(19) Let A be an Abelian group of order 24 whose elements have orders as shown in Table 30.1. Suppose $\varphi : A \to \mathbb{Z}_{22}$ is a non-trivial homomorphism. What can you say about $|\text{Ker}(\varphi)|$? What specific group is $\text{Im}(\varphi)$? Explain.

Element	a_1	a_2	a_3	a_4	a_5	a_6	a_7	a_8	a_9	a_{10}	a_{11}	a_{12}
Order	1	12	12	6	2	4	6	3	12	4	3	12
Element	a_{13}	a_{14}	a_{15}	a_{16}	a_{17}	a_{18}	a_{19}	a_{20}	a_{21}	a_{22}	a_{23}	a_{24}
Order	12	6	4	12	6	6	4	2	6	12	12	2

Table 30.1
Orders of elements in A.

(20) Let $G = U_{44}$, $N = \langle [3] \rangle$, and $K = \langle [9] \rangle$.

(a) Find $G/K, G/N, N/K$, and $(G/K)/(N/K)$.

(b) Explain why $(G/K)/(N/K) \cong G/N$.

(21) We can use the Second Isomorphism Theorem to derive a relationship between the greatest common divisor and the least common multiple of two integers. Let a and b be nonzero integers. Let $K = a\mathbb{Z}$ and $N = b\mathbb{Z}$.

(a) What is $K \cap N$? Prove your answer.

(b) What is $K + N$? Prove your answer.

(c) Let r and s be nonzero integers such that r divides s. What is $[r\mathbb{Z} : s\mathbb{Z}]$? Prove your answer.

(d) Explain how the Second Isomorphism Theorem and the result of part (c) tell us that

$$\text{lcm}(a, b) = \frac{ab}{\gcd(a, b)}.$$

(22) In this exercise, we will prove an interesting fact about the subgroups of the symmetric groups. Let $n \geq 2$ be an integer. Show that every subgroup H of S_n contains either all even permutations or exactly one-half of the elements in H are even permutations. (Hint: If H is a subgroup of S_n, consider the function $\varphi : H \to \{1, -1\}$ defined by

$$\varphi(\alpha) = \begin{cases} 1, & \text{if } \alpha \text{ is even,} \\ -1, & \text{if } \alpha \text{ is odd} \end{cases},$$

and then apply the First Isomorphism Theorem.)

(23) Let G be a group. Let $a \in G$, and define $\varphi_a : G \to G$ by $\varphi_a(g) = aga^{-1}$.

(a) Completely describe the mappings φ_a for all $a \in D_3$. Then construct an operation table for the set $\{\varphi_a : a \in D_3\}$, using the operation of function composition. What do you notice about this table?

(b) Show that φ_a is an automorphism of G. (See Exercise 33 on page 416 of Investigation 29.) The automorphism φ_a is called the **inner automorphism of G induced by** a. All other automorphisms of a group are called **outer** automorphisms.

(c) Let Inn(G) denote the set of all inner automorphisms of G. Prove that Inn(G) is a subgroup of Aut(G).

(d) Let G be an Abelian group. Which specific subgroup of Aut(G) is Inn(G)? Explain.

(24) In this exercise, we will determine a specific fact about inner automorphisms (see Exercise 23) and then find Inn(D_n).

(a) Let G be a group. We can associate to any element $a \in G$ the inner automorphism φ_a defined by $\varphi_a(g) = aga^{-1}$ for each $g \in G$. Define a function $\Phi : G \to \text{Inn}(G)$ by

$$\Phi(a) = \varphi_a.$$

(i) Show that Φ is a homomorphism.

(ii) Is Φ an epimorphism? Prove your answer.

(iii) Find Ker(Φ). Is Φ a monomorphism? Explain.

(iv) Explain why the following theorem is true.
Theorem 30.25. *If G is any group, then $G/Z(G) \cong$ Inn(G).*

(b) To what familiar group is Inn(D_n) isomorphic? (Hint: The result of Exercise 2 on page 330 of Investigation 24 is useful.)

(25) Let $n \geq 2$ be an integer.

(a) Find Inn(S_3). (Hint: Use the results of Exercise 24 in this investigation and Exercise 10 on page 330 of Investigation 25.)

(b) Show that Aut(S_3) $\cong$ Inn(S_3). Note that this result—namely, that Aut(S_n) $\cong$ Inn(S_n)—is almost true in general, but fails when $n = 6$, providing another fascinating property of the number 6. (Other interesting properties of the number 6 are that 6 is the smallest order a non-Abelian group can have, 6 is the smallest perfect number, and 6 is the smallest positive integer that is the product of two distinct primes.) Segal* gives a reasonably accessible proof of the fact that Aut(S_n) $\cong$ Inn(S_n) for *every* n not equal to 6. (Segal states that the first proof is due to Hölder.†) A good project would be to find an outer automorphism of S_6.

Connections

A homomorphism from a group G to a group H is a structure-preserving map. The isomorphism theorems for groups show us different ways that homomorphisms can determine isomorphisms. If

*I. Segal, "The automorphisms of the symmetric group," *Bulletin of the American Mathematical Society*, **46**(6), p. 565.
†*Mathematische Annalen*, **46**, 1895, pp. 340–345.

you studied ring theory before group theory, you should notice the connections between group homomorphisms in this investigation and ring homomorphisms in Investigation 16. The major difference between a group homomorphism and a ring homomorphism is that a ring has two operations, and so there is a bit more structure to preserve. In spite of this difference, a careful perusal of the group isomorphism theorems shows that they are completely analogous to the corresponding ring isomorphism theorems.

Investigation 31

The Fundamental Theorem of Finite Abelian Groups

Focus Questions

By the end of this investigation, you should be able to give precise and thorough answers to the questions listed below. You may want to keep these questions in mind to focus your thoughts as you complete the investigation.

- What is a p-group?
- In what sense is each finite Abelian group made up of p-subgroups?
- What does the Fundamental Theorem of Finite Abelian Groups tell us, and why is the theorem considered "fundamental"?

Preview Activity 31.1. The orders of the elements of the group $U_{35} = \{[1], [2], [3], [4], [6], [8], [9], [11], [12], [13], [16], [17], [18], [19], [22], [23], [24], [26], [27], [29], [31], [32], [33], [34]\}$ are given in Table 31.1.

$\|[1]\| = 1$	$\|[9]\| = 6$	$\|[18]\| = 12$	$\|[27]\| = 4$
$\|[2]\| = 12$	$\|[11]\| = 3$	$\|[19]\| = 6$	$\|[29]\| = 2$
$\|[3]\| = 12$	$\|[12]\| = 12$	$\|[22]\| = 4$	$\|[31]\| = 6$
$\|[4]\| = 6$	$\|[13]\| = 4$	$\|[23]\| = 12$	$\|[32]\| = 12$
$\|[6]\| = 2$	$\|[16]\| = 3$	$\|[24]\| = 6$	$\|[33]\| = 12$
$\|[8]\| = 4$	$\|[17]\| = 12$	$\|[26]\| = 6$	$\|[34]\| = 2$

Table 31.1
Orders of elements in U_{35}.

(a) The group U_{35} contains subgroups K and N of orders 12 and 2, respectively, so that U_{35} is the internal direct product $K \times N$. Find such groups and explain why $U_{35} = K \times N$.

(b) Find two cyclic groups $\mathbb{Z}_m$ and $\mathbb{Z}_n$ (for some integers m and n) so that U_{35} is isomorphic to the external direct product $\mathbb{Z}_m \oplus \mathbb{Z}_n$. Explain your reasoning.

(c) Notice that $|U_{35}| = 2^3 \times 3$. Let $G(2)$ be the subset of all of the elements a of U_{35} such that $|a|$ is a power of 2. Does $G(2)$ form a subgroup of U_{35}? Explain.

(d) Let $G(3)$ be the set of all elements of U_{35} whose order is a power of 3. Does $G(3)$ form a subgroup of U_{35}? Explain.

(e) Explain why $U_{35} = G(2) \times G(3)$.

(f) Explain why $G(2) \cong \mathbb{Z}_4 \oplus \mathbb{Z}_2$. (Hint: What is the maximal order an element in $G(2)$ has?) Why does this imply $U_{35} \cong \mathbb{Z}_4 \oplus \mathbb{Z}_2 \oplus \mathbb{Z}_3$? Does this contradict your answer to part (b)? Explain.

Introduction

Classification theorems are among the most difficult in mathematics. To classify something is to understand it completely. As a result, it is a significant accomplishment that there is a classification theorem for all finite Abelian groups. It is that classification that we will study in this investigation.

The Components: p-Groups

Preview Activity 31.1 suggests that certain subgroups of a finite Abelian group—namely, those whose orders are powers of a prime number p—might provide factors of the group. This brings us to the important idea of what we call p-*groups*.

Definition 31.2. A p-**group**, where p is a prime, is a group whose order is a power of p.

The p-groups are important because they will form the building blocks of all finite Abelian groups. As we will see, a finite Abelian group can be broken down into a direct sum of p-subgroups. Preview Activity 31.1 gave us a hint about how we might begin with such a decomposition when we wrote the subgroup $G(2)$ as an internal direct product using an element of maximal order (as a power of 2) in the group. The next theorem shows us that we can always take this step for a p-group. Its proof is quite technical, so we will present it in its entirety and then consider an example to illustrate.

Theorem 31.3. *Let p be a prime, and let G be a finite Abelian p-group. Let $a \in G$ be an element of maximal order in G. Then there exists a subgroup K of G so that $G = \langle a \rangle \times K$.*

Proof. Let p be a prime, let G be a p-group with identity e, and let $a \in G$ have maximal order in G. To find a candidate for the subgroup K of G, we will look for a group K that is large (to account for the elements not in $\langle a \rangle$) and that also satisfies $\langle a \rangle \cap K = \{e\}$. There is at least one subgroup H of G that has the property that $\langle a \rangle \cap H = \{e\}$ (namely, $H = \{e\}$). Since G is a finite group, G must contain such a subgroup K of largest order.

To show that $G = \langle a \rangle \times K = \{a^m k : m \in \mathbb{Z}, k \in K\}$, we must show that any element $g \in G$ can be written in the form $a^m k$ for some nonnegative integer m and some element $k \in K$. We will proceed by contradiction and assume there is some element $b \in G$ so that b cannot be written in this form. Since $e = ee = a^0 e$, we know that $b \neq e$. Since G is a finite Abelian p-group, there is a positive integer r so that $b^{p^r} = e$ and, consequently, $b^{p^r} \in \langle a \rangle \times K$. Let s be the smallest positive integer so that $b^{p^s} \in \langle a \rangle \times K$. Let $c = b^{p^{s-1}}$. Then

$$c \notin \langle a \rangle \times K \quad \text{and} \quad c^p = \left(b^{p^{s-1}}\right)^p = b^{p^s} \in \langle a \rangle \times K. \tag{31.1}$$

Let m be a nonnegative integer and $k \in K$ so that

$$c^p = a^m k. \tag{31.2}$$

Since $a \in G$ and G is a p-group, there is a positive integer n so that $|a| = p^n$.

Claim 1. If $g \in G$, then $g^{p^n} = e$ for every $g \in G$.

To verify Claim 1, let $g \in G$. Since G is a p-group, we must have $|g| = p^t$ for some integer t. Since a has maximal order in G, we know $|g| \leq p^n$. It follows that $|g|$ divides $|a|$, and so $g^{p^n} = e$ for every $g \in G$.

By Claim 1, we know $c^{p^n} = e$, and so

$$e = c^{p^n} = (c^p)^{p^{n-1}} = (a^m k)^{p^{n-1}}.$$

Thus $(a^m)^{p^{n-1}} = k^{-p^{n-1}}$ is an element of both $\langle a \rangle$ and K. Since $\langle a \rangle \cap K = \{e\}$, it follows that

$$(a^m)^{p^{n-1}} = k^{-p^{n-1}} = e.$$

Therefore, the order of a must divide mp^{n-1}—that is, $p^n \mid mp^{n-1}$. Since $n > n-1$, Euclid's Lemma (see page 36) implies that $p \mid m$. So there is an integer w such that $pw = m$. Thus,

$$c^p = a^m k = a^{pw} k,$$

or

$$k = c^p a^{-pw} = (ca^{-w})^p. \tag{31.3}$$

Let

$$d = ca^{-w}. \tag{31.4}$$

Claim 2. $d \notin K$, but $d^p \in K$.

To verify Claim 2, note that if $d = ca^{-w} = k_0 \in K$, then $c = a^w k_0 \in \langle a \rangle \times K$, which contradicts (31.1). Thus, $d \notin K$. However, (31.3) shows $d^p \in K$.

Claim 3. The subgroup $H = K\langle d \rangle$ of G contains K as a proper subgroup.

To verify Claim 3, note that $H = K\langle d \rangle = \{xd^y : x \in K, y \in \mathbb{Z}\}$ is a subgroup of G since K is a normal subgroup of G. (See Activity 27.22 on page 372.) Since $xd^0 = x$ is in H for every $x \in K$, we see that K is a subset of H. Since $d \notin K$ and $d = ed^1 \in H$, we see that K is a proper subset of H.

Recall that K is the largest subgroup of G with the property that $\langle a \rangle \cap K = \{e\}$. Therefore,

$\langle a \rangle \cap H \neq \{e\}$. Let z be a non-identity element in $\langle a \rangle \cap H$. Then there exist a nonnegative integer j, an element k_1 in K, and an integer q so that

$$z = a^j = k_1 d^q. \tag{31.5}$$

Recall that $d^p \in K$ by Claim 2. So if $p \mid q$, then for some integer t, we have $z = k_1 d^q = k_1(d^p)^t \in K$. Thus, $z \in \langle a \rangle \cap K$. Since z is not the identity element, this cannot happen. So p cannot divide q. Therefore, $\gcd(p, q) = 1$, and Corollary 3.11 (see page 29) implies that there exist integers u, v so that

$$pu + qv = 1.$$

Substituting from (31.2) and (31.4) gives us

$$c = c^{pu+qv} = (c^p)^u(c^q)^v = (a^m k)^u((da^w)^q)^v = (a^m k)^u(d^q a^{wq})^v.$$

Substituting from (31.5) now yields

$$c = (a^m k)^u(d^q a^{wq})^v = (a^m k)^u(a^j k_1^{-1} a^{wq})^v = (a^{mu} a^{jv} a^{wqv})(k^u k_1^{-v}).$$

Now $c = (a^{mu} a^{jv} a^{wqv})(k^u k_1^{-v}) \in \langle a \rangle \times K$, which contradicts (31.1). Therefore, we can conclude that every element in G is in $\langle a \rangle \times K$. Since every element in $\langle a \rangle \times K$ is also in G, it follows that $G = \langle a \rangle \times K$, as desired. ■

There are two key steps in applying Theorem 31.3:

- First, we must find an element $a \in G$ of maximal order.
- Next, we must find a maximal subgroup K of G satisfying $\langle a \rangle \cap K = \{e\}$. Then, for this K, we will have
$$G = \langle a \rangle K = \{a^m k : m \in \mathbb{Z}, k \in K\}.$$

Activity 31.4. We will illustrate the proof of Theorem 31.3 using $G = U_{60}$ as an example. Note that

$$U_{60} = \{[1], [7], [11], [13], [17], [19], [23], [29], [31], [37], [41], [43], [47], [49], [53], [59]\},$$

and so $|U_{60}| = 16 = 2^4$. The orders of the elements of U_{60} are given in Table 31.2.

$\|[1]\| = 1$	$\|[17]\| = 4$	$\|[31]\| = 2$	$\|[47]\| = 4$
$\|[7]\| = 4$	$\|[19]\| = 2$	$\|[37]\| = 4$	$\|[49]\| = 2$
$\|[11]\| = 2$	$\|[23]\| = 4$	$\|[41]\| = 2$	$\|[53]\| = 4$
$\|[13]\| = 4$	$\|[29]\| = 2$	$\|[43]\| = 4$	$\|[59]\| = 2$

Table 31.2
Orders of elements in U_{60}.

(a) Find an element a in U_{60} of maximal order. Compute $\langle a \rangle$.

(b) Find a maximal subgroup K of G so that $\langle a \rangle \cap K = \{[1]\}$. (Hint: Recall that we want to have $\langle a \rangle \times K = G$, so what must the order of K be? Also note that $[7]^2 = [13]^2 = [17]^2 = [23]^2 = [37]^2 = [43]^2 = [47]^2 = [53]^2 = [49] \in U_{60}$.)

(c) It will necessarily be the case that every element in U_{60} can be written uniquely in the form $a^m k$ for some nonnegative integer m and some element $k \in K$. Verify that this does in fact happen for the choices of a and K you found in parts (a) and (b).

The group K in Theorem 31.3 is itself a p-group, so we can apply Theorem 31.3 to K. Continuing this process will allow us to write the original group as an internal direct product of cyclic subgroups.

Activity 31.5. Let p be a prime, and let G be a finite Abelian p-group with order p^n for some nonnegative integer n. In this activity, we will show that G can be written as an internal direct product of cyclic groups.

(a) Explain why we are done if $n = 1$.

(b) Now assume that any p-group of order less than p^n is an internal direct product of cyclic p-groups.

(i) What does Theorem 31.3 tells us about G?

(ii) If K is the subgroup as described in Theorem 31.3, what can we say about $|K|$?

(iii) By considering cases and using Theorem 31.3, complete the proof of the following corollary to Theorem 31.3:

Corollary 31.6. *Let p be a prime and G a finite Abelian p-group. Then G is an internal direct product of cyclic p-groups.*

Next we will apply Corollary 31.6 to an example to see a complete decomposition of a finite Abelian p-group into an internal direct product of cyclic subgroups.

Activity 31.7. Consider the group U_{60} whose decomposition as an internal direct product of cyclic groups we began in Activity 31.4. In the inductive step of Corollary 31.6, we applied Theorem 31.3 to the subgroup K. Illustrate this process by completing the steps below for the subgroup K you found in part (b) of Activity 31.4.

(a) Choose an element a_K of maximal order in K.

(b) Find a maximal subgroup K_1 of K so that $\langle a_K \rangle \cap K_1 = \{[1]\}$.

(c) Construct U_{60} from these cyclic subgroups.

Example 31.8. As an illustration of Corollary 31.6, we will determine all Abelian groups of order 8. Let G be an Abelian group of order $8 = 2^3$. By Corollary 31.6, we know that G is isomorphic to a direct product of cyclic p-groups for $p = 2$. The only possibilities for G are then

$$\mathbb{Z}_8, \qquad \mathbb{Z}_4 \oplus \mathbb{Z}_2, \qquad \text{or} \qquad \mathbb{Z}_2 \oplus \mathbb{Z}_2 \oplus \mathbb{Z}_2.$$

Since $\mathbb{Z}_8$ contains one element of order 2, $(\mathbb{Z}_4 \oplus \mathbb{Z}_2)$ contains three elements of order 2, and $\mathbb{Z}_2 \oplus \mathbb{Z}_2 \oplus \mathbb{Z}_2$ contains seven elements of order 2, we see that these three groups form three distinct isomorphism classes of groups of order 8. So there are exactly 3 Abelian groups of order 8, up to isomorphism.

An interesting question to ask is, given a prime power q^n, how many isomorphism classes are there of finite Abelian groups of order q^n? (The reason why we are switching to use q as our prime instead of p should become clear shortly.) The answer is connected to integer partitions. An **integer**

n	Partitions of n	Isomorphism classes of groups of order q^n
1	1	$\mathbb{Z}_q$
2	2	$\mathbb{Z}_{q^2}$
	1+1	$\mathbb{Z}_q \oplus \mathbb{Z}_q$
3	3	$\mathbb{Z}_{q^3}$
	2+1	$\mathbb{Z}_{q^2} \oplus \mathbb{Z}_q$
	1+1+1	$\mathbb{Z}_q \oplus \mathbb{Z}_q \oplus \mathbb{Z}_q$
4	4	$\mathbb{Z}_{q^4}$
	3+1	$\mathbb{Z}_{q^3} \oplus \mathbb{Z}_q$
	2+2	$\mathbb{Z}_{q^2} \oplus \mathbb{Z}_{q^2}$
	2+1+1	$\mathbb{Z}_{q^2} \oplus \mathbb{Z}_q \oplus \mathbb{Z}_q$
	1+1+1+1	$\mathbb{Z}_q \oplus \mathbb{Z}_q \oplus \mathbb{Z}_q \oplus \mathbb{Z}_q$
5	5	$\mathbb{Z}_{q^5}$
	4+1	$\mathbb{Z}_{q^4} \oplus \mathbb{Z}_q$
	3+2	$\mathbb{Z}_{q^3} \oplus \mathbb{Z}_{q^2}$
	3+1+1	$\mathbb{Z}_{q^3} \oplus \mathbb{Z}_q \oplus \mathbb{Z}_q$
	2+2+1	$\mathbb{Z}_{q^2} \oplus \mathbb{Z}_{q^2} \oplus \mathbb{Z}_q$
	2+1+1+1	$\mathbb{Z}_{q^2} \oplus \mathbb{Z}_q \oplus \mathbb{Z}_q \oplus \mathbb{Z}_q$
	1+1+1+1+1	$\mathbb{Z}_q \oplus \mathbb{Z}_q \oplus \mathbb{Z}_q \oplus \mathbb{Z}_q \oplus \mathbb{Z}_q$

Table 31.3
Isomorphism classes of Abelian groups of order q^n, for $1 \leq n \leq 5$.

partition of a positive integer n is a non-increasing sequence of positive integers whose sum is n. As an example,

$$\begin{aligned} 4 &= 4 \\ &= 3 + 1 \\ &= 2 + 2 \\ &= 2 + 1 + 1 \\ &= 1 + 1 + 1 + 1, \end{aligned}$$

so there are 5 partitions of 4. The **integer partition function** is denoted p, where $p(n)$ is the number of partitions of the positive integer n. Our example above shows that $p(4) = 5$. Some values of $p(n)$, along with connections to various branches of mathematics, can be found in the *On-Line Encyclopedia of Integer Sequences*.* Table 31.3 suggests that there is a connection between the number of isomorphism classes of finite Abelian groups of order q^n and the number $p(n)$ of partitions of the power n. You will formalize this connection in Exercise 5.

*This is sequence A000041 at `oeis.org`. The OEIS is a fascinating site and you can learn a lot of interesting mathematics there. We highly recommend that you visit the site.

The Fundamental Theorem

Preview Activity 31.9. Let $G = U_{28}$ where

$$U_{28} = \{[1], [3], [5], [9], [11], [13], [15], [17], [19], [23], [25], [27]\}.$$

The orders of the elements in U_{28} are shown in Table 31.4.

$\|[1]\| = 1$	$\|[11]\| = 6$	$\|[19]\| = 6$
$\|[3]\| = 6$	$\|[13]\| = 2$	$\|[23]\| = 6$
$\|[5]\| = 6$	$\|[15]\| = 2$	$\|[25]\| = 3$
$\|[9]\| = 3$	$\|[17]\| = 6$	$\|[27]\| = 2$

Table 31.4
Orders of elements in U_{28}.

(a) Let $G(2)$ be the subset of G consisting of the elements whose orders are powers of 2. Find the elements of $G(2)$.

(b) Let $G(3)$ be the subset of G consisting of the elements whose orders are powers of 3. Find the elements of $G(3)$.

(c) Are $G(2)$ and $G(3)$ subgroups of G? Explain

(d) What group is $G(2) \times G(3)$?

We will now extend the results of the last section to all finite Abelian groups, using p-groups as our building blocks. Let G be a finite Abelian group of order n. By the Fundamental Theorem of Arithmetic, we can find distinct primes $p_1, p_2, \ldots, p_k$ and positive integers $m_1, m_2, \ldots, m_k$ so that

$$n = p_1^{m_1} p_2^{m_2} \cdots p_k^{m_k}.$$

In this section, we will show that we can break G up into an internal direct product of p-groups,

$$G = G(p_1) \times G(p_2) \times \cdots \times G(p_k), \tag{31.6}$$

where $|G(p_i)| = p_i^{m_i}$ for each i from 1 to k. We can then use Corollary 31.6 to write each $G(p_i)$ as a direct product of cyclic groups. When we put this all together, we have G as an internal direct product of cyclic groups as well, which will be isomorphic to an external direct product of cyclic groups.

In what follows, we will see how to decompose G into a direct product of p-groups as in (31.6). First, we will define the groups $G(p_i)$. Preview Activity 31.9 indicates that the groups we want are exactly those groups whose elements have orders that are powers of the primes. The next definition formalizes this idea.

Definition 31.10. Let G be a finite Abelian group of order n, and let p be a prime factor of n. The **p-primary component** of G is the set

$$G(p) = \{a \in G : |a| = p^t \text{ for some nonnegative integer } t\}.$$

In other words, $G(p)$ is the set of all elements $a \in G$ whose order is a power of p.

In our example in Preview Activity 31.9, the p-primary components of G were subgroups of G. In the next activity, we will determine if this is always the case.

Activity 31.11. Let G be a finite Abelian group with identity e, and let p be a prime divisor of $|G|$.

(a) Is $e \in G(p)$? Explain.

(b) Is $G(p)$ closed? Explain.

(c) If $a \in G(p)$, must a^{-1} be in $G(p)$? Explain.

(d) Is $G(p)$ a subgroup of G? Explain.

The next lemma formalizes the result of Activity 31.11 and also shows that $G(p)$ is a p-group.

Lemma 31.12. *If p is a prime factor of $|G|$, then $G(p)$ is a subgroup of G whose order is a power of p.*

Proof. Let p be a prime factor of $|G|$. The proof that $G(p)$ is a subgroup of G is contained in Activity 31.11. We will show that $G(p)$ has order a power of p. Suppose q is a prime, with $q \neq p$, such that q divides $|G(p)|$. By Cauchy's Theorem for Finite Abelian Groups, we know that $G(p)$ contains an element g of order q. This contradicts the fact that every element of $G(p)$ has order p^t for some positive integer t. Therefore, the only prime divisor of $|G(p)|$ is p, and so the order of $G(p)$ is a power of p. ∎

In other words, Lemma 31.12 shows that $G(p)$ is a p-group. It is these p-primary components of G that will form the factors of G, as in (31.6).

Theorem 31.13 (The Fundamental Theorem of Finite Abelian Groups)**.** *Let G be a finite Abelian group. Then*

$$G = G(p_1) \times G(p_2) \times \cdots \times G(p_k), \tag{31.7}$$

where $p_1, p_2, \ldots, p_k$ are the distinct prime factors of $|G|$.

Proof. Let G be a finite Abelian group with identity e, and let $p_1, p_2, \ldots, p_k$ be the distinct prime factors of $|G|$. We will show that any element $a \in G$ can be written in the form

$$a = a_1 a_2 \cdots a_k, \tag{31.8}$$

with $a_i \in G(p_i)$ for each i.

Let $a \in G$. Since $|a|$ divides $|G|$ and $p_1, p_2, \ldots, p_k$ are the distinct prime factors of $|G|$, the order of a must have the form

$$|a| = q_1^{r_1} q_2^{r_2} \cdots q_s^{r_s},$$

where $Q = \{q_1, q_2, \ldots, q_s\}$ is a subset of $\{p_1, p_2, \ldots, p_k\}$ and $r_i > 0$ for each i. We will assume, without loss of generality, that $q_1 < q_2 < \cdots < q_s$. If we show that a can be factored as

$$a = b_1 b_2 \cdots b_s,$$

where $b_i \in G(q_i)$ for each i, then to verify (31.8) we can take $a_i = b_i$ when $p_i \in Q$ and $a_i = e$ when $p_i \notin Q$.

To show $a = b_1 b_2 \cdots b_s$, where $b_i \in G(q_i)$ for each i, we will proceed by induction on s, the number of distinct prime factors of $|a|$. If $s = 1$, then there is only one prime factor of $|a|$. Thus, we can choose $b_1 = a \in G(q_1)$ and we will be done. Now suppose that for any positive integer $t < s$, whenever $q_1, q_2, \ldots, q_t$ are the distinct prime factors of an element $g \in G$, then we can write $g = b_1 b_2 \cdots b_t$, where $b_i \in G(q_i)$ for each i between 1 and t.

Now suppose $q_1, q_2, \ldots, q_s$ are the distinct prime factors of $|a|$. Let $|a| = q_1^{r_1} q_2^{r_2} \cdots q_s^{r_s}$ for some positive integers $r_1, r_2, \ldots, r_s$. Let $u = q_1^{r_1}$ and $v = q_2^{r_2} \cdots q_s^{r_s}$. Then $\gcd(u, v) = 1$, and so we can find integers x, y such that $xu + yv = 1$. (See Corollary 3.11 on page 29.) Now

$$a = a^{xu+yv} = (a^{xu})(a^{yv}). \tag{31.9}$$

Note that

$$(a^{xu})^v = (a^{uv})^x = (a^{|a|})^x = e,$$

and so $|a^{xu}|$ must be a divisor of v. (See Theorem 23.5 on page 320.) Therefore, the only primes dividing $|a^{xu}|$ are $q_2, q_3, \ldots, q_s$. Thus, we can apply the induction hypothesis to a^{xu} and write

$$a^{xu} = b_2 b_3 \cdots b_s, \tag{31.10}$$

where $b_i \in G(q_i)$ for each i between 2 and s.

Similarly,

$$(a^{yv})^u = (a^{uv})^y = (a^n)^y = e,$$

and so the only prime factor of $|a^{yv}|$ is q_1. We can again apply the induction hypothesis, this time to a^{yv}, and write

$$a^{yv} = b_1, \tag{31.11}$$

where $b_1 \in G(q_1)$. Combining (31.9), (31.10), and (31.11), gives us

$$a = a^{yv} a^{xu} = b_1 b_2 \cdots b_s,$$

where $b_i \in G(q_i)$ for each i between 1 and s. This completes our proof that a has the form from (31.8).

Finally, to show that G is an internal direct product of $G(p_1), G(p_2), \ldots, G(p_k)$, we will show that the decomposition of an element in G into a product of elements in the p-primary components of G is unique. (Activity 31.16 establishes the validity of this approach.) Let $a \in G$, and suppose

$$a = a_1 a_2 \cdots a_k = b_1 b_2 \cdots b_k,$$

where $a_i, b_i \in G(p_i)$ for each i. Then, for each j between 1 and k we have

$$a_j b_j^{-1} = (b_1 a_1^{-1})(b_2 a_2^{-1}) \cdots (b_{j-1} a_{j-1}^{-1})(b_{j+1} a_{j+1}^{-1}) \cdots (b_k a_k^{-1}).$$

The only prime that can divide the order of $a_j b_j^{-1}$ is p_j. If p_j divides $|a_j b_j^{-1}|$, then p_j divides

$$|(b_1 a_1^{-1})(b_2 a_2^{-1}) \cdots (b_{j-1} a_{j-1}^{-1})(b_{j+1} a_{j+1}^{-1}) \cdots (b_k a_k^{-1})|. \tag{31.12}$$

But for each $i \neq j$, we have $b_i a_i^{-1} \in G(p_i)$. So the only primes that can divide (31.12) are $p_1, p_2, \ldots, p_{j-1}, p_{j+1}, \ldots, p_k$. Therefore, $|a_j b_j^{-1}| = 1$, and so $a_j = b_j$. Since j was chosen arbitrarily, we can conclude that $a_j = b_j$ for each j from 1 to k. Thus, the decomposition of the element a as $a = a_1 a_2 \cdots a_k$, where $a_i \in G(p_i)$ for each i, is unique.

It follows that if G is a finite Abelian group and $p_1, p_2, \ldots, p_k$ are the distinct prime divisors of $|G|$, then

$$G = G(p_1) \times G(p_2) \times \cdots \times G(p_k).$$

■

Activity 31.14. In this activity, we will illustrate Theorem 31.13 with $G = U_{28}$.

(a) Determine specific groups $G(p_1), G(p_2), \ldots, G(p_k)$ so that $G = U_{28}$ is the internal direct product

$$G = G(p_1) \times G(p_2) \times \cdots \times G(p_k).$$

(b) Complete Table 31.5 to show that

$$U_{28} = G(p_1) \times G(p_2) \times \cdots \times G(p_k).$$

If these groups $G(p_i)$ are different than the ones you found in part (a), then explicitly identify $G(p_1), G(p_2), \ldots, G(p_k)$ in this case.

	[1]	[9]	[25]
[1]			
[13]			
[15]			
[27]			

Table 31.5
U_{28} as an internal direct product of p-groups.

Example 31.15. To illustrate the Fundamental Theorem of Finite Abelian Groups, we will find all Abelian groups of order 360. If G is such a group, then since $360 = 2^3 \times 3^2 \times 5$, it follows that

$$G = G(2) \times G(3) \times G(5).$$

By Corollary 31.6, we know:

- $G(2)$ is isomorphic to $\mathbb{Z}_8$, $\mathbb{Z}_4 \oplus \mathbb{Z}_2$, or $\mathbb{Z}_2 \oplus \mathbb{Z}_2 \oplus \mathbb{Z}_2$.
- $G(3)$ is isomorphic to $\mathbb{Z}_9$ or $\mathbb{Z}_3 \oplus \mathbb{Z}_3$.
- $G(5)$ is isomorphic to $\mathbb{Z}_5$.

Therefore, G is isomorphic to one of the following:

$$\begin{array}{rl} \mathbb{Z}_8 \oplus \mathbb{Z}_9 \oplus \mathbb{Z}_5 & \mathbb{Z}_8 \oplus \mathbb{Z}_3 \oplus \mathbb{Z}_3 \oplus \mathbb{Z}_5 \\ \mathbb{Z}_4 \oplus \mathbb{Z}_2 \oplus \mathbb{Z}_9 \oplus \mathbb{Z}_5 & \mathbb{Z}_4 \oplus \mathbb{Z}_2 \oplus \mathbb{Z}_3 \oplus \mathbb{Z}_3 \oplus \mathbb{Z}_5 \\ \mathbb{Z}_2 \oplus \mathbb{Z}_2 \oplus \mathbb{Z}_2 \oplus \mathbb{Z}_9 \oplus \mathbb{Z}_5 & \mathbb{Z}_2 \oplus \mathbb{Z}_2 \oplus \mathbb{Z}_2 \oplus \mathbb{Z}_3 \oplus \mathbb{Z}_3 \oplus \mathbb{Z}_5. \end{array}$$

One question remains. Let G be a finite Abelian group whose order has prime power factorization

$$|G| = p_1^{m_1} p_2^{m_2} \cdots p_k^{m_k}.$$

How do we know that the groups we obtain using the method in Example 31.15 are all different? It seems obvious, but of course the obvious isn't always true. However, it is true in this case. Exercise 5 tells us how to determine all of the p_i-primary components, but it doesn't tell us what happens when we put them together. If we can show that different choices for $G(p_i)$ yield non-isomorphic groups G when combined as in (31.7), then we will have our answer. You are asked to supply the details in Exercise 10.

Concluding Activities

Activity 31.16. In the proof of the Fundamental Theorem of Finite Abelian Groups, we showed that the decomposition of an element in G into a product of elements in the p-primary components of G is unique. We then concluded that G is the internal direct product of the p-primary components of G. In this activity, we will verify the validity of this conclusion.

Assume that N_1, N_2, ..., N_k are normal subgroups of a group G with $|N_1||N_2|\cdots|N_k| = |G|$ and that the decomposition of an element in $N_1N_2\cdots N_k$ is unique. That is, if $a_i, b_i \in N_i$ for each i and

$$a_1a_2\cdots a_k = b_1b_2\cdots b_k,$$

then it follows that $a_i = b_i$ for all $1 \leq i \leq k$.

(a) Let e be the identity in G. Prove that

$$N_i \cap (N_1N_2\cdots N_{i-1}N_{i+1}\cdots N_k) = \{e\}$$

for each i between 1 and k. (You may want to look back at Activity 28.16.) Conclude that

$$N_1N_2\cdots N_k = N_1 \times N_2 \times \cdots \times N_k.$$

(b) Use part (a) and the fact that $|N_1||N_2|\cdots|N_k| = |G|$ to explain why

$$G = N_1 \times N_2 \times \cdots \times N_k.$$

Activity 31.17. There is an alternate version of the Fundamental Theorem of Finite Abelian Groups that says that if G is a finite Abelian group, then G is isomorphic to a direct product of cyclic groups

$$\mathbb{Z}_{c_1} \oplus \mathbb{Z}_{c_2} \oplus \cdots \oplus \mathbb{Z}_{c_t}, \tag{31.13}$$

where $c_i \geq 2$ for all i and $c_t \mid c_{t-1} \mid c_{t-2} \mid \cdots \mid c_1$. The integers c_1, c_2, ..., c_t are called the **invariant factors** of G because two finite Abelian groups with the same invariant factors are isomorphic. (See Exercise 13.) In this activity, we will prove the alternate, invariant factor version of the Fundamental Theorem of Finite Abelian Groups.

Let G be a finite Abelian group so that $|G| = p_1^{m_1}p_2^{m_2}\cdots p_k^{m_k}$ for some distinct primes $p_1, p_2, \ldots, p_k$ and some positive integers $m_1, m_2, \ldots m_k$. By the original version of the Fundamental Theorem of Finite Abelian Groups, we know that

$$G = G(p_1) \times G(p_2) \times \cdots \times G(p_k),$$

where $G(p_i)$ is a p_i group for each i. Also, for each i we have

$$G(p_i) \cong \mathbb{Z}_{p_i^{r_{i,1}}} \oplus \mathbb{Z}_{p_i^{r_{i,2}}} \oplus \cdots \oplus \mathbb{Z}_{p_i^{r_{i,t_i}}} \tag{31.14}$$

for some $t_i \geq 1$, where $r_{i,1} \geq r_{i,2} \geq \cdots \geq r_{i,t_i}$.

(a) The group U_{1001} has order $720 = 2^4 \times 3^2 \times 5$. Also,

$$U_{1001}(2) = \langle[34]\rangle \times \langle[274]\rangle \times \langle[428]\rangle$$
$$U_{1001}(3) = \langle[100]\rangle \times \langle[144]\rangle$$
$$U_{1001}(5) = \langle[92]\rangle.$$

(i) Identify all of the variables in (31.14) for the group U_{1001}.

(ii) Find the invariant factors of U_{1001}.

(b) To prove this alternate formulation of the Fundamental Theorem of Finite Abelian Groups, we will need some additional notation. Let $t = \max\{t_i : 1 \leq i \leq k\}$. Note that $\mathbb{Z}_1 = \{[0]\}$, so we can extend $G(p_i)$ to

$$G(p_i) \cong \mathbb{Z}_{p_i^{r_{i,1}}} \oplus \mathbb{Z}_{p_i^{r_{i,2}}} \oplus \cdots \oplus \mathbb{Z}_{p_i^{r_{i,t_i}}} \oplus \cdots \oplus \mathbb{Z}_{p_i^{r_{i,t}}},$$

where $r_{i,j} = 0$ for $j > t_i$. Using this notation, determine the invariant factors of G, and show that G can be written in the form of equation (31.13).

Activity 31.18. Write a short summary that describes the important concepts, and the relationships between these concepts, that were introduced in this investigation. Explain how the material in this investigation is connected to the material in Investigations 20, 22, 28, and 29.

Exercises

(1) Find all of the isomorphism classes of finite Abelian groups of the following orders.

(a) 12 (b) 16 (c) 30

(d) 36 (e) 252 (f) 8600

(2) Write the group U_{120} as an internal direct product of cyclic groups. (Hint: No element in U_{120} has order greater than 4.)

(3) For this exercise, let A be an Abelian group of order 24 whose elements have orders as shown in Table 31.6.

Element	a_1	a_2	a_3	a_4	a_5	a_6	a_7	a_8	a_9	a_{10}	a_{11}	a_{12}
Order	1	12	12	6	2	4	6	3	12	4	3	12
Element	a_{13}	a_{14}	a_{15}	a_{16}	a_{17}	a_{18}	a_{19}	a_{20}	a_{21}	a_{22}	a_{23}	a_{24}
Order	12	6	4	12	6	6	4	2	6	12	12	2

Table 31.6
Orders of elements in A.

(a) Find all of the isomorphism classes of Abelian groups of order 24. Use the Fundamental Theorem of Finite Abelian Groups to determine which class the group A belongs to. Explain. (Note that $a_6^2 \neq a_{20}$.)

(b) It is known that there are 15 groups (up to isomorphism) of order 24. How many of these groups are non-Abelian? Explain.

(c) List six potentially different non-Abelian groups of order 15. Explain why at least two of the non-Abelian groups that you listed are not isomorphic.

(4) Prove that the integer partition function p is a strictly increasing function.

⋆ (5) Let p be prime. Table 31.3 indicates a connection between the number of distinct isomorphism classes of a finite Abelian group of order q^m and the partition function p. Find this connection and explain it. Be sure to provide an explanation for why there are *exactly* the number of isomorphism classes you claim.

(6) (a) Classify the orders of all finite Abelian groups for which there are exactly 4 distinct isomorphism classes. (Hint: Exercise 5 might be useful.)

(b) Classify the orders of all finite Abelian groups for which there are exactly 5 distinct isomorphism classes.

(7) (a) Suppose the order of some finite Abelian group is divisible by 6. Prove that the group has a cyclic subgroup of order 6.

(b) Is there something special about the number 6 in part (a)? That is, if k is a positive integer and the order of a finite Abelian group G is divisible by k, must G contain a cyclic subgroup of order k? If yes, prove it. If no, find a counterexample and then determine conditions on k that make the statement true.

(8) We know that a finite cyclic group of order n contains exactly one subgroup of order m for each divisor m of n. Is the converse true if our group is Abelian? That is, suppose G is a finite Abelian group that has exactly one subgroup for each divisor of $|G|$. Must G be cyclic? Prove your answer.

(9) Let $G = \mathbb{Z}_{720} \oplus \mathbb{Z}_{120} \oplus \mathbb{Z}_{15}$. To which groups are the p-primary components of G isomorphic? Explain.

⋆ (10) Let G be a finite Abelian group of order $p_1^{m_1} p_2^{m_2} \cdots p_k^{m_k}$, where $p_1, p_2, \ldots, p_k$ are distinct primes and $m_1, m_2, \ldots, m_k$ are positive integers. The Fundamental Theorem of Finite Abelian Groups tells us that

$$G = G(p_1) \times G(p_2) \times \cdots \times G(p_k).$$

Since $|G(p_i)|$ is a power of p_i, the Fundamental Theorem of Arithmetic tells us that $|G(p_i)| = p_i^{m_i}$ for each i. In this exercise, we will show that different choices for $G(p_i)$ yield non-isomorphic groups G. To do so, let H_1, H_2, ..., H_k and Q_1, Q_2, ..., Q_k be groups with $|H_i| = |Q_i| = p_i^{m_i}$ for each i. If

$$G \cong (H_1 \oplus H_2 \oplus \cdots \oplus H_k) \cong (Q_1 \oplus Q_2 \oplus \cdots \oplus Q_k),$$

prove that $H_i \cong Q_i$ for each i.

(11) (a) Use the results of Exercises 5 and 10 to determine the number of isomorphism classes of Abelian groups of order $8318750000 = 2^4 \times 5^8 \times 11^3$. Use the fact that $p(8)$, the number of partitions of 8, is 22. Please do not attempt to write out all representatives of all of the isomorphism classes.

(b) Let G be an Abelian group of order $q_1^{m_1} q_2^{m_2} \cdots q_k^{m_k}$, where $q_1, q_2, \ldots, q_k$ are distinct primes and $m_1, m_2, \ldots, m_k$ are positive integers. Find a formula, in terms of the partition function, for the number of isomorphism classes of Abelian groups of order $|G|$.

(12) Let G be a finite Abelian group, and let a be an element of maximum order in G. Show that $|b|$ divides $|a|$ for any element b in G. (Hint: Activity 31.17 might be useful.)

⋆ (13) Let G and H be finite Abelian groups, and let $\nu_G(k)$ and $\nu_H(k)$ denote the number of elements of order k in G and H, respectively.

(a) Prove that if $\nu_G(k) = \nu_H(k)$ for all $k \in \mathbb{Z}^+$, then G and H have the same invariant factors. (See Activity 31.17.)

(b) Prove that $G \cong H$ if and only if G and H have the same invariant factors.

(c) In Exercise 9 of Investigation 29 (see page 413), we showed that there exists a finite Abelian group K and a finite non-Abelian group L so that $|K| = |L|$ and $\nu_K(k) = \nu_L(k)$ for all $k \in \mathbb{Z}^+$, but $K \not\cong L$. Is this same property possible if both G and H are finite Abelian groups? That is, if $\nu_G(k) = \nu_H(k)$ for all $k \in \mathbb{Z}^+$, must G and H be isomorphic? Prove your answer.

(14) Let G be a group, and define $t(G)$ to be the set of all elements of G of finite order.

(a) Is $t(G)$ always a subgroup of G? If the answer is yes, prove it. If the answer is no, illustrate with an example and then find at least two different types of groups for which the answer is yes. Explain. (Hint: Refer to Exercise 5 on page 323 of Investigation 23.)

(b) Let S be the set of all complex numbers with norm 1. (See Exercise 6 on page 311 of Investigation 22.) Determine the elements of $t(S)$.

(c) Let G be an Abelian group. Then $t(G)$ is a normal subgroup of G called the **torsion subgroup** of G. What can we say about the orders of the elements in $G/t(G)$? Prove your answer.

(d) Prove that if G is an Abelian group, then $G \cong t(G) \oplus G/t(G)$.

(e) Describe as best you can the elements of the group $S/t(S)$, where S is the group of complex numbers of norm 1. Your description must involve more than just the definition of a quotient group.

Investigation 32

The First Sylow Theorem

Focus Questions

By the end of this investigation, you should be able to give precise and thorough answers to the questions listed below. You may want to keep these questions in mind to focus your thoughts as you complete the investigation.

- What does it mean for two elements to be conjugate in a group? What are some important properties of conjugacy?
- How is the conjugacy class of an element in a group G related to a special subgroup of G?
- How does the Class Equation work, and what are two important consequences of the Class Equation?
- What is the general formulation of Cauchy's Theorem, and why is it important?
- What are the three Sylow theorems, and what do they tell us about finite groups?
- What is a Sylow p-subgroup of a group G? Why are these subgroups important?

Preview Activity 32.1. In this activity, we will introduce three important theorems known as the Sylow theorems. In order to understand the Sylow theorems, we will first need to understand *conjugacy*, defined formally as follows:

Definition 32.2. Let G be a group. The element $a \in G$ is **conjugate** to $b \in G$ if there is an element $g \in G$ so that $a = gbg^{-1}$.

The element $b = gag^{-1}$ is called a *conjugate* of a or the *conjugate of a by g*.

(a) Find all of the conjugates of the element r in D_4.

(b) Note that conjugacy can be used to define a relation on a group. In particular, let $a \sim b$ if a is conjugate to b. It seems natural to ask what properties this conjugacy relation satisfies.

  (i) Is $\sim$ a reflexive relation? Prove your answer.

  (ii) Is $\sim$ a symmetric relation? Prove your answer.

  (iii) Is $\sim$ a transitive relation? Prove your answer.

(c) Is the relation $\sim$ defined in part (b) an equivalence relation? If so, find the distinct equivalence classes of this relation (also known as *conjugacy classes*) for the group D_4.

Introduction

After Lagrange's Theorem, the most important results in finite group theory are arguably the Sylow Theorems. These theorems are named after Norwegian mathematician Peter Ludvig Mejdell Sylow (1832–1918).* The Sylow theorems are among the most important and profound results in the theory of finite groups.

As we have discussed, the subgroup structure of a group tells us a lot about the group. Our first big theorem in studying the subgroup structure of finite groups was Lagrange's Theorem. Recall that Lagrange's Theorem told us that if G is a finite group and H is a subgroup of G, then $|H|$ divides $|G|$. While this theorem tells us something about the subgroup structure of finite groups, it is not as strong a result as we might like. It would be ideal if the converse of Lagrange's Theorem were true—that is, if whenever a positive integer k divides the order of a finite group G, then G must have a subgroup of order k. Unfortunately, this is not the case, as we have seen that the alternating group A_4 has order 12 but has no subgroup of order 6. The Sylow theorems are important because they give us some information about which subgroups a finite group *must* have.

In this investigation, we will state the three Sylow theorems and prove the first one. The second and third Sylow Theorems will be discussed in more detail in the next investigation. To begin, we will develop a counting technique that depends on conjugacy.

Conjugacy and the Class Equation

The results of Preview Activity 32.1 show that conjugacy is an equivalence relation on a group G. As such, the conjugacy relation partitions a group G into disjoint equivalence classes, called *conjugacy classes*.

Definition 32.3. Let G be a group and $a \in G$. The **conjugacy class of** a in G is the equivalence class of a under the conjugacy relation.

We denote the conjugacy class of an element a in a group G as $\text{cl}(a)$. So

$$\text{cl}(a) = \{gag^{-1} : g \in G\}.$$

Activity 32.4. Find $\text{cl}(a)$ for each $a \in S_3$.

It is important to note that, in general, $\text{cl}(a)$ is **NOT** a subgroup of G.

The notion of a relation on a group is not new to us. Recall that if G is a group and H a subgroup of G, then we can define an equivalence relation $\sim_H$ on G by letting $a \sim_H b$ if $b^{-1}a \in H$. The equivalence classes under $\sim_H$ are the left cosets of H in G. The $\sim_H$ relation was useful in that it allowed us to write any finite group G as a disjoint union of left cosets and then count the elements in the distinct left cosets to obtain Lagrange's Theorem. What made this argument work was that any two equivalence classes (left cosets of H in G) had the same number of elements. As our previous activities have shown, two conjugacy classes can contain different numbers of elements,

*The paper in which the Sylow Theorems appeared was titled "Théorèmes sur les groupes de substitutions" and was published in *Mathematische Annalen*, vol. 5 (pp. 584–594) in 1872.

so counting with conjugacy classes will be a bit more complicated than counting with left cosets. However, many useful results will follow from the conjugacy relation.

Since conjugacy is an equivalence relation on G, we know that G is the union of the disjoint conjugacy classes. So we can write

$$G = \bigcup_{a \in G} \text{cl}(a).$$

If we want to use conjugacy classes to count elements, we will need to know how to determine the number of elements in $\text{cl}(a)$ for a given $a \in G$. The answer, of course, depends on the element a. In some cases, it is easy to determine the number of elements in a conjugacy class, as the next activity demonstrates.

Activity 32.5. Let G be a group, and let $a \in G$.

(a) In Preview Activity 32.1 we saw that it is possible to have $\text{cl}(a) = \{a\}$. Determine specific conditions on a for this to happen.

(b) If $a \in Z(G)$, what is $\text{cl}(a)$? Explain.

(c) Complete the following lemma

Lemma 32.6. *Let G be a group and $a \in G$. Then $cl(a) = \{a\}$ if and only if* ___________.

Lemma 32.6 seems to imply that conjugacy classes are not very interesting for Abelian groups (where $G = Z(G)$). However, if G is non-Abelian, then conjugacy classes are very helpful in studying the subgroup structure of G.

Let G be a group, and let $a \in G$. Since $\text{cl}(a)$ is not in general a subgroup of G, there is no obvious relationship between the size of $\text{cl}(a)$ and the order of G. (For example, we can't use Lagrange's Theorem to make any conclusions about the relationship between $|\text{cl}(a)|$ and $|G|$.) If we dig a little deeper, however, we will find that there is a relationship that will make it easier for us to count elements using conjugacy classes. Suppose we have two conjugates, $x = gag^{-1}$ and $y = hah^{-1}$, of an element $a \in G$. In order to count only distinct conjugates, we would like to have a way to tell when these conjugates are the same. In other words, we would like to determine the circumstances in which

$$gag^{-1} = hah^{-1}?$$

If $gag^{-1} = hah^{-1}$, then we have

$$\left(g^{-1}h\right) a = a \left(g^{-1}h\right).$$

Thus, $g^{-1}h$ commutes with a. So, to count the number of distinct conjugates of an element a, we need to understand how many elements of G commute with a. This observation motivates the following definition.

Definition 32.7. Let G be a group and $a \in G$. The **centralizer** of $a \in G$ is the set $C(a) = \{g \in G : ga = ag\}$.

In other words, $C(a)$ is the set of elements in G that commute with a.

Activity 32.8. Let $G = S_3$.

(a) Find $C(a)$ for each $a \in G$.

(b) Compare the sets $C(a)$ in G for choices of a that are in the same conjugacy class. (See Activity 32.4.) What do you notice?

Our observations about the sets $C(a)$ in Activity 32.8 seem to suggest that, unlike the conjugacy class of an element, the centralizer of an element might always be a group. In the next activity, we will see if this is actually the case.

Activity 32.9. Let G be a group with identity e, and let $a \in G$.

(a) Show that $C(a)$ is not empty.

(b) Is $C(a)$ closed under the operation in G? Verify your answer.

(c) If $g \in C(a)$, must g^{-1} also be in $C(a)$? Verify your answer.

(d) Is $C(a)$ a subgroup of G?

Recall that the number of distinct conjugates of an element a in a finite group G is the number of elements that commute with a, or the order of $C(a)$. So we now have a relationship between conjugacy classes and subgroups. We can exploit this relationship to count the number of elements in a group. Recall that two conjugates gag^{-1} and hah^{-1} of a are equal when $g^{-1}h$ is in $C(a)$. We also know that $g^{-1}h \in C(a)$ implies $gC(a) = hC(a)$. So $gag^{-1} = hah^{-1}$ implies $gC(a) = hC(a)$. Conversely, if $gC(a) = hC(a)$, then $g^{-1}h$ is in $C(a)$ and $gag^{-1} = hah^{-1}$. Therefore, the number of elements in the conjugacy class of $a \in G$ is the number of distinct right (or left) cosets of $C(a)$ in G. These observations establish the following theorem:

Theorem 32.10. *Let G be a finite group, and let $a \in G$. The number of elements in the conjugacy class of $a \in G$ is*

$$[G : C(a)] = \frac{|G|}{|C(a)|}.$$

The next activity asks you to verify this theorem for a particular example.

Activity 32.11. Verify Theorem 32.10 for the elements in S_3.

The Class Equation

Counting techniques are important throughout finite group theory. We saw one example of the use of a counting technique when we proved Lagrange's Theorem. In this section, we will develop another counting tool known as the *Class Equation*.

As we observed earlier, we can write G as the union of the disjoint conjugacy classes of elements in G. In other words,

$$G = \bigcup_{a \in G} \text{cl}(a).$$

By Theorem 32.10, we know how many elements are in each conjugacy class. This leads to the following result:

Theorem 32.12 (The Class Equation). *Let G be a finite group, and let $cl(a_1)$, $cl(a_2)$, ..., $cl(a_k)$ be the distinct conjugacy classes of G. Then*

$$|G| = \sum_{i=1}^{k} \frac{|G|}{|C(a_i)|}. \tag{32.1}$$

The next activity uses our prior calculations and provides a quick example of the Class Equation.

Activity 32.13. Verify the Class Equation for the group S_3.

There are several variations on the Class Equation. For one variation, recall that the single element classes correspond exactly to those elements in the center of G. Therefore, if $\text{cl}(a_1), \text{cl}(a_2), \ldots, \text{cl}(a_r)$ are the distinct conjugacy classes of G, then

$$|G| = |Z(G)| + \sum_{i=1}^{r} \frac{|G|}{|C(a_i)|}. \tag{32.2}$$

Note also that $a \in Z(G)$ if and only if $C(a) = G$. So if $a \notin Z(G)$, then, since both the identity and a commute with a, we have $|C(a)| \geq 2$.

The Class Equation has many applications in group theory. Later on, we will use the Class Equation to prove the First Sylow Theorem. The Class Equation can also be used to prove other results about groups, such as the following:

Theorem 32.14. *Let p be a prime and n a positive integer. If G is a group of order p^n, then $Z(G)$ is non-trivial.*

Proof. Let p be a prime and n a positive integer. Let G be a group of order p^n. Let $z = |Z(G)|$. Since every element in $Z(G)$ has a one-element conjugacy class, there are z distinct one-element conjugacy classes. Let $a \in G$ such that $a \notin Z(G)$. Then $|C(a)|$ is larger than 1, less than $|G|$, and divides p^n. So $|C(a)| = p^{k_a}$ for some positive integer k_a less than n. Then $\frac{|G|}{|C(a)|} = p^{n-k_a} \geq p$. The Class Equation (32.2) then implies that

$$p^n = z + \sum_a \frac{|G|}{|C(a)|} = z + \sum_a p^{n-k_a}.$$

A little algebra shows that

$$z = p^n - \sum_a p^{n-k_a} = p\left(p^{n-1} - \sum_a p^{n-k_a-1}\right).$$

Therefore, $p \mid z$. Thus, z is larger than 1, and so $Z(G)$ is non-trivial. ∎

The next activity establishes another useful result.

Activity 32.15. Let p be a prime, and let G be a group of order p^2. In this activity, we will determine what type of group G must be.

(a) What does Theorem 32.14 tell us about $Z(G)$? What are the possible orders of $Z(G)$?

(b) What can we say about G if $|Z(G)| = p^2$?

(c) If $|Z(G)| \neq p^2$, then there is an element $g \in G \setminus Z(G)$. What can we say about $|C(g)|$? What conclusion can we draw?

(d) Explain how we have proved the following theorem:

Corollary 32.16. *Let p be a prime. If G is a group of order p^2, then G is Abelian.*

Corollary 32.16 allows us to classify all groups of order p^2, where p is prime. (See Exercise 3 for more details.)

Cauchy's Theorem

In Investigation 27, we proved Cauchy's Theorem for Abelian Groups, which states that if a prime p divides the order of a finite Abelian group G, then G contains a subgroup of order p. As it turns out, Cauchy's Theorem is true for any finite group G, and we can use the Class Equation to prove it. Cauchy's Theorem is an important result that we will use in our proofs of the Sylow theorems.

Theorem 32.17 (Cauchy's Theorem)**.** *Let G be a finite group, and let p be a prime dividing the order of G. Then G has an element of order p.*

Proof. Let G be a finite group with identity element e. We will proceed by using strong induction on the order of G. Since we have completely classified the groups of order 1, 2, and 3, we know the theorem is true in those cases. Now let $n \in \mathbb{Z}^+$, and assume that, for any group of order less than n, if a prime p divides the order of the group, then the group has an element of order p.

Let G be a group of order n, and assume p is a prime factor of n. Then $n = pm$ for some positive integer m. We know that G is the union of finitely many conjugacy classes and that $Z(G)$ contains all of the single element conjugacy classes. Let $\text{cl}(a_1)$, $\text{cl}(a_2)$, $\ldots$ $\text{cl}(a_r)$ be the distinct multi-element conjugacy classes. The Class Equation tells us that

$$pm = |Z(G)| + \sum_{i=1}^{r} \frac{|G|}{|C(a_i)|}.$$

We will consider two cases: (1) if p divides $|C(a_i)|$ for some i; and (2) if p divides no $|C(a_i)|$.

Case 1: Suppose there is an integer i between 1 and r such that p divides $|C(a_i)|$. Since $1 < |C(a_i)| < |G|$, we can use the induction hypothesis to conclude that $C(a_i)$ contains an element of order p. Thus, G contains an element of order p.

Case 2: Now suppose there is no integer i such that p divides $|C(a_i)|$. Then $\frac{|G|}{|C(a_i)|} = \frac{pm}{|C(a_i)|}$ must be a multiple of p for each value of i. The Class Equation then tells us that p divides $pm - \sum_{i=1}^{r} \frac{|G|}{|C(a_i)|} = |Z(G)|$. Therefore, the order of $Z(G)$ is at least p. Since $Z(G)$ is an Abelian group whose order is a multiple of p, Cauchy's Theorem for Abelian Groups shows us that $Z(G)$ contains an element of order p.

In either case, we can conclude that G contains an element of order p. ∎

With these preliminary results out of the way, we can now turn our attention to the First Sylow Theorem.

The First Sylow Theorem

Preview Activity 32.18. We have seen that the group A_4 of order 12 does not contain a subgroup of order 6. Find all of the non-trivial proper subgroups of A_4, and determine their orders. What can you say about the orders of these subgroups in relation to the order of A_4?

The result of Activity 32.18 is that, while A_4 contains no subgroup of order 6, it does contain subgroups of every order p^n, where p is prime and p^n divides $|A_4|$. The First Sylow Theorem guarantees that this happens in general. In order to prove it, we will need one more piece of information about subgroups of quotient groups.

Lemma 32.19. *Let G be a group and N a normal subgroup of G. Every subgroup H' of G/N has the form H/N for some subgroup H of G containing N. Moreover, $H \lhd G$ if and only if $H' \lhd G/N$.*

The proof of Lemma 32.19 is left as Exercise 23.

We are now ready to prove the First Sylow Theorem.

Theorem 32.20 (First Sylow Theorem)**.** *Let G be a finite group. If p is a prime number and $k \in \mathbb{Z}^+$ such that p^k divides $|G|$, then G has a subgroup of order p^k.*

Proof. Let G be a finite group. We will proceed by using strong induction on the order n of $|G|$. We have already classified all of the groups of orders 1 through 6, and so we know that the theorem is true in those cases. Now let $n \in \mathbb{Z}^+$, and assume that for any group H of order less than n and for any prime p and any positive integer k satisfying the property that p^k divides $|H|$, we know that H contains a subgroup of order p^k.

Let G be a group of order n. Let p be a prime number, and let $k \in \mathbb{Z}^+$ such that p^k divides $|G|$. Then there is a positive integer m such that $n = p^k m$. We will show that G contains a subgroup of order p^k.

We know that G is the union of finitely many conjugacy classes. We also know that $Z(G)$ contains all of the single-element conjugacy classes. Let $\text{cl}(a_1)$, $\text{cl}(a_2)$, $\ldots$, $\text{cl}(a_r)$ be the distinct multi-element conjugacy classes of G. The Class Equation tells us that

$$|G| = |Z(G)| + \sum_{i=1}^{r} \frac{|G|}{|C(a_i)|}.$$

As in the proof of Cauchy's Theorem, we will now consider two cases.

Case 1: Suppose there is an i between 1 and r such that p does not divide $\frac{|G|}{|C(a_i)|}$. Since $|C(a_i)|$ divides $|G| = p^k m$, we must then have that p^k divides $|C(a^i)|$. Recall that $1 < |C(a_i)| < |G|$, so we can apply our induction hypothesis to $C(a_i)$ to conclude that $C(a_i)$ contains a subgroup of order p^k. Hence, G contains a subgroup of order p^k.

Case 2: Suppose p divides $\frac{|G|}{|C(a_i)|}$ for each i. Then p divides

$$|G| - \sum_{i=1}^{r} \frac{|G|}{|C(a_i)|} = |Z(G)|.$$

So $Z(G)$ has order at least p. By Cauchy's theorem we know that $Z(G)$ has an element, x, of order p. Let $N = \langle x \rangle$. Since $x \in Z(G)$, we know $N \lhd G$. Now consider the group $K = G/N$. Since $|K| = \frac{|G|}{|N|} = p^{k-1}m < |G|$, by our induction hypothesis the group K has a subgroup X of order p^{k-1}. By Lemma 32.19, we know this subgroup X is of the form H/N for some subgroup H of G that contains N. We also know that

$$p^{k-1} = |X| = |H/N| = \frac{|H|}{|N|} = \frac{|H|}{p}.$$

Therefore, $|H| = p^k$.

In either case, we have found a subgroup of G of order p^k. ■

The First Sylow Theorem tells us that every finite group G contains subgroups of every prime power divisor of the order of G. This provides significant information about the subgroup structure of G. Recall that if G is an Abelian group with $p_1^{m_1}p_2^{m_2}\cdots p_k^{m_k}$ as the prime power factorization of $|G|$, then

$$G = G(p_1) \times G(p_2) \times \cdots \times G(p_k),$$

where $G(p_i)$ is the p_i-primary component of G and has order $p_i^{m_i}$. If G is a non-Abelian group with $p_1^{m_1}p_2^{m_2}\cdots p_k^{m_k}$ as the prime power factorization of $|G|$, the First Sylow Theorem tells us that G contains a subgroup of order $p_i^{m_i}$ for each i. These maximal p-subgroups are given a special name.

Definition 32.21. Let G be a finite group, and let p be a prime that divides $|G|$. Let $|G| = p^k m$ for some positive integers k and m with $\gcd(k, m) = 1$, so that p^k is the highest power of p that divides $|G|$. A subgroup S of G is a **Sylow p-subgroup** if $|S| = p^k$.

So a Sylow p-subgroup of an Abelian group is just the p-primary component of the group. A non-Abelian group may contain one or more Sylow p-subgroups for a given prime p, as the next activity illustrates.

Activity 32.22.

(a) Find a Sylow p-subgroup of D_6 for each prime divisor of $|D_6|$.

(b) Does D_6 contain more than one Sylow 2-subgroup? Does D_6 contain more than one Sylow 3-subgroup? Explain.

The second and third Sylow theorems provide additional information about the Sylow p-subgroups of a group. In particular, they tell us about the relationship between the different Sylow p-subgroups of a group, as well as the number of Sylow p-subgroups a group can have.

The Second and Third Sylow Theorems

In Activity 32.22, we showed that D_6 contains exactly one Sylow 3-subgroup, but has more than one Sylow 2-subgroup. From Exercise 36 of Investigation 27 (see page 378), we know that the only normal subgroups of D_6 are $\{I\}$, $\langle R^3\rangle$ and D_6. So the sole Sylow 3-subgroup of D_6 is a normal subgroup. In general, this is one of the conclusions we can draw from the Second Sylow Theorem. In particular, the Second Sylow Theorem tells us that any two Sylow p-subgroups (for the same value of p) are *conjugate*. We first encountered conjugates of subgroups in Activity 27.20 (see page 372), and we will formally define them here.

Definition 32.23. Let G be a group and H a subgroup of G. The **conjugate** of H in G by $a \in G$ is the set

$$aHa^{-1} = \{aha^{-1} : h \in H\}.$$

Activity 27.20 shows that aHa^{-1} is always a subgroup of G. We say that the subgroup H is **conjugate** to the subgroup K if $H = aKa^{-1}$ for some $a \in G$. We will explore the conjugacy relation in more detail in the next investigation. For now, however, we can use it to formally state the Second Sylow Theorem.

Theorem 32.24 (Second Sylow Theorem). *Let G be a finite group, and let p be a prime divisor of $|G|$. If H and K are Sylow p-subgroups of G, then there exists $g \in G$ such that $H = gKg^{-1}$.*

One consequence of the Second Sylow Theorem is the following corollary:

Corollary 32.25. *Let G be a finite group, and let p be a prime divisor of $|G|$. A Sylow p-subgroup S of G is normal in G if and only if S is the only Sylow p-subgroup of G.*

The proof of Corollary 32.25 is outlined in Activity 32.30. The proof of the subsequent corollary is outlined in Activity 32.31.

The Third Sylow Theorem provides information—although it is not at all obvious—about the number of Sylow p-subgroups of a group G.

Theorem 32.26 (Third Sylow Theorem). *Let G be a finite group, and let p be a prime divisor of $|G|$. The number of Sylow p-subgroups of G divides $|G|$ and is of the form $1 + pk$ for some nonnegative integer k.*

Corollary 32.27. *Let G be a finite group, and let p be a prime such that $|G| = p^r m$ for some $r, m \in \mathbb{Z}^+$ with $\gcd(p, m) = 1$. Then the number of Sylow p-subgroups of G divides m and is of the form $1 + pk$ for some nonnegative integer k.*

Note that the First Sylow Theorem tells us about the *existence* of certain types of subgroups of a finite group, while the second and third Sylow theorems give us information about those subgroups. We will postpone the proofs of the second and third Sylow theorems to the next investigation, but the three theorems work together as a group (no pun intended), so we will freely use all three theorems from this point on. (You should do the same for the exercises.)

We will conclude this investigation with two examples that illustrate some of the power of the Sylow theorems.

Example 32.28. Let p be an odd prime, and let G be a non-Abelian group of order $2p$ with identity e. Then Theorem 29.16 (see page 407) shows that $G \cong D_p$. We can also prove this using the Sylow theorems. The Third Sylow Theorem tells us that the number n_p of Sylow p-subgroups of G is congruent to 1 mod p, and Corollary 32.27 shows that n_p divides 2. Since $p > 2$, the only way this can happen is if $n_p = 1$. So there is one Sylow p-subgroup N of G, and Corollary 32.25 tells us that N is normal in G. Since $|N| = p$, N is cyclic and is therefore generated by some element n. Let K be a Sylow 2-subgroup of G. By Activity 27.22 (see page 372), we know that NK is a subgroup of G. Furthermore, Exercise 2 shows that $N \cap K = \{e\}$. Now we will show that $G = NK$ by demonstrating that $|NK| = 2p$. By Exercise 7, we know that the representation of an element of NK in the form nk (where $n \in N$ and $k \in K$) is unique. Thus, $|NK| = |N||K| = 2p$, and so $G = NK$. If $K \triangleleft G$, then $G = (N \times K) \cong (N \oplus K) \cong Z_{2p}$. Therefore, we can assume that K is not normal in G.

Since $|K| = 2$, we know that K is cyclic and is generated by some element k. Since $G = NK$, every element in G is of the form $n^i k^j$ for some $0 \leq i < p$ and $0 \leq j < 2$. The operation table for G, and hence the structure of G, will be determined by the product kn. Since G is non-Abelian, it must be the case that $kn \neq nk$. Given that $N \triangleleft G$, we must have $knk = knk^{-1} \in N$. So $knk = n^t$

for some $1 \leq t < p$. Therefore,

$$\begin{aligned} n &= k^2nk^2 \\ &= k(knk)k \\ &= kn^tk \\ &= (knk)^t \\ &= \left(n^t\right)^t \\ &= n^{t^2}. \end{aligned}$$

Thus, $n^{t^2-1} = e$, and it follows that p divides $t^2 - 1 = (t+1)(t-1)$. So it must be the case that p divides $t+1$ or $t-1$. Recall that $1 \leq t < p$, so if p divides $t+1$, then $t = p-1$. Also, if p divides $t-1$, then $t = 1$. Since $kn \neq nk$, we can rule out $t = 1$, which leaves us with the conclusion that $t = p-1$. This tells us that $knk = n^{p-1} = n^{-1}$. Therefore, G has presentation

$$\langle n, k \mid n^p = k^2 = 1, knk^{-1} = n^{-1} \rangle \cong D_p.$$

Example 32.29. As a second example, we will classify all groups of order 66. Note that $66 = 2 \times 3 \times 11$. Let G be a group of order 66, and let n_p be the number of Sylow p-subgroups of G for each prime divisor of $|G|$. The Third Sylow Theorem tells us that $n_{11} \equiv 1 \pmod{11}$ and that n_{11} divides 6. This can only happen if $n_{11} = 1$. So there is a unique Sylow 11-subgroup N of G. Let K be a Sylow 3-subgroup of G. Since $N \triangleleft G$ (by Corollary 32.25), we know that NK is a subgroup of G (by Activity 27.22 on page 372). Also, Exercise 2 shows that $K \cap N = \{e\}$, so $|NK| = 33$. The result of Activity 32.32 shows that NK is a cyclic group and therefore is generated by some element x. Note that $[G : NK] = 2$, so $NK \triangleleft G$ (by Exercise 19 on page 375 of Investigation 27).

Now let H be any Sylow 2-subgroup of G. Since NK is normal and $\gcd(|NK|, |H|) = 1$, Exercise 2 implies that $|NK \cap H| = \{e\}$. Exercise 7 then implies that $|NKH| = |NK||H|$. Since $|H|$ is a power of 2, it must be that $|H| = 2$ and $|NKH| = 66$. From this, we can conclude that $H = \langle h \rangle$ for some $h \in G$ and that $NKH = G$. Since NK is generated by x and H is generated by h, it follows that every element of G can be written as x^th^j for some integers t and j with $0 \leq t < 33$ and $0 \leq j < 2$. In particular, since hx is an element of G, $hx = x^th^j$ for some $0 \leq t < 33$ and $0 \leq j < 2$. However, if $j = 0$, then $hx = x^t$, which implies that $h \in \langle x \rangle$, a contradiction. Thus, $j = 1$ and $hx = x^th$—or, equivalently, $hxh^{-1} = x^t$. Furthermore, since $h \neq e$, we know that $1 \leq t \leq 32$. But since $|H| = 2$, we also know that $h^{-1} = h$, and so $hxh = x^t$. Therefore, by the same argument as in Example 32.28, $x^{t^2-1} = e$, and so 33 divides $t^2 - 1$. Since $1 \leq t \leq 32$, the possibilities are $t = 1$, $t = 10$, $t = 23$, and $t = 32$. Each of these four values of t determines a group with presentation

$$\langle x, h \mid x^{33} = h^2 = 1, hx = x^th \rangle.$$

In Exercise 12, you will show that none of these four groups are isomorphic to each other. This establishes that, up to isomorphism, there are exactly four groups of order 66.

Concluding Activities

Activity 32.30. Use the Second Sylow Theorem to prove Corollary 32.25.

Activity 32.31. Use the Third Sylow Theorem to prove Corollary 32.27.

Activity 32.32. We have spent a significant amount of time classifying groups of various orders. The Sylow Theorems are critical in this process except for very small groups. In addition, the Sylow Theorems can help classify entire families of groups. In this activity, we will see how to explicitly determine the groups of order pq, where p and q are primes, $p < q$, and p does not divide $q - 1$. To begin, let G be such a group.

(a) How many Sylow q-subgroups must G have?

(b) How many Sylow p-subgroups must G have?

(c) Let K be a Sylow q-subgroup of G and N a Sylow q-subgroup of G. Explain why K and N are both normal subgroups of G.

(d) Prove that G is a cyclic group.

Activity 32.33. Write a short summary that describes the important concepts, and the relationships between these concepts, that were introduced in this investigation. Explain how the material in this investigation is connected to the material in Investigations 20, 22, and 29.

Exercises

(1) Describe the conjugacy classes of elements in an Abelian group G.

* (2) Let G be a group with identity e, and let H and K be finite subgroups of G.

(a) Suppose that $\gcd(|H|, |K|) = 1$. Determine the elements of $H \cap K$.

(b) Suppose that H and K are cyclic groups of the same prime order. What is $H \cap K$?

* (3) (a) Classify all groups of order p^2 when p is prime.

(b) Is it true that every group of order n^2, where $n \in \mathbb{Z}^+$, is Abelian? Explain.

(4) Let A be an Abelian group of order 24 whose elements have orders as shown in Table 32.1. Find all the Sylow p-subgroups of A. Clearly explain your work.

Element	a_1	a_2	a_3	a_4	a_5	a_6	a_7	a_8	a_9	a_{10}	a_{11}	a_{12}
Order	1	12	12	6	2	4	6	3	12	4	3	12
Element	a_{13}	a_{14}	a_{15}	a_{16}	a_{17}	a_{18}	a_{19}	a_{20}	a_{21}	a_{22}	a_{23}	a_{24}
Order	12	6	4	12	6	6	4	2	6	12	12	2

Table 32.1
Orders of elements in A.

(5) Recall that the Second Sylow Theorem tells us that any two Sylow p-subgroups are conjugate. Illustrate this theorem by finding a Sylow 2-subgroup H of S_3. Show specifically how all other Sylow 2-subgroups of S_3 are conjugate to H.

(6) Let G be a group of order 200. What are the sizes of the Sylow p-subgroups of G? Explain why at least one of the Sylow p-subgroups of G is normal in G.

⋆ (7) Let G be a group with identity e, and let N and K be subgroups of G with $N \lhd G$. If $N \cap K = \{e\}$, show that the representation of an element of the form nk in the subgroup NK (where $n \in N$ and $k \in K$) is unique. Conclude that $|NK| = |N||K|$.

(8) Let N be a normal subgroup of a finite group G, and let $a \in G$ such that $\gcd(|a|, |G/N|) = 1$. Show that $a \in N$.

⋆ (9) Let G be a group and H a subgroup of G.

(a) Show by example that not all conjugates of H must be the same.

(b) Show that every conjugate of H in G is isomorphic to H.

(10) Consider the group $\mathbf{Q} = \{1, -1, i, -i, j, -j, k, -k\}$ of quaternions.

(a) Determine the distinct conjugacy classes in $\mathbf{Q}$.

(b) Find $Z(\mathbf{Q})$. Use the Class Equation to verify the order of $Z(\mathbf{Q})$.

(11) Consider the group $D_4 = \{r^i R^j : 0 \leq i \leq 1, 0 \leq j \leq 3\}$ of symmetries of the square.

(a) Determine the distinct conjugacy classes in D_4.

(b) Find $Z(D_4)$. Use the Class Equation to verify the order of $Z(D_4)$.

⋆ (12) Demonstrate that the four groups $\mathbb{Z}_{66}$, D_{33}, $D_{11} \oplus \mathbb{Z}_3$, and $D_3 \oplus \mathbb{Z}_{11}$ form four different isomorphism classes of groups of order 66. Then determine the values of t from Example 32.29 that correspond to each group.

(13) We know that for each prime p there is exactly one group of order p. Is this true for any composite numbers? If so, find the smallest composite number n so that there is exactly one group of order n (up to isomorphism). Verify your result.

(14) Let G be a finite group and K a Sylow p-subgroup of G. If φ is an automorphism of G, must $\varphi(K) = K$? If not, what can we say about $\varphi(K)$?

(15) Let n and k be positive integers with $k \leq n$, and let p be a prime number.

(a) Determine the number of k-cycles in S_n.

(b) Describe the Sylow p-subgroups of S_p. How many distinct Sylow p-subgroups does S_p contain?

(16) Let G be a group of order $p^m n$, where p is prime, $p > n$, and $\gcd(p, n) = 1$. Let S be a Sylow p-subgroup of G. Must S be normal in G? Prove your answer.

(17) Suppose G is a group of order pq, where p and q are distinct primes with $p < q$.

(a) Must G contain a normal subgroup of order p? Explain.

(b) Must G contain a normal subgroup of order q? Explain.

(18) (a) Classify all groups of order 45.

(b) Let p and q be prime with $p < q$ and $q \not\equiv 1 \pmod{p}$. Classify all groups of order $p^2 q$.

(19) Find all of the Sylow p-subgroups of D_{12}.

(20) We have shown that normality is not transitive in general. However, let G be a finite group with subgroups P and N such that $P \lhd N$, $N \lhd G$, and P is a Sylow subgroup of N. Show that $P \lhd G$.

(21) Show that if all of the Sylow subgroups of a finite group G are normal in G, then G is isomorphic to the direct sum of its Sylow subgroups. Is the converse true? Explain.

(22) In this exercise, we will classify all groups of order 21.

(a) How many Abelian groups of order 21 are there? Explain.

(b) Now let G be a non-Abelian group of order 21.

(i) Determine the number of Sylow subgroups of each order in G.

(ii) Let $N = \langle n \rangle$ be the Sylow 7-subgroup of G and let $K = \langle k \rangle$ be one Sylow 3-subgroup. Use the fact that $n = k^{-3}nk^3$ to show that $n^{t^3-1} = e$, where e is the identity in G. Conclude that 7 divides $t^3 - 1$.

(iii) Determine the possible values of t from the previous part of this problem, and use your result to complete the classification of all groups of order 21.

$\star$ (23) Explain how the Fourth Isomorphism Theorem verifies Lemma 32.19.

(24) Classify all groups of order 30.

$\star$ (25) Let p be a prime, and let G be a p-group of order p^m. The First Sylow Theorem tells us that G contains subgroups of each order p^k for $0 \leq k \leq m$, demonstrating that the converse of Lagrange's Theorem is true for p-groups. We can show something more, though. Prove that G has *normal* subgroups of all orders p^k for $0 \leq k \leq m$.

Connections

The Sylow theorems provide a number of tools that we can use to make conclusions about the subgroup structure of an arbitrary group. In the case of a finite Abelian group G, the p-primary components of G that we defined in Investigation 31 are exactly the Sylow p-subgroups of G. While we can use the Fundamental Theorem of Finite Abelian Groups to completely classify all Abelian groups of a given order, the problem of classifying non-Abelian groups is much more difficult. The Sylow theorems play an important role in the classification of such groups.

Investigation 33

The Second and Third Sylow Theorems

Focus Questions

By the end of this investigation, you should be able to give precise and thorough answers to the questions listed below. You may want to keep these questions in mind to focus your thoughts as you complete the investigation.

- If G is a group and H is a subgroup of G, what does it mean for two subgroups of G to be H-conjugate? How is this different than two subgroups being conjugate?
- What properties does the H-conjugacy relation satisfy, and why are these properties important?
- What is the normalizer of a subgroup? Why are normalizers important, and what properties do they satisfy?
- What role do H-conjugacy and normalizers play in the proofs of the second and third Sylow theorems?
- How can the Sylow theorems be used to classify finite groups?

Preview Activity 33.1. In Investigation 32, we utilized conjugacy of elements in a group to develop an important counting tool, the Class Equation. In this investigation, we will take our work a step further and investigate the idea of conjugate subgroups. The definition is similar to that of conjugate elements.

Definition 33.2. Let G be a group and H a fixed subgroup of G. A subgroup K of G is H**-conjugate** to a subgroup N of G if there is an element $h \in H$ so that $K = hNh^{-1} = \{hnh^{-1} : n \in N\}$.

(a) Let $G = D_6$, and let $H = \langle R \rangle$. Is the subgroup $K = \langle rR \rangle$ H-conjugate to the subgroup $N = \langle r \rangle$? Find all of the subgroups N' of G that are H-conjugate to N.

(b) Conjugation is a relation on the collection of subgroups of a group. If H is a fixed subgroup of a group G, we will write $K \sim_H N$ to denote that the subgroup K of G is H-conjugate to the subgroup N of G. We will also say that two subgroups K and N of G are conjugate (without reference to a subgroup) if K and N are G-conjugate. It is natural at this point to ask what kind of relation $\sim_H$ is.

(i) Show that $\sim_H$ is a reflexive relation.

(ii) Show that $\sim_H$ is a symmetric relation.

(iii) Show that $\sim_H$ is a transitive relation.

(iv) What can you conclude about $\sim_H$ from your work in parts (i) – (iii)?

Introduction

Recall that the First Sylow Theorem tells us that if p is a prime and p^k divides the order of a finite group G, then G contains a subgroup of order p^k. In particular, if p is prime and the order of a finite group G is $p^k m$, where $k, m \in \mathbb{Z}+$ with $\gcd(p, m) = 1$, then any subgroup of G of order p^k is called a Sylow p-subgroup of G. In this investigation, we will prove the second and third Sylow theorems. These theorems tell us about the relationship between Sylow p-subgroups and the number of Sylow p-subgroups in a group.

Conjugate Subgroups and Normalizers

In Preview Activity 33.1, we began our exploration of conjugate subgroups. Recall that if G is a group and H is a subgroup of G, then two subgroups K and N of G are H-conjugate if $K = hNh^{-1}$ for some $h \in H$. As we saw in the activity, H-conjugacy defines an equivalence relation on the set of all subgroups of G, which we denote by $\sim_H$. Conjugates play an important role in group theory, and the lemma below provides some useful properties of conjugates.

Lemma 33.3. *Let G be a group, let K and N be subgroups of G, and let $a, b \in G$.*

(i) *If $K = aNa^{-1}$, then $N = a^{-1}Ka$.*

(ii) $(ab)K((ab)^{-1}) = a(bKb^{-1})a^{-1}$.

(iii) *If $a \in K$, then $aKa^{-1} = K$.*

We proved the first part of Lemma 33.3 in Preview Activity 33.1, and the proof of remaining parts of Lemma 33.3 are left for the exercises. (See Exercise 2.)

In order to prove the second and third Sylow theorems, we will also need to understand the normalizer of a subgroup. Recall that a subgroup N is normal in a group G if $aNa^{-1} = N$ for every $a \in G$. It is also possible to have $aNa^{-1} = N$ for just some elements $a \in G$, as the next activity demonstrates.

Activity 33.4. Let $K = \langle r \rangle$ in D_6.

(a) Is $RKR^{-1} = K$?

(b) Find all $a \in D_6$ so that $aKa^{-1} = K$. What kind of set do you obtain?

Activity 33.4 suggests that the set of elements a for which $aKa^{-1} = K$ might be interesting. This leads us to the following definition.

Definition 33.5. Let G be a group and K a subgroup of G. The **normalizer** of K in G is the set

$$N(K) = \{a \in G : aKa^{-1} = K\}.$$

You may have encountered the normalizer of a subgroup in Exercise 18 of Investigation 27. (See page 375.) If so, you already know that $N(K)$ is a subgroup of G, and you can omit parts (a) – (c) of the following activity.

Activity 33.6. Let G be a group with identity e, and let K be a subgroup of G.

(a) Can $N(K)$ be empty?

(b) Is $N(K)$ closed? You may use Lemma 33.3 if it helps.

(c) Does $N(K)$ contain the inverse of each of its elements? What conclusion can you draw about $N(K)$?

(d) Prove the following lemma. You may use Lemma 33.3 if it helps.

Lemma 33.7. *Let G be a group and K a subgroup of G. Then $N(K)$ is a subgroup of G and K is a normal subgroup of $N(K)$.*

As with Lagrange's Theorem and the relation that led us to left cosets, as well as the conjugacy relation on elements in a group that led us to the Class Equation, we will use the relation $\sim_H$ to count elements in finite groups. Doing so will allow us to prove the second and third Sylow theorems, but first we will need to learn a bit more about H-conjugacy. The next example suggests an important connection that will help us in this regard.

Activity 33.8. Let $K = \langle r \rangle$ in D_6, and let $H = \langle R \rangle$. We have seen that $N(K) = \{I, r, R^3, rR^3\}$ and that the distinct H-conjugates of K in D_6 are N, $\langle rR^2 \rangle$, and $\langle rR^4 \rangle$. There is a useful way to count the number of distinct H-conjugates of K in G. This number depends on $|H|$ and $|H \cap N(K)|$. Find $H \cap N(K)$ for this example, and use your work to conjecture a relationship between $|H|$, $|H \cap N(K)|$, and the number of distinct H-conjugates of K in G.

Although it is not wise to extrapolate from just one example, the result of Activity 33.8 might lead us to wonder if the relationship we observed there holds in general. The next lemma shows that it does.

Lemma 33.9. *Let G be a finite group and H a fixed subgroup of G. Let K be a subgroup of G. The number of distinct H-conjugates of K in G is the index of $H \cap N(K)$ in H. In particular, the number of distinct H-conjugates of K in G divides $|H|$.*

Proof. Let G be a group and H a fixed subgroup of G. Let K be a subgroup of G. To count the number of H conjugates of K, let aKa^{-1} and bKb^{-1} be H conjugates of K. Suppose $aKa^{-1} = bKb^{-1}$. Parts (a) and (b) of Lemma 33.3 imply that

$$K = b^{-1}(aKa^{-1})b = (b^{-1}a)K(b^{-1}a)^{-1},$$

so $b^{-1}a \in N(K)$. Since a and b are in H, we also have $b^{-1}a \in (H \cap N(K))$. Thus, $a(H \cap N(K)) = b(H \cap N(K))$. So different left cosets of $H \cap N(H)$ in G correspond to different H-conjugates of K in G. Therefore, the index of $H \cap N(K)$ in G is less than or equal to the number of distinct H-conjugates of K in G.

To prove equality, suppose $a(H \cap N(K)) = b(H \cap N(K))$ for some $a, b \in H$. Then $b^{-1}a \in (H \cap N(K))$ and

$$K = (b^{-1}a)K(b^{-1}a)^{-1} = b^{-1}(aKa^{-1})b,$$

from which it follows that

$$aKa^{-1} = bKb^{-1}.$$

Therefore, the number of left cosets of $H \cap N(K)$ in H is equal to the number of distinct H-conjugates of K in G. Since $H \cap N(K)$ is a subgroup of H, we know that the number of distinct H-conjugates of K in G divides $|H|$. ■

There is one other result we will use in our proof of the Second Sylow Theorem. The H-conjugation relation is a tool that will allow us to count elements in a finite group. In order to do so, we will need to determine the circumstances in which a Sylow p-subgroup group can be H-conjugate to itself. The next lemma provides the details.

Lemma 33.10. *Let Q be a Sylow p-subgroup of a finite group G, and let $g \in G$. If $|g| = p^k$ for some nonnegative integer k and $gQg^{-1} = Q$, then $g \in Q$.*

Proof. Let Q be a Sylow p-subgroup of a finite group G with identity e. Since Q is a p-group, we know $|Q| = p^n$ for some positive integer n. Let $g \in G$ with $|g| = p^k$ for some nonnegative integer k such that $gQg^{-1} = Q$. Then $g \in N(Q)$. Lemma 33.7 shows that Q is a normal subgroup of $N(Q)$, so $N(Q)/Q$ is defined. Consider $gQ \in N(Q)/Q$. By the Fourth Isomorphism Theorem (see page 426), $\langle gQ \rangle = H/Q$ for some subgroup H of G that contains Q. Now $(gQ)^{p^k} = g^{p^k}Q = eQ = Q$, so $|gQ| = p^m$ for some nonnegative integer $m \leq k$. Therefore,

$$p^m = |\langle gQ \rangle| = |H/Q| = \frac{|H|}{|Q|} = \frac{|H|}{p^n},$$

so $|H| = p^{n+m}$. Since Q is a Sylow p-subgroup, n is the largest power of p that divides $|G|$. It follows that $m = 0$ and $H = Q$. Thus, $g \in H$ implies $g \in Q$, as desired. ■

The Second Sylow Theorem

We are now in position to prove the Second Sylow Theorem, which tells us that any two Sylow p-subgroups of a group are conjugate. The next activity will help us to better understand the ideas behind the proof.

Preview Activity 33.11. Let $G = D_6$.

(a) Find all of the Sylow 2-subgroups of G. (You can use the result of the Second Sylow Theorem here.)

(b) Label the Sylow 2-subgroups of G as $K_1, K_2, \ldots, K_t$, and let $S = \{K_1, K_2, \ldots, K_t\}$. Find t and identify $K_1, K_2, \ldots, K_t$.

(c) Let H be one of the Sylow 2-subgroups of D_6. Let $[K_i]_H$ be the collection of all subgroups of G that are H-conjugate to K_i. Find $[K_i]_H$ for each i.

(d) Since $K_i \in [K_i]_H$ for each i, we can write S as a union of the sets $[K_i]_H$. However, there may be some overlap. To be more concise, we can write S as a disjoint union of the sets $[K_i]_H$, where we discard any repetition. Relabeling if necessary, identify s and the $K_1, K_2, \ldots, K_s$ so that

$$S = \bigcup_{v=1}^{s} [K_v]_H$$

is a disjoint union of sets of the form $[K_v]_H$.

The Second Sylow Theorem should not be all that surprising. Exercise 9 in Investigation 32 (see page 458) shows that if K is a subgroup of a group G, then any conjugate of K is isomorphic to K. This means that any conjugate of a Sylow p-subgroup is again a Sylow p-subgroup. The Second Sylow Theorem tells us that these are the *only* Sylow p-subgroups. Its proof uses H-conjugacy to count elements and is the first time we have used this technique.

Theorem 33.12 (Second Sylow Theorem). *Let G be a finite group, and let p be a prime divisor of $|G|$. If H and K are Sylow p-subgroups of G, then there is an element $g \in G$ so that $H = gKg^{-1}$.*

Proof. Let G be a finite group. Let p be a prime number, and let $k \in \mathbb{Z}^+$ such that $|G| = p^k m$ with $\gcd(p, m) = 1$. Let H and K be Sylow p-subgroups of G. Since K is a Sylow p-subgroup of G, we know $|K| = p^k$. Let $S = \{K_1, K_2, \ldots, K_t\}$ be the set of distinct conjugates of K in G. Lemma 33.9 shows that the number of conjugates of K in G is

$$t = [G : G \cap N(K)] = \frac{|G|}{|N(K)|}.$$

Since K is a subgroup of $N(K)$, we know p^k divides $|N(K)|$. Let $|N(K)| = p^k l$ for some positive integer l. So $t = \frac{p^k m}{p^k l} = \frac{m}{l}$. Since $\gcd(p, m) = 1$, it follows that $\gcd(p, t) = 1$. We will now show that $H = K_i$ for some i between 1 and t.

Let i and j be integers with $1 \leq i, j \leq t$. Note that both K_i and K_j are conjugate to K, and, by transitivity, K_i and K_j are conjugate to each other. Let $[K_i]_H$ be the collection of all subgroups of G that are H-conjugate to K_i. Since any H-conjugate of K_i is a G-conjugate of K_i, every H-conjugate of K_i is equal to K_r for some $1 \leq r \leq t$. Thus, $[K_i]_H \subseteq S$. Also, $K_i \in [K_i]_H$, so S is a union of the H-conjugacy classes of the form $[K_i]_H$. Since H-conjugacy is an equivalence relation, S can be written as a disjoint union of H-conjugacy classes with representatives from S. After relabeling, we can assume

$$S = \bigcup_{v=1}^{s} [K_v]_H, \tag{33.1}$$

for some integer s between 1 and t. By Lemma 33.9, the number of distinct H-conjugates of K_i in G is equal to $[H : H \cap N(K_i)] = \frac{|H|}{|H \cap N(K_i)|}$, which is a divisor of $|H| = p^k$. Let $\frac{|H|}{|H \cap N(K_i)|} = p^{k_i}$ for some positive integer $k_i \leq k$. Then $|[K_i]_H| = p^{k_i}$.

We will now count the number of distinct conjugates of K in G in two different ways. On the one hand, S contains the distinct conjugates of K in G. On the other hand, we can add up the number of elements in each distinct conjugacy class in (33.1). Equating the two totals yields

$$t = |S| = \sum_{v=1}^{s} |[K_v]_H| = \sum_{v=1}^{s} p^{k_v}.$$

If all of the k_v are positive, then $p \mid t$. This contradicts the fact that $\gcd(t, p) = 1$. Therefore, there is an index w so that $k_w = 0$. Then $\frac{|H|}{|H \cap N(K_w)|} = 1$, or $|H| = |H \cap N(K_w)|$. This means $H \subseteq N(K_w)$. So, if $x \in H$, then x has order p^m for some nonnegative integer m and $xK_wx^{-1} =$

K_w. Lemma 33.10 then shows that $x \in K_w$. Thus, $H \subseteq K_w$. Since both K_w and H are Sylow p-subgroups of G, they have the same order. Thus, $H = K_w$, and by the transitivity of conjugacy, H and K are conjugate. ■

The Third Sylow Theorem

The Third Sylow Theorem provides us with information about the number of Sylow p-subgroups in a group. More specifically:

Theorem 33.13 (Third Sylow Theorem)**.** *Let G be a finite group, and let p be a prime dividing the order of G. The number of Sylow p-subgroups of G divides $|G|$ and is of the form $1 + pm$ for some nonnegative integer m.*

Lemma 33.9 proves part of this result—namely, that the number of distinct conjugates of a Sylow p-subgroup in G divides $|G|$. The other conclusion, however, might seem mysterious. Activity 33.16 outlines a proof of the Third Sylow Theorem, and in doing so, sheds some light on why its second conclusion makes sense.

To complete this investigation, we will consider two more examples that illustrate the use of the Sylow theorems.

Example 33.14. Let G be a group of order 36. We will show that the Sylow theorems guarantee that G contains a normal subgroup of order 3 or of order 9. If G is Abelian, then any subgroup of G is normal in G, so we will assume that G is non-Abelian. Since $36 = 2^2 \times 3^2$, the First Sylow Theorem tells us that G contains a Sylow 2-subgroup and a Sylow 3-subgroup. Let n_2 be the number of Sylow 2-subgroups of G and n_3 the number of Sylow 3-subgroups of G. The Third Sylow Theorem and its corollary (see Corollary 32.27 on page 455) tell us that $n_3 = 1 + 3t$ for some nonnegative integer t, and n_3 divides 4. These conditions can only be satisfied if $n_3 = 1$ or $n_3 = 4$. If $n_3 = 1$, then the Sylow 3-subgroup of G is normal in G. But what can we say if $n_3 = 4$?

Let H be a Sylow 3-subgroup of G. Recall that $P(S)$ is the group of permutations of the set S and G/H is the set of left cosets of H in G. The mapping $\varphi : G \to \mathcal{P}(G/H)$ defined by $\varphi(g) = \pi_g$ is a non-trivial homomorphism with $|\text{Im}(\varphi)| \geq [G : H]$, where $\pi_g(aH) = (ga)H$. (See Exercise 7.) Now $[G : H] = 4$, so $|P(G/H)| = 4! = 24$, and φ cannot be a monomorphism. Let $K = \text{Ker}(\varphi)$. Note that $|G/K| = \frac{|G|}{|K|} = \frac{36}{|K|}$ divides both $|G|$ and $|\text{Im}(\varphi)|$, so $|G/K| = 4$, $|G/K| = 6$, or $|G/K| = 12$. So $|K| = 3$, $|K| = 6$, or $|K| = 9$. If $|K| = 3$ or $|K| = 9$, then we are done. If $|K| = 6$, then the Third Sylow Theorem tells us that K contains a unique Sylow 3-subgroup N of order 3. Since $|N| = 3$, there is an element $n \in K$ such that $N = \langle n \rangle$. We will now show that N is normal in G. Let $g \in G$ and $m \in N$. Then $gmg^{-1} \in K$ since $K \lhd G$. If m is the identity e in K, then $gmg^{-1} = e \in N$. If m is not the identity, then $|gmg^{-1}| = 3$. Since N is the unique subgroup of K of order 3, N contains all of the elements in K of order 3, and so $gmg^{-1} \in N$. Thus, N is a normal subgroup of G of order 3. In each case, we have that G contains a normal subgroup of order 9 or a normal subgroup of order 3.

Example 33.15. Let p and q be primes, and let G be a group of order p^2q^2. We will show that G is not a simple group—that is, G has a non-trivial proper normal subgroup. If G is Abelian, then any subgroup of G of order p is a non-trivial proper normal subgroup. So assume G is non-Abelian. Now consider the case where $p = q$. Then G is a p-group, and Theorem 32.14 shows that $Z(G)$ is non-trivial and is therefore a non-trivial proper normal subgroup of G. Thus, we can also assume that p and q are distinct primes with $p < q$. Let n_p be the number of Sylow p-subgroups of G, and let n_q be the number of Sylow q-subgroups of G. The Third Sylow Theorem and its corollary (Corollary 32.27) tell us that $n_q = 1 + qt$ for some nonnegative integer t, and n_q divides p^2. Since $p < q$, we cannot have $n_q = p$, so the only possibilities are $n_q = 1$ or $n_q = p^2$. If $n_q = 1$, then the Sylow q-subgroup of G is a non-trivial proper normal subgroup of G. So assume $n_q = p^2$. Then $1 + qt = p^2$, and $qt = p^2 - 1 = (p+1)(p-1)$. Since q is prime, it follows that q divides $p - 1$ or q divides $p + 1$. The former is impossible because $p < q$, so we conclude that q divides $p + 1$. The only way this can happen is if $p = 2$, $q = 3$, and $|G| = 36$. But Example 33.14 shows that any group of order 36 contains a normal subgroup of order 3 or 9, which completes the argument that G is not simple.

Concluding Activities

Activity 33.16. In this activity, we will provide the framework for a proof of the Third Sylow Theorem. Let G be a finite group, and let p be a prime divisor of $|G|$. Let K_1 K_2, ..., K_t be the distinct Sylow p-subgroups of G, and let $S = \{K_1, K_2, \ldots, K_t\}$. Let $H = K_1$, and for each i, let $[K_i]_H$ be the collection of all subgroups of G that are H-conjugate to K_i (as in our proof of the Second Sylow Theorem).

(a) In Activity 33.11, we saw that the sets $[K_i]_H$ did not all have the same number of elements. In particular, there was one case in which $[K_i]_H$ contained only one element. Use Lemma 33.10 to show that $|[K_i]_H| = 1$ if and only if $K_i = H$. (As we will see, this is where the 1 in $1 + pm$ comes from.)

(b) Suppose $K_i \neq H$. If we want to use H-conjugacy to count the elements in S (as we did in the proof of Theorem 33.12), then we will need to understand what $|[K_i]_H|$ can be. Review the proof of Theorem 33.12 and use part (a) of this activity to explain why p divides $|[K_i]_H|$. (This is where the pm part comes from.)

(c) As in the proof of Theorem 33.12, we can write S as a disjoint union of H-conjugacy classes—that is,

$$S = \bigcup_{v=1}^{s} [K_v]_H$$

for some integer s between 1 and t. Use this equation to explain why $t = 1 + pm$ for some nonnegative integer m.

(d) Combine the previous pieces of this activity into a complete and formal proof of the Third Sylow Theorem.

Activity 33.17. Let G be a finite group and p a prime divisor of $|G|$. Let H be a subgroup of G of order p^m for some m. Prove that H is a subgroup of a Sylow p-subgroup. (Hint: Refer to the proof of the Second Sylow Theorem.)

Activity 33.18. Write a short summary that describes the important concepts, and the relationships between these concepts, that were introduced in this investigation. Explain how the material in this investigation is connected to the material in Investigations 20, 22, and 32.

Exercises

(1) Let $G = S_3$.

(a) For $H = \langle(1\ 2)\rangle$, find all of the H-conjugates of $K = \langle(1\ 2\ 3)\rangle$ in G.

(b) For $H = \langle(1\ 2\ 3)\rangle$, find all of the H-conjugates of $K = \langle(1\ 2)\rangle$ in G.

⋆ (2) Let G be a group, let H and K be subgroups of G, and let $a, b \in G$.

(a) Prove part (b) of Lemma 33.3—that is, prove that

$$(ab)H(ab)^{-1} = a(bHb^{-1})a^{-1}.$$

(b) Prove part (c) of Lemma 33.3—that is, prove that if $a \in H$, then $aHa^{-1} = H$.

(3) Find all of the Sylow p-subgroups of S_4. (Hint: See Exercise 37 on page 378 of Investigation 27.)

(4) In this exercise, we will investigate the question of whether subgroups inherit Sylow p-subgroups under intersections. Let G be a finite group, let K be a Sylow p-subgroup of G, and let N be any subgroup of G.

(a) Must $K \cap N$ be a Sylow p-subgroup of N? (Hint: See Exercise 3.)

(b) Must it be true that $K \cap N$ is a Sylow p-subgroup of N if N is normal in G? Prove your answer.

(5) Let p be a prime divisor of the order of a group G, and let N be a p-subgroup of G.

(a) Must N be contained in every Sylow p-subgroup of G? Explain.

(b) Are there any conditions on N so that N is contained in *every* Sylow p-subgroup of G? Prove your answer.

(6) **Classifying simple groups.** The Sylow theorems are especially useful in classifying simple groups, as we will explore in this exercise. We have already argued that the only simple Abelian groups are the groups $\mathbb{Z}_p$ for a prime p. Therefore, we will focus here on non-Abelian groups. Let G be a non-Abelian group of order n. Let p, q, and r be primes with $p < q < r$. At this point, we know the following:

- One result of Activity 32.32 (see page 457) is that any group of order pq has a normal Sylow q-subgroup.
- Exercise 25 of Investigation 32 (see page 459) shows that any non-Abelian p-group has non-trivial normal subgroups.
- Exercise 18 of Investigation 32 (see page 458) shows that any group of order p^2q with $q \not\equiv 1 \pmod{p}$ has a normal Sylow q-subgroup.
- Exercise 36 of Investigation 27 (see page 378) shows that D_n contains the non-trivial normal subgroup $\langle R\rangle$ of index 2 and also the normal subgroups $\langle r, R^2\rangle$ and $\langle rR, r^2\rangle$ when n is even.

In this exercise, we will show that the order of the smallest non-Abelian simple group is 60.

(a) Show that if $|G| = pq^k$ with $p < q$ and $k \in \mathbb{Z}^+$, then G has a non-trivial proper normal subgroup.

(b) Suppose $|G| = pqr$. Show that G has a non-trivial proper normal subgroup.

(c) Explain why if G is a non-Abelian simple group of order less than or equal to 60, then $|G| = 60$.

⋆ (7) This exercise was used in Example 33.15 and will be used in Exercise 8. Let G be a finite group and K a subgroup of G. Let G/K denote the collection of left cosets of K in G. (Note that we are not assuming the K is normal in G, so G/K may not be a group.) Recall that $P(S)$ denotes the collection of permutations of a set S.

(a) Let $a \in G$. Show that the function π_a defined by $\pi_a(gK) = (ag)K$ is a permutation of G/K.

(b) Define $\varphi : G \to P(G/K)$ by $\varphi(a) = \pi_a$. Show that φ is a homomorphism.

(c) If K is a proper subgroup of G, show that φ is not the trivial homomorphism.

(d) If K is a proper subgroup of G, show that $|\text{Im}(\varphi)| \geq [G : K]$.

(8) **Simple groups of order 60.** In this exercise, we will use the result of Exercise 7 to show that the only simple group of order 60 is A_5. We know that there is no simple Abelian group of order 60, so for this exercise, assume that G is a non-Abelian simple group of order 60.

(a) For each prime divisor p of $|G|$, let n_p denote the number of Sylow p-subgroups of G. Determine the possible values of n_p for each p.

(b) Use Exercise 7 with K a Sylow 2-subgroup of G to show that $n_2 \neq 3$. (Hint: What can you say about $|\text{Ker}(\varphi)|$?)

(c) If K and N are distinct Sylow 2-subgroups of G, what is the maximum value $|K \cap N|$ can have? Use this idea and count elements to show that $n_2 \neq 15$.

(d) Assume now that $n_2 = 5$, and let K be a Sylow 2-subgroup of G.

(i) Let $\varphi : G \to P(G/K)$ be defined by $\varphi(a) = \pi_a$ (as in Exercise 7). Explain why φ must be a monomorphism.

(ii) Show that G is isomorphic to a subgroup G' of S_5.

(iii) Show that $G' = A_5$, and conclude that $G \cong A_5$. (Hint: First determine what group $G'A_5$ is if G' is not contained in A_5. Then use the Second Isomorphism Theorem to calculate $|G' \cap A_5|$.)

Part VII

Special Topics

Investigation 34

RSA Encryption

Focus Questions

By the end of this investigation, you should be able to give precise and thorough answers to the questions listed below. You may want to keep these questions in mind to focus your thoughts as you complete the investigation.

- What is public key encryption, and what are some of its applications?
- How does RSA encryption work, and how is the security of RSA encryption related to prime factorization of integers?
- What mathematical results are necessary to establish the validity of RSA encryption, and how do these results follow from previously established properties of the integers?

Preview Activity 34.1. In Investigations 1 – 4, we learned about divisibility, greatest common divisors, and prime factorization in the integers. In this investigation, we will tie all of these ideas together to investigate an interesting and important application: encryption. The system we will study is one that is commonly used to transmit sensitive data over the internet. Its security rests on an important observation that should become apparent as you attempt to complete the following tasks.

(a) For each of the parts below, find a number whose prime factors are exactly the numbers listed:

(i) 4861 and 2621

(ii) 7907 and 619

(iii) 1753 and 1759

(b) Each of the numbers below has exactly two prime factors. Using any mathematically correct method, find these two factors.

(i) 13494211

(ii) 3902233

(iii) 1776977

(c) Which was easier: part (a) or part (b)? Explain why.

Introduction

Throughout history, secret codes have been used to send private messages from one person to another. From the primitive Caeser cipher, which simply shifts each letter in the original message by a fixed amount, to the fascinating Enigma machines used during World War II, numerous encryption schemes and devices have been invented in an attempt to keep sensitive data out of the hands of unauthorized (and potentially malicious!) third parties. In this investigation, we will study one of the most commonly used modern encryption schemes, named *RSA encryption* after its three inventors: Rivest, Shamir, and Asleman. We will first explain the details of the RSA method and consider a specific example to illustrate how it works. Then, throughout the remainder of the investigation, you will establish the validity of RSA encryption by completing a sequence of activities and proving several intermediate results.

RSA encryption was first proposed in 1978, and since then it has become the gold standard for what is now known as *public-key encryption*. The basic idea behind a public-key scheme is that anyone should be able to send a message, but only the intended recipient should be able to read it. Thus, the key to *encrypt* a message is made public, but the key to *decrypt* the message is kept secret and distributed only to those who are authorized to view the encrypted text. This type of encryption is particularly useful for tasks such as sending data over the internet. For instance, a banking web site might want to allow any user to send information (such as a user name and password) over a secure connection. However, for the transaction to be truly secure, only the bank should be able to decode the information sent. In order for such a system to work, it has to be very difficult or even impossible for a potential attacker to determine the *private* decryption key just by knowing the *public* encryption key. RSA encryption achieves this design feature by using the properties of prime factorizations suggested in Preview Activity 34.1. In particular, RSA encryption exploits the fact that it is relatively easy to multiply two prime numbers together (even if they are large primes), but nearly impossible to efficiently factor a large number into its prime factors. As we will see shortly, the RSA scheme translates this theoretical fact into a very practical and secure encryption method.

Congruence and Modular Arithmetic

Before we dive into the details of RSA encryption, a few notes regarding congruence and modular arithmetic are in order.

First, we will use "mod" notation in two distinct but related ways. In particular, we will write

$$a \equiv b \bmod n$$

when we want to specify a *relationship* between the integers a and b—namely, that n divides $(a-b)$, or equivalently, a and b have the same remainder when divided by n. This is the standard usage that we are familiar with from Investigation 2, except that we have omitted the parentheses that typically surround the "mod n" portion of the notation. This omission is common, and it will often make our notation easier to read, especially when we are working with expressions that already contain several sets of parentheses.

We will also sometimes write $a = b \bmod n$ (note the use of $=$ instead of $\equiv$) for the purposes of *defining* a to be the unique remainder guaranteed by the Division Algorithm when b is divided by

n. This usage treats "mod" as an operation, like addition or multiplication. For instance, when we write

$$x = 546 \bmod 17,$$

we are defining x to be the unique remainder that results from dividing 546 by 17. This remainder happens to be 2, so we could also write

$$x = 546 \bmod 17 = 2.$$

The Basics of RSA Encryption

The first step in any encryption scheme is to decide what *alphabet* will be used and how the elements of the alphabet will be assigned numerical representations. The most basic and natural choice for an alphabet is the standard A through Z alphabet, with A represented by 0, B represented by 1, and so on. As we will see, however, there are many security advantages to using a larger alphabet that consists not only of letters, but entire words. So, for instance, we might use a two-digit numerical representation of each letter (00 to 25) and encode our messages using blocks of letters. If we were to do so, the word "SECRET" would be encoded as 180402170419. Keep in mind that at this point, we haven't actually encrypted anything. We have simply developed a way of translating letters or blocks of letters into the numerical representations that will be used by our encrypting function. Note that we could have also included numerical codes for spaces, numbers, punctuation, and the like.

Once we have established our alphabet, it takes only a few simple steps to encrypt a message:

(1) First, we choose two different prime numbers, p and q, and calculate the quantity $m = pq$ (called the *modulus*). For the resulting system to be secure, p and q need to be extremely large, perhaps having hundreds or even thousands of digits.* The value of m becomes public—it can be shared with anyone—but p and q must be kept secret.

(2) Next, we choose a positive integer e such that

$$\gcd(e, (p-1)(q-1)) = 1.$$

This number e is called the *encryption key*, and the value of e is made public. The quantity $(p-1)(q-1)$ is often called the *totient*, denoted t. The totient must be kept secret, as it plays an essential role in the decrypting process.

(3) Finally, to encrypt a message, we input the numerical representation of each block of letters into the encoding function f defined by

$$f(x) = x^e \bmod m.$$

Once we have encrypted a message, the decryption process is similar and can be described as follows:

*As we will see later on, in order for RSA encryption to work, p and q each need to be larger than the number of elements in the given alphabet. This is usually not a problem since the security of RSA encryption relies on choosing primes that are huge, and certainly larger than the size of any reasonable alphabet.

(1) First, we find a positive integer d (the *decryption key*) such that

$$ed \equiv 1 \bmod (p-1)(q-1).$$

(2) To decrypt an encoded message, we input each block of data into the decoding function g defined by

$$g(x) = x^d \bmod m.$$

It should be noted that not all of the steps described above are straightforward or easy to complete. For instance, the decryption key is defined to be an integer that satisfies a particular congruence relation. It is not immediately obvious that such an integer will always exist. We will have to use what we have learned about the integers to prove not only that a suitable decryption key exists, but also that it can be found in a relatively straightforward manner, and that the corresponding decoding function actually returns encrypted messages to their original, unencrypted state.

An Example

Before we go any further into investigating the details of RSA encryption and why it works, let's look at a simple example.

Example 34.2. Suppose we want to use RSA encryption to encode the highly sensitive, top-secret message, "JOHNNY LOVES SALLY". We will begin by choosing p and q. In practice, p and q are often hundreds of digits long, but for simplicity, we will choose two smaller primes,

$$p = 4000433442120074580 13 \text{ and } q = 500030066366269001203.$$

With these choices of p and q, our modulus m and totient t are

$$m = pq = 200033699955714283345172521584008468989639$$

and

$$t = (p-1)(q-1) = 200033699955714283344272448173430192530424.$$

Recall that m will be made public, but p, q, and t will be kept secret. This is significant, since in order to calculate t from m, we would have to first factor m, a task that a computer could fairly easily complete for this example, but not for examples involving larger primes. In fact, one website on RSA cryptography notes that "if p and q are each 1024 bits long, the sun will burn out before the most powerful computers presently in existence can factor the modulus into p and q."[†]

The next step in encoding our message is to choose an encryption key. We need to choose a number e such that $\gcd(e, (p-1)(q-1)) = 1$. Note that any prime number that does not divide $t = (p-1)(q-1)$ will suffice here; a common choice is $e = 2^{16} + 1 = 65537$, and this is what we will use.

To perform the actual encryption, we will first break our message into three 6-character blocks. We will use the standard 00 – 25 encoding for the letters A – Z, and we will also use the code 99 to denote a space. Thus, the numerical representation of our message is:

091407131324 991114210418 991800111124

[†] `http://fringe.davesource.com/Fringe/Crypt/RSA/Algorithm.html`

We then apply our encoding function to each of these 12-digit numbers:

$$\begin{aligned} f(091407131324) &= (091407131324)^e \bmod m \\ &= 009505729493564929202343371764084584555016 \\ f(991114210418) &= (991114210418)^e \bmod m \\ &= 012196119237767316793050190360104919489384 \\ f(991800111124) &= (991800111124)^e \bmod m \\ &= 124080637343749317837866219863773135637684 \end{aligned}$$

Note that we have appended zeros to the beginning of each encoded block so that each has exactly 42 digits. This would allow the entire message to be sent as a single string and then unambiguously decomposed into its three distinct parts prior to decryption.

To decrypt the message, we would first need to find the decryption key, d. It is at this step that some of the theory we have been studying in previous investigations is particularly useful (and in fact necessary). For now, however, we will skip over the "how" and simply assume that we have been able to find an integer, say

$$d = 921894177863251936178098635065731653140\,81,$$

such that

$$ed \equiv 1 \bmod (p-1)(q-1).$$

A computer algebra system can then readily verify that

$$\begin{aligned} &g(009505729493564929202343371764084584555016) \\ &\quad = (009505729493564929202343371764084584555016)^d \bmod m \\ &\quad = 091407131324, \end{aligned}$$

$$\begin{aligned} &g(012196119237767316793050190360104919489384) \\ &\quad = (012196119237767316793050190360104919489384)^d \bmod m \\ &\quad = 991114210418, \end{aligned}$$

and

$$\begin{aligned} &g(124080637343749317837866219863773135637684) \\ &\quad = (124080637343749317837866219863773135637684)^d \bmod m \\ &\quad = 991800111124. \end{aligned}$$

In other words, the decoding function g returns each block of the encrypted message to its original, unencrypted state, as desired. It is interesting to note that, even in this relatively simple example, the computations in the decryption process involve raising one 40-digit number to another 40-digit exponent. Fortunately, there are efficient algorithms for performing such exponentiations, even when the numbers involved are much larger (as would be the case if larger, and hence more realistic, values of p and q were chosen.)

Why RSA Works

Now that we've seen an example, we are ready to get to work and show that RSA encryption actually works the way it is intended. In particular, we must show three important facts:

- First, we must show that no matter what primes we choose for p and q, it will always be possible to find an encryption key, e, that satisfies
$$\gcd(e, (p-1)(q-1)) = 1.$$
- Next, we must show that it is always possible to find a decryption key, d, such that
$$ed \equiv 1 \bmod (p-1)(q-1).$$
- Finally, we must show that the decoding function $g(x) = x^d \bmod m$ is the inverse of the encoding function $f(x) = x^e \bmod m$. In other words, we must show that for all x in our alphabet,
$$(x^e)^d \bmod m = x^{ed} \bmod m = x.$$

The goal for the rest of this investigation is to establish these three facts, and the activities below suggest a series of steps that will accomplish exactly that goal.

Task 1: Finding the Encryption Key

Activity 34.3. Let p and q be any prime numbers.

(a) Explain why it is always possible to find a prime number e that does not divide $(p-1)(q-1)$.

(b) Explain why the number found in part (a) would always satisfy
$$\gcd(e, (p-1)(q-1)) = 1.$$

Task 2: Finding the Decryption Key

Activity 34.4. Consider the fact that the encryption key e is chosen specifically so that
$$\gcd(e, (p-1)(q-1)) = 1.$$

(a) Use Bezout's Identity (Theorem 3.9 on page 28) to write down a linear combination corresponding to $\gcd(e, (p-1)(q-1))$.

(b) Use your answer to part (a) to explain why there must exist an integer d such that
$$ed \equiv 1 \bmod (p-1)(q-1).$$

(c) What process or algorithm would allow you to actually determine the value of d that is guaranteed to exist by part (b)? (Hint: We have studied this algorithm in a previous investigation.)

(d) Suppose that we were able to find an integer d' such that

$$ed' \equiv 1 \bmod (p-1)(q-1),$$

but $d' < 0$. Explain how we could use d' to find a positive integer d that also satisfies

$$ed \equiv 1 \bmod (p-1)(q-1).$$

(Hint: Add a convenient quantity to d'.)

Task 3: Proving an Inverse Relationship Between f and g

In order to prove that the decoding function g actually undoes the work of the encoding function f, we must show that for all x in our alphabet,

$$x^{ed} \bmod m = x.$$

Doing so will require the following three intermediate results:

Theorem 34.5 (Binomial Theorem). *Let n be a positive integer, and let a and b be any real numbers. Then*

$$(a+b)^n = \sum_{k=0}^{n} \binom{n}{k} a^{n-k} b^k,$$

where

$$\binom{n}{k} = \frac{n!}{k!(n-k)!}.$$

Theorem 34.6 (Freshman's Dream). *Let p be a prime number. Then for all integers a and b,*

$$(a+b)^p \equiv (a^p + b^p) \bmod p.$$

Theorem 34.7 (Fermat's Little Theorem). *Let p be a prime number. Then for every positive integer a,*

$$a^p \equiv a \bmod p.$$

Proofs of the Binomial Theorem and the Freshman's Dream are outlined in Exercises 9 and 10 of Investigation 8. (See page 103.) The proof of Fermat's Little Theorem follows from the Freshman's Dream. (If you have completed Investigation 26, an alternative proof using group theory is suggested in Exercise 20 on page 356.)

Activity 34.8. Use induction on a, along with the Freshman's Dream, to prove Fermat's Little Theorem.

Activity 34.9. Explain why the conclusion of Fermat's Little Theorem (namely, that $a^p \equiv a \bmod p$) is equivalent to

$$a^{p-1} \equiv 1 \bmod p$$

as long as $p \nmid a$.

We can now use the Freshman's Dream and Fermat's Little Theorem to establish an inverse relationship between the RSA encoding and decoding functions.

Activity 34.10. Let p, q, m, d, and e be as stated previously. Note that, by definition,

$$ed \equiv 1 \bmod (p-1)(q-1).$$

(a) Use the definition of congruence to explain why $ed \equiv 1 \bmod (p-1)$ and $ed \equiv 1 \bmod (q-1)$.

(b) Use part (a), Activity 34.9, and the fact that p and q were chosen to be very large (and, in particular, much larger than the number of letters in the alphabet, $\mathbb{A}$) in order to prove that

$$x^{ed} \equiv x \bmod p \text{ and } x^{ed} \equiv x \bmod q$$

for all $x \in \mathbb{A}$.

(c) Use part (b) to explain why for all $x \in \mathbb{A}$, $x^{ed} \equiv x \bmod m$.

(d) Explain how your answer to part (c) actually implies that $x^{ed} \bmod m = x$. (Hint: It again matters that p and q, and thus m, are larger than the number of elements in $\mathbb{A}$. Remember that if two numbers are both less than m and yet congruent modulo m, then they must be equal.)

(e) Deduce from your answer to part (d) that the decoding function used in RSA encryption always returns an encrypted message to its original, unencrypted state.

Concluding Thoughts and Notes

Before we conclude our investigations of RSA encryption, a few additional observations are worth mentioning.

First, note that since p and q are chosen to be very large, it is important that e be large enough so that x^e is typically greater than the modulus, $m = pq$. If x^e is not greater than m, then messages can be easily decrypted by simply taking the eth root of the encrypted data, since the encryption function in this case does not involve a reduction modulo m.

Next, note that RSA encryption schemes are deterministic, meaning that they have no random component to them. Because of this, potential attackers could use the public encryption key to develop a dictionary of likely words and their encryptions. This dictionary could then be used to try to decipher encrypted messages by comparing the encrypted words to the entries in the dictionary. Encrypting larger blocks of data (instead of individual letters) reduces this security vulnerability.

One aspect of public key encryption that we have not considered is that of *signing* or *authentication*. Since RSA schemes enable anyone to encrypt a message, it is important to be able to verify that encrypted messages are actually from who they claim to be from. Several methods are available for this purpose, many of which involve introducing an additional private key that identifies the sender.

Finally, most experts agree that with large enough primes, RSA encryption seems to be secure for the near future. However, advances in computer science, and especially in the study of quantum computing, have the potential to one day render RSA encryption obsolete.

Exercises

(1) Caeser's cipher is an example of a **shift cipher**, in that it encrypts messages by simply shifting each letter in the message by a fixed amount. (For example A $\to$ E, B $\to$ F, etc.) The encoding

functions associated with shift ciphers always have the form

$$f(x) = x + a \bmod n,$$

where n is the number of letters in the alphabet (typically 26).

(a) Assuming that the message below was encrypted using a shift cipher, decrypt the message. Make sure you explain how you determined which shift cipher was used.

```
XLMW QIWWEKI AEW IRGVCTXIH YWMRK E WLMJX GMTLIV.
```

(b) Was the following message encrypted using a shift cipher? Why or why not?

```
AJMZ YBZZGLB EGZ QCA BQPSOUABD
KZMQL G ZJMHA PMUJBS.
```

(2) In contrast to a shift cipher (see Exercise 1), a **stretch cipher** uses multiplication instead of addition to encode messages. That is, instead of using an encoding function of the form

$$f(x) = x + a \bmod 26,$$

it uses one of the form

$$f(x) = ax \bmod 26,$$

where a is some integer.

(a) Use a stretch cipher of your choosing to encode this message: ABSTRACT ALGEBRA MAKES ME SMILE.

(b) Does the stretch cipher you used in part (a) have a corresponding decoding function? In other words, is there a rule that can be used to decode any message encoded by the cipher?

(c) Is it always possible to decode a message that has been encoded using a stretch cipher? If so, explain how. Otherwise, determine the values of a (assuming a 26 letter alphabet) for which the corresponding stretch cipher is decode-able.

(3) The frequency with which each letter in the alphabet occurs in ordinary English is given in Table 34.1.[‡] Explain in a precise way how this table could be used to break shift and stretch ciphers.

(4) **Hill ciphers** use matrices to encode and decode messages. For instance, using the 2×2 matrix

$$A = \begin{bmatrix} 2 & 5 \\ 1 & 4 \end{bmatrix},$$

the message "ATTACK" would be encrypted by multiplying A by the vector representation of each pair of consecutive letters. Doing so, we obtain

$$A \begin{bmatrix} 0 \\ 19 \end{bmatrix} = \begin{bmatrix} 95 \\ 76 \end{bmatrix}, \quad A \begin{bmatrix} 19 \\ 0 \end{bmatrix} = \begin{bmatrix} 38 \\ 19 \end{bmatrix}, \quad \text{and} \quad A \begin{bmatrix} 2 \\ 10 \end{bmatrix} = \begin{bmatrix} 54 \\ 42 \end{bmatrix}.$$

(a) Reduce these vectors modulo 26 in order to finish encrypting the message.

[‡]This table originally appeared in *Applications of Abstract Algebra with Maple and MATLAB* (2nd ed.) by Klima, Stitzinger, and Sigmon, CRC Press, 2006.

Letter	Frequency (%)	Letter	Frequency (%)
A	8.167	N	6.749
B	1.492	O	7.507
C	2.782	P	1.929
D	4.253	Q	0.095
E	12.702	R	5.987
F	2.228	S	6.327
G	2.015	T	9.056
H	6.094	U	2.758
I	6.966	V	0.978
J	0.153	W	2.360
K	0.772	X	0.150
L	4.025	Y	1.974
M	2.406	Z	0.074

Table 34.1
Frequency of each letter in the English language.

(b) Are the two A's in the original message encrypted to the same letter? What about the two T's?

(c) Are Hill ciphers more or less susceptible to frequency analysis (that is, the analysis suggested in Exercise 3) than shift and stretch ciphers? Clearly explain your answer.

(d) What conditions must be placed on the encrypting matrix in order to guarantee that the resulting Hill cipher will be decode-able? Give a convincing argument to justify your answer.

(e) Use a Hill cipher with a matrix different than A to encrypt the message, "CRYPTOGRAPHY IS FUN". Then find the corresponding decrypting matrix, and verify that it does in fact return the encrypted message to its original form.

(5) Are any of the systems mentioned in Exercises 1 – 4 public-key schemes? That is, is it ever possible to encode messages using one of these schemes without also knowing how to decode messages?

Investigation 35

Check Digits

Focus Questions

By the end of this investigation, you should be able to give precise and thorough answers to the questions listed below. You may want to keep these questions in mind to focus your thoughts as you complete the investigation.

- What are check digits, and how are they used?
- What are some common check digit schemes for credit card and ISBN numbers?
- What is Verhoeff's check digit scheme, and how is it related to the dihedral group of order 10?

Preview Activity 35.1. This investigation will involve both congruence of integers and the dihedral group D_5. In this activity, we will review a few of the basics.

(a) Determine the value of x that satisfies the congruence equation

$$2(1) + 3(2) + 4(2) + 5(6) + 6x \equiv 0 \pmod 7.$$

(b) Consider the permutation $\pi = \begin{pmatrix} 0 & 1 & 2 & 3 & 4 & 5 & 6 & 7 & 8 & 9 \\ 1 & 5 & 7 & 6 & 2 & 8 & 3 & 0 & 9 & 4 \end{pmatrix}$ in S_{10}. (Note that we are permuting the digits 0 through 9 instead of 1 through 10.)

(i) Explain why π is an even permutation.

(ii) What is the order of π?

(iii) Find $\pi^4(5)$, where π^4 denotes the composition of 4 copies of π (that is, $\pi \circ \pi \circ \pi \circ \pi$).

(iv) If x is any integer between 0 and 9, what is $\pi^8(x)$? Explain.

Introduction

In this investigation, we will discuss check digits and explore various check digit schemes, including one that uses the dihedral group D_5. Check digits are important because nowadays most information is transmitted electronically. When we use an ATM or pay with a credit card, there is always the possibility that an error in transmission can occur. For instance, noise can be introduced in a message, information can be lost, and data can be altered during transmission. Another important source of error is human error. Data can be confused when humans enter numbers into machines or communicate to others. Some mistakes are more prevalent than others. For example, according to Richard Hamming,* the two most common human errors when dealing with information are interchanging digits (e.g., typing 12 instead of 21) and changing one of a string of three digits when two adjacent digits are the same (e.g., using 112 instead of 122). Clearly, problems can arise when data is not encoded or transmitted properly. Fortunately, there are ways to compensate for these errors. The first step is to determine when they occur, and that is where check digits come into play.

Check Digits

A check digit is a digit appended to a string, usually at the end, to make the sum of the digits in that string congruent to a specific number modulo a given integer. For example, the 10 digit identification number 2361068754 might have an extra digit d appended to the end so that the digit sum is congruent to 0 modulo 9. In this case, the check digit would be 3 and the ID number would be 23610687543. As their name suggests, check digits perform a check to ensure that the number received is a valid number. However, just because the number appears to be valid, that does not necessarily make it legitimate. For instance, a credit card company may choose to only use a small subset of the set of valid credit card numbers. We should also note that even when check digits are used to detect errors, they do not necessarily provide a way to correct the errors they find. (There are other methods for doing that.)

Check digits are used in UPC codes, credit card numbers, ISBN numbers, and most other identification systems. We will now consider a few of the more common and/or mathematically interesting check digit schemes.

Credit Card Check Digits

Different credit cards have account numbers, or ID codes, of different lengths and with different prefixes. Each code consists of a string of numbers, with each digit between 0 and 9. The prefix of a card is the one or two digit block at the beginning (the leftmost digits) of the ID code. In particular:

- MasterCard codes have 16 digits and use prefixes of 51, 52, 53, 54, and 55.
- VISA codes have either 13 or 16 digits and use a prefix of 4.
- American Express codes have 15 digits and use prefixes 34 or 37.
- Discover codes have 16 digits and use a prefix of 6011.

The prefix is the Major Industry Identifier (MII) and indicates the type of industry that issues the card. For example, VISA and MasterCard are issued by the banking and financial sector (with prefix

* *Coding and Information Theory* (2nd ed.), Prentice-Hall, 1986, p. 27.

numbers 4 or 5) while American Express is in the travel and entertainment category (prefix number 3). All credit cards compute a check digit modulo 10.† To find the check digit, we can use a process known as the *Luhn algorithm*:‡

(1) Beginning with the second digit from the right (in other words, don't include the check digit) and moving from right to left, double every other digit. Add the individual *digits* of these numbers (e.g., if the doubled number is 16, add 1 and 6).

(2) Sum the digits (but not the check digit) not considered in step (1).

(3) Add the results of steps (1) and (2). Call this result s.

(4) The check digit d is the solution to $s + d \equiv 0 \pmod{10}$.

Activity 35.2.

(a) Find the correct check digit d for the sample VISA card with number 4417 1234 5678 911d.

(b) Create your own valid American Express card number.

ISBN Check Digits

The acronym ISBN is an abbreviation for the International Standard Book Number, which is used to identify books. An ISBN-10 has the form

$$X_1X_2X_3X_4X_5X_6X_7X_8X_9X_{10},$$

where each X_i is a digit between 0 and 9. In an ISBN, the first digit, X_1, represents the language of the book (0 is English), the next block (2 or 3 digits) identifies the publisher, the third block (5 or 6 digits) is a publisher's number for the book, and the last digit is a check digit.

The check digit in an ISBN is determined by first attaching a weight to each digit, with the leftmost digit (X_1) having a weight of 1, the next digit (X_2) having a weight of 2, and so on. (In general, the weight of the digit X_k is k.) Next, we multiply each digit (except the check digit) by its weight and compute the weighted sum. The check digit is congruent to the weighted sum modulo 11, with X representing a check digit of 10.

A quick way to implement this scheme is through the use of weight vectors and dot products. Recall that the dot product of vectors $[v_1, v_2, \ldots, v_n]$ and $[w_1, w_2, \ldots, w_n]$ is the scalar

$$[w_1, w_2, \ldots, w_n] \cdot [v_1, v_2, \ldots, v_n] = \sum_{i=1}^{n} w_i v_i.$$

By taking the dot product of the first 9 digits of an ISBN with the weight vector $[1, 2, 3, 4, 5, 6, 7, 8, 9]$, we can easily determine what the check digit should be.

Activity 35.3. The ISBN-10 for a very interesting book is 0-471-33193-?. Find the check digit.

†Some books and web sites state that MasterCard and VISA use the prefix digits when determining the check digit, while American Express and Discover do not. To the best of our knowledge, all companies use the prefix digits in their calculations, and so we will do the same.

‡The Luhn Algorithm was created by Hans Peter Luhn, who worked for IBM. It is patented in U.S. Patent No. 2.950.048.

This ISBN-10 check digit scheme will find all single digit errors, but will also catch errors obtained by interchanging digits (for example, typing 12 instead of 21). However, the ISBN-10 scheme is restricted to ID numbers with 10 digits, and we have to introduce the extra symbol X to represent the digit 10 as a possible check digit. In the next section, we will examine a check digit scheme that works for ID numbers with any number of digits.

Verhoeff's Dihedral Group D_5 Check

In the late 1960s, Dutch mathematician Jacobus Verhoeff proposed a check digit scheme based on the dihedral group D_5.[§] This scheme is an improvement on others in that it works for any length number, and it detects all single digit errors and all transposition errors involving two adjacent digits. However, the Verhoeff scheme is a little more complicated to implement.

We begin with the operation table for D_5 given in Table 35.1. (Note that the elements in this table are listed in a different order than usual; this is done to match Verhoeff's labeling.)

	I	R	R^2	R^3	R^4	rR^4	rR^3	rR^2	rR	r
I	I	R	R^2	R^3	R^4	rR^4	rR^3	rR^2	rR	r
R	R	R^2	R^3	R^4	I	rR^3	rR^2	rR	r	rR^4
R^2	R^2	R^3	R^4	I	R	rR^2	rR	r	rR^4	rR^3
R^3	R^3	R^4	I	R	R^2	rR	r	rR^4	rR^3	rR^2
R^4	R^4	I	R	R^2	R^3	r	rR^4	rR^3	rR^2	rR
rR^4	rR^4	r	rR	rR^2	rR^3	I	R^4	R^3	R^2	R
rR^3	rR^3	rR^4	r	rR	rR^2	R	I	R^4	R^3	R^2
rR^2	rR^2	rR^3	rR^4	r	rR	R^2	R	I	R^4	R^3
rR	rR	rR^2	rR^3	rR^4	r	R^3	R^2	R	I	R^4
r	r	rR	rR^2	rR^3	rR^4	R^4	R^3	R^2	R	I

Table 35.1
Operation table for D_5.

We then replace the elements in the D_5 table with the digits 0 to 9 (keeping the elements in the same order as in Table 35.1) to obtain Table 35.2:

Verhoeff's check digit scheme requires an ID number of the form $a_{n-1}a_{n-2}\cdots a_1a_0$ (note that the digits are indexed from right to left, starting with an index of 0) to satisfy the equation

$$\pi^0(a_0)\cdot\pi^1(a_1)\cdot\pi^2(a_2)\cdots\pi^{n-1}(a_{n-1}) = 0,$$

where

$$\pi = \begin{pmatrix} 0 & 1 & 2 & 3 & 4 & 5 & 6 & 7 & 8 & 9 \\ 1 & 5 & 7 & 6 & 2 & 8 & 3 & 0 & 9 & 4 \end{pmatrix}$$

[§] J. Verhoeff, "Error detecting decimal codes," *Mathematical Centre* Tract 29, The Mathematical Centre, Amsterdam, 1969.

$\cdot$	0	1	2	3	4	5	6	7	8	9
0	0	1	2	3	4	5	6	7	8	9
1	1	2	3	4	0	6	7	8	9	5
2	2	3	4	0	1	7	8	9	5	6
3	3	4	0	1	2	8	9	5	6	7
4	4	0	1	2	3	9	5	6	7	8
5	5	9	8	7	6	0	4	3	2	1
6	6	5	9	8	7	1	0	4	3	2
7	7	6	5	9	8	2	1	0	4	3
8	8	7	6	5	9	3	2	1	0	4
9	9	8	7	6	5	4	3	2	1	0

Table 35.2
Operation table for the Verhoeff check digit scheme.

is a permutation, $\pi^i = \pi \circ \pi \circ \cdots \circ \pi$ is the composition of i copies of π, and the $\cdot$ operation is that which arises from D_5 as indicated in Table 35.2. Note that the permutation applied to each digit depends on the position of the digit in the ID number. For example, if 4 is the digit in the third position from the right, then we apply π^2 to 4 to obtain 7. The result of applying the powers of π to any position can be described in the *permutation table* shown in Table 35.3, where the ith row shows the result of applying the permutation π^i to each possible digit. Note that the powers of π are periodic, so the rows repeat after row 8. In other words, $\pi^{i+8}(k) = \pi^i(k)$ for all k.

	0	1	2	3	4	5	6	7	8	9
π^0	0	1	2	3	4	5	6	7	8	9
π^1	1	5	7	6	2	8	3	0	9	4
π^2	5	8	0	3	7	9	6	1	4	2
π^3	8	9	1	6	0	4	3	5	2	7
π^4	9	4	5	3	1	2	6	8	7	0
π^5	4	2	8	6	5	7	3	9	0	1
π^6	2	7	9	3	8	0	6	4	1	5
π^7	7	0	4	6	9	1	3	2	5	8

Table 35.3
Permutation table for the Verhoeff check digit scheme.

Now that we understand some of the mechanics involved, the Verhoeff scheme can be implemented as follows:

(1) Start with the n digit number

$$a_{n-1}a_{n-2}a_{n-3}\cdots a_1a_0,$$

with the digits labeled from right to left, starting with a_0.

(2) Let c denote the *checksum*, and set $c = 0$ initially.

(3) Step through the n-digit number digit by digit, each time replacing c with $c \cdot \pi^i(a_i)$ (where the operation $\cdot$ is indicated in Table 35.2).

The original number has a valid check digit if and only if, at the end of the process, the checksum c is equal to 0.

As an example, Table 35.4 shows how the steps described above can be followed to validate the checksum for the ID number 4134705. Since the final value of c is 0, we have a valid ID number.

i	a_i	$\pi^i(a_i)$	Old c	New c: Old $c \cdot \pi^i(a_i)$
0	5	5	0	5
1	0	1	5	9
2	7	1	9	8
3	4	0	8	8
4	3	3	8	5
5	1	2	5	8
6	4	8	8	0

Table 35.4
A Verhoeff check digit example.

The Verhoeff algorithm detects all single digit errors and all transposition errors made by interchanging two adjacent digits. Also, the Verhoeff algorithm detects over 95% of twin errors (where aa is changed to bb), over 94% of jump transpositions (abc replaced with cba) and jump twin errors (aca replaced with bcb), and most phonetic errors ($a0$ replaced with $1a$—for example, 40 replaced with 14; notice how these two numbers sound similar when read aloud). Gallian¶ describes an interesting application in which the German government used a slight modification of the Verhoeff scheme to append check digits to serial numbers on their banknotes.

Activity 35.4. How do we find the check digit for an ID number using the Verhoeff scheme? Consider the ID number $1023857d$, where d is the check digit.

(a) Determine the value of the checksum c for the related ID number 10238570.

(b) Let $d = c^{-1}$. Show that $1023857d$ is a valid ID number. (As it turns out, this technique will always yield a correct check digit. Exercise 5 asks you to prove this result.)

Concluding Activities

Activity 35.5. Since 2007, books have been identified with 13 digit ISBNs (the ISBN-13). The first 12 digits of the ISBN-13 contain the identifying code and the 13th digit is the check digit. The

¶*Contemporary Abstract Algebra* (5th ed.), Houghton Mifflin Company, 2002.

check digit scheme is similar to the one used in ISBN-10 IDs. Research the ISBN-13 check digit scheme and explain how it works. Be sure to cite all of your sources in your explanation. Then find the check digit d for the ISBN-13 978-082183798d.

Activity 35.6. Write a short summary that describes the important concepts, and the relationships between these concepts, that were introduced in this investigation. Explain how the material in this investigation is connected to the material in Investigations 19, 20, and 24.

Exercises

(1) Determine if each of the following is a valid ID number. Assume that the last digit of each number is the check digit. If the check digit is not correct, fix it.

(a) 10513485, using the Luhn algorithm

(b) The ISBN-10 ID number 0-131-87718-4

(c) 2401346782, using the Verhoeff scheme.

(2) **Airline ticket ID numbers.** Airlines tickets have a 15-digit identification number. The first digit (reading left to right) is the coupon number and identifies the leg of the trip. (Coupon number 1 indicates the first flight in the trip, 2 the second, and so on, while coupon number 0 is the customer's receipt.) The next three digits identify the airline, and the next 10 digits comprise the document number, while the last digit is the check digit. Airlines use a simple mod 7 system to determine the check digit. If the airline ticket has digits

$$d_1d_2d_3d_4d_5d_6d_7d_8d_9d_{10}d_{11}d_{12}d_{13}d_{14}d_{15},$$

then the check digit d_{15} satisfies

$$d_{15} \equiv d_1d_2d_3d_4d_5d_6d_7d_8d_9d_{10}d_{11}d_{12}d_{13}d_{14} \pmod{7}.$$

(a) Verify that 1-101-2134601379-2 is a valid airline ticket. Use a computer algebra system if necessary.

(b) Calculating the remainder when dividing a 14-digit number by 7 is easy for a computer algebra system but a bit time-consuming for humans. Here we will investigate one method that can be done by hand: the method of *casting out sevens*. This method works by determining the remainders when dividing powers of 10 by 7.

(i) Determine a general formula for calculating $10^n \pmod{7}$ for nonnegative integers n.

(ii) Expand 11012134601379 in powers of 10, and then use your answer to part (a) to calculate the remainder when 11012134601379 is divided by 7.

(c) Another algorithm that can be used to determine if a large number is divisible by 7 is the following:

- Remove the last (rightmost) digit from the number.
- Subtract twice the value of the removed digit from the remaining number.
- Repeat until you can tell if the number obtained is divisible by 7.

(i) Apply this algorithm to show that 11012134601377 is divisible by 7. Then explain why the algorithm works.

(ii) Can a method similar to this one, where we remove the last digit and subtract some single-digit multiple of that digit from the new number, be used to test for divisibility by any other single digit integer? If so, find and explain all of the cases in which such a method can be used.

(3) In this exercise, we will examine the types of errors that are detected (or not detected) by the Luhn algorithm.

(a) Does the Luhn algorithm find all single digit errors? That is, if

$$a_{n-1} \cdots a_{i+1} a_i a_{i-1} \cdots a_0 d$$

and

$$a_{n-1} \cdots a_{i+1} b_i a_{i-1} \cdots a_0 d'$$

are valid IDs with $a_i \neq b_i$ and check digits d and d', must d and d' be different? If the answer is yes, prove it. If no, provide a counterexample.

(b) Will the Luhn algorithm detect all errors obtained by interchanging digits (e.g., typing 12 instead of 21)? If the answer is yes, prove it. If no, provide a counterexample.

(4) In this exercise, we will examine the types of errors that are detected (or not detected) by the ISBN-10 scheme.

(a) Show that the ISBN-10 scheme detects all single digit errors. (See part (a) of Exercise 3.)

(b) Show that the ISBN-10 scheme detects all errors obtained by interchanging digits (for example, typing 12 instead of 21).

(c) Another common family of errors are twin errors. One example of a twin error is changing aa to bb. Does the ISBN-10 scheme detect all twin errors? If the answer is yes, prove it. If no, provide a counterexample.

(d) Two other types of common errors are jump transposition errors (for example, abc replaced with cba) and jump twin errors (for example, aca replaced with bcb). Does the ISBN-10 scheme detect these errors? If the answer is yes, prove it. If no, provide a counterexample.

⋆ (5) Activity 35.4 provided a method for finding a correct check digit using the Verhoeff scheme. Prove that this method works in general. That is, show that if the original ID number ends in 0 and has checksum c, then c^{-1} (from Table 35.2) is the correct value to use as the check digit in place of the final 0.

(6) In this exercise, we will examine the types of errors that are detected (or not detected) by the Verhoeff scheme.

(a) Show that the Verhoeff scheme detects all single digit errors. (See part (a) of Exercise 3.)

(b) Complete the following steps to show that the Verhoeff scheme detects all errors obtained by interchanging digits (for example, typing 12 instead of 21).

(i) Show by direct calculation that if $a \neq b$ in D_5, then $a \cdot \pi(b) \neq \pi(a) \cdot b$.

(ii) Extend the result of the previous part to show that if $a \neq b$ in D_5, then $\pi^{i-1}(a)\pi^i(b) \neq \pi^i(a)\pi^{i-1}(b)$ for all $i \in \mathbb{Z}^+$.

(iii) Now show that the Verhoeff scheme detects all errors obtained by interchanging digits.

Connections

Abstract algebra has many practical applications, and this investigation considered one of these applications. In particular, we saw how congruence of integers (from Investigation 5) and a dihedral group (from Investigation 24) can be used to create check digit schemes. These check digit schemes help make transfers of information more reliable and are therefore important components of our electronic world.

Investigation 36

Games: NIM and the 15 Puzzle

Focus Questions

By the end of this investigation, you should be able to give precise and thorough answers to the questions listed below. You may want to keep these questions in mind to focus your thoughts as you complete the investigation.

- How is the game of NIM related to group theory? What is a good strategy for playing NIM?
- What is the 15 Puzzle, and how can the symmetric groups tell us if a 15 Puzzle is solvable?

Games can be fun to play—and, as it turns out, to study. In fact, many games involve mathematical ideas or can be analyzed using mathematics. In this investigation, we will learn how group theory can be used to determine winning strategies in the game of NIM and to determine if a 15 Puzzle is solvable.

Preview Activity 36.1. Go to any online version of NIM and play the game a few times. Search for a winning strategy.

The Game of NIM

To play the game of NIM, one begins with a number of sets, or stacks, of objects. We can think of these stacks as piles of stones, as shown in Table 36.1. (Note that we have displayed the piles horizontally to save space.) In this example, the first pile has 6 stones, the second has 2, and the third has 3.

•	•	•	•	•	•
•	•				
•	•	•			

Table 36.1
A NIM game.

The game is played by two players alternating turns. At each turn, a player can take as many stones from a single pile as he or she wants, but the player must remove at least one stone. The object is to be the last player to remove stones.

The number of stones in each pile in a NIM game is an element in the set $\mathbb{W}$ of whole numbers. So we can think of a particular state of a NIM game with three piles as an element in the Cartesian product $\mathbb{W} \times \mathbb{W} \times \mathbb{W}$. Recall that $\mathbb{W}$ is not a group under standard addition of integers. Therefore, to relate this game to group theory, we will need to define a relevant operation under which $\mathbb{W}$ is a group. As we will see, using binary representations of whole numbers will help us to define such an operation.

To study binary representations, it will be helpful to first review how the decimal representation of a whole number works. Recall that we typically think of the digits of a whole number as representing place value. For example, in the integer 1234, the digit 4 is located in the ones place, the digit 3 is in the 10's place, 2 is in the 100's place, and 1 is in the 1000's place. In other words, the integer 1234 can also be represented by the sum

$$(1 \times 10^3) + (2 \times 10^2) + (3 \times 10^1) + (4 \times 10^0).$$

This is the *decimal representation* of the number 1234. In the decimal system, we add two whole numbers digit-by-digit, from right to left, reducing modulo 10 and carrying a 1 to the next digit whenever the sum of two digits is 10 or greater.

There is nothing particularly special about the base 10 used in the decimal representation of a number, other than the fact that it is convenient. After all, most people have 10 fingers and 10 toes, so a base 10 system is natural. However, we could just as easily replace the base 10 with any other base. In the *binary* system, we replace the base 10 with the base 2. There is an adjustment we must make, though. With the decimal system, each individual digit can be anywhere between 0 and 9 (because these integers are less than 10). With the binary system, we will only use the digits 0 and 1. For example, the binary number 10110 represents the decimal number $(1 \times 2^4) + (0 \times 2^3) + (1 \times 2^2) + (1 \times 2^1) + (0 \times 2^0) = 22$. In fact, any whole number can be represented in binary format.

To represent a given whole number in the binary system, we first look for the highest power of 2 that is less than the number. Then we subtract this highest power of 2 and repeat the process with the difference. To illustrate, let's convert 219 to binary. First note that $2^7 = 128$ and $2^8 = 256$. So 2^7 is the highest power of 2 less than 219. Now $219 - 1 \times 2^7 = 91$, and so $219 = (1 \times 2^7) + 91$. Repeating the process with the integer 91, we note that the highest power of 2 in 91 is 2^6. Since $91 - 2^6 = 27$, we see that $219 = (1 \times 2^7) + (1 \times 2^6) + 27$. We then continue reducing the differences until we no longer have any powers of 2 remaining. This leaves us with

$$\begin{aligned} 219 = {} & (1 \times 2^7) + (1 \times 2^6) + (0 \times 2^5) + (1 \times 2^4) \\ & + (1 \times 2^3) + (0 \times 2^2) + (1 \times 2^1) + (1 \times 2^0). \end{aligned}$$

Therefore, the binary representation of 219 is 11011011. The standard sum of two whole numbers $a = a_n a_{n-1} a_{n-2} \cdots a_1 a_0$ and $b = b_n b_{n-1} b_{n-2} \cdots b_1 b_0$ in binary is similar to the decimal sum: we add digit-by-digit, from right to left, reducing modulo 2 and carrying a 1 to the next digit whenever the sum of two digits is 2 or greater.

Remember that $\mathbb{W}$ is not a group under standard addition of integers. If, however, we convert each whole number into its binary representation, then we can define a special operation under which $\mathbb{W}$ is a group. Let $x = x_n x_{n-1} \dots x_1 x_0$ and $y = y_n y_{n-1} \dots y_1 y_0$ be whole numbers in binary form. We can assume both integers have the same number of digits in their binary representations by simply appending zeros to the left end of one number if necessary. We define the "NIM sum" of x and y to be the binary number $x \oplus y = s_n s_{n-1} \dots s_1 s_0$, where $s_i = (x_i + y_i) \pmod 2$. Note that the NIM sum of two binary numbers is the same as the normal binary sum, except that we don't allow carrying. For example, the NIM sum $101101 \oplus 00111$ is 101010. Of course, we can also add more than two numbers this way. For example, $10111 \oplus 1110 \oplus 111 = 11110$.

Activity 36.2. Let $\mathcal{B}$ be the set of binary representations of the whole numbers.

(a) Is the NIM sum operation $\oplus$ well-defined in $\mathcal{B}$? Explain.

(b) Is $\mathcal{B}$ closed under the NIM sum $\oplus$? Explain.

(c) Is there an identity element in $\mathcal{B}$ with respect to the NIM sum $\oplus$? If yes, what is the identity? If no, why not?

(d) Does $\mathcal{B}$ contain an inverse for each of its elements with respect to the NIM sum $\oplus$? If yes, what is the inverse of a given element? If no, why not?

(e) Is the NIM sum associative in $\mathcal{B}$? Prove your answer.

(f) What conclusion can we draw about $\mathcal{B}$?

Now that we have an operation under which the whole numbers in binary form are a group, we can also make a group out of $\mathcal{B}^n = \underbrace{\mathcal{B} \oplus \mathcal{B} \oplus \cdots \oplus \mathcal{B}}_{n \text{ factors}}$. The group $\mathcal{B}^n$ forms the playing field for all NIM games with n piles of stones—that is, each element in $\mathcal{B}^n$ corresponds to a particular stage in a NIM game. As such, we will call any element in $\mathcal{B}^n$ a *configuration*. If the ith pile of stones contains N_i stones, then the NIM game has the configuration $(N_1, N_2, \ldots, N_n)$.

A *legal move* in a NIM game consists of removing some number of stones (at least one) from a single pile. Note that we can view a move as a configuration as well; in particular, the move that takes m stones from pile i can be thought of as the configuration $M = (0, 0, \ldots, 0, m, 0, \ldots, 0)$, with m in the ith component. Since every element in $\mathcal{B}^n$ is its own inverse, the result of performing move M on configuration X is the configuration $X \oplus M$. For example, let $X = (011, 100, 001)$ be the configuration in $\mathcal{B}^3$ with 3 stones in the first pile, 4 in the second, and 1 in the third, as shown on the left in Table 36.2. Let $M = (010, 000, 000)$ be the move that takes 2 stones from the first pile. The result of applying the move M to the configuration X is the configuration $X \oplus M = (001, 100, 001)$, as shown on the right in Table 36.2.

Table 36.2
A NIM move.

In any NIM game, the last move will result in the configuration $(0, 0, \ldots, 0)$. This configuration has the special property that the NIM sum of all of the components is 0. In general, a configuration that satisfies this property is called an *even* configuration—that is, $(N_1, N_2, \ldots, N_n)$ is an *even* configuration if

$$\bigoplus_{i=1}^{n} N_i = 0.$$

Any other configuration is called an *odd* configuration. Note that every move is an odd configuration.

Activity 36.3. Let n be a positive integer, and let $\mathcal{E}^n$ be the subset of $\mathcal{B}^n$ consisting of the even configurations. Is $\mathcal{E}^n$ a subgroup of $\mathcal{B}^n$? Prove your answer.

We will now explore some strategy behind the game of NIM.

Activity 36.4.

(a) If $X \in \mathcal{E}^n$ is not the identity, how many nonzero components must X have?

(b) If $X \in \mathcal{E}^n$ is not the identity, how many nonzero components must its inverse have? Explain what this observation tells us about the possibility of winning a NIM game from a nonzero even configuration.

The result of Activity 36.4 reveals a strategy for playing NIM defensively. In particular, if we can always present our opponent with an even configuration, then he or she cannot win. The question now is how that can be done.

Activity 36.5. Let n be a positive integer, and let $X \in \mathcal{E}^n$. Determine and describe all moves M so that $(X \oplus M) \in \mathcal{E}^n$. Relate your answer to playing the game of NIM.

Activity 36.5 tells us that if we present our opponent with an even configuration, we will always be confronted with an odd configuration on our next turn. So the final question is whether we can convert an odd configuration into an even one. This is a bit more complicated, and so we will describe the process using a NIM game with three piles.

Let $X = (N_1, N_2, N_3)$ be an odd configuration in $\mathcal{B}^3$. We want to find a move $M = (M_1, M_2, M_3)$ with exactly one of M_1, M_2, M_3 nonzero so that $X \oplus M$ is even. (It is not true that the NIM sum of two odd configurations is even, and you should find a simple example to convince yourself of this.) Let $N_1 = a_m a_{m-1} \ldots a_1 a_0$, $N_2 = b_m b_{m-1} \ldots b_1 b_0$, and $N_3 = c_m c_{m-1} \ldots c_1 c_0$ (all in binary). Since $N_1 \oplus N_2 \oplus N_3 \neq 0$, there is some index i so that $a_i + b_i + c_i \equiv 1 \pmod 2$. Let k be the largest index for which this happens. At least one of a_k, b_k, c_k must be 1. Without loss of generality, assume $a_k = 1$. This means that $b_k + c_k \equiv 0 \pmod 2$. For $0 \leq i < k$, let

$$a_i' = \begin{cases} 0, & \text{if } (b_i + c_i) \equiv 0 \pmod 2 \\ 1, & \text{otherwise.} \end{cases}$$

So $a_i' + b_i + c_i \equiv 0 \pmod 2$ for $i < k$. Let M be the move that takes stones from pile 1 so that $N_1' = a_m a_{m-1} \ldots a_{k+1} 0 a_{k-1}' a_{k-2}' \ldots a_1' a_0'$ remain. The result of applying move M to X is the configuration $X' = (X \oplus M) = (N_1', N_2, N_3)$. Recall that $(b_k + c_k) \equiv 0 \pmod 2$, and $(a_i + b_i + c_i) \equiv 0 \pmod 2$ if $i > k$ (by the definition of k). Since $a_i' + b_i + c_i \equiv 0 \pmod 2$ for $i < k$, we can therefore conclude that $N_1' \oplus N_2 \oplus N_3 = 0$, and $X' \in \mathcal{E}^n$.

In terms of the NIM game, this result tells us that if we are confronted with an odd configuration, we can always change it to an even configuration.

Activity 36.6. Apply the algorithm provided above to find a move that converts the NIM configuration in Table 36.3 to an even configuration.

Based on our work up to this point, we have proved the following theorem (with $n = 3$ for part (iii), but you are asked to extend this to $\mathcal{B}^n$ for any n in Exercise 2).

Theorem 36.7. *Let n be a positive integer.*

(i) *If $X \in \mathcal{E}^n$ is nonzero, then there is no move M so that $X \oplus M = 0$.*

Table 36.3
A NIM configuration.

(ii) *If* $X \in \mathcal{E}^n$ *is nonzero, then* $(X \oplus M) \notin \mathcal{E}^n$ *for any move* M.

(iii) *If* $X \in (\mathcal{B}^n \setminus \mathcal{E}^n)$*, then there exists a move* M *so that* $(X \oplus M) \in \mathcal{E}^n$.

Interpreted in more natural language, Theorem 36.7 presents us with the following strategy for playing the NIM game:

(i) Since it is not possible to win from a nonzero even configuration, always present our opponent with an even configuration if possible.

(ii) If we can present our opponent with an even configuration, any move our opponent makes will present us with an odd configuration.

(iii) If we have an odd configuration at our turn, we can always turn it into an even configuration.

So our strategy is to always present our opponent with an even configuration. If the configuration is the 0 configuration, we have won. If not, then (ii) shows that our opponent cannot win because he or she will be forced to present us with an odd configuration. By part (iii), we can turn that odd configuration into an even configuration so that our opponent cannot win on the next turn. Since at least one stone is removed at each turn, our opponent will eventually have to present us with an odd configuration from which we can win.

One final note: if we are ever presented with even configuration, we cannot win the game unless our opponent makes a mistake. For this reason, it is always advantageous to be able to decide whether to move first or second after seeing the initial configuration.

We will illustrate the strategy described above with the game from Table 36.3.

Table 36.4
Our first move (left), our opponent's move (middle), and the coup de gras (right).

As we argued before, if we move first, we should remove stones from pile 1 to leave exactly 2 stones as shown on the left of Table 36.4. Now, whatever move our opponent makes, we will be left with an odd configuration. Suppose our opponent's move leaves us with the configuration shown in the middle of Table 36.4. This configuration gives us $N_1 = 2 = 10$, $N_2 = 4 = 100$, and $N_3 = 4 = 100$. Now $10 \oplus 100 \oplus 100 = 010$, so we must change the 1 in the 2's place. We can do this by removing all of the stones from pile 1, leaving us with the NIM sum $100 \oplus 100 = 000$, which corresponds to the configuration on the right of Table 36.4. Whatever moves our opponent makes, we can now win the game by keeping both piles of the same size. This will always result in a 0 NIM sum. Thus, the game is ours.

The 15 Puzzle

Preview Activity 36.8. Go to any on-line version of the 15 Puzzle and play the game a few times. Find one that allows you to create your own 15 Puzzle. Are all 15 Puzzles solvable? If not, search for a pattern that determines which puzzles are solvable.

The classic 15 Puzzle was made famous in the 19th century by puzzleist Sam Loyd.* The game consists of a starting position, which is a 4×4 array of the integers between 1 and 15 along with a symbol # (which we interpret as a blank space), as shown in Table 36.5. We will call each entry of the array a *cell*.

2	9	7	3
10	15	12	8
1	4	#	14
6	13	5	11

Table 36.5
15 Puzzle: Configuration 1.

The game is played with one type of legal move: interchanging the blank cell with a cell either to the left or right or the cell above or below. (The children's game is usually made of sliding tiles that are numbered 1 to 15. The interchange mentioned here is done by sliding a tile to the empty cell.) In this example, we can interchange the blank with the 12, 4, 14 or 5. Interchanging the blank and the 5 leaves us with the configuration in Table 36.6. We can then interchange the blank with the 5, 13, or 11.

2	9	7	3
10	15	12	8
1	4	5	14
6	13	#	11

Table 36.6
15 Puzzle: Configuration 2.

The object of the game is to interchange the blank with other cells and transform the starting

*In 2006, Jerry Slocum and Dic Sonneveld published their book, *The 15 Puzzle* (Slocum Puzzle Foundation), in which they write, "Sam Loyd did not invent the 15 puzzle and had nothing to do with promoting or popularizing it. The puzzle craze that was created by the 15 Puzzle began in January 1880 in the US and in April in Europe. The craze ended by July 1880 and Sam Loyd's first article about the puzzle was not published until sixteen years later, January 1896. Loyd first claimed in 1891 that he invented the puzzle, and he continued until his death a 20 year campaign to falsely take credit for the puzzle. The actual inventor was Noyes Chapman, the Postmaster of Canastota, New York, and he applied for a patent in March 1880."

position to the standard position in Table 36.7. It is important to note that each interchange is reversible. So an equivalent game is to begin with the standard position and move to obtain a specified position. It is this latter version of the game that we will analyze.

1	2	3	4
5	6	7	8
9	10	11	12
13	14	15	#

Table 36.7
15 Puzzle: Standard configuration.

Permutations and the 15 Puzzle

To analyze this game, we will construct a correspondence between possible configurations of the 4×4 array and elements in the symmetric group S_{16}. To do this, we will let the number 16 represent the blank cell. Recall that any permutation can be written as a product of transpositions. We will apply a transposition $(a\ b)$ to a configuration A by interchanging the labels of the cells in *positions* a and b in A. (Note that this may be different than interchanging the cells *labeled* a and b.) For example, the transposition $(12\ 16)$ applied to the standard configuration I will correspond to Table 36.8 in which the cells in positions 12 and 16 have been interchanged.

1	2	3	4
5	6	7	8
9	10	11	#
13	14	15	12

Table 36.8
The transposition $(12\ 16)$ applied to I.

To apply a permutation $\sigma \in S_{16}$ to a configuration A, we can first write σ as a product $\tau_m \tau_{m-1} \cdots \tau_2 \tau_1$ of transpositions and then apply the transpositions, in order, to A. The resulting configuration is denoted by $\sigma(A)$. It is important to note that only certain permutations in S_{16} can be applied to a given configuration. In particular, the transpositions involved must always interchange two adjacent cells. So, for example, while we could apply the transposition $(12\ 16)$ to the standard configuration I (as we did in Table 36.8), we could not apply the transposition $(12\ 13)$ to I, since cells 12 and 13 are not adjacent. For convenience, we will refer to permutations in S_{16} that can be applied to I as *valid* permutations.

Activity 36.9. Find a permutation that converts the standard configuration to the configuration shown in Table 36.9.

We can construct a one-to-one correspondence between possible configurations and valid elements in S_{16} by assigning to each valid $\sigma \in S_{16}$ the configuration $\sigma(I)$. Because of this correspon-

1	#	2	4
5	6	3	8
9	10	7	12
13	14	11	15

Table 36.9
A target configuration.

dence, we will from this point on refer to valid elements of S_{16} as configurations, and vice versa. Note that the standard configuration is represented by the identity permutation.

Solving the 15 Puzzle

To determine if a 15 Puzzle is solvable, we are interested in answering the following equivalent questions:

- From which initial configuration can we obtain the standard configuration?
- Which configurations can be obtained starting from the standard configuration?

We will answer the second of these questions and, consequently, obtain the answer to the first question at the same time. To make our work a little easier, first observe that any configuration can be reduced to one in which the blank square is at location 16.

Activity 36.10. As an example, let A be the configuration shown in Table 36.5. Find a permutation σ for which $\sigma(A)$ produces the configuration shown in Table 36.10.

2	9	7	3
10	15	12	8
1	4	14	11
6	13	5	#

Table 36.10
Moving the blank to cell 16.

With this in mind, we need only to determine the configurations that have the blank in position 16 and can be obtained from the standard position I. The next theorem, which we will prove throughout the remainder of this investigation, tells us exactly which configurations meet these conditions.

Theorem 36.11. *Let H be the subset of S_{16} corresponding to all configurations that have the blank in position 16 and can be obtained from the standard position I. Then $H = A_{15}$.*

Note that even though S_{15} is not a subset of S_{16}, we can consider S_{15} to be contained in S_{16} as the set of all permutations that fix 16.

Activity 36.12.

(a) Explain how Theorem 36.11 tells us exactly which 15 Puzzles are solvable.

(b) Let σ_0 represent the configuration shown in Table 36.5. Can we find a sequence of allowable moves to transform σ_0 to I? Why or why not?

It is probably not surprising that H, the set of all permutations with the blank in position 16 that can be obtained from I, is a subgroup of S_{16}. This is left for you to prove in Exercise 4. It is also the case that every element in H is an even permutation. (See Exercise 5.) Thus, $H \subseteq A_{15}$. To complete the proof of Theorem 36.11, we need to show that $A_{15} \subseteq H$. Two facts will be useful in our argument:

- Lemma 27.16 (see page 368) shows that for $n \geq 3$, any permutation in A_n can be written as a product of 3-cycles.
- Exercise 16 of Investigation 25 (see page 344) shows that for any $\alpha \in S_{16}$, any $x, y, z \in \{1, 2, 3, \ldots, 16\}$, and any $n \in \mathbb{Z}^+$, we have

$$\alpha^n(x\ y\ z)\alpha^{-n} = (\alpha^n(x)\ \alpha^n(y)\ \alpha^n(z)). \tag{36.1}$$

Now we can show that H contains every 3-cycle of the form $(a\ b\ c)$ for $a, b, c \in S = \{1, 2, 3, \ldots, 15\}$. Since A_{15} is generated by three cycles, we will be able to conclude that $H = A_{15}$, as desired.

Lemma 36.13. *The group H contains every 3-cycle of the form $(a\ b\ c)$ for $a, b, c \in \{1, 2, 3, \ldots, 15\}$.*

Proof. Let $(a\ b\ c)$ be a 3-cycle with $a, b, c \in S = \{1, 2, 3, \ldots, 15\}$. If $\alpha \in S_{16}$ is in H and $(a\ b\ c) \in H$, then equation (36.1) shows that

$$\alpha^n(a\,b\,c)\alpha^{-n} = (\alpha^n(a)\ \alpha^n(b)\ \alpha^n(c)) \in H \tag{36.2}$$

for any $n \in \mathbb{Z}^+$.

To complete the proof of Lemma 36.13, we will apply this idea to some specific elements in H, as indicated in the next activity, to show the following:

(1) Every 3-cycle of the form $(11\ 7\ b)$ is in H, where $b \in S$ and $b \neq 7, 11, 16$.

(2) Every 3-cycle of the form $(a\ b\ 11)$ is in H, where $a, b \in S$, $a \neq 11, 16$, and $b \neq 7, 11, 16$.

(3) Every 3-cycle is in H.

Once we have established these facts, we will have proved the lemma. ■

Activity 36.14.

(a) Let $b \in S$ with $b \neq 7, 11, 16$. Here we will show that every 3-cycle of the form $(11\ 7\ b)$ is in H.

(i) Explain why

$$\alpha_1 = (11\ 15\ 12) = (16\ 12)(12\ 11)(11\ 15)(15\ 16)$$

(as shown in Table 36.11) is in H.

1	2	3	4
5	6	7	8
9	10	12	15
13	14	11	#

Table 36.11
α_1.

1	2	3	4
5	7	8	12
9	6	11	15
13	10	14	#

Table 36.12
α_2.

(ii) Explain why

$$\begin{aligned}\alpha_2 &= (16\ 12)(12\ 8)(8\ 7)(7\ 6)(6\ 10)(10\ 14)(14\ 15)(15\ 16)\\ &= (6\ 10\ 14\ 15\ 12\ 8\ 7)\end{aligned}$$

(as shown in Table 36.12) is an element of H. Then use equation (36.2) to show that $(11\ 7\ 6)$ is an element of H.

(iii) Next, note that the element α_3 defined by

$$\begin{aligned}\alpha_3 &= (16\ 12)(12\ 8)(8\ 4)(4\ 3)(3\ 2)(2\ 1)(1\ 5)(5\ 6)\\ &\qquad (6\ 10)(10\ 9)(9\ 13)(13\ 14)(14\ 15)(15\ 16)\\ &= (1\ 5\ 6\ 10\ 4\ 13\ 14\ 15\ 12\ 8\ 4\ 3\ 2)\end{aligned}$$

(as shown in Table 36.13) is another element of H.

2	3	4	8
1	5	7	12
10	6	11	15
9	13	14	#

Table 36.13
α_3.

Use equation (36.2) and the fact that α_3 fixes 7, 11, and 16 to complete the argument that every 3-cycle of the form $(11\ 7\ b)$, where $b \in S$ and $b \neq 7, 11, 16$, is in H.

(b) Let $a, b \in S$ with $a \neq 11, 16$ and $b \neq 7, 11, 16$. Show that every 3-cycle of the form $(a\ b\ 11)$ is in H. (Hint: Why is $(11\ b\ 7) \in H$?)

(c) Finally, show that every 3-cycle is in H. How does this show that $H = A_{15}$?

The work we did in Activity 36.14 completes our proof of Lemma 36.13. To summarize, we can determine if a particular 15 Puzzle is solvable by first transforming it to a puzzle A with the blank in position 16. Then if there is an even permutation σ so that $\sigma(I) = A$, we can conclude that our original 15 Puzzle is solvable.

Activity 36.15. Find a 15 Puzzle that is not solvable and that is also not easily seen to be unsolvable. Explain how you know your puzzle is not solvable.

One final note: Our analysis of the 15 Puzzle completely classifies which games can be won but does not tell us how to win. There are strategies for winning, but trial and error is often the best bet.

Concluding Activities

Activity 36.16. Go to any online site that has a NIM game and play it using the strategies we have described in this activity. Make sure you win!

Activity 36.17. Write a short summary that describes the important concepts, and the relationships between these concepts, that were introduced in this investigation. Explain how the material in this investigation is connected to the material in Investigations 20, 22, 25, and 27.

Exercises

(1) Let m be a positive integer, and let $\mathcal{B}_m$ be the set of whole numbers, in binary, that are less than 2^m.

(a) Show that $\mathcal{B}_m$ is a subgroup of $\mathcal{B}$.

(b) As a finite Abelian group, the group $\mathcal{B}_m$ is isomorphic to a direct sum of cyclic groups. To what familiar direct sum is $\mathcal{B}_m$ isomorphic?

$\star$ (2) Let n be a positive integer. Show that if $X \notin \mathcal{E}^n$, then there is a move M so that $(X + M) \in \mathcal{E}^n$.

(3) There is a story that when Sam Loyd first distributed his game, the configuration of the puzzle was the standard configuration but with the 14 and 15 pieces in reversed position. Loyd offered a prize of 1000 dollars for a correct solution to that puzzle. Did he ever pay the 1000 dollar prize? Explain.

$\star$ (4) Let H be the subset of S_{16} consisting of all permutations with the blank in position 16 that can be obtained from the standard configuration I. Prove that H is a subgroup of S_{16}.

$\star$ (5) Prove that if σ is in H, the set of all permutations with the blank in position 16 that can be obtained from I, then σ is an even permutation.

Connections

Abstract algebra has many applications—some serious, and some more fun. In this investigation, we saw how group theory could be used to analyze two games: NIM and the 15 Puzzle. The structure of the game of NIM is related to the group $\mathcal{B}^n$ and its subgroup $\mathcal{E}^n$ of the even configurations, while the symmetric groups are important in determining which 15 Puzzles can be solved.

Investigation 37

Finite Fields, the Group of Units in $\mathbb{Z}_n$, and Splitting Fields

Focus Questions

By the end of this investigation, you should be able to give precise and thorough answers to the questions listed below. You may want to keep these questions in mind to focus your thoughts as you complete the investigation.

- What is a finite field? How many elements can a finite field have?
- How are any two finite fields of the same order related?
- What is the structure of the group U_n of units in $\mathbb{Z}_n$?
- What is the splitting field of a polynomial? Why do we refer to "the" splitting field instead of "a" splitting field?
- How can one construct the splitting field of a polynomial?

Preview Activity 37.1. Let $p(x) = x^2 + x + [1]$ in $\mathbb{Z}_2[x]$.

(a) Find all of the elements of $K = \mathbb{Z}_2[x]/p(x)$.

(b) Create the addition and multiplication tables for K.

(c) What special kind of ring is K? Explain.

(d) What kind of group is $U(K)$, the group of units in K? Explain.

(e) Explain why K contains all of the roots of the polynomial $f(x) = x^4 - x$ over $\mathbb{Z}_2$. In fact, show that every element of K is a root of $f(x) = x^4 - x$ over $\mathbb{Z}_2$.

Introduction

The structure of the groups U_n, which consist of the units in $\mathbb{Z}_n$, is not obvious. However, with a bit of work we can determine the structure of these groups. To do so, we will first show that the group U_p is cyclic when p is prime. We will then use this result to describe every group U_n, whether n

is prime or not. In the process, we will classify all finite fields and also explore splitting fields of polynomials.

Finite Fields

When p is a prime, the ring $\mathbb{Z}_p$ is a field. The field $\mathbb{Z}_p$ is different than the more familiar fields like $\mathbb{Q}$ and $\mathbb{R}$ in that $\mathbb{Z}_p$ contains only a finite number of elements. The first question we will address in this investigation is whether there are finite fields other than $\mathbb{Z}_p$ for p prime. Activity 37.1 shows that there are in fact such fields by demonstrating the existence of a field with 4 elements. As we will see, we can actually explicitly determine all of the finite fields.

Activity 37.2. Let F be a finite field.

(a) Every field has a characteristic. Explain why the characteristic of F must be prime. (Hint: In Exercise 7 on page 103 of Investigation 8, we showed that the characteristic of an integral domain is either zero or prime.)

(b) Explain why F must contain a copy of $\mathbb{Z}_p$, and so, without loss of generality, we can consider F to be an extension of $\mathbb{Z}_p$.

(c) Since F is finite, F must be a finite dimensional vector space over $\mathbb{Z}_p$.* Let $\{x_1, x_2, \ldots, x_n\}$ be a basis for F over $\mathbb{Z}_p$. Explain why F contains p^n elements.

The result of Activity 37.2 is that the order of any finite field is a power of a prime. We can arrive at this result in another way by observing that if F is a field of prime characteristic p, then as an Abelian group under addition, every nonzero element has order p. Thus, by the Fundamental Theorem of Finite Abelian Groups (see page 440), F must be a p-group, and so $|F| = p^n$ for some integer n.

We now have the conditional statement that if F is a finite field, then the order of F is a power of a prime. It is natural to ask if the converse is true—that is, given a prime p and a positive integer n, is there a field of order p^n? This question is a little more complicated.

Let p be a prime, and let n be a positive integer. In Activity 37.1, we saw that the field F whose elements are the roots of the polynomial $f(x) = x^4 - x$ over $\mathbb{Z}_2$ is a field of order 4. As we will see, this construction works in general. If F is a finite field of order p^n, then the group $U(F)$ of units in F has order $p^n - 1$. So every nonzero element in F satisfies the polynomial equation $x^{p^n-1} - 1 = 0$. If we toss in the 0 element, then every element in F is a root of the polynomial $x^{p^n} - x$. What we will ultimately show is that the set of all roots of the polynomial $x^{p^n} - x$ over $\mathbb{Z}_p$ is actually a field and has order p^n. Let $m = p^n$, and let $f(x) = x^m - x \in \mathbb{Z}_p[x]$. Kronecker's Theorem (see page 204) tells us that there is an extension field K of $\mathbb{Z}_p$ in which $f(x)$ *splits* into a product of linear factors. In other words, in $K[x]$, we have

$$f(x) = c(x - r_1)(x - r_2)\cdots(x - r_m)$$

for some c, r_1, r_2, …, r_m in K.

We will show that the set $A = \{k \in K : f(k) = 0\}$ of roots of $f(x)$ in K is a subfield of K of

*If F is a field and K is a subfield of F, then F can be viewed as a vector space over K by simply taking scalar multiplication to be the same as multiplication within F, restricted to the set K of scalars.

order p^n, thus demonstrating that there is a field of order p^n. To do this, we first need to show that the set of roots of $f(x)$ is a subfield of K.

Activity 37.3. Let s and t be two roots of $f(x)$.

(a) Recall the Freshman's Dream from Exercise 10 of Investigation 8 (see page 103), which states that $(a+b)^p = a^p + b^p$ whenever a and b are elements of a commutative ring with prime characteristic p. A similar result is that $(a \pm b)^m = a^m \pm b^m$ for all $a \in K$. (See Exercise 4.) Use this fact to show that

$$f(s-t) = (s^m - s) - (t^m - t).$$

(b) Use the fact that s and t are roots of $f(x)$ to show that $s-t$ is a root of $f(x)$.

(c) Now assume that $t \neq 0$. Show that st^{-1} is a root of $f(x)$.

(d) Let $A = \{k \in K : f(k) = 0\}$. Explain why A is a subfield of K.

Activity 37.3 shows that the set A of roots of $f(x) = x^{p^n} - x$ in K is a subfield of K. By definition, all of the elements of A are roots of $f(x)$. What remains is for us to determine the order of A. We know that A contains only $r_1, r_2, \ldots, r_m$, the roots of $f(x)$ in K. This means that $|A| \leq m = p^n$, but this inequality could be strict if $f(x)$ has a repeated root. So to show that $|A| = p^n$, we need to demonstrate that $f(x)$ has no repeated root in K. To do so, let r be a root of $f(x)$ in K. We need to show that $(x-r)^2$ does not divide $f(x)$. We will begin by rewriting $f(x)$ as follows:

$$\begin{aligned} f(x-r) &= (x-r)^m - (x-r) \\ &= x^m - r^m - x + r \\ &= x^m - r - x + r \\ &= x^m - x \\ &= f(x). \end{aligned}$$

Therefore,

$$f(x) = f(x-r) = (x-r)^m - (x-r) = (x-r)\left[(x-r)^{m-1} - 1\right].$$

But $(x-r)$ does not divide $(x-r)^{m-1} - 1$ (do you see why?) and so $(x-r)^2$ cannot divide $f(x)$. Thus, r is a root of $f(x)$ of multiplicity 1. Since our choice of r was arbitrary, we have shown that $f(x)$ does not have any repeated roots in K. Therefore, $|A| = p^n$, and A is the desired field.

We have now demonstrated that every finite field of characteristic p has order p^n for some n. We have also shown that there is a finite field of order p^n for each prime p and each positive integer n. The next theorem summarizes these results.

Theorem 37.4. *Every finite field has order p^n for some prime p and some positive integer n. Moreover, if p is a prime and n is a positive integer, then the set of all roots of the polynomial $x^{p^n} - x$ over $\mathbb{Z}_p$ is a field with p^n elements.*

One important question remains: How many finite fields are there of order p^n? We will answer this question later on in this investigation. In the meantime, we will return to one of the problems posed at the start of this investigation—namely, what can we say about the group U_p of units in $\mathbb{Z}_p$, where p is prime?

The Group of Units of a Finite Field

Preview Activity 37.5. Recall that $U_n = U(\mathbb{Z}_n)$ is the group of units in the ring $\mathbb{Z}_n$. There is a common structure to the groups U_p, where p is prime. Determine the structure of the groups U_2, U_3, U_5, U_7, and U_{11}. What property do they all have in common? Is this property satisfied even when p is not prime?

Activity 37.5 indicates that there may be something special about the group of units in $\mathbb{Z}_p$, where p is prime. In fact, the observations we made about the units in $\mathbb{Z}_p$ are consequences of a more general result—namely, that if F is a finite field of order p^n for some prime p, then F contains an element with multiplicative order $p^n - 1$. A tool we will need to prove this result is the somewhat technical next lemma.

Lemma 37.6. *Let G be a group, and let $a, b \in G$ such that $ab = ba$. If $|a| = k$ and $|b| = n$, then there is an element $c \in G$ with $|c| = lcm(k, n)$.*

A note about Lemma 37.6: It is tempting to think that if $ab = ba$ in a finite group, then $|ab| = \text{lcm}(|a|, |b|)$. Unfortunately, this is not true. For example, $|[4]| = 6$ and $|[16]| = 3$ in U_{35}, but $|[29]| = 2$. So there is some work to do to prove this lemma.

Proof of Lemma 37.6. Let G be a group, and let $a, b \in G$ such that $ab = ba$. Assume $|a| = k$ and $|b| = n$. Let $d = \gcd(k, n)$. Furthermore, let $p_1, p_2, \ldots p_m$ be the distinct primes that appear in either of the prime factorizations of k and n, and let

$$k = p_1^{\alpha_1} p_2^{\alpha_2} \cdots p_m^{\alpha_m} \quad \text{and} \quad n = p_1^{\beta_1} p_2^{\beta_2} \cdots p_m^{\beta_m},$$

where α_i and β_i are nonnegative integers for each i. Then

$$d = p_1^{\gamma_1} p_2^{\gamma_2} \cdots p_m^{\gamma_m},$$

where $\gamma_i = \min\{\alpha_i, \beta_i\}$ for each i. Break d into factors d_1 and d_2 as follows:

$$d_1 = p_1^{\sigma_1} p_2^{\sigma_2} \cdots p_m^{\sigma_m} \quad \text{where } \sigma_i = \begin{cases} \gamma_i & \text{if } \gamma_i = \alpha_i \text{ and } \alpha_i \neq 0 \\ 0 & \text{otherwise} \end{cases}$$

and

$$d_2 = p_1^{\mu_1} p_2^{\mu_2} \cdots p_m^{\mu_m} \quad \text{where } \mu_i = \begin{cases} \gamma_i & \text{if } \gamma_i = \beta_i \text{ and } \beta_i \neq \alpha_i \\ 0 & \text{otherwise.} \end{cases}$$

To see why $d = d_1 d_2$, we will examine $\sigma_i + \mu_i$. Consider two cases:

- If $\alpha_i \leq \beta_i$, then $\gamma_i = \alpha_i$ and $\sigma_i = \gamma_i$. So $\mu_i = 0$ and $\sigma_i + \mu_i = \gamma_i + 0 = \gamma_i$.
- If $\alpha_i > \beta_i$, then $\gamma_i = \beta_i$. So $\sigma_i = 0$ and $\mu_i = \gamma_i$. Again, $\sigma_i + \mu_i = 0 + \gamma_i = \gamma_i$.

It follows that

$$d_1 d_2 = p_1^{\sigma_1+\mu_1} p_2^{\sigma_2+\mu_2} \cdots p_m^{\sigma_m+\mu_m} = p_1^{\gamma_1} p_2^{\gamma_2} \cdots p_m^{\gamma_m} = d.$$

Next, we will compute $|a^{d_1}|$ and $|b^{d_2}|$. First note that

$$|a^{d_1}| = \frac{k}{\gcd(k, d_1)} = \frac{k}{d_1} = p_1^{\alpha_1 - \sigma_1} p_2^{\alpha_2 - \sigma_2} \cdots p_m^{\alpha_m - \sigma_m}$$

and

$$|b^{d_2}| = \frac{n}{\gcd(n, d_2)} = \frac{n}{d_2} = p_1^{\beta_1-\mu_1} p_2^{\beta_2-\mu_2} \cdots p_m^{\beta_m-\mu_m}.$$

For each i, if $\alpha_i - \sigma_i > 0$, then $\alpha_i > 0$ and we must have $\sigma_i = 0$. Then $\gamma_i = \beta_i = \mu_i$, and so $\beta_i - \mu_i = 0$. Thus, $|a^{d_1}|$ and $|b^{d_2}|$ have no prime factors in common, and $\gcd(|a^{d_1}|, |b^{d_2}|) = 1$. Exercise 18 of Investigation 23 (see page 324; this result states that if a and b are elements in a group G with $ab = ba$ and $\gcd(|a|, |b|) = 1$, then $|ab| = |a||b|$) now implies that

$$\begin{aligned}
|a^{d_1} b^{d_2}| &= |a^{d_1}||b^{d_2}| \\
&= (p_1^{\alpha_1-\sigma_1} p_2^{\alpha_2-\sigma_2} \cdots p_m^{\alpha_m-\sigma_m})(p_1^{\beta_1-\mu_1} p_2^{\beta_2-\mu_2} \cdots p_m^{\beta_m-\mu_m}) \\
&= p_1^{(\alpha_1+\beta_1)-(\sigma_1+\mu_1)} p_2^{(\alpha_2+\beta_2)-(\sigma_2+\mu_2)} \cdots p_m^{(\alpha_m+\beta_m)-(\sigma_m+\mu_m)} \\
&= \frac{p_1^{(\alpha_1+\beta_1)} p_2^{(\alpha_2+\beta_2)} \cdots p_m^{(\alpha_m+\beta_m)}}{p_1^{(\sigma_1+\mu_1)} p_2^{(\sigma_2+\mu_2)} \cdots p_m^{(\sigma_m+\mu_m)}} \\
&= \frac{kn}{p_1^{\gamma_1} p_2^{\gamma_2} \cdots p_m^{\gamma_m}} \\
&= \frac{kn}{d} \\
&= \operatorname{lcm}(k, n).
\end{aligned}$$

Therefore, we have found an element in G of order $\operatorname{lcm}(k, n)$, as desired. ■

The result of Lemma 37.6 is that, given commuting elements a and b of orders k and n in a group G, we will always be able to find an element in G of order $\operatorname{lcm}(a, b)$. We will use this fact in the next activity to prove that the group of units of any finite field is a cyclic group. Since $\mathbb{Z}_p$ is a field when p is prime, it naturally follows that the groups U_p are cyclic when p is prime.

Activity 37.7. Let F be a finite field, and let $G = U(F) = \{x \in F : x \neq 0_F\}$ be the group of units in F. Since F is finite, we can enumerate the elements in G. Let $G = \{a_1, a_2, \ldots, a_m\}$, and also let $|a_i| = n_i$ for each i.

(a) Explain why we can find an element $b_1 \in G$ with $|b_1| = \operatorname{lcm}(n_1, n_2)$.

(b) How can we show that for each j with $1 \leq j < m$, there exists an element b_j in G such that $|b_j| = \operatorname{lcm}(n_1, n_2, \ldots, n_{j+1})$?

(c) Let $k = |b_{m-1}|$. How is k related to each n_i? How is k related to $|G|$?

(d) What can we say about a_i^k for each i?

(e) How many roots can the polynomial $p(x) = x^k - 1$ in $F[x]$ have? What does this tell us about the relationship between k and m? Explain.

(f) Explain how we have proved the following theorem.

Theorem 37.8. *Let F be a finite field. The group $G = U(F)$ of units in F is a cyclic group.*

Now that we have proved Theorem 37.8, we can determine the groups U_n of the units in $\mathbb{Z}_n$, as we will see in the next section.

The Group of Units of $\mathbb{Z}_n$

Let n be a positive integer. The group U_n of units of $\mathbb{Z}_n$ is a finite Abelian group. The Fundamental Theorem of Finite Abelian Groups (see page 440) tells us that U_n must be isomorphic to a direct sum of finite cyclic groups. Recall that to find the summands in such a direct sum, we broke up the group in question into an internal direct product of its p-primary components. We then decomposed each p-primary component into internal direct products of cyclic subgroups of largest order. Even without knowing the order of U_n directly, we can do something similar here.

First, the result of Exercise 11 tells us that if $m = n_1 n_2 \cdots n_k$, where n_i and n_j are relatively prime for $i \neq j$, then

$$U_m \cong (U_{n_1} \oplus U_{n_2} \oplus \cdots \oplus U_{n_k}).$$

So once we decompose m into a product of primes, we have reduced our problem to that of determining the groups of the form U_{p^n} for each prime p.

To tackle this problem, we will first determine the order of U_{p^n}. Recall that the elements of U_{p^n} are the congruence classes $[a]$ such that $\gcd(a, p^n) = 1$. The only way $\gcd(a, p^n) > 1$ is if a is a multiple of p. The multiples of p between 1 and p^n are

$$p, 2p, 3p, \ldots, (p^{n-1})p,$$

so there are a total of p^{n-1} multiples of p between 1 and p^n. Thus, the number of integers a between 1 and p^n that are relatively prime to p^n is $p^n - p^{n-1}$. Therefore, $|U_{p^n}| = p^n - p^{n-1}$.

We will now proceed to the decomposition of U_{p^n} as a product of cyclic groups. As is often the case, the prime 2 exhibits different behavior than the odd primes. One can experiment with many odd primes and notice that U_{p^n} seems to always be a cyclic group. That conjecture is verified by the following theorem. The proof is a bit technical, so we present it in its entirety.

Theorem 37.9. *If p is an odd prime and n a positive integer, then $U_{p^n} \cong \mathbb{Z}_{p^n - p^{n-1}}$.*

Proof. Recall that $[a]_k$ denotes the class of the integer a in U_k. We already know that $|U_{p^n}| = p^n - p^{n-1} = (p-1)p^{n-1}$. To show that U_{p^n} is cyclic, we will demonstrate the existence of an element of order $p-1$ and an element of order p^{n-1} in U_{p^n}. First we will find an element of order $p-1$.

Since U_p is cyclic, there is a generator $[a]_p$ for U_p with $|[a]_p| = p-1$. Since $\gcd(a,p) = 1$, we also know that $\gcd(a, p^n) = 1$, and so $[a]_{p^n} \in U_{p^n}$. Suppose $|[a]_{p^n}| = k$. Then k divides $|U_{p^n}| = p^{n-1}(p-1)$, and so

$$kr = p^{n-1}(p-1) \tag{37.1}$$

for some $r \in \mathbb{Z}^+$. Also, $([a]_{p^n})^k = [1]_{p^n}$, and so p^n divides $a^k - 1$, which implies that p divides $a^k - 1$. Thus, $[a]_p^k = [1]_p$ (in $\mathbb{Z}_p$), and so $|[a]_p| = p-1$ divides k. Let $s \in \mathbb{Z}^+$ such that

$$k = (p-1)s. \tag{37.2}$$

Equations (37.1) and (37.2) combine to give us $rs = p^{n-1}$, and so $s = p^j$ for some j with $0 \leq j \leq n-1$. Thus,

$$|[a]_{p^n}| = k = (p-1)s = (p-1)p^j,$$

and

$$|[a]_{p^n}^{p^j}| = \frac{|[a]_{p^n}|}{\gcd(p^j, |[a]_{p^n}|)} = \frac{(p-1)p^j}{p^j} = p-1.$$

Therefore, $[a]_{p^n}^{p^j}$ is an element of order $p-1$ in U_{p^n}.

Next we will show that $|[p+1]_{p^n}| = p^{n-1}$. First, we will prove by induction that for each integer $k \geq 2$, $(p+1)^{p^{k-2}} - 1$ is divisible by p^{k-1} but not p^k. For our base case, note that $(p+1)^1 - 1 = p$ is divisible by p but not p^2, so our statement is true when $k = 2$. Now assume that, for some integer $k \geq 2$, $(p+1)^{p^{k-2}} - 1$ is divisible by p^{k-1} but not p^k. Then $(p+1)^{p^{k-2}} - 1 = qp^{k-1}$ for some integer q with $\gcd(q, p) = 1$. So $(p+1)^{p^{k-2}} = 1 + qp^{k-1}$, and raising both sides to the pth power gives us

$$\begin{aligned}(p+1)^{p^{k-1}} &= (1+qp^{k-1})^p \\ &= 1 + p(qp^{k-1}) + \sum_{i=2}^{p} \binom{p}{i} (qp^{k-1})^i \\ &= 1 + qp^k + p^{k+1}r\end{aligned}$$

for some integer r. Therefore p^k divides $(p+1)^{p^{k-1}} - 1$ but p^{k+1} does not (because $\gcd(q,p) = 1$). From this, we can conclude that $[p+1]_{p^n}^{p^{n-1}} = [1]_{p^n}$ but $[p+1]_{p^n}^{p^{k-1}} \neq [1]_{p^n}$ for any smaller positive value of k. This completes our proof that $|[p+1]_{p^n}| = p^{n-1}$.

Finally, since $|[a]_{p^n}^{p^j}| = p-1$ and $|[p+1]_{p^n}| = p^{n-1}$, it follows that $\gcd(|[a]_{p^n}^{p^j}|, |[p+1]_{p^n}|) = 1$. Thus,

$$|[a]_{p^n}^{p^j}[p+1]_{p^n}| = |[a]_{p^n}^{p^j}||[p+1]_{p^n}| = (p-1)p^{n-1} = |U_{p^n}|,$$

which implies that U_{p^n} is a cyclic group of order $(p-1)p^{n-1}$. We have therefore shown that $U_{p^n} \cong \mathbb{Z}_{p^n - p^{n-1}}$ when p is an odd prime. ■

Next, we will turn to the case when $p = 2$. We know that $U_2 = \{[1]\}$ is a cyclic group of order 1 and that $U_4 = \{[1], [3]\} = \langle [3] \rangle$ is a cyclic group of order 2. Let us now consider the groups U_{2^n} for $n \geq 3$.

Recall that when we decomposed finite Abelian p-groups as internal direct products of cyclic groups (see Investigation 31), we began by finding an element of largest order. We have already shown that $|U_{2^n}| = 2^n - 2^{n-1} = 2^{n-1}$. Some routine calculations show that in U_8 the largest order of any element is 2, in U_{16} the largest order is 4, in U_{32} the largest order is 8, and the pattern seems to continue. This might lead us to conjecture that $U_{2^n} \cong (\mathbb{Z}_2 \oplus \mathbb{Z}_{2^{n-2}})$. It is also curious that, in each of the above examples, $[5]$ appears to have order 2^{n-2}. This observation turns out to be a key idea in the proof of our conjecture, which is stated formally as Theorem 37.10.

Theorem 37.10. *Let $n \geq 3$ be an integer. Then*

$$U_{2^n} \cong (\mathbb{Z}_2 \oplus \mathbb{Z}_{2^{n-2}}).$$

Proof. First, we will demonstrate that $[5]$ has order 2^{n-2} in U_{2^n}. Our proof will rely on the fact that $5^{2^{n-3}} \equiv 1 + 2^{n-1} \pmod{2^n}$ for each $n \geq 3$, which we will verify by induction. The case where $n = 3$ is straightforward, since $5^1 = 1 + 2^2$. Now assume that $5^{2^{n-3}} \equiv 1 + 2^{n-1} \pmod{2^n}$ for some $n \geq 3$. Then

$$5^{2^{n-3}} = 1 + 2^{n-1} + t2^n$$

for some integer t. Dividing t by 2, we obtain integers q and r with $0 \leq r < 2$ and $t = 2q + r$. Then

$$5^{2^{n-3}} = 1 + 2^{n-1} + (2q+r)2^n = 1 + 2^{n-1} + r2^n + q2^{n+1}.$$

So

$$5^{2^{n-3}} \equiv 1 + 2^{n-1} + r2^n \pmod{2^{n+1}}.$$

We can now square both sides of this equation to obtain

$$\begin{aligned} 5^{2^{n-2}} &\equiv \left(1+2^{n-1}+r2^n\right)^2 \\ &\equiv 1+2^n+r2^{n+1}+2^{2(n-1)}+r2^{2n}+r^2 2^{2n} \\ &\equiv 1+2^n \pmod{2^{n+1}}. \end{aligned}$$

So, we have $[5]^{2^{n-3}} = [1+2^{n-1}]$ in U_{2^n}. Now $[1+2^{n-1}]^2 = [1+2^n+2^{2(n-1)}] = [1]$ in U_{2^n}, so $|[1+2^{n-1}]| = 2$ in U_{2^n}. This means $|[5]^{2^{n-3}}| = 2$ or that $|[5]| = 2^{n-2}$. Thus, in U_{2^n} we have that $\langle[5]\rangle \cong \mathbb{Z}_{2^{n-2}}$. Since $|U_{2^n}| = 2^{n-1}$, to complete our decomposition of U_{2^n} as an internal direct product of subgroups we only need to find an element of order 2 that is not in $\langle[5]\rangle$. We know that $[-1]^2 = [1]$ in U_{2^n}, and so $[-1]$ has order 2 in U_{2^n}. Recall that a cyclic group of order m has a unique subgroup of each order that divides m. (See Theorem 23.7 on page 321.) We have already seen that $\langle[5]\rangle$ has a subgroup $\langle[1+2^{n-1}]\rangle$ of order 2. Since $1+2^{n-1} \not\equiv -1 \pmod{2^n}$, we also see that $\langle[-1]\rangle \neq \langle[1+2^{n-1}]\rangle$. Therefore, $[-1] \notin \langle[5]\rangle$ and so

$$U_{2^n} = (\langle[-1]\rangle \times \langle[5]\rangle) \cong (\mathbb{Z}_2 \oplus \mathbb{Z}_{2^{n-2}})$$

when $n \geq 3$. ■

Example 37.11. As an example, we will decompose U_{200} into a sum of cyclic groups. Note that $200 = 2^3 \times 5^2$, and so

$$U_{200} \cong (U_{2^3} \oplus U_{5^2}).$$

Theorem 37.10 shows that $U_{2^3} \cong (\mathbb{Z}_2 \oplus \mathbb{Z}_2)$, and Theorem 37.9 tells us that $U_{5^2} \cong \mathbb{Z}_{20}$. Combining these results shows that

$$U_{200} \cong (\mathbb{Z}_2 \oplus \mathbb{Z}_2 \oplus \mathbb{Z}_{20}).$$

Splitting Fields

In this section, we will return to the question of how many fields there are of a given order p^n. To address this question, we will in fact answer a much more general question about a special class of field extensions known as *splitting fields*.

Preview Activity 37.12. Splitting fields, which we will define shortly, are related to roots of polynomials. To begin our study of splitting fields, we will begin by investigating the structure of sets of polynomials with a common root. Let F be a field, and let K be an extension of F containing an element a that is a root of some nonzero polynomial in $F[x]$. Let $I = \{f(x) \in F[x] : f(a) = 0\}$.

(a) Show that I is an ideal of $F[x]$.

(b) We will now determine more specifically the structure of the ideal I.

  (i) Explain why I must contain a polynomial $p(x)$ of smallest degree that has a as a root. (Hint: What principle have we used to find smallest elements?)

  (ii) Show that if $f(x) \in I$, then $p(x)$ divides $f(x)$. Conclude that $I = \langle p(x)\rangle$. (Hint: How can we divide one polynomial into another?)

  (iii) Show that $p(x)$ is an irreducible polynomial.

Elements that are roots of polynomials (as in Activity 37.12) are similar to the numbers we defined to be *algebraic* in Investigation 15. The next definition builds on this important similarity by extending the definition of an algebraic number.

Definition 37.13. Let F be a field, and let K be an extension field of F. An element $a \in K$ is **algebraic** over F if a is the root of some non-constant polynomial in $F[x]$.

We have many familiar examples of algebraic elements. For example, $\sqrt{2}$ is algebraic over $\mathbb{Q}$ (being a root of $x^2 - 2$) and i is algebraic over $\mathbb{R}$ (as a root of $x^2 + 1$). Not all numbers are algebraic. Elements that are not algebraic are called **transcendental**.[†] Joseph Liouville[‡] was the first to prove the existence of transcendental numbers, while Hermite[§] showed that the number e is transcendental over $\mathbb{Q}$ and Lindemann[¶] proved that π is also transcendental over $\mathbb{Q}$. In general, it is very difficult to prove that a number is transcendental, and we won't say anything more about such numbers.

Let's now return to the question of how many fields there are of a given finite order. Let p be a prime, and let K and L be fields of order p^n for some positive integer n. As vector spaces over $\mathbb{Z}_p$, the two fields K and L have the same dimension. Therefore, K and L are isomorphic as vector spaces. In addition, $U(K)$ and $U(L)$ are both cyclic groups of the same order and are therefore isomorphic as groups. With this much structure in common, we might think that K and L are isomorphic as fields.

In fact, there is another very good reason to suspect that any two fields of order p^n are closely related. If K is any field of order p^n, then $U(K)$ is a cyclic group of order $p^n - 1$, and so $a^{p^n - 1} = 1$ for every nonzero $a \in K$. Thus, $a^{p^n} = a$ for every $a \in K$, which means that every element of K is a root of the polynomial $f(x) = x^{p^n} - x \in \mathbb{Z}_p[x]$. Since we also know that $|K| = p^n$ and that $f(x)$ has at most p^n roots (see Theorem 13.9 on page 171), it follows that K must be a field of smallest order that contains all of the roots of $f(x)$. So any two fields of order p^n are fields of the same size that contain all of the roots of the polynomial $f(x) = x^{p^n} - x$ over $\mathbb{Z}_p$. (Note the connection to Theorem 37.4, which states that the set of all roots of $f(x) = x^{p^n} - x$ is a field of order p^n. The above observations indicate that *every* field of order p^n contains this set of roots.) For these reasons, we might expect that any two fields of order p^n are isomorphic. We will verify this conjecture by proving a more general result.

As we have just discussed, a field K of order $m = p^n$ contains exactly the m roots of the polynomial $f(x) = x^m - x$ over $\mathbb{Z}_p$. Therefore, we can factor the polynomial $f(x)$ in $K[x]$ as a product of linear terms. In other words,

$$f(x) = c(x - r_1)(x - r_2) \cdots (x - r_n)$$

for some elements c, r_1, r_2, ..., r_n in K. In this case, we say that $f(x)$ *splits* over the field K. Of course, this idea works in other fields as well. For example, $x^2 + 1 = (x + i)(x - i)$ in $\mathbb{C}[x]$. A smallest field that contains all of the roots of a polynomial so that we can factor the polynomial into linear factors is called a *splitting field*.

Definition 37.14. Let F be a field, and let $f(x) \in F[x]$.

(i) The polynomial $f(x)$ **splits** in an extension K of F if

$$f(x) = c(x - r_1)(x - r_2) \cdots (x - r_n)$$

[†]Leibniz appears to have first used the word transcendental (as "transcendens") in the fall of 1673 in *Progressio figurae segmentorum circuli aut ei sygnotae*. Leibniz's papers can be viewed at an archive of publications by Leibniz at `http://www.nlb-hannover.de/Leibniz/Leibnizarchiv/Veroeffentlichungen/`.

[‡]In the *Comptes-rendus*, **18**, 1844, p. 883, p. 910 (reproduced in *Journal de mathématiques pures et appliquées*, **16**, 1851).

[§]C. Hermite, "Sur la fonction exponentielle," *C. R. Acad Sei.*, **77**, 1873, pp. 18–24, 74–79, 226–233, and 285–293; also in *Œuvres*, **3**, Gauthier-Villars, Paris, 1912, pp. 150–181.)

[¶]F. Lindemann, "Über die Zahl π," *Mathematische Annalen*, **20**, 1882, pp. 213–225. This also seems to be the place where the term "transcendental number" appears (as "transsc endente Zahl").

for some elements c, $r_1, r_2, \ldots, r_n$ in K.

(ii) An extension field K of F is a **splitting field** for $f(x)$ over F if $f(x)$ splits in K but does not split in any proper subfield of K.

In other words, a splitting field for a polynomial $f(x)$ in $F[x]$ is an extension of F of smallest order that contains all of the roots of $f(x)$. It may not be surprising that any two splitting fields of a polynomial $f(x)$ over a field F are isomorphic, and we will finish this investigation by verifying this fact. Since any fields of order p^n for a prime p and a positive integer n are splitting fields for $f(x) = x^{p^n} - x$, this will show that any two finite fields of the same order are isomorphic.

We will need some results about extensions and irreducible polynomials to arrive at our final result. Recall that irreducible polynomials in polynomial rings behave like the primes do in the integers. In particular, irreducible polynomials form the building blocks of all polynomials in the sense that every polynomial in a polynomial ring over a field can be factored in some unique way as a product of irreducible polynomials. Since irreducible polynomials cannot be factored in a non-trivial way, if an element a is a root of a polynomial $f(x)$ in some polynomial ring $F[x]$, then a must be a root of some irreducible polynomial over $F[x]$. We can say a bit more than this, as the result of Activity 37.12 shows. The next theorem states that result formally.

Theorem 37.15. *Let F be a field, and let K be an extension of F containing an element a that is algebraic over F. Then there exists an irreducible polynomial $p(x) \in F[x]$ that has a as a root. Moreover, any polynomial in $F[x]$ that has a as a root is divisible by $p(x)$.*

The irreducible polynomial $p(x)$ in Theorem 37.15 need not be unique. In fact, any nonzero scalar multiple of $p(x)$ will work. If we want to specify a specific polynomial to use, we often select a monic polynomial—that is, one with a leading coefficient of 1. This polynomial is given a special name.

Definition 37.16. The **minimal polynomial** for an algebraic element a over a field F is the monic irreducible polynomial in $F[x]$ having a as a root.

So, for example, $x^2 + 1$ is the minimal polynomial for i over $\mathbb{R}$.

In Investigation 15 (specifically, Theorem 15.11 on page 204), we showed that the quotient ring $F[x]/\langle p(x)\rangle$ is a field if and only if $p(x)$ is irreducible in $F[x]$. There is something even more we can say about this quotient field—namely, that it is (up to isomorphism) the smallest extension of F that contains a. We will explore this idea in the next activity.

Activity 37.17. Let F be a field, let a be an element that is algebraic over F, and let $p(x)$ be an irreducible polynomial in $F[x]$ having a as a root. Denote by $F(a)$ the smallest extension of F that contains a. Define the function $\varphi : F[x] \to F(a)$ by $\varphi(f(x)) = f(a)$.

(a) Explain why φ is a ring homomorphism.

(b) Show that $\text{Ker}(\varphi) = \langle p(x)\rangle$.

(c) Show that $\text{Im}(\varphi) = F(a)$.

(d) Use the First Isomorphism Theorem for rings to conclude the following theorem:

Theorem 37.18. *Let F be a field, let a be an algebraic element over F, and let $p(x) \in F[x]$ be an irreducible polynomial having a as a root. Then*

$$F(a) \cong F[x]/\langle p(x)\rangle.$$

A useful corollary of Theorem 37.18 is the following:

Corollary 37.19. *Let F be a field, and let $p(x) \in F[x]$ be an irreducible polynomial. If a is a root of $p(x)$ in some extension E of F and b is a root of $p(x)$ in some extension K of F, then $F(a) \cong F(b)$.*

Proof. Let a be a root of $p(x)$ in some extension E of F, and let b be a root of $p(x)$ in some extension K of F. Theorem 37.18 shows that $F(a) \cong F[x]/\langle p(x)\rangle$ and also that $F[x]/\langle p(x)\rangle \cong F(b)$. Therefore, $F(a) \cong F(b)$. ■

As a reminder, our goal in this section is to show that any two splitting fields for a polynomial are isomorphic. Corollary 37.19 seems to have us heading in the right direction. However, Corollary 37.19 only helps us if our polynomial is irreducible. Since any polynomial is the product of irreducible polynomials, we will need to do a bit more work to reach our destination. Our next step is to generalize Corollary 37.19 to the case where we have isomorphic—but not necessarily equal—base fields. (In Corollary 37.19, F is the base field.) Before we can do that, though, we need to discuss how to extend a field isomorphism to an isomorphism of polynomial rings.

Suppose F and E are fields and $\varphi : F \to E$ is an isomorphism. There is a natural function that maps $F[x]$ into $E[x]$ as follows: assign to the polynomial

$$f(x) = a_n x^n + a_{n-1}x^{n-1} + \cdots + a_1 x + a_0 \in F[x]$$

the polynomial

$$\varphi(a_n)x^n + \varphi(a_{n-1})x^{n-1} + \cdots + \varphi(a_1)x + \varphi(a_0) \in E[x].$$

Although it is a bit of an abuse of notation, we will denote this image of $f(x)$ as $\varphi(f(x))$. Note that $\varphi : F[x] \to E[x]$ has the property that $\varphi(c) = c$ for any $c \in F$. So this mapping between the polynomial rings agrees with the corresponding isomorphism between the coefficient fields. For this reason, we say that $\varphi : F[x] \to E[x]$ **extends** the function φ, and it makes sense to use the same notation for both maps. Since φ is an isomorphism, we might expect that the extension of φ to the corresponding polynomial rings preserves much of the structure of the polynomial rings. In fact, in Exercise 5, you are asked to show that $\varphi : F[x] \to E[x]$ is a ring isomorphism. Knowing this, we can now proceed to the major result that will ultimately allow us to prove that any two splitting fields for a polynomial are isomorphic.

Theorem 37.20. *Let F be a field, and let $p(x) \in F[x]$ be an irreducible polynomial. Let a be a root of $p(x)$ in some extension F' of F. Let E be a field, and let $\varphi : F \to E$ be a field isomorphism. If b is a root of $\varphi(p(x))$ in some extension E' of E, then there is an isomorphism Φ from $F(a)$ to $E(b)$ so that*

(i) $\Phi(c) = \varphi(c)$ *for all $c \in F$, and*

(ii) $\Phi(a) = b$.

Proof. Let b be a root of $\varphi(p(x))$ in some extension E' of E. Since φ is an isomorphism, the polynomial $\varphi(p(x))$ is irreducible in $E[x]$. Let $\beta : E[x]/\langle\varphi(p(x))\rangle \to E(b)$ be the isomorphism whose existence was demonstrated in Theorem 37.18. Recall that $\beta(g(x) + \langle\varphi(p(x))\rangle) = g(b) \in E(b)$ and so $\beta(x + \langle\varphi(p(x))\rangle) = b$. Let α be the inverse of the corresponding isomorphism from $F[x]/\langle p(x)\rangle$ to $F(a)$. If we can show that there is an isomorphism $\psi : F[x]/\langle p(x)\rangle \to E[x]/\langle\varphi(p(x))\rangle$, then we can combine these three isomorphisms

$$F(a) \xrightarrow{\alpha} F[x]/\langle p(x)\rangle \xrightarrow{\psi} E[x]/\langle\varphi(p(x))\rangle \xrightarrow{\beta} E(b)$$

to obtain an isomorphism $\Psi : F(a) \to E(b)$.

Define $\psi : F[x]/\langle p(x)\rangle \to E[x]/\langle \varphi(p(x))\rangle$ by

$$\psi(f(x) + \langle p(x)\rangle) = \varphi(f(x)) + \langle \varphi(p(x))\rangle.$$

That ψ is a well-defined isomorphism is left for you to prove in Exercise 3. What remains is to demonstrate that $\Phi = \beta\psi\alpha$ satisfies conditions (i) and (ii). Let $c \in F$. Then $\alpha^{-1}(c + \langle p(x)\rangle) = c$, and so $\alpha(c) = c + \langle p(x)\rangle$. Thus,

$$\Phi(c) = (\beta\psi)(\alpha(c)) = \beta(\psi(c + \langle p(x)\rangle)) = \beta(\varphi(c) + \langle \varphi(p(x))\rangle) = \varphi(c),$$

and Φ satisfies condition (i). Moreover, $\alpha^{-1}(x + \langle p(x)\rangle) = a$, and so $\alpha(a) = x + \langle p(x)\rangle$. That, and the fact that $\varphi(x) = x$, gives us

$$\begin{aligned}\Phi(a) &= (\beta\psi)(\alpha(a))\\ &= \beta(\psi(x + \langle p(x)\rangle))\\ &= \beta(\varphi(x) + \langle \varphi(p(x))\rangle)\\ &= \beta(x + \langle \varphi(p(x))\rangle)\\ &= b,\end{aligned}$$

and so Φ satisfies condition (ii) as well. Therefore, Φ is the desired isomorphism. ■

We now need to extend the result of Theorem 37.20 to any polynomial, and then we can show that any two splitting fields for a polynomial are isomorphic.

Corollary 37.21. *Let φ be an isomorphism from a field F to a field E, and let $f(x) \in F[x]$. Let F' be a splitting field for $f(x)$ over F, and let E' be a splitting field of $\varphi(f(x))$ over E. Then there is an isomorphism Φ from F' to E' so that $\Phi(c) = \varphi(c)$ for all $c \in F$.*

Since we don't know that $f(x)$ is irreducible, we cannot directly use Theorem 37.20 to prove Corollary 37.21. However, we can split $f(x)$ into irreducible factors and use Theorem 37.20 on each factor. This will allow us to build up extension fields with the desired properties from polynomials of smaller degree than $f(x)$. Mathematical induction is the tool that formalizes this idea in the proof that follows.

Proof of Corollary 37.21. We will induct on $n = \deg(f(x))$. If $n = 1$, then $F' = F$, $E' = E$, and $\Phi = \varphi$ is the desired isomorphism. Now assume $n > 1$. For our inductive hypothesis, assume that Corollary 37.21 is true for any polynomial of degree less than n.

Let $p(x)$ be an irreducible factor of $f(x)$, a a root of $p(x)$ in F', and b a root of $\varphi(p(x))$ in E'. By Theorem 37.20, there is an isomorphism α from $F(a)$ to $E(b)$ that agrees with φ on F and maps a to b. By the Division Algorithm,

$$f(x) = (x - a)g(x)$$

for some $g(x) \in F(a)[x]$ with $\deg(g(x)) < \deg(f(x))$. It follows that F' is a splitting field for $f(x)$ over $F(a)$ and E' is a splitting field for $\varphi(f(x))$ over $E(b)$. By our induction hypothesis (using $g(x)$ as our polynomial in $F(a)[x]$, and α an isomorphism from $F(a)$ to $E(b)$), there is an isomorphism Φ from F' to E' so that $\Phi(r) = \alpha(r)$ for all $r \in F(a)$. Note that if $c \in F$, then

$$\Phi(c) = \alpha(c) = \varphi(c),$$

and so Φ is the desired isomorphism. ■

We can now establish our main result that splitting fields are unique.

Activity 37.22. Explain how Corollary 37.21 proves the following theorem:

Theorem 37.23. *Let F be a field, and let $f(x) \in F[x]$. If K and L are splitting fields for $f(x)$ over F, then $K \cong L$.*

Because any two splitting fields of a polynomial are isomorphic, we usually just refer to "the" splitting field of a polynomial. In particular, since a finite field of order p^n is the splitting field of $x^{p^n} - x$ over $\mathbb{Z}_p$, we have just proved the next result.

Corollary 37.24. *Let K and L be finite fields of order p^n for some prime p and some positive integer n. Then $K \cong L$.*

In other words, there is exactly one finite field (up to isomorphism) of any given order p^n, where p is prime and n is a positive integer. Different notations are used to denote finite fields, but we will denote the field of order p^n as $\mathbb{F}_{p^n}$. This field is called the *Galois field* of order p^n in honor of the mathematician Evariste Galois, whose work on the solvability of polynomial equations formed the basis of much of group theory.

Concluding Activities

Activity 37.25. Theorem 9.10 showed us the form of a simple or primitive quadratic extension of a field. Prove the following generalization of this theorem:

Theorem 37.26. *Let F be a field and $p(x)$ an irreducible polynomial of degree n in $F[x]$. Let E be an extension of F containing a root r of $p(x)$. The smallest extension of F containing field containing r has the form*

$$F(r) = \{a_0 + a_1 r + a_2 r^2 + \cdots + a^{n-1} r^{n-1} : a_0, a_1, \ldots, a_{n-1} \in F\}. \tag{37.3}$$

The field $F(r)$ is called a **simple** or **primitive** extension of F and is obtained from F by attaching the root r to F. Note that $F(r)$ is essentially the field $F[x]/\langle p(x)\rangle$.

Activity 37.27. Although we have studied the properties of splitting fields, we have not yet discussed a general method for finding the splitting field of a polynomial. In this activity, we will illustrate such a method with the example of $f(x) = x^4 - 5x^2 + 6 \in \mathbb{Q}[x]$.

(a) Show that $(x^2-2)(x^2-3)$ is a factorization of $f(x)$ into a product of irreducible polynomials in $\mathbb{Q}[x]$.

(b) The extension $\mathbb{Q}(\sqrt{2})$ contains both $\sqrt{2}$ and $-\sqrt{2}$, so $\mathbb{Q}(\sqrt{2})$ is the splitting field for $x^2 - 2$ over $\mathbb{Q}$. But is $\mathbb{Q}(\sqrt{2})$ the splitting field for $x^2 - 3$ over $\mathbb{Q}$? If yes, then we have found the splitting field for $f(x)$ over $\mathbb{Q}$. If no, then we will need to do more work. Show that $\sqrt{3} \notin \mathbb{Q}(\sqrt{2})$. How does that answer our question about whether $\mathbb{Q}(\sqrt{2})$ is the splitting field for $f(x)$ over $\mathbb{Q}$? (Hint: Use Theorem 37.26.)

(c) To completely factor $f(x)$ over $\mathbb{Q}$, we need to add a root of $x^2 - 3$ to $\mathbb{Q}(\sqrt{2})$. Denote by $\mathbb{Q}(\sqrt{2}, \sqrt{3})$ the field $\mathbb{Q}(\sqrt{2})(\sqrt{3})$. Explain why $\mathbb{Q}(\sqrt{2}, \sqrt{3})$ is the splitting field for $f(x)$ over $\mathbb{Q}$.

(d) Since $\sqrt{2}\sqrt{3} = \sqrt{6}$, it is tempting to think that $\mathbb{Q}(\sqrt{6}) = \mathbb{Q}(\sqrt{2}, \sqrt{3})$. Is this the case? Prove your answer.

As this activity illustrates, to find the splitting field of a polynomial $f(x)$ over a field F, we first factor $f(x)$ into a product of irreducible polynomials. If $p(x)$ is an irreducible factor of $f(x)$, then we know that the field $F[x]/\langle p(x)\rangle$ contains roots of $p(x)$. We attach these roots to F to obtain an extension field E_1. Then we divide $f(x)$ by $p(x)$ in $E_1[x]$ and repeat the process on the quotient in $E_1[x]$. Eventually, we arrive at an extension K of F that is obtained by adding roots of $f(x)$ to F so that K contains all of the roots of $f(x)$. This field K is the splitting field for $f(x)$ over F.

Activity 37.28. Write a short summary that describes the important concepts, and the relationships between these concepts, that were introduced in this investigation. Explain how the material in this investigation is connected to the material in Investigations 4, 7, 10, 15, 20, 23, and 31.

Exercises

(1) The splitting field of the polynomial $f(x) = x^9 - x$ over $\mathbb{Z}_3$ is a field of order 9. We will explicitly construct this field in this exercise.

(a) Note that $f(x)$ is not irreducible in $\mathbb{Z}_3[x]$. Factor $f(x)$ into a product of irreducible polynomials in $\mathbb{Z}_3[x]$.

(b) Let i be a root of $x^2 + [1]$ in the extension $\mathbb{Z}_3(i)$ of $\mathbb{Z}_3$. Show that $f(x)$ factors completely over $\mathbb{Z}_3(i)$.

(c) Write the operation tables for $\mathbb{Z}_3(i)$.

(d) Find a generator for $U(\mathbb{Z}_3(i))$.

(2) Explicitly construct a field with 16 elements.

$\star$ (3) Let F be a field, and let $p(x) \in F[x]$ an irreducible polynomial. Let a be a root of $p(x)$ in some extension F' of F. Let E be a field and $\varphi : F \to E$ a field isomorphism. Define $\psi : F[x]/\langle p(x)\rangle \to E[x]/\langle \varphi(p(x))\rangle$ by

$$\psi(f(x) + \langle p(x)\rangle = \varphi(f(x)) + \langle \varphi(p(x))\rangle.$$

Show that ψ is a well-defined isomorphism.

$\star$ (4) Let K be a field of characteristic p, and let n be a positive integer.

(a) Prove that $(a \pm b)^p = a^p \pm b^p$ for any $a, b \in K$.

(b) Prove that $(a \pm b)^{p^n} = a^{p^n} \pm b^{p^n}$ for any $a, b \in K$ and any positive integer n.

$\star$ (5) Let F and E be fields with $\varphi : F \to E$ an isomorphism. If $f(x) = a_n x^n + a_{n-1}x^{n-1} + \cdots + a_1 x + a_0 \in F[x]$, define $\varphi(f(x))$ to be the polynomial

$$\varphi(a_n)x^n + \varphi(a_{n-1})x^{n-1} + \cdots + \varphi(a_1)x + \varphi(a_0) \in E[x].$$

(a) Prove that $\varphi : F[x] \to E[x]$, as defined above, is an isomorphism.

(b) Show that $p(x)$ is irreducible in $F[x]$ if and only if $\varphi(p(x))$ is irreducible in $E[x]$.

(6) Find the splitting field of $x^2 + 1$ over $\mathbb{Q}$. Explain why this splitting field is not $\mathbb{C}$.

(7) Show that $x^2 + x + 1$ and $x^2 + 3$ have the same splitting fields over $\mathbb{Q}$.

(8) What are the possible splitting fields of a polynomial in $\mathbb{R}[x]$? Explain.

(9) Find the splitting field of $x^4 + 1$ over $\mathbb{Q}$. Would your answer be different if you had found the splitting field over $\mathbb{R}$ instead?

(10) Find the splitting field of $x^2 + x + 2$ over $\mathbb{Z}_3$. Then find all of the roots of $f(x)$ in this splitting field.

* (11) (a) Let s and t be positive integers with $\gcd(s, t) = 1$. Prove that $U_{st} \cong (U_s \oplus U_t)$.

(b) Let $m = n_1 n_2 \cdots n_k$, where $\gcd(n_i, n_j) = 1$ for all $i \neq j$. Prove that

$$U_m \cong (U_{n_1} \oplus U_{n_2} \oplus \cdots \oplus U_{n_k}).$$

(12) Write each of U_{4752} and U_{114244} as a direct sum of cyclic groups.

(13) Let n be an integer with $n \geq 2$.

(a) Use the result of Exercise 11 to calculate $|U_n|$ in terms of the prime factors of n.

(b) Recall that the Euler phi function φ gives us the number of positive integers less than a given integer that are relatively prime to that integer. Explain why $\varphi(mn) = \varphi(m)\varphi(n)$ if m and n are positive integers and $\gcd(m, n) = 1$. (Such functions are called **multiplicative** functions and are important in number theory.) Then find a formula for $\varphi(n)$ in terms of the prime factors of n. Use this formula to find $\varphi(6860)$.

Connections

In this investigation, we used material from both ring theory and group theory to study the structure of finite fields. Of course, any finite field is a ring, as introduced in Investigation 7, and we saw that facts about polynomials and polynomial rings (Investigation 11), field extensions (Investigation 9), ring isomorphisms (Investigation 10), roots of polynomials (Investigation 13), irreducible polynomials (Investigation 13 and 14), quotients of polynomial rings (Investigation 15), and ideals (Investigation 16) all came up in our classification of finite fields. Moreover, when determining the group of units of a finite field, we also used information about groups (Investigation 20), cyclic groups (Investigation 22), direct products of groups (Investigation 28), and the classification of finite Abelian groups (Investigation 31). As should be obvious from this lengthy list of investigations—all of which were necessary to answer the questions in this investigation—there are many important connections between ring theory and group theory.

Investigation 38

Groups of Order 8 and 12: Semidirect Products of Groups

Focus Questions

By the end of this investigation, you should be able to give precise and thorough answers to the questions listed below. You may want to keep these questions in mind to focus your thoughts as you complete the investigation.

- What is a semidirect product of groups? In what ways are semidirect products of groups useful?
- Up to isomorphism, how many groups are there of order 8?
- Up to isomorphism, how many groups are there of order 12?

Preview Activity 38.1. Summarize our previous work in classifying groups, and list all of the isomorphism classes of groups of orders from 1 to 7, 9 to 11, and 13 to 15. Provide whatever information you can about groups of orders 8 and 12.

Introduction

As Activity 38.1 demonstrates, we have classified all groups of order 1 through 15, with the exception of groups of order 8 and 12. We have also determined all groups of prime order (see Activity 26.17 on page 352 and Theorem 29.18 on page 409), all groups of order $2p$, where p is prime (see Corollary 27.13 on page 367 and Theorem 29.16 on page 407), all groups of order pq where p and q are distinct primes and p does not divide $q-1$ (see Activity 32.32 on page 457), and all groups of order p^2, where p is prime (see Activity 32.15 on page 451). In this investigation, we will fill in the gaps and classify all groups of order 8 and 12, which will complete our classification of groups of order less than 16. (Since there are 14 groups of order 16, this seems like a reasonable place to stop.) In the process of classifying the groups of order 8 and 12, we will introduce and use the idea of the semidirect product of two groups. We will also see that when classifying groups of a given order, it is not the order itself that determines the difficulty of the classification, but rather how large the powers of the prime divisors of the order are.

Groups of Order 8

In this section, we will determine the distinct isomorphism classes of groups of order 8. Since the Fundamental Theorem of Finite Abelian Groups tells us about the Abelian groups of order 8, we will concentrate on non-Abelian groups. Recall that we already know two non-Abelian groups of order 8—namely, D_4 and the quaternions $\mathbf{Q}$. We will now determine if there are, up to isomorphism, any other non-Abelian groups of order 8, and we will classify all such groups.

Activity 38.2. Let G be a non-Abelian group of order 8 with identity e.

(a) Use Exercise 2 of Investigation 21 (see page 301) to explain why G must contain an element b of order 4.

(b) Let $N = \langle b \rangle$. Explain why N is normal in G.

(c) Let $a \in G$ with $a \notin N$. Then $G = N \cup aN = \{e, b, b^2, b^3, a, ab, ab^2, ab^3\}$. The operation table for G will be determined once we know how to represent the elements ba and a^2 in the form $a^i b^j$, with $0 \leq i \leq 1$ and $0 \leq j \leq 3$. Given that $N \lhd G$, we know that $aba^{-1} \in N$, so $aba^{-1} = b^t$ for some t with $0 \leq t \leq 3$

(i) Explain why t cannot be 0 or 1.

(ii) Now explain why t cannot be equal to 2. (Hint: What is $|aba^{-1}|$?)

(iii) Assume $aba^{-1} = b^3 = b^{-1}$. Since $a \in G$ and $a \neq e$, we must have $|a| = 2$ or $|a| = 4$. What can we say about G if $|a| = 2$?

(iv) What can we say about G if $aba^{-1} = b^{-1}$ and $|a| = 4$?

(d) How many groups are there of order 8?

Activity 38.2 tells us about the groups of order 8, so we will now turn our attention to groups of order 12. It is possible (and not that difficult) to classify the groups of order 12 directly, but in the next section we will introduce a new tool, the semidirect product, that is very helpful in the general context of classifying groups of a given order. We will then use semidirect products to classify groups of order 12 and of order p^3, where p is an odd prime.

Semi-direct Products of Groups

Preview Activity 38.3. Let G be the set of ordered pairs

$$\{([a]_3, [b]_2) : [a]_3 \in \mathbb{Z}_3, [b]_2 \in \mathbb{Z}_2\}.$$

Define an operation $\cdot$ on G by

$$([a]_3, [b]_2) \cdot ([c]_3, [d]_2) = ([a + (-1)^b c]_3, [b + d]_2)$$

(a) Explain why this operation is well-defined on G.

(b) Construct the operation table for the set G with the operation defined above.

(c) Is G a group under this operation? If no, why not? If yes, to what familiar group is G isomorphic? Explain.

While we can decompose some groups into an internal direct product of normal subgroups, we cannot do this for every group. For example, if we try to write the group D_3 as an internal direct product of two proper subgroups, we run into a problem. Recall that the group $H = \langle R \rangle$ is a normal subgroup of D_3 of order 3, so we would need a normal subgroup K of order 2 to be able to write D_3 as the product $H \times K$. However, D_3 has no normal subgroup of order 2. So we cannot decompose D_3 into an internal direct product of nontrivial normal subgroups. It turns out, however, that we can decompose D_3 into what is called a semidirect product of subgroups. Activity 38.3 gives an example of such a decomposition. (Don't worry if the construction there does not seem obvious or natural to you at the moment.)

Up to this point, we have seen several different products of groups:

- The direct product $H \oplus K$ of two groups is again a group, external to both H and K.
- If H and K are subgroups of a group G with $H \lhd G$, then the product

$$HK = \{hk : h \in H, K \in K\}$$

is a subgroup of G. (See Activity 27.22 on page 372.) Moreover, if $H \cap K = \{e\}$, where e is the identity in G, then $|HK| = |H||K|$. To see this, suppose $h_1k_1 = h_2k_2$ in HK. Then $h_2^{-1}h_1 = k_2k_1^{-1} \in (H \cap K)$, so $h_2^{-1}h_1 = k_2k_1^{-1} = e$. Thus, $h_1 = h_2$, and $k_1 = k_2$. It follows that if $H \cap K = \{e\}$ and $|H||K| = |G|$, then $G = HK$.

- If H and K are normal subgroups of a group G with $H \cap K = \{e\}$, then $HK = H \times K$ is the internal direct product of H and K.

In this section, we will construct another type of product called a *semidirect product*. To understand how semidirect products work, recall that the construction of the internal product HK requires that H and K be subgroups of some group G that is already known. The question we want to answer now is if we can generalize this construction. In other words, given any two arbitrary groups H and K, can we find a group G so that G contains copies of both H and K—that is, subgroups H' and K' that are isomorphic to H and K, respectively—with $H' \lhd G$ and $H' \cap K' = \{e\}$, where e is the identity in G? If so, then $G = H'K'$, and we will denote this special decomposition as $H \rtimes K$. (We will say more about this notation later.)

To explore this construction more, let H and K be groups with identities e_H and e_K, respectively. We want to find a group G that contains isomorphic copies of H and K satisfying the conditions described above. A natural place to start is to let $G = \{(h, k) : h \in H, k \in K\}$. Certainly, G will contain a copy $H' = \{(h, e_K) : h \in H\}$ of H and a copy $K' = \{(e_H, k) : k \in K\}$ of K. The key to constructing the group G is to define an appropriate operation. We want to make $G = H'K'$ with $H' \lhd G$. This will mean that if $k \in K'$ and $h \in H'$, then $khk^{-1} \in H'$. If $a = h_1k_1$ and $b = h_2k_2$ are elements of $H'K'$, then it will follow that

$$ab = (h_1k_1)(h_2k_2) = h_1\left(k_1h_2k_1^{-1}\right)(k_1k_2) \tag{38.1}$$

is also an element of $H'K'$. So the operation we define in G will need to mimic the product in (38.1).

What makes the product in (38.1) work is that $k_1h_2k_1^{-1}$ is in H. In fact, conjugation by the element k_1 is an automorphism of H (an inner automorphism to be specific). We also saw this idea in Activity 38.3, where left multiplication by $(-1)^b$ for $b = 0$ or $b = 1$ is an automorphism on $\mathbb{Z}_3$. In other words, we had a mapping φ with domain $\mathbb{Z}_2$ that assigned to each $[b] \in \mathbb{Z}_2$ an automorphism on $\mathbb{Z}_3$. More specifically, we had $\varphi([0]_2)$ as the identity automorphism and $\varphi([1]_2)$ as the automorphism that sends $[a]_3$ to $[2a]_3$ for all $[a]_3 \in \mathbb{Z}_3$. Expressed another way, $\varphi : \mathbb{Z}_2 \to \text{Aut}(\mathbb{Z}_3)$ is the mapping for which $\varphi([0]_2)([a]_3) = [a]_3$ (for all $[a]_3 \in \mathbb{Z}_3$) and $\varphi([1]_2)([a]_3) = [2a_3]$ (for all $[a]_3 \in \mathbb{Z}_3$). (It turns out that φ is a homomorphism as well, and you should verify that for yourself.)

Before we proceed, a word of caution is in order: the above notation can be very confusing since we are dealing with functions whose images are functions as well. As you work through this section, it is vitally important to distinguish between the elements of a given group and the functions that act on these elements. In the next activity, we will explore these ideas in a more general context.

Activity 38.4. Let H and K be groups with identities e_H and e_K, respectively, let $\varphi : K \to \text{Aut}(H)$ be a homomorphism, and let $G = H \times K$ be the Cartesian product of H and K. Then we can define a product on G as follows:

$$(h_1, k_1)(h_2, k_2) = (h_1\varphi(k_1)(h_2), k_1k_2). \tag{38.2}$$

Note that this product has the same form as the products in (38.1) and Activity 38.3. In Activity 38.3, our example turned out to be a group, so it seems reasonable to ask if G will always be a group with the product defined by (38.2).

(a) Is G closed under the operation from (38.2)?

(b) Does G contain an identity element? If so, what is it? Explain.

(c) Is the operation defined by (38.2) associative? Prove your answer.

(d) Does G contain an inverse for each of its elements? If so, what is the form of an inverse of an element in G?

Activity 38.4 tells us that G, as defined above, is a group under the operation from (38.2). We can actually say more about the group G, as stated in the following theorem.

Theorem 38.5. *Let H and K be groups with identities e_H and e_K, respectively, and let $\varphi : K \to$ Aut(H) be a homomorphism. Then $G = \{(h, k) : h \in H, k \in K\}$ with the operation*

$$(h_1, k_1)(h_2, k_2) = (h_1\varphi(k_1)(h_2), k_1k_2)$$

is a group. Moreover,

(i) $H' = \{(h, e_K) : h \in H\}$ *is a normal subgroup of G isomorphic to H;*

(ii) $K' = \{(e_H, k) : k \in K\}$ *is a subgroup of G isomorphic to K; and*

(iii) $H' \cap K' = \{(e_H, e_K)\}$.

Proof. Since Activity 38.4 shows that G is a group, we will focus here on parts (i) – (iii). In particular, we will prove part (i) and leave the remaining parts for the reader in Exercise 4.

We will show that $H' = \{(h, e_K) : h \in H\}$ is a normal subgroup of G isomorphic to H. The element (e_H, e_K) is in H', so H' contains the identity element in G. Let (h_1, e_K) and (h_2, e_K) be in H'. Then

$$(h_1, e_K)(h_2, e_K) = (h_1, \varphi(e_K)(h_2), e_K) = (h_1h_2, e_K) \in H',$$

and so H' is closed under the operation in G.

Also,

$$(h_1, e_K)^{-1} = \left(\varphi\left(e_K^{-1}\right)\left(h_1^{-1}\right), e_K^{-1}\right) = \left(h_1^{-1}, e_K\right) \in H',$$

and so H' is a subgroup of G by the Subgroup Test. (See page 305.)

To show that H' is a normal subgroup of G, let $(h, e_K) \in H'$ and let $g = (a, b) \in G$. Then

$$\begin{aligned}(a, b)^{-1}(h, e_K)(a, b) &= \left(\varphi\left(b^{-1}\right)\left(a^{-1}\right), b^{-1}\right)(h\varphi(e_K)(a), b) \\ &= \left(\varphi\left(b^{-1}\right)\left(a^{-1}\right)\varphi\left(b^{-1}\right)(ha), b^{-1}b\right) \\ &= \left(\varphi\left(b^{-1}\right)\left(a^{-1}\right)\varphi\left(b^{-1}\right)(ha), e_K\right)\end{aligned}$$

and $(a, b)^{-1}(h, e_K)(a, b) \in H'$. Therefore, $g^{-1}H'g = H'$ and $H' \lhd G$.

That H' is isomorphic to H can be shown by considering the mapping $\alpha : H \to H'$ defined by $\alpha(h) = (h, e_K)$. Let $h_1, h_2 \in H$. Then

$$\alpha(h_1h_2) = (h_1h_2, e_K) = (h_1, e_K)(h_2, e_K) = \alpha(h_1)\alpha(h_2),$$

and α is a homomorphism. If $\alpha(h_1) = \alpha(h_2)$, then $(h_1, e_K) = (h_2, e_K)$ and $h_1 = h_2$. Thus, α is a monomorphism. If $(x, e_K) \in H'$, then $\alpha(x) = (x, e_K)$, and so α is an epimorphism. We have therefore shown that α is an isomorphism and $H \cong H'$.

■

The group G described in Theorem 38.5 is called the **semidirect product** of H and K and is denoted $H \rtimes_\varphi K$ to indicate its dependence on the particular homomorphism $\varphi : K \to \text{Aut}(H)$. When the homomorphism φ is clear from the context, we will simply write $H \rtimes K$.

Definition 38.6. Let H and K be groups, and let $\varphi : K \to \text{Aut}(H)$ be a homomorphism. The **semidirect product** of H and K with respect to φ is the group

$$H \rtimes_\varphi K = \{(h, k) : h \in H, k \in K\}$$

with the operation

$$(h_1, k_1)(h_2, k_2) = (h_1\varphi(k_1)(h_2), k_1k_2).$$

Why are semidirect products important? First, the direct product $H \oplus K$ is an example of a semidirect product (see Exercise 3), so we can think of the semidirect product as an extension of the direct product. (This is also the main motivation for the notation $\rtimes$ for the semidirect product, as it is more general than the internal direct product.) Second, if φ is a nontrivial homomorphism, then the semidirect product $H \rtimes_\varphi K$ is a non-Abelian group (see Exercise 6), so semidirect products provide a method for constructing non-Abelian groups. Semidirect products are also useful in classifying groups. For example, if H and K are subgroups of a group G so that H is normal in G, $H \cap K$ is trivial, and $|HK| = |G|$, then G will be isomorphic to a semidirect product $H \rtimes_\varphi K$ for some φ. The next activity makes this last point clear.

Activity 38.7. Let G be a group with identity e and subgroups H and K such that

(i) $H \lhd G$ and

(ii) $H \cap K = \{e\}$.

Let $\varphi : K \to \text{Aut}(H)$ be the mapping that sends each element $k \in K$ to the inner automorphism defined by conjugation by k. That is, let $\varphi(k) = \pi_k$, where $\pi_k(h) = khk^{-1}$. There is a natural function $\Phi : HK \to (H \rtimes_\varphi K)$. Define this function Φ and show that it is an isomorphism. Then explain how we have proved the following theorem:

Theorem 38.8. *Let G be a group with identity e and subgroups H and K with*

(i) $H \lhd G$ *and*

(ii) $H \cap K = \{e\}$.

Then $HK \cong (H \rtimes_\varphi K)$ *for some homomorphism* $\varphi : K \to \text{Aut}(H)$.

In the next section, we will use Theorem 38.8 to classify all groups of order 12 and all groups of order p^3, where p is a prime.

Groups of Order 12 and p^3

We will begin this section with groups of order 12. The Fundamental Theorem of Finite Abelian Groups tells us that the groups $\mathbb{Z}_{12}$ and $\mathbb{Z}_6 \oplus \mathbb{Z}_2$ are the distinct Abelian groups of order 12. We already know at least three non-Abelian groups of order 12—namely, D_6, A_4, and T, where T is described in Exercise 18 of Investigation 25 (see page 344) as having presentations $\langle s, t \mid s^6 = 1, s^3 = t^2, sts = t\rangle$ and $\langle x, y \mid x^4 = y^3 = 1, yxy = x\rangle$. We will now show that these three groups are, up to isomorphism, the only non-Abelian groups of order 12.

Let G be a non-Abelian group of order 12 with identity e. Since $12 = 2^2 \times 3$, G has a Sylow 3-subgroup H of order 3 and a Sylow 2-subgroup K of order 4. Since H is cyclic, $H = \langle h \rangle$ for some $h \in G$.

Define $\varphi : G \to P(G/H)$ by $\varphi(a) = \pi_a$ where $\pi_a(gH) = (ag)H$. Recall that G/H denotes the collection of left cosets of H in G (even if H is not normal in G) and $P(G/H)$ denotes the group of permutations of G/H. In Exercise 7 of Investigation 33 (see page 469), we showed that φ is a homomorphism. Now $[G : H] = 4$, and so $P(G/H) \cong S_4$. If φ is a monomorphism, then G is isomorphic to a subgroup of order 12 in S_4. The only such subgroup is A_4, and so $G \cong A_4$ in this case.

Now assume that $|\text{Ker}(\varphi)| > 1$. We will next show that $\text{Ker}(\varphi) \subseteq H$. Let $a \in \text{Ker}(\varphi)$. Then $\varphi(a) = \pi_a$ is the identity permutation in $P(G/H)$, so $H = \pi_a(H) = aH$ and $a \in H$. Thus, $\text{Ker}(\varphi) \subseteq H$. Because $|H| = 3$ and $|\text{Ker}(\varphi)| > 1$, it follows that $\text{Ker}(\varphi) = H$, and so $H \lhd G$.

We know that $H \cap K = \{e\}$ and that $|G| = |H||K|$, so $G = HK$. Theorem 38.8 shows us that $G \cong (H \rtimes_\varphi K)$ for some φ. Next we will determine the different groups of this form.

Since $|H| = 3$, we know that $H \cong \mathbb{Z}_3$. Recall from Exercise 34 of Investigation 29 (see page 416) that $\text{Aut}(\mathbb{Z}_3) \cong U_3$, so $|\text{Aut}(H)| = 2$. The two automorphisms of H are the identity automorphism π_0 and the automorphism π_1 defined by $\pi_1(h) = h^{-1} = h^2$. Since $|K| = 4$, there are two possibilities for K: $K \cong \mathbb{Z}_4$ and $K \cong (\mathbb{Z}_2 \oplus \mathbb{Z}_2)$.

Case 1: $K \cong \mathbb{Z}_4$**.** In this case, we can identify K with $\mathbb{Z}_4$, and so any homomorphism φ from K to

$\text{Aut}(H)$ is determined by its action on $[1]$. Thus, there are two possibilities: $\varphi([1]) = \pi_0$ and $\varphi([1]) = \pi_1$. In the first case, G will be Abelian and isomorphic to $\mathbb{Z}_3 \oplus \mathbb{Z}_4$. In the second case, the product in HK will have the form

$$(h_1k_1)(h_2k_2) = (h_1h_2^{-1})(k_1k_2).$$

In particular, we will have

$$(h^ik^j)(h^uk^v) = h^{i-j}k^{u+v}$$

for all i, j, u, and v. More specifically, $hkh = k$. So G has the presentation

$$\langle h, k \mid h^3 = k^4 = 1, hkh = k \rangle,$$

and $G \cong T$.

Case 2: $K \cong (\mathbb{Z}_2 \oplus \mathbb{Z}_2)$. In this case, we will identify K with $\mathbb{Z}_2 \oplus \mathbb{Z}_2$, and so any homomorphism from K to $\text{Aut}(H)$ will be determined by its actions on the generators $([1],[0])$ and $([0],[1])$.

- Let $\varphi_0 : K \to \text{Aut}(H)$ be defined by

$$\varphi_0(([1],[0])) = \pi_0 \text{ and } \varphi_0(([0],[1])) = \pi_0.$$

Then G is Abelian and isomorphic to $\mathbb{Z}_3 \oplus (\mathbb{Z}_2 \oplus \mathbb{Z}_2)$.

- Let $\varphi_1 : K \to \text{Aut}(H)$ be defined by

$$\varphi_1(([1],[0])) = \pi_0 \text{ and } \varphi_1(([0],[1])) = \pi_1.$$

Then

$$\varphi_1(([a],[b])) = \varphi_1(a([1],[0]) + b([0],[1])) = \pi_0^a\pi_1^b = \pi_1^b.$$

Note that $\pi_1^b(t) = t^{2^b}$. In this case, the product on HK is

$$\begin{aligned}(h^i([a],[b]))(h^u([c],[d])) &= (h^i\pi_1^b(h^u))([a+c],[b+d]) \\ &= h^{i+u2^b}([a+c],[b+d]).\end{aligned}$$

Let $x = ([1],[0])$ in K. Then $(hx)^2 = (hx)(hx) = h^2$, $(hx)^3 = x$, $(hx)^4 = h$, $(hx)^5 = h^2x$, and $(hx)^6 = e$. So $|hx| = 6$, and $N = \langle hx \rangle$ is a subgroup of G of order 6. Since $[G : N] = 2$, it follows that $N \lhd G$. (See Exercise 19 on page 375.) Let $y = ([0],[1]) \in K$. Then $y^2 = e$, and so $G = N\langle y \rangle$. Note that

$$y(hx)y^{-1} = y(hx)y = h^2([1],[1])y = h^2x = (hx)^{-1}.$$

So G has presentation $\langle hx, y \mid (hx)^6 = y^2 = 1, y(hx)y^{-1} = (hx)^{-1} \rangle$. But this is exactly the presentation for D_6, and so $G \cong D_6$ in this case.

- Let $\varphi_2 : K \to \text{Aut}(H)$ be defined by

$$\varphi_2(([1],[0])) = \pi_1 \text{ and } \varphi_2(([0],[1])) = \pi_0.$$

In this case, we also have $G \cong D_6$, which is left as an exercise for you to verify. (See Exercise 5.)

- Let $\varphi_3 : K \to \text{Aut}(H)$ be defined by

$$\varphi_3(([1],[0])) = \pi_1 \text{ and } \varphi_3(([0],[1])) = \pi_1.$$

In this case, we again have $G \cong D_6$, which is left as an exercise for you to verify. (See Exercise 5.)

We can therefore conclude that the only non-Abelian groups of order 12 are D_6, A_4, and T.

Our classification of groups of order 12 demonstrates that two groups $H \rtimes_\alpha K$ and $H \rtimes_\beta K$ can be isomorphic even if $\alpha \neq \beta$. In general, it can be difficult to determine if two semidirect products with the same underlying groups are isomorphic or not. One tool is given in the next lemma (which we will use in our classification of groups of order p^3), and others are presented in Activity 38.10.

Lemma 38.9. *Let K be a finite cyclic group, and let H be any group. Let $\varphi_1 : K \to Aut(H)$ and $\varphi_2 : K \to Aut(H)$ be homomorphisms so that $\varphi_1(K)$ and $\varphi_2(K)$ are conjugate subgroups of $Aut(H)$. Then $(H \rtimes_{\varphi_1} K) \cong (H \rtimes_{\varphi_2} K)$.*

Proof. Since K is cyclic, there is an element k so that $K = \langle k \rangle$. The fact that $\varphi_1(K)$ and $\varphi_2(K)$ are conjugate subgroups of $\mathrm{Aut}(H)$ means that there exists $\delta \in \mathrm{Aut}(H)$ so that $\delta^{-1}\varphi_2(K)\delta = \varphi_1(K)$. So

$$\varphi_1(k) = \delta^{-1}\varphi_2(k^t)\delta$$

for some integer t. Then

$$\varphi_1(k)^i = \left(\delta^{-1}\varphi_2(k^t)\delta\right)^i,$$

and so

$$\begin{aligned}\varphi_1(k^i) &= \delta^{-1}\varphi_2(k^t)^i\delta \\ &= \delta^{-1}\varphi_2(k^i)^t\delta\end{aligned}$$

for any integer i. Thus,

$$\varphi_1(x) = \delta\varphi_2(x^t)\delta^{-1} \tag{38.3}$$

for all $x \in K$.

Since $K = \langle k \rangle$, it follows that $\mathrm{Im}(\varphi_1) = \langle \varphi_1(k) \rangle$ and $\mathrm{Im}(\varphi_2) = \langle \varphi_2(k) \rangle$. Since $\varphi_1(k)$ and $\varphi_2(k)$ are conjugate, it also follows that $|\mathrm{Im}(\varphi_1)| = |\mathrm{Im}(\varphi_2)|$. Thus, $|\varphi_1(k)| = |\varphi_2(k)|$. Equation (38.3) shows that $\varphi_1(k) = \delta^{-1}\varphi_2(k)^t\delta$, and so $|\varphi_2(k)| = |\varphi_1(k)| = |\varphi_2(k)^t|$. Thus, $\gcd(|\varphi_1(K)|, t) = 1$. Since $|\varphi_1(K)|$ divides $|K|$, Exercise 12 shows that there is an integer t' with $t' \equiv t \pmod{|\varphi_1(K)|}$ and $\gcd(t', |K|) = 1$. Since $\varphi_2(x^t) = \varphi_2(x)^t = \varphi_2(x)^{t'} = \varphi_2(x^{t'})$ for all $x \in K$, we can replace t with t' in (38.3). This allows us to assume without loss of generality that $\gcd(t, |K|) = 1$, and so there exist integers x and y with $x|K| + yt = 1$.

Define $\Psi : (H \rtimes_{\varphi_1} K) \to (H \rtimes_{\varphi_2} K)$ by $\Psi((a,b)) = (\delta(a), b^t)$. To show that Ψ is a homomorphism, let (a_1, b_1) and (a_2, b_2) be in $H \times K$. Let $\cdot_1$ denote the operation in $H \rtimes_{\varphi_1} K$, and let $\cdot_2$ denote the operation in $H \rtimes_{\varphi_2} K$. Then

$$\begin{aligned}\Psi((a_1,b_1)\cdot_1(a_2,b_2)) &= \Psi((a_1\varphi_1(b_1)(a_2), b_1b_2)) \\ &= (\delta(a_1\varphi_1(b_1)(a_2)), (b_1b_2)^t) \\ &= (\delta(a_1(\delta^{-1}\varphi_2(b_1^t)\delta)(a_2)), b_1^tb_2^t) \\ &= (\delta(a_1)\delta((\delta^{-1}\varphi_2(b_1^t)\delta)(a_2)), b_1^tb_2^t) \\ &= (\delta(a_1)\varphi_2(b_1^t)(\delta(a_2)), b_1^tb_2^t) \\ &= (\delta(a_1), b_1^t)\cdot_2(\delta(a_2), b_2^t) \\ &= \Psi((a_1,b_1))\cdot_2\Psi((a_2,b_2)),\end{aligned}$$

and Ψ is a homomorphism.

To show that Ψ is a monomorphism, suppose $(a, k^r) \in \mathrm{Ker}(\Psi)$. Let e_H be the identity in H, and let e_K be the identity in K. Then

$$(e_H, e_K) = \Psi((a, k^r)) = (\delta(a), k^{rt}),$$

and so $\delta(a) = e_H$ and $k^{rt} = e_K$. Since $\delta \in \text{Aut}(H)$, we can conclude that $a = e_H$. That $k^{rt} = e_K$ means $|K|$ divides rt. But $\gcd(t, |K|) = 1$ implies that $|K|$ divides r. Thus, $k^r = e_K$, and so (a, k^r) is the identity in $H \rtimes_{\varphi_1} K$. Therefore, Ker(Ψ) is trivial and Ψ is a monomorphism.

Finally, we will demonstrate that Ψ is an epimorphism. Let $(w, z) \in (H \rtimes_{\varphi_2} K)$. Since δ^{-1} is a surjection, there exists $a \in H$ such that $\delta^{-1}(a) = w$. Recall that $x|K| + yt = 1$, so

$$z = z^{x|K|+yt} = \left(z^{|K|}\right)^x (z^y)^t = (z^y)^t.$$

Then

$$\Psi\left((\delta^{-1}(a), z^y)\right) = (w, z),$$

and we have shown that Ψ is an epimorphism. Therefore, Ψ is an isomorphism, and $(H \rtimes_{\varphi_1} K) \cong (H \rtimes_{\varphi_2} K)$.

■

We will end this investigation by classifying all groups of order p^3, where p is a prime. Doing so will illustrate the fact that it is not necessarily the size of a group that makes classification difficult, but rather how large the powers of the prime divisors of the order are. In working out the classification, we will leave a number of details for you to complete in the exercises. Since we have already classified the groups of order 8, we will restrict ourselves to odd primes.

Let G be a group of order p^3, where p is an odd prime. We know the Abelian groups of order p^3 by the Fundamental Theorem of Finite Abelian Groups: $\mathbb{Z}_{p^3}$, $\mathbb{Z}_{p^2} \oplus \mathbb{Z}_p$, and $\mathbb{Z}_p \oplus \mathbb{Z}_p \oplus \mathbb{Z}_p$. Therefore, we will now focus on the non-Abelian groups of order p^3.

Let G be a non-Abelian group of order p^3. Since G is a p-group, we know that $Z = Z(G)$ is nontrivial. Since G is non-Abelian, there are two possibilities for $|Z|$: $|Z| = p$ or $|Z| = p^2$. If $|Z| = p^2$, then $|G/Z| = p$ and so G/Z is cyclic. It follows then that G is Abelian (see Theorem 27.10 on page 364), a contradiction. We can therefore conclude that $|Z| = p$. Thus, $|G/Z| = p^2$, and so $G/Z \cong \mathbb{Z}_{p^2}$ or $G/Z \cong (\mathbb{Z}_p \oplus \mathbb{Z}_p)$. Since G/Z is not cyclic, it must be that $G/Z \cong (\mathbb{Z}_p \oplus \mathbb{Z}_p)$.

Recall from Exercise 9 of Investigation 24 (see page 332) that the commutator of elements $x, y \in G$ is the element $[x, y] = x^{-1}y^{-1}xy$. Since G is non-Abelian, the subgroup G' generated by the commutators of pairs of elements is nontrivial. Let $x, y \in G$. Then $xZ, yZ \in G/Z \cong (\mathbb{Z}_p \oplus \mathbb{Z}_p)$ (an Abelian group), and so

$$Z = (xZ)^{-1}(yZ)^{-1}(xZ)(yZ) = (x^{-1}y^{-1}xy)Z.$$

This, however, implies that $x^{-1}y^{-1}xy \in Z$. Thus, $G' \subseteq Z$, and so $G' = Z$.

Define $\psi : G \to G$ by $\psi(g) = g^p$. That ψ is a homomorphism can be demonstrated as follows. Let $a, b \in G$. Since $G' = Z$, the commutators commute with every element in G. Exercise 8 and the fact that every nonidentity element in Z has order p show that

$$\begin{aligned}
\psi(a)\psi(b) &= a^p b^p \\
&= (ab)^p [a, b]^{p(p-1)/2} \\
&= (ab)^p \left([a, b]^{(p-1)/2}\right)^p \\
&= (ab)^p \\
&= \psi(ab),
\end{aligned}$$

and ψ is a homomorphism. Moreover, we can again use the fact that every nonidentity element in G/Z has order p to see that

$$Z = (aZ)^p = a^p Z,$$

and so $a^p \in Z$ for every $a \in G$. Thus, $\text{Im}(\psi) \subseteq Z$.

Now consider $\text{Ker}(\psi)$. If every nonidentity element in G has order p, then $\text{Ker}(\psi) = G$. Otherwise, G contains an element of order p^2. We will consider each of these cases. Let e be the identity in G.

Case 1: There is an element h of order p^2 in G. Let $H = \langle h \rangle$. It follows that $H \cap Z = \langle h^p \rangle$. By Exercise 7, we have that H is normal in G. If we can find a subgroup K of G so that $G = HK$, then we will have $G \cong (H \rtimes_\varphi K$ for some φ.

Since $h^p \neq e$, we have $\text{Ker}(\psi) \neq G$. So $|\text{Im}(\psi)| > 1$, and it follows that $\text{Im}(\psi) = Z$. Thus, $|\text{Ker}(\psi)| = p^2$. Since $h \notin \text{Ker}(\psi)$, we have that $\text{Ker}(\psi) \neq H$. Let k be any element of $\text{Ker}(\psi)$ such that $k \notin H$, and let $K = \langle k \rangle$. Note that $e = \psi(k) = k^p$, so $|k| = p$. Since every nonidentity element in K generates K, it follows that $H \cap K = \{e\}$, and so $G = HK$. Thus, $G \cong (\mathbb{Z}_{p^2} \rtimes_\varphi \mathbb{Z}_p)$ for some nontrivial $\varphi : \mathbb{Z}_p \to \text{Aut}(\mathbb{Z}_{p^2})$. Now

$$\text{Aut}(\mathbb{Z}_{p^2}) \cong U_{p^2} \cong \mathbb{Z}_{p^2-p}$$

by Theorem 37.9 (see page 510), so $\text{Aut}(\mathbb{Z}_{p^2})$ contains a unique subgroup of order p. (In fact, $\text{Aut}(\mathbb{Z}_{p^2}) = \langle \gamma \rangle$, where $\gamma([x]) = (1+p)[x]$.) Lemma 38.9 then shows that the mappings $\varphi : \mathbb{Z}_p \to \text{Aut}(\mathbb{Z}_{p^2})$ will all produce isomorphic groups, and so there is just one non-Abelian group G of order p^3 if G contains an element of order p^2.

Case 2: Every nonidentity element in G has order p. Every p-group contains normal subgroups of any order dividing the order of the group (see Exercise 25 on page 459 of Investigation 32), so let H be a normal subgroup of G of order p^2. Since no element in G has order p^2, it must be the case that $H \cong (\mathbb{Z}_p \oplus \mathbb{Z}_p)$. Let a and b be generators for H so that $|a| = |b| = p$ and $H = \langle a \rangle \langle b \rangle$. Let $k \in G$ with $k \notin H$, and let $K = \langle k \rangle$. Again, $H \cap K = \{e\}$ and so $G = HK$. Thus, $G \cong (H \rtimes_\varphi K)$ for some nontrivial $\varphi : K \to \text{Aut}(H)$. (Note that $|k| = p$.) Now $\text{Aut}(H) \cong \text{Aut}(\mathbb{Z}_p \oplus \mathbb{Z}_p) \cong \text{GL}_2(\mathbb{Z}_p)$, and $|\text{Aut}(H)| = p^4 - p^3 - p^2 + p = p(p^3 - p^2 - p + 1)$. (See Exercise 10.) The Sylow p-subgroups of $\text{Aut}(H)$ all have order p, so any two subgroups of $\text{Aut}(H)$ of order p are conjugate.

Define $\gamma \in \text{Aut}(H) = \langle a \rangle \langle b \rangle$ by $\gamma(a) = ab$ and $\gamma(b) = b$. Since every non-identity element in H has order p, it follows that $|\gamma| = p$, and so any Sylow p-subgroup of $\text{Aut}(H)$ is conjugate to $\gamma(K)$. (We can also represent γ as an element in $\text{GL}_2(\mathbb{Z}_p)$ as $\begin{bmatrix} 1 & 1 \\ 0 & 1 \end{bmatrix}$.) Lemma 38.9 again shows that there is exactly one non-Abelian group of order p^3 of this type. This group is the Heisenberg group we introduced in Exercise 9 of Investigation 29. (See page 413.)

So the groups of order p^3, for p an odd prime, are:

- $\mathbb{Z}_{p^3}$;
- $\mathbb{Z}_{p^2} \oplus \mathbb{Z}_p$;
- $\mathbb{Z}_p \oplus \mathbb{Z}_p \oplus \mathbb{Z}_p$;
- $\mathbb{Z}_{p^2} \rtimes_\varphi \mathbb{Z}_p$, where $\varphi([1]) = \gamma$ with $\gamma([x]) = (1+p)[x]$; and
- $(\mathbb{Z}_p \oplus \mathbb{Z}_p) \rtimes_\varphi \mathbb{Z}_p$, where $\varphi([1]) = \gamma$ with γ represented by the matrix transformation $\begin{bmatrix} 1 & 1 \\ 0 & 1 \end{bmatrix}$.

Concluding Activities

Activity 38.10. Let H and K be finite groups, and let φ_1 and φ_2 be homomorphisms from K to $\text{Aut}(H)$. It can be a difficult task to determine if $H \rtimes_{\varphi_1} K$ is isomorphic to $H \rtimes_{\varphi_2} K$.

(a) If φ_1 and φ_2 act in a similar way on elements of K, then we might expect the corresponding semidirect products to be isomorphic. Show that if $\varphi_1 = \varphi_2\theta$ for some $\theta \in \text{Aut}(K)$, then $(H \rtimes_{\varphi_1} K) \cong (H \rtimes_{\varphi_2} K)$.

(b) If φ_1 and φ_2 are significantly different in some way, we should expect that $H \rtimes_{\varphi_1} K$ is not isomorphic to $H \rtimes_{\varphi_2} K$. In this part we will show that if $\gcd(|H|, |K|) = 1$ and $\text{Ker}(\varphi_1) \ncong \text{Ker}(\varphi_2)$, then $H \rtimes_{\varphi_1} K$ is not isomorphic to $H \rtimes_{\varphi_2} K$. Let e_H be the identity in H, and let e_K be the identity in K. Then let $H' = \{(h, e_K) : h \in H\}$ and $K' = \{(e_H, k) : k \in K\}$ be subgroups of $H \rtimes_{\varphi_1} K$ that are copies of H and K. For $h \in H$, let $h' = (h, e_K)$ and for $k \in K$, let $k' = (e_H, k)$.

(i) Show that $\text{Ker}(\varphi_1) = \{x \in K : x'h'(x')^{-1} = h' \text{ for all } h' \in H'\}$. In other words, $\text{Ker}(\varphi_1) \cong C_{K'}(H')$.

(ii) Assume that $\gcd(|H|, |K|) = 1$ throughout the remainder of this exercise. Show that H' is the only subgroup of order $|H|$ in $G_1 = H \rtimes_{\varphi_1} K$.

(iii) Prove that if $H \rtimes_{\varphi_1} K$ is isomorphic to $H \rtimes_{\varphi_2} K$, then $\text{Ker}(\varphi_1) \cong \text{Ker}(\varphi_2)$. This will show that if $\text{Ker}(\varphi_1) \ncong \text{Ker}(\varphi_2)$, then $(H \rtimes_{\varphi_1} K) \ncong (H \rtimes_{\varphi_2} K)$.

Activity 38.11. Write a short summary that describes the important concepts, and the relationships between these concepts, that were introduced in this investigation. Explain how the material in this investigation is connected to the material in Investigations 20, 28, and 29.

Exercises

(1) Let $H = \mathbb{Z}_n$ and $K = \mathbb{Z}_2$. Let $\varphi : K \to \text{Aut}(H)$ be defined by $\varphi([1]_2) = \pi$, where $\pi([x]) = -[x]$ for all $x \in H$. Show that $(H \rtimes_\varphi K) \cong D_n$.

(2) Can the group $\mathbf{Q} = \{\pm 1, \pm i, \pm j, \pm k\}$ of quaternions be written as a semidirect product of proper subgroups? Explain.

* (3) **Direct sums and semidirect products.** The direct sum of two groups is a special case of the semidirect product, as this exercise illustrates. Let H and K be groups with identities e_H and e_K, respectively, and let $\varphi : K \to \text{Aut}(H)$ be a homomorphism. Show that the following are equivalent.

(a) The map $\Phi : (H \rtimes_\varphi K) \to (H \oplus K)$ defined by $\Phi((h, k)) = hk$ is an isomorphism.

(b) The homomorphism φ is the trivial homomorphism.

(c) The subgroup $K' = \{(e_H, k) : k \in K\}$ is normal in $H \rtimes_\varphi K$.

* (4) Prove the remaining items from Theorem 38.5. That is, prove that

(ii) $K' = \{(e_H, k) : k \in K\}$ is a subgroup of G isomorphic to K and

(iii) $H' \cap K' = \{(e_H, e_K)\}$.

⋆ (5) Complete the classification of groups of order 12 by showing that $(\mathbb{Z}_3 \rtimes_{\varphi_2} (\mathbb{Z}_2 \oplus \mathbb{Z}_2)) \cong D_3$ and that $(\mathbb{Z}_3 \rtimes_{\varphi_3} (\mathbb{Z}_2 \oplus \mathbb{Z}_2)) \cong D_3$, where φ_2 and φ_3 are as described in that section.

⋆ (6) When is $H \rtimes_\varphi K$ an Abelian group? Prove your answer. (Hint: Exercise 3 might be helpful.)

⋆ (7) Let G be a finite group of order n, and let p be the smallest prime divisor of n. Prove that any subgroup of index p in G is normal in G. (Hint: Consider the homomorphism $\pi : G \to P(G/N)$ defined by $\pi(a)(gN) = (ag)N$, where N is a subgroup of G of index p.)

⋆ (8) Let G be any group, and let $x, y \in G$ so that x and y commute with $[x, y] = x^{-1}y^{-1}xy$. Prove that

$$x^n y^n = (xy)^n [x, y]^{n(n-1)/2}$$

for every nonnegative integer n.

(9) **Groups of order pq.** We have previously shown that any group of order pq is cyclic if p and q are distinct primes and p does not divide $q - 1$. We will now classify the rest of the groups of order pq. Let p and q be primes with $p < q$ so that p divides $q - 1$, and let G be a group of order pq. Determine all of the groups to which G could be isomorphic.

⋆ (10) Let p be a prime. Explain why $\text{Aut}(\mathbb{Z}_p \oplus \mathbb{Z}_p) \cong \text{GL}_2(\mathbb{Z}_p)$. Then show that the order of $\text{GL}_2(\mathbb{Z}_p)$ is $p^4 - p^3 - p^2 + p$. (Hint: Use the result from linear algebra that a 2×2 matrix is invertible if and only if no row is a multiple of the other.)

(11) Our definition of a semidirect product requires two groups. In this exercise, we will introduce a useful construction of a semidirect product that uses only one group. Let H be any group, and let $K = \text{Aut}(H)$. Define $\varphi : K \to \text{Aut}(H)$ by $\varphi(\pi) = \pi$. The resulting semidirect product $H \rtimes_\varphi K$ is called the **holomorph** of H and is denoted $\text{Hol}(H)$. Holomorphs provide a context in which to study elements of a group and their automorphisms together.

(a) Let $n \geq 2$ be an integer. Find $|\text{Hol}(\mathbb{Z}_n)|$ in terms of the prime factors of n. (Hint: See Exercise 13 on page 519 of Investigation 37.)

(b) Create the operation table for $\text{Hol}(\mathbb{Z}_3)$. To which familiar group is $\text{Hol}(\mathbb{Z}_3)$ isomorphic?

(c) Let $H = \mathbb{Z}_2 \oplus \mathbb{Z}_2$, $K = \text{Aut}(H)$, and $G = \text{Hol}(H)$. Show that $G \cong S_4$ using the following steps.

(i) Determine the order of $\text{Hol}(\mathbb{Z}_2 \oplus \mathbb{Z}_2)$

(ii) For ease of notation, we will identify K with the subgroup of G that is isomorphic to K. Note that $[G : K] = 4$, so the permutation group $P(G/K)$ of the left cosets of K in G is isomorphic to S_4. Now define $\theta : G \to P(G/K)$ by $\theta(g)(aK) = (ga)K$. We have shown that θ is a homomorphism, so if we can show that θ is a bijection, then we will have that $G \cong S_4$. Prove that θ is a monomorphism. How can we conclude that $G \cong S_4$?

⋆ (12) Let t, m, and n be integers such that m divides n and $\gcd(t, m) = 1$. Prove that there exists an integer t' such that $t' \equiv t \pmod{m}$ and $\gcd(t', n) = 1$.

(13) (a) There are two distinct semidirect products of the form $\mathbb{Z}_4 \rtimes \mathbb{Z}_2$. What are they? Construct the operation table for the non-Abelian one. To which familiar group is this non-Abelian semidirect product isomorphic?

(b) Let $n \geq 3$ be an integer. Show that $D_n \cong (\mathbb{Z}_n \rtimes_{\varphi} \mathbb{Z}_2)$ for some appropriate choice of φ.

(14) **Groups of order pq^2.**

(a) Construct a non-Abelian group of order 75.

(b) Determine all non-Abelian groups of order 75.

(c) Determine all groups of order pq^2, where p and q are primes with $p < q$ such that p does not divide $q - 1$.

(15) Classify all groups of order 20.

(16) **Groups of order $2p^2$.** Let p be an odd prime. In this exercise, we will classify all groups of order 18 as part of the more general problem of classifying all groups of order $2p^2$.

(a) Use the steps suggested below to determine the conjugacy classes of the elements of order 2 in $\mathrm{GL}_2(\mathbb{F}_p)$.

(i) Show that every element of order 2 or less in $\mathrm{GL}_2(\mathbb{F}_p)$ is conjugate to a diagonal matrix with 1's and/or -1's along the diagonal.

(ii) If A is an element of order 2 in $\mathrm{GL}_2(\mathbb{F}_p)$, show that A is conjugate to either $-I$ or the matrix $\begin{bmatrix} 1 & 0 \\ 0 & -1 \end{bmatrix}$.

(b) Classify all groups of order $2p^2$. Note that there are three non-Abelian groups, and you may not be able to easily distinguish them. You might try to show that they have different centers. Part (a) of this exercise and Activity 38.10 should be useful.

Connections

This investigation continued our classification of groups of various orders that we began in Investigation 29. The tools we used in this investigation were mostly familiar, with the exception of the new construction of the semidirect product, an extension of the direct product first discussed in Investigation 28.

Appendix A

Functions

Focus Questions

By the end of this investigation, you should be able to give precise and thorough answers to the questions listed below. You may want to keep these questions in mind to focus your thoughts as you complete the investigation.

- What is a function?
- What does it mean to say that a function is an injection? How can we prove that a function is (or is not) an injection?
- What does it mean to say that a function is a surjection? How can we prove that a function is (or is not) a surjection?
- What is a bijection?
- What is the composition of two functions, and what is a composite function? What are some important theorems about composite functions?
- What is the inverse of a function? Under what conditions is the inverse of a function $f : A \to B$ a function from B to A?
- What are some important theorems about functions and their inverses?

Functions are frequently used in mathematics to define and describe certain relationships between sets and other mathematical objects. In this appendix, we will first study special types of functions known as injections and surjections. Before defining these types of functions, we will review the definition of a function and explore certain functions with finite domains.

Definition A.1. A function f from a set A to a set B is a collection of ordered pairs

$$\{(a, b) : a \in A \text{ and } b \in B\}$$

such that for each element a in A, there is one and only one element in B such that (a, b) is in f.

There is a special notation, called *functional notation*, that is commonly used to describe functions and the way they act on sets. In particular, if (a, b) is in the function f, we write $f(a) = b$ (read as "f of a equals b"). It is important to note the dual use of the symbol f here; we use f to represent a collection of ordered pairs and also to describe an action (pairing a with b in $f(a) = b$). In general practice, we use functional notation and think of a function as assigning to an element a a unique element b. In this context, we think of the elements in A as the input of the assignment and the elements in B as the output. In this way, we can consider f as a *mapping* from A to B and write

$f : A \to B$ to indicate this mapping action. How we read this notation depends on the context in which it appears. For instance, the statement

Consider the function $f : A \to B$

would be read as, "Consider the function f from A to B." On the other hand, if we write

Let $f : A \to B$,

then this statement would be read as "Let f be a function from A to B" or "Let f map from A to B."

There is some familiar terminology and notation associated with functions. Let f be a function from a set A to a set B.

- The set A is called the **domain** of f, and we write $\text{dom}(f) = A$.
- The set B is called the **codomain** of f, and we write $\text{codom}(f) = B$.
- The subset $\{f(a) : a \in A\}$ of B is called the **range** of f, which we denote by range(f). Note that the range of f could equivalently be defined as follows:

$$\text{range}(f) = \{y \in B \mid y = f(x) \text{ for some } x \in A\}.$$

- If $a \in A$, then $f(a)$ is the **image** of a under f.
- If $b \in B$ and $b = f(a)$ for some $a \in A$, then a is called a **pre-image** of b.

Notice that, according to these definitions, $\text{range}(f) \subseteq \text{codom}(f)$, but it is not necessarily the case that $\text{range}(f) = \text{codom}(f)$. Whether we have this set equality or not depends on the function f, as we will see in the next section.

Special Types of Functions: Injections and Surjections

Preview Activity A.2. Let $A = \{1, 2, 3\}$, $B = \{a, b, c, d\}$, and $C = \{s, t\}$. Define

$f : A \to B$ by	$g : A \to B$ by	$h : A \to C$ by
$f(1) = a$	$g(1) = a$	$h(1) = s$
$f(2) = b$	$g(2) = b$	$h(2) = t$
$f(3) = c$	$g(3) = a$	$h(3) = s$

(a) Consider the following property, defined for an arbitrary function F:

For all $x, y \in \text{dom}(F)$, if $x \neq y$, then $F(x) \neq F(y)$.

Which of the functions defined above satisfy this property?

(b) Which of the functions defined above satisfy the following property (defined in terms of an arbitrary function F)?

For all $x, y \in \text{dom}(F)$, if $F(x) = F(y)$, then $x = y$.

(c) Determine the range of each of the functions f, g, and h.

(d) Which of these functions have their range equal to their codomain?

(e) Which of the these functions satisfy the following property (again, defined in terms of an arbitrary function F)?

For all y in the codomain of F, there exists an $x \in \text{dom}(F)$ such that $F(x) = y$.

(f) Let F be a function from a set S to a set T.

(i) Is it possible to have two elements x_1 and x_2 in S with $x_1 \neq x_2$ and $F(x_1) = F(x_2)$? If no, explain why not. If yes, give an example and explain why this does not violate the definition of a function.

(ii) Are the range and codomain of a function the same or different? If they are the same, explain why. If different, give an example to illustrate the difference and explain any relationships that must exist between the two sets.

Preview Activity A.3. Let A and B be nonempty sets, and let $f : A \to B$. In Preview Activity A.2, we determined whether or not certain functions satisfied some specified properties. These properties were written in the form of statements, and we will now examine these statements in more detail.

(a) Consider the following statement:

For all $x, y \in A$, if $x \neq y$, then $f(x) \neq f(y)$.

(i) Write the contrapositive of this conditional statement.

(ii) Write the negation of this conditional statement.

(b) Now consider the following statement:

For all $y \in B$, there exists an $x \in A$ such that $f(x) = y$.

Write the negation of this statement.

(c) Let $g : \mathbb{R} \to \mathbb{R}$ be defined by $g(x) = 5x + 3$, for all $x \in \mathbb{R}$. Complete the proofs of the following propositions about the function g.

Proposition 1. For all $a, b \in \mathbb{R}$, if $g(a) = g(b)$, then $a = b$.

Proof. Let $a, b \in \mathbb{R}$, and assume that $g(a) = g(b)$. We will prove that $a = b$. Since $g(a) = g(b)$, we know that

$$5a + 3 = 5b + 3.$$

(Now prove that in this situation, $a = b$.)

Proposition 2. For all $b \in \mathbb{R}$, there exists an $a \in \mathbb{R}$ such that $g(a) = b$.

Proof. Let $b \in \mathbb{R}$. We will prove that there exists an $a \in \mathbb{R}$ such that $g(a) = b$ by constructing such an a in $\mathbb{R}$. In order for this to happen, we need $g(a) = 5a + 3 = b$.

(Now solve the equation for a, and then show that for this real number a, $g(a) = b$.)

Injections

We have now seen examples of functions for which there exist different inputs that produce the same output. Using more formal notation, this means that there are functions $f : A \to B$ for which there exist $x_1, x_2 \in A$ with $x_1 \neq x_2$ and $f(x_1) = f(x_2)$. The work in the preview activities was intended to motivate the following definition.

Definition A.4. Let $f : A \to B$ be a function from the set A to the set B. The function f is called an **injection** provided that

$$\text{for all } x_1, x_2 \in A, \text{ if } x_1 \neq x_2, \text{ then } f(x_1) \neq f(x_2).$$

When f is an injection, we also say that f is a **one-to-one function**, or that f is an **injective function**.

Notice that the condition that specifies that a function f is an injection is given in the form of a conditional statement. As we will see, in proofs it is usually easier to use the contrapositive of this conditional statement. Although we did not define the term then, we have already written the contrapositive for the conditional statement in the definition of an injection in part (a) of Preview Activity A.3. In that preview activity, we also wrote the negation of the definition of an injection. The box below summarizes this work by giving the conditions that are equivalent to f being an injection or not being an injection.

Let $f : A \to B$.

"The function f is an injection" means that

- for all $x_1, x_2 \in A$, if $x_1 \neq x_2$, then $f(x_1) \neq f(x_2)$; or
- for all $x_1, x_2 \in A$, if $f(x_1) = f(x_2)$, then $x_1 = x_2$.

"The function f is not an injection" means that

- there exist $x_1, x_2 \in A$ such that $x_1 \neq x_2$ and $f(x_1) = f(x_2)$.

Activity A.5. Now that we have defined what it means for a function to be an injection, we can see that in part (c) of Preview Activity A.3, we proved that the function $g : \mathbb{R} \to \mathbb{R}$, where $g(x) = 5x+3$ for all $x \in \mathbb{R}$, is an injection. Use the definition (or its negation) to determine whether or not the following functions are injections.

(a) $k : A \to B$, where $A = \{a, b, c\}$, $B = \{1, 2, 3, 4\}$, and $k(a) = 4$, $k(b) = 1$, and $k(c) = 3$

(b) $f : A \to C$, where $A = \{a, b, c\}$, $C = \{1, 2, 3\}$, and $f(a) = 2$,$f(b) = 3$, and $f(c) = 2$

(c) $F : \mathbb{Z} \to \mathbb{Z}$ defined by $F(m) = 3m + 2$ for all $m \in \mathbb{Z}$

(d) $h : \mathbb{R} \to \mathbb{R}$ defined by $h(x) = x^2 - 3x$ for all $x \in \mathbb{R}$

(e) $s : \mathbb{Z}_5 \to \mathbb{Z}_5$ defined by $s(x) = x^3$ for all $x \in \mathbb{Z}_5$

Surjections

In previous mathematics courses and in Preview Activity A.2, we have seen that there exist functions $f : A \to B$ for which the codomain and range of f are equal—that is, range$(f) = B$. This means that every element of B is an output of the function f for some input from the set A. Using quantifiers, this means that for every $y \in B$, there exists an $x \in A$ such that $f(x) = y$. One of the objectives of the preview activities was to motivate the following definition:

Definition A.6. Let $f : A \to B$ be a function from the set A to the set B. The function f is called a **surjection** provided that the range of f equals the codomain of f. This means that

$$\text{for every } y \in B, \text{ there exists an } x \in A \text{ such that } f(x) = y.$$

When f is a surjection, we also say that f is an **onto function**, that f maps A **onto** B, or that f is a **surjective function**.

Note that the main condition defining what it means for a function f to be a surjection is given in the form of a universally quantified statement. Although we did not define the term then, we have already written the negation of this statement defining a surjection in part (b) of Preview Activity A.3. The box below summarizes the conditions for f being a surjection or not being a surjection.

Let $f : A \to B$.

"The function f is a surjection" means that

- $\text{range}(f) = \text{codom}(f) = B$; or
- for every $y \in B$, there exists an $x \in A$ such that $f(x) = y$.

"The function f is not a surjection" means that

- $\text{range}(f) \neq \text{codom}(f)$; or
- there exists a $y \in B$ such that for all $x \in A$, $f(x) \neq y$.

Activity A.7. Now that we have defined what it means for a function to be a surjection, we can see that in part (c) of Preview Activity A.3, we proved that the function $g : \mathbb{R} \to \mathbb{R}$, where $g(x) = 5x+3$ for all $x \in \mathbb{R}$, is a surjection. Determine whether or not the following functions are surjections. Are any of these functions injections?

(a) $k : A \to B$, where $A = \{a, b, c\}$, $B = \{1, 2, 3, 4\}$, and $k(a) = 4$, $k(b) = 1$, and $k(c) = 3$.

(b) $f : \mathbb{R} \to \mathbb{R}$ defined by $f(x) = 3x + 2$ for all $x \in \mathbb{R}$.

(c) $F : \mathbb{Z} \to \mathbb{Z}$ defined by $F(m) = 3m + 2$ for all $m \in \mathbb{Z}$.

(d) $s : \mathbb{Z}_5 \to \mathbb{Z}_5$ defined by $s(x) = x^3$ for all $x \in \mathbb{Z}_5$.

Another important class of functions are those that are both injective and surjective. Any such function is called a *bijection*.

Definition A.8. A **bijection** is a function that is both an injection and a surjection. If the function f is a bijection, we also say that f is **one-to-one and onto** and that f is a **bijective function**.

Activity A.9. Which of the functions in Activity A.7 are bijections?

The Importance of the Domain and Codomain

The functions in the next activity will illustrate why the domain and the codomain are just as important as the rule defining the outputs when we are trying to determine if a given function is injective and/or surjective.

Activity A.10. Let $f : \mathbb{R} \to \mathbb{R}$ be defined by $f(x) = x^2 + 1$. Notice that

$$f(2) = 5 \text{ and } f(-2) = 5.$$

This observation is enough to prove that the function f is not an injection since we can see that there exist two different inputs that produce the same output.

Since $f(x) = x^2 + 1$, we know that $f(x) \geq 1$ for all $x \in \mathbb{R}$. This implies that the function f is not a surjection. For example, -2 is in the codomain of f and $f(x) \neq -2$ for all x in the domain of f.

(a) Now let $T = \{y \in \mathbb{R} \mid y \geq 1\}$, and define $F : \mathbb{R} \to T$ by $F(x) = x^2 + 1$. Notice that the function F uses the same formula as the function f and has the same domain as f, but has a different codomain than f.

  (i) Explain why F is not an injection.

  (ii) Is F a surjection? Justify your conclusion.

(b) Let $\mathbb{Z}^* = \{x \in \mathbb{Z} \mid x \geq 0\} = \mathbb{N} \cup \{0\}$. Define $g : \mathbb{Z}^* \to \mathbb{N}$ by $g(x) = x^2 + 1$. (Notice that this is the same formula used in part (a).)

  (i) Calculate $g(0), g(1), g(2), g(3), g(4)$, and $g(5)$. Based on this information, does the function g appear to be an injection? Does the function g appear to be a surjection?

  (ii) Is the function g an injection? Justify your conclusion with a proof or a counterexample.

  (iii) Is the function g a surjection? Justify your conclusion with a proof of a counterexample.

In Activity A.10, the same mathematical formula was used to determine the outputs for the functions. However:

- One of the functions was neither an injection nor a surjection.
- Another one of the functions was not an injection but was a surjection.
- The third function was an injection but was not a surjection.

This illustrates the important fact that whether a function is injective or surjective not only depends on the formula that defines the output of the function but also on the domain and codomain of the function.

Composition of Functions

The basic idea of function composition is that, when possible, the output of a function f is used as the input of a function g. The resulting function can be referred to as "f followed by g" and is called the composition of f with g. For example, if

$$f(x) = 3x^2 + 2 \quad \text{and} \quad g(x) = \sin(x),$$

then we can compute $g(f(x))$ as follows:

$$\begin{aligned} g(f(x)) &= g\left(3x^2+2\right) \\ &= \sin\left(3x^2+2\right). \end{aligned}$$

In this case, $f(x)$, the output of the function f, was used as the input for the function g. This idea motivates the formal definition of the composition of two functions.

Definition A.11. Let A, B, and C be nonempty sets, and let $f : A \to B$ and $g : B \to C$ be functions. The **composition of f and g** is the function $g \circ f : A \to C$ defined by

$$(g \circ f)(x) = g\left(f(x)\right)$$

for all $x \in A$. We often refer to the function $g \circ f$ as a **composite function**.

Activity A.12. Let $A = \{1, 2, 3\}$, $B = \{a, b, c, d\}$, and $C = \{s, t\}$. Define $f : A \to B$ by

$$f(1) = a, f(2) = b, f(3) = c,$$

$g : A \to B$ by

$$g(1) = c, g(2) = d, g(3) = c,$$

and $h : B \to C$ by

$$h(a) = s, h(b) = s, h(c) = t, h(d) = s.$$

(a) Find the images of the elements in A under the function $f \circ h$.

(b) Find the images of the elements in A under the function $g \circ h$.

(c) Is $f \circ h$ an injection? Is $f \circ h$ a surjection? Explain.

(d) Is $g \circ h$ an injection? Is $g \circ h$ a surjection? Explain.

In Activity A.12, we asked questions about whether certain composite functions were injections and/or surjections. In mathematics, it is typical to explore whether certain properties of an object transfer to related objects. In particular, we might want to know whether or not the composite of two injective functions is also an injection. (Of course, we could ask a similar question for surjections.) These types of questions are explored in the next activity.

Activity A.13. Let the sets A, B, C, and D be as follows:

$$A = \{a, b, c\}, \quad B = \{p, q, r\}, \quad C = \{u, v, w, x\}, \quad \text{and} \quad D = \{u, v\}.$$

(a) Construct a function $f : A \to B$ that is an injection and a function $g : B \to C$ that is an injection. In this case, is the composite function $g \circ f : A \to C$ an injection? Explain.

(b) Construct a function $f : A \to B$ that is a surjection and a function $g : B \to D$ that is a surjection. In this case, is the composite function $g \circ f : A \to D$ a surjection? Explain.

(c) Construct a function $f : A \to B$ that is a bijection and a function $g : B \to A$ that is a bijection. In this case, is the composite function $g \circ f : A \to A$ a bijection? Explain.

In Activity A.13, we explored some properties of composite functions related to injections, surjections, and bijections. The following theorem summarizes the results that these explorations were intended to illustrate.

Theorem A.14. *Let A, B, and C be nonempty sets, and assume that $f : A \to B$ and $g : B \to C$.*

(i) *If f and g are both injections, then $(g \circ f) : A \to C$ is an injection.*

(ii) *If f and g are both surjections, then $(g \circ f) : A \to C$ is a surjection.*

(iii) *If f and g are both bijections, then $(g \circ f) : A \to C$ is a bijection.*

The proof of part (i) is Exercise 4, and part (iii) is a direct consequence of the first two parts. Therefore, we will focus here on constructing a proof of part (ii). Our goal is to prove that $g \circ f$ is a surjection. Since $g \circ f : A \to C$, this is equivalent to proving that

$$\text{for all } c \in C, \text{ there exists an } a \in A \text{ such that } (g \circ f)(a) = c.$$

Thus, we need to find an $a \in A$ such that $(g \circ f)(a) = c$.

Now we can look at the hypotheses. In particular, we are assuming that both $f : A \to B$ and $g : B \to C$ are surjections. Since we have chosen $c \in C$, and $g : B \to C$ is a surjection, we know that there exists a $b \in B$ such that $g(b) = c$. Now, $b \in B$ and $f : A \to B$ is a surjection. Therefore, there exists an $a \in A$ such that $f(a) = b$. If we now compute $(g \circ f)(a)$, we will see that

$$(g \circ f)(a) = g(f(a)) = g(b) = c.$$

We can now write the complete proof as follows:

Proof of Theorem A.14, part (ii). Let A, B, and C be nonempty sets, and assume that $f : A \to B$ and $g : B \to C$ are both surjections. We will prove that $g \circ f : A \to C$ is a surjection.

Let c be an arbitrary element of C. We will prove there exists an $a \in A$ such that $(g \circ f)(a) = c$. Since $g : B \to C$ is a surjection, it follows that there exists a $b \in B$ such that $g(b) = c$. Now $b \in B$ and $f : A \to B$ is a surjection. Hence, there exists an $a \in A$ such that $f(a) = b$. We now see that

$$\begin{aligned}(g \circ f)(a) &= g(f(a)) \\ &= g(b) \\ &= c.\end{aligned}$$

We have therefore shown that for every $c \in C$, there exists an $a \in A$ such that $(g \circ f)(a) = c$. This proves that $g \circ f$ is a surjection. ■

Inverse Functions

Now that we have studied composite functions, we will move on to consider another important idea: the inverse of a function. In order to study inverse functions, we will need to use the concept of the

Cartesian product of two sets A and B, denoted by $A \times B$, which is the set of all ordered pairs (x, y) where $x \in A$ and $y \in B$. That is,

$$A \times B = \{(x, y) : x \in A \text{ and } y \in B\}.$$

In previous mathematics courses, you probably learned that the exponential function (with base e) and the natural logarithm functions are inverses of each other. You may have seen this relationship expressed as follows:

$$\text{For each } x \in \mathbb{R} \text{ with } x > 0 \text{ and for each } y \in \mathbb{R},$$
$$y = \ln(x) \text{ if and only if } x = e^y.$$

Notice that x is the input and y is the output for the natural logarithm function if and only if y is the input and x is the output for the exponential function. In essence, the inverse function (in this case, the exponential function) reverses the action of the original function (in this case, the natural logarithm function). In terms of ordered pairs (input-output pairs), this means that if (x, y) is an ordered pair for a function, then (y, x) is an ordered pair for its inverse. The idea of reversing the roles of the first and second coordinates is the basis for our definition of the inverse of a function.

Definition A.15. Let $f : A \to B$ be a function. The **inverse** of *f*, denoted by f^{-1}, is the set of ordered pairs $\{(b, a) \in B \times A \mid f(a) = b\}$. That is,

$$f^{-1} = \{(b, a) \in B \times A : f(a) = b\}.$$

If we use the ordered pair representation for *f*, we could also write

$$f^{-1} = \{(b, a) \in B \times A : (a, b) \in f\}.$$

Notice that this definition does not state that f^{-1} is a function. Rather, f^{-1} is simply a subset of $B \times A$. In Activity A.16, we will explore the conditions under which the inverse of a function $f : A \to B$ is itself a function from B to A.

Activity A.16. Let $A = \{a, b, c\}$, $B = \{a, b, c, d\}$, and $C = \{p, q, r\}$. Define

$f : A \to C$ by	$g : A \to C$ by	$h : B \to C$ by
$f(a) = r$	$g(a) = p$	$h(a) = p$
$f(b) = p$	$g(b) = q$	$h(b) = q$
$f(c) = q$	$g(c) = p$	$h(c) = r$
		$h(d) = q$

(a) Determine the inverse of each function as a set of ordered pairs.

(b) (i) Is f^{-1} a function from C to A? Explain.

(ii) Is g^{-1} a function from C to A? Explain.

(iii) Is h^{-1} a function from C to B? Explain.

(c) Make a conjecture about what conditions on a function $F : S \to T$ will ensure that its inverse is a function from T to S.

We will now consider a general argument suggested by the explorations in Activity A.16. By definition, if $f : A \to B$ is a function, then f^{-1} is a subset of $B \times A$. However, f^{-1} may or may

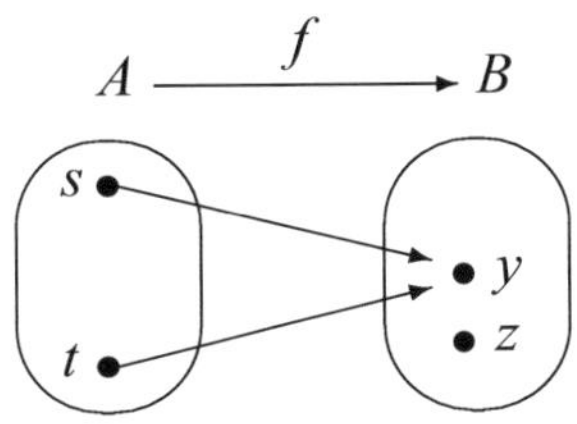

Figure A.1
The inverse is not a function.

not be a function from B to A. For example, suppose that $s, t \in A$ with $s \neq t$ and $f(s) = f(t)$ (as illustrated in Figure A.1).

In this case, if we try to reverse the arrows, we will not get a function from B to A. This is because $(y, s) \in f^{-1}$ and $(y, t) \in f^{-1}$ with $s \neq t$. Consequently, f^{-1} is not a function. This observation suggests that if f is not an injection, then f^{-1} is not a function.

Also, if f is not a surjection, then there exists a $z \in B$ such that $f(a) \neq z$ for all $a \in A$, as in the diagram in Figure A.1. In other words, there is no ordered pair in f with z as the second coordinate. This means that there would be no ordered pair in f^{-1} with z as a first coordinate. Consequently, f^{-1} cannot be a function from B to A.

Theorem A.17 formalizes these observations. In the proof of the theorem, we will use both the input-output representation and the ordered pair representation of a function. The idea is that if $G : S \to T$ is a function, then for $s \in S$ and $t \in T$,

$$G(s) = t \text{ if and only if } (s, t) \in G.$$

When we use the ordered pair representation of a function, we will also use the ordered pair representation of its inverse. In this case, we know that

$$(s, t) \in G \text{ if and only if } (t, s) \in G^{-1}.$$

Theorem A.17. *Let A and B be nonempty sets, and let $f : A \to B$. The inverse of f is a function from B to A if and only if f is a bijection.*

Proof. Let A and B be nonempty sets, and let $f : A \to B$. We will first assume that f is a bijection and prove that f^{-1} is a function from B to A. To do this, we will show that f^{-1} satisfies the conditions of Definition A.1.

Let $b \in B$. Since the function f is a surjection, there exists an $a \in A$ such that $f(a) = b$. This implies that $(a, b) \in f$ and hence that $(b, a) \in f^{-1}$. Thus, each element of B is the first coordinate of an ordered pair in f^{-1}. We must now prove that each element of B is the first coordinate of exactly one ordered pair in f^{-1}. So let $b \in B$, $a_1, a_2 \in A$ and assume that

$$(b, a_1) \in f^{-1} \text{ and } (b, a_2) \in f^{-1}.$$

This means that $(a_1, b) \in f$ and $(a_2, b) \in f$. We can then conclude that

$$f(a_1) = b \text{ and } f(a_2) = b.$$

But this means that $f(a_1) = f(a_2)$. Since f is a bijection, f is by definition an injection, and we can conclude that $a_1 = a_2$. This proves that b is the first element of only one ordered pair in f^{-1}. Consequently, we have proved that f^{-1} satisfies the conditions of Definition A.1 and hence f^{-1} is a function from B to A.

We will now assume that f^{-1} is a function from B to A and prove that f is a bijection. First, to prove that f is an injection, we will assume that $a_1, a_2 \in A$ and that $f(a_1) = f(a_2)$. We wish to show that $a_1 = a_2$. If we let $b = f(a_1) = f(a_2)$, we can conclude that

$$(a_1, b) \in f \text{ and } (a_2, b) \in f.$$

But this means that

$$(b, a_1) \in f^{-1} \text{ and } (b, a_2) \in f^{-1}.$$

Since we have assumed that f^{-1} is a function, we can conclude that $a_1 = a_2$. Hence, f is an injection.

Now to prove that f is a surjection, we will choose an arbitrary $b \in B$ and show that there exists an $a \in A$ such that $f(a) = b$. Since f^{-1} is a function, b must be the first coordinate of some ordered pair in f^{-1}. Consequently, there exists an $a \in A$ such that

$$(b, a) \in f^{-1}.$$

Now this implies that $(a, b) \in f$, and so $f(a) = b$. This proves that f is a surjection. Since we have also proved that f is an injection, we can conclude that f is a bijection, as desired. ■

Theorems about Inverse Functions

In the situation where $f : A \to B$ is a bijection and f^{-1} is a function from B to A, we can write $f^{-1} : B \to A$. In this case, we frequently say that f is an **invertible function**, and we usually do not use the ordered pair representation for either f or f^{-1}. Instead of writing $(a, b) \in f$, we write $f(a) = b$, and instead of writing $(b, a) \in f^{-1}$, we write $f^{-1}(b) = a$. Using the fact that $(a, b) \in f$ if and only if $(b, a) \in f^{-1}$, we can now write $f(a) = b$ if and only if $f^{-1}(b) = a$. Theorem A.18 formalizes this observation.

Theorem A.18. *Let A and B be nonempty sets, and let $f : A \to B$ be a bijection. Then $f^{-1} : B \to A$ is a function, and for every $a \in A$ and $b \in B$,*

$$f(a) = b \textit{ if and only if } f^{-1}(b) = a.$$

The next two results are two important theorems about inverse functions. The first can be considered to be a corollary of Theorem A.18.

Corollary A.19. *Let A and B be nonempty sets, and let $f : A \to B$ be a bijection. Then*

(i) *For every x in A, $\left(f^{-1} \circ f\right)(x) = x$.*

(ii) *For every y in B, $\left(f \circ f^{-1}\right)(y) = y$.*

Activity A.20. Prove Corollary A.19. For the first part, let $x \in A$, write $f(x) = y$, and then use the result in Theorem A.18.

We will now consider the case where $f : A \to B$ and $g : B \to C$ are both bijections. In this case, $f^{-1} : B \to A$ and $g^{-1} : C \to B$. Figure A.2 illustrates this situation.

By Theorem A.14, $g \circ f : A \to C$ is also a bijection. Hence, by Theorem A.17, $(g \circ f)^{-1}$ is a function and, in fact, $(g \circ f)^{-1} : C \to A$. Notice that we can also form the composition of g^{-1} followed by f^{-1} to get $f^{-1} \circ g^{-1} : C \to A$. Figure A.2 helps illustrate the result of the next theorem.

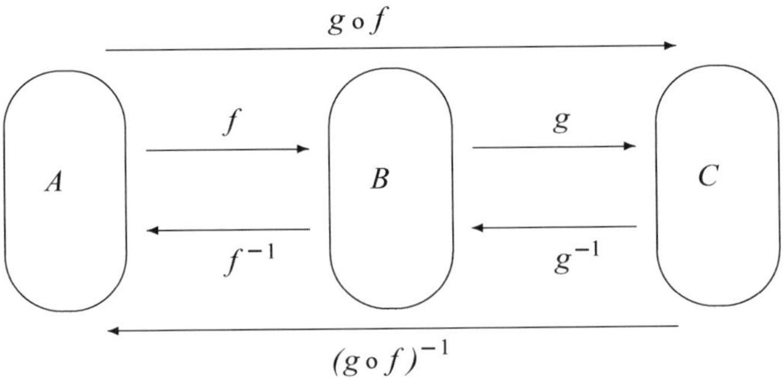

Figure A.2
Composition of two bijections.

Theorem A.21. *Let $f : A \to B$ and $g : B \to C$ be bijections. Then $g \circ f$ is a bijection and $(g \circ f)^{-1} = f^{-1} \circ g^{-1}$.*

Proof. Let $f : A \to B$ and $g : B \to C$ be bijections. Then $f^{-1} : B \to A$ and $g^{-1} : C \to B$. Hence, $f^{-1} \circ g^{-1} : C \to A$. Also, by Theorem A.14, $g \circ f : A \to C$ is a bijection, and hence $(g \circ f)^{-1} : C \to A$. We will now prove that for each $z \in C$, $(g \circ f)^{-1}(z) = (f^{-1} \circ g^{-1})(z)$.

Let $z \in C$. Since the function g is a surjection, there exists a $y \in B$ such that

$$g(y) = z. \tag{A.1}$$

Also, since f is a surjection, there exists an $x \in A$ such that

$$f(x) = y. \tag{A.2}$$

Now equations (A.1) and (A.2) can be written in terms of the respective inverse functions as

$$g^{-1}(z) = y \text{ and} \tag{A.3}$$
$$f^{-1}(y) = x. \tag{A.4}$$

Using equations (A.3) and (A.4), we see that

$$\begin{aligned}\left(f^{-1} \circ g^{-1}\right)(z) &= f^{-1}\left(g^{-1}(z)\right)\\ &= f^{-1}(y)\\ &= x. \end{aligned} \tag{A.5}$$

Using equations (A.1) and (A.2) again, we see that $(g \circ f)(x) = z$. However, in terms of the inverse function, this means that

$$(g \circ f)^{-1}(z) = x. \tag{A.6}$$

Comparing equations (A.5) and (A.6), we have shown that for all $z \in C$, $(g \circ f)^{-1}(z) = (f^{-1} \circ g^{-1})(z)$. This proves that $(g \circ f)^{-1} = f^{-1} \circ g^{-1}$. ■

Concluding Activities

Activity A.22. Prove the following:

If $f : A \to B$ is a bijection, then $f^{-1} : B \to A$ is also a bijection.

Activity A.23. Write a short summary that describes the important concepts, and the relationships between these concepts, that were introduced in this investigation.

Exercises

(1) For each of the following functions, determine if the function is an injection, a surjection, a bijection, or none of these. Justify all of your conclusions.

(a) $F : \mathbb{R} \to \mathbb{R}$ defined by $F(x) = 5x + 3$, for all $x \in \mathbb{R}$.

(b) $G : \mathbb{Z} \to \mathbb{Z}$ defined by $G(x) = 5x + 3$, for all $x \in \mathbb{Z}$.

(c) $f : (\mathbb{R} - \{4\}) \to \mathbb{R}$ defined by $f(x) = \dfrac{3x}{x-4}$, for all $x \in (\mathbb{R} - \{4\})$.

(d) $g : (\mathbb{R} - \{4\}) \to (\mathbb{R} - \{3\})$ defined by $g(x) = \dfrac{3x}{x-4}$, for all $x \in (\mathbb{R} - \{4\})$.

(2) Define $f : \mathbb{N} \to \mathbb{Z}$ as follows: For each $n \in \mathbb{N}$,

$$f(n) = \frac{1 + (-1)^n(2n - 1)}{4}.$$

Is the function f an injection? Is the function f a surjection? Justify your conclusions.

Suggestions: Start by calculating several outputs for the function before you attempt to write a proof. In exploring whether or not the function is an injection, it might be a good idea to use cases based on whether the inputs are even or odd. In exploring whether f is a surjection, consider using cases based on whether the output is positive or less than or equal to zero.

(3) An operation $*$ on a set S is a function from $S \times S$ to S that assigns to the pair $(x, y) \in S \times S$ the element $x * y$ in S. For example, addition of integers can be defined as a function $f : \mathbb{Z} \times \mathbb{Z} \to \mathbb{Z}$ that maps the pair $(a, b) \in \mathbb{Z} \times \mathbb{Z}$ to the integer $f(a, b) = a + b$.

(a) Is the function f an injection? Justify your conclusion.

(b) Is the function f a surjection? Justify your conclusion.

(4) Prove Part (i) of Theorem A.14:

Let A, B, and C be nonempty sets, and let $f : A \to B$ and $g : B \to C$. If f and g are both injections, then $g \circ f : A \to C$ is an injection.

(5) Suppose $f : A \to B$ and $g : B \to C$ are functions.

(a) Is it true that if $f \circ g$ is an injection, then both f and g are injections? If the answer is no, are there any conditions that f or g must satisfy to make $f \circ g$ an injection? Prove your answers.

(b) Is it true that if $f \circ g$ is a surjection, then both f and g are surjections? If the answer is no, are there any conditions that f or g must satisfy to make $f \circ g$ a surjection? Prove your answers.

(6) Is composition of functions a commutative operation? Prove your answer.

(7) Is composition of functions an associative operation? Prove your answer.

(8) (a) Define $f : \mathbb{Z}_5 \to \mathbb{Z}_5$ by $f([x]) = [x^2 + 4]$ for all $[x] \in \mathbb{Z}_5$. Write the inverse of f as a set of ordered pairs, and explain why f^{-1} is not a function.

(b) Define $g : \mathbb{Z}_5 \to \mathbb{Z}_5$ by $f([x]) = [x^3 + 4]$ for all $[x] \in \mathbb{Z}_5$. Write the inverse of g as a set of ordered pairs, and explain why g^{-1} is a function.

(c) Is it possible to write a formula for $g^{-1}([y])$, where $[y] \in \mathbb{Z}_5$? The answer to this question depends on whether or not it is possible to define a cube root of elements of $\mathbb{Z}_5$. Recall that for a real number x, we define the cube root of x to be the real number y such that $y^3 = x$. That is,

$$y = \sqrt[3]{x} \text{ if and only if } y^3 = x.$$

Using this idea, is it possible to define the cube root of each element of $\mathbb{Z}_5$? If so, what is $\sqrt[3]{[0]}$, $\sqrt[3]{[1]}$, $\sqrt[3]{[2]}$, $\sqrt[3]{[3]}$, and $\sqrt[3]{[4]}$.

(d) Now answer the question posed at the beginning of part (c). If possible, determine a formula for $g^{-1}([y])$ where $g^{-1} : \mathbb{Z}_5 \to \mathbb{Z}_5$.

Appendix B

Mathematical Induction and the Well-Ordering Principle

Focus Questions

By the end of this investigation, you should be able to give precise and thorough answers to the questions listed below. You may want to keep these questions in mind to focus your thoughts as you complete the investigation.

- What does the Principle of Mathematical Induction say? What do we need to verify in order to prove a statement using the Principle of Mathematical Induction?
- How do the extended and strong forms of induction differ from the Principle of Mathematical Induction? How are all of these different versions of induction similar?
- What does the Well-Ordering Principle say? What do we need to verify in order to prove a statement using the Well-Ordering Principle?
- How are the Principle of Mathematical Induction, the Extended Principle of Mathematical Induction, the Strong Form of Mathematical Induction, and the Well-Ordering Principle all related?

Preview Activity B.1. Suppose you are on a game show called *Let's Make a Great Deal*. You have reached the final round and will be asked one question. If you answer the question correctly, you will win a key to open door number 1. Behind door number 1 is a prize and a key to open door number 2. Behind door number 2 is a prize and a key to open door number 3. Behind door number 3 is a prize and a key to open door number 4, and so on.

(a) How many prizes will you win if you fail to answer the question correctly?

(b) Which prizes will you win if you answer the question correctly?

Introduction

Mathematical induction is an important tool in mathematics. Induction helps us prove that certain types of statements are true for *all* positive integers. This is quite a feat, since there are infinitely many positive integers! Mathematical induction comes in more than one flavor. There is the basic principle, the extended principle, and the strong (or second, or complete) principle. An equivalent version of the Principle of Mathematical Induction is the Well-Ordering Principle. We will study each of these principles in this investigation.

The Principle of Mathematical Induction

Activity B.1 demonstrated the basic idea behind induction. Just as no prize comes for free (in our game, we needed to answer the question in order to win anything), to verify a statement using induction, we will need to prove something. In other words, we will need to unlock the door that has our first statement behind it. But unlocking the first door is not enough to unlock every other door, unless we are able to establish—as was specified in the rules of our game show—that each door, when opened, contains the key to the next door. If this condition also holds, then once we open the first door—that is, once we prove the first statement—we will be able to open every other door, thus proving our statement for every positive integer.

To illustrate this process in a more concrete way, consider the example in the following activity.

Activity B.2. Let n be a positive integer. Complete Table B.2. What do you notice?

n	$1+2+3+\cdots+n$	$\frac{n(n+1)}{2}$
1		
2		
3		
4		
5		

The calculations in Activity B.2 show that

$$1+2+3+\cdots+n=\frac{n(n+1)}{2} \tag{B.1}$$

for all integers n between 1 and 5. A few more calculations might convince you that equation (B.1) is actually true for many more integers, and perhaps for all positive integers. Although we cannot physically evaluate both sides of equation (B.1) to determine if it is true for every positive integer, we can use mathematical induction to accomplish the same goal.

To return to our game show analogy, think of verifying equation (B.1) as the goal of the game. Each door corresponds to one instance of equation (B.1). The first door corresponds to equation (B.1) with $n = 1$, the second door to equation (B.1) with $n = 2$, and so on. In general, for each positive integer m, the mth door corresponds to equation (B.1) with $n = m$.

In this context, the question we need to answer to open door number one is whether equation (B.1) is true when $n = 1$.

Activity B.3. Is equation (B.1) is true when $n = 1$? Why?

Opening the first door is important, but it does not complete the problem of proving equation (B.1) for *all* positive integers. In the game, behind each door was a key to opening the next door. Of course, these keys were a critical part of the game. If one of the doors did not have a key to the next, then we wouldn't necessarily win all of the prizes just by opening the first door. In the same way, to complete our verification of equation (B.1), we will need to show that the first door ($n = 1$) contains the key to the second door ($n = 2$), the second door ($n = 2$) contains the key to the third door ($n = 3$), and so on.

Our calculations in Activity B.2 show that equation (B.1) is true for n from 1 to 5, but they don't demonstrate that each holds the key to the next. In other words, if equation (B.1) is true for $n = 1$, must it also be true when $n = 2$? And if equation (B.1) is true when $n = 2$, must it also be true when $n = 3$? And so on. In a nutshell, what we need to demonstrate is that if equation (B.1) is true for some arbitrary positive integer n, then it must also be true for the integer $n + 1$. This shows that each instance when equation (B.1) is true for a given positive integer n provides the key to proving that the equation is also true for the integer $n + 1$.

As an example, let's show that if equation (B.1) is true for $n = 1$, then it must also be true for $n = 2$. To do so, we will assume equation (B.1) is true when $n = 1$. That is, we will assume that

$$1 = \frac{(1)(2)}{2}. \tag{B.2}$$

Assuming equation (B.2) is true, we need to prove that equation (B.1) is true when $n = 2$, or that

$$1 + 2 = \frac{(2)(3)}{2}.$$

Since we are assuming equation (B.1) to be true when $n = 1$, we can begin with the true statement (B.2). Adding 2 to both sides of equation (B.2) yields

$$\begin{aligned} 1 + 2 &= \frac{(1)(2)}{2} + 2 \\ &= \frac{(1)(2) + 2(2)}{2} \\ &= \frac{(2)(1 + 2)}{2} \\ &= \frac{(2)(3)}{2}, \end{aligned}$$

which shows that equation (B.1) is true when $n = 2$ (assuming that the same equation is true when $n = 1$).

Activity B.4. To complete our proof of equation (B.1) for all positive integers n, we need to verify that whenever equation (B.1) for some arbitrary positive integer n, then it is also true for the integer $n + 1$.

(a) Continue, as above, to show that if equation (B.1) is true for $n = 2$, then it is also true for $n = 3$.

(b) Of course, we cannot continue showing each specific implication in turn, as that would take an infinite amount of time. To be more efficient, we really want to show that if n is any positive integer and equation (B.1) is true for n, then equation (B.1) is also true for $n + 1$. To do this, we let n be an arbitrary positive integer and assume that equation (B.1) is true for n. That is, we assume that

$$1 + 2 + 3 + \cdots + n = \frac{(n)(n + 1)}{2}.$$

Use this assumption to show that equation (B.1) is true for $n+1$.

To summarize, we needed to prove two things to show that equation (B.1) is true for all positive integers n:

(1) Equation (B.1) is true when $n = 1$; and

(2) whenever equation (B.1) is true for a positive integer n, then it is also true for the integer $n+1$.

Step 1 is equivalent to answering the question in our game show and opening the first door. The prize is that equation (B.1) is true when $n = 1$. Step 2 verifies that each door holds the key to opening the next—that is, if equation (B.1) is true for the integer n, then it is also true for $n+1$. Completing both steps shows that equation (B.1) is true for every positive integer n.

Let's now formalize the ideas from the previous example. In doing so, we will develop the Principle of Mathematical Induction, which can be used when we have a family of statements, one for each positive integer, that we want to prove. For example, for each positive integer n, let $P(n)$ be the statement that

$$1 + 2 + 2^2 + 2^3 + \cdots + 2^{n-1} = 2^n - 1 \tag{B.3}$$

To prove that $P(n)$ is true for all $n \in \mathbb{N}$, we have seen that we need to prove that:

(1) $P(1)$ is true (we call this the *base case*); and

(2) for every positive integer n, if $P(n)$ is true, then $P(n+1)$ is also true. (This second step is called the *inductive step*.)

When we prove the inductive step, we assume $P(n)$ is true for some arbitrary positive integer n. This assumption is called the *induction hypothesis* or *inductive hypothesis*. We then show, using this assumption, that $P(n+1)$ is also true.

Rephrasing this process in a slightly different form leads us to the formal statement of the Principle of Mathematical Induction. Let S be the set of positive integers for which $P(n)$ is true. Proving $P(1)$ is true is the same as showing that 1 is in S. Likewise, showing that $P(n)$ implies $P(n+1)$ is equivalent to showing that $n+1 \in S$ whenever $n \in S$. Combining these two observations, we arrive at the following axiom:

Axiom B.5 (Principle of Mathematical Induction). *Let S be a subset of the set of natural numbers $\mathbb{N}$. If*

(i) *S contains 1 and*

(ii) *S contains the positive integer $n+1$ whenever S contains n,*

then $S = \mathbb{N}$.

In essence, the Principle of Mathematical Induction tells us that if we have a set $S \subseteq \mathbb{N}$ containing 1, and if S contains the integer $n+1$ whenever S contains n, then S must contain $1 + 1 = 2$. But then S must also contain $2 + 1 = 3$, and $3 + 1 = 4$, and so on. Therefore, S will contain *all* natural numbers. By using the Principle of Mathematical Induction, we can prove infinitely many statements in only two steps.

To illustrate, let's formally apply the Principle of Mathematical Induction to establish equation (B.3). Let

$$S = \{n \in \mathbb{N} : 1 + 2 + 2^2 + 2^3 + \cdots + 2^{n-1} = 2^n - 1\}.$$

To use the Principle of Mathematical Induction, we need to show that $1 \in S$ and that $n+1 \in S$ whenever $n \in S$. First we will show that $1 \in S$ (the base case). Notice that when $n = 1$, we have

$$1 + 2 + 2^2 + 2^3 + \cdots + 2^{n-1} = 2^0 = 1$$

and

$$2^n - 1 = 2^1 - 1 = 2 - 1 = 1.$$

So equation (B.3) is true when $n = 1$, which means that $1 \in S$. For the inductive step, we need to show that $n+1 \in S$ whenever $n \in S$. To do so, we will assume that $n \in S$ for some integer $n \geq 1$ (the inductive hypothesis). In this case, we will assume that

$$1 + 2 + 2^2 + 2^3 + \cdots + 2^{n-1} = 2^n - 1. \tag{B.4}$$

We then need to prove that $n+1 \in S$. So we need to show that

$$1 + 2 + 2^2 + 2^3 + \cdots + 2^{n-1} + 2^{(n+1)-1} = 2^{n+1} - 1,$$

or, equivalently,

$$1 + 2 + 2^2 + 2^3 + \cdots + 2^{n-1} + 2^n = 2^{n+1} - 1. \tag{B.5}$$

To prove (B.5), we can substitute from (B.4) in the left hand side of (B.5) to obtain

$$\begin{aligned} 1 + 2 + 2^2 + 2^3 + \cdots + 2^{n-1} + 2^n &= \left(1 + 2 + 2^2 + 2^3 + \cdots + 2^{n-1}\right) + 2^n \\ &= (2^n - 1) + 2^n \\ &= 2(2^n) - 1 \\ &= 2^{n+1} - 1, \end{aligned}$$

which shows that $n+1 \in S$. Therefore, $S = \mathbb{N}$ by the Principle of Mathematical Induction, which means that (B.3) is true for all $n \in \mathbb{N}$.

Activity B.6. Let r be a real number with $r > 1$. Consider the statements

$$1 + r + r^2 + r^3 + \cdots r^{n-1} = \frac{r^n - 1}{r - 1} \tag{B.6}$$

for all $n \geq 1$. (Note that you have probably seen equation (B.6) in a previous class. Do you remember where it comes from?) We will use induction to prove that (B.6) is true for all $n \in \mathbb{N}$.

(a) Identify an appropriate set S on which to apply our induction argument.

(b) What is the base case? Give a precise statement, and then verify your statement.

(c) What is the induction hypothesis? What is the inductive step?

(d) Complete the inductive step to verify that equation (B.6) is true for all $n \in \mathbb{N}$.

The Extended Principle of Mathematical Induction

Preview Activity B.7. Let's now return to our *Let's Make a Great Deal* game. Suppose you fail to correctly answer the question that allows you to open the first door. But suppose also that, after doing so, the host gives you the opportunity to answer a second question. If you correctly answer question number 2, then you win a key to open door number 2. As before, behind door number 2 is a prize and a key to open door number 3. Behind door number 3 is a prize and a key to open door number 4, and so on.

(a) Which prizes would you win if you answered the second question correctly?

(b) Suppose you fail to answer the second question correctly. The host then gives you an option of answering a third question. If you correctly answer question number 3, then you win a key to open door number 3. Which prizes would you win if you answered the third question correctly?

(c) You probably see the pattern by now. Suppose you fail to answer the first $m - 1$ questions correctly for some integer $m \geq 2$. The host then gives you an option of answering an mth question. If you correctly answer question number m, then you win a key to open door number m. Which prizes would you win if you answered the mth question correctly?

(d) Compare this version of the game to the version described in Activity B.1. How are the two versions alike, and how are they different?

The Principle of Mathematical Induction that we stated in the previous section is a method for proving an entire family of statements, one for each positive integer n. There are, however, instances when we need to modify our induction arguments slightly. For example, consider the statement

$$2^n < n!.$$

If we try to prove this statement for all $n \in \mathbb{N}$, we immediately encounter a problem in that the statement is not true for $n = 1$. In fact, the statement is also false for $n = 2$ and $n = 3$. However, the statement does appear to be true for $n \geq 4$. If we want to use the Principle of Mathematical Induction to prove that this statement is true for $n \geq 4$, we will need to somehow translate it to an equivalent statement that is true for *all* positive integers. One way to do this is by re-indexing the statement to say

$$2^{n+3} < (n+3)!$$

for all positive integers n. Although this would solve the problem, the resulting statement is more complicated, and if we started with a more involved result, re-indexing could potentially make the situation appear more difficult than it really is. This is where the result of Activity B.7 is useful; in fact, proving that $2^n < n!$ for all $n \geq 4$ is analogous to answering the 4th question correctly. The point of Activity B.7 is that it really shouldn't make any difference what our starting point is, as long as we have one. If we can open door number n_0 for some integer n_0, then we will be able to open all doors with numbers higher than n_0 as well—even though we may not be able to open the doors numbered lower than n_0. This is the idea behind the *Extended Principle of Mathematical Induction.*

Axiom B.8 (Extended Principle of Mathematical Induction). *Let S be a subset of the set of the integers $\mathbb{Z}$. If there is an integer n_0 such that*

(i) *S contains n_0 and*

(ii) *for all $n \geq n_0$, S contains the positive integer $n + 1$ whenever S contains n,*

then S contains every integer greater than or equal to n_0.

When applying the Extended Principle of Mathematical Induction, our base case is when $n = n_0$ instead of $n = 1$, but the inductive step is still the same. Note that our goal in this situation is to show that S contains every integer greater than or equal to n_0, which establishes that our statement is true for all $n \geq n_0$. Doing so doesn't rule out the possibility that S could contain other integers as well, but we are only interested in the integers greater than or equal to n_0.

Activity B.9. To illustrate the Extended Principle of Mathematical Induction, we will continue with our example of proving that $2^n < n!$ for all $n \geq 4$.

(a) State and verify the base case for this inductive proof.

(b) What is the inductive hypothesis in this proof? Give a precise statement, and then complete the inductive step.

(c) What conclusion can you draw from your work in parts (a) and (b)?

Notice that the only difference between the Principle of Mathematical Induction and the Extended Principle of Mathematical Induction is the base case. In fact, letting $n_0 = 1$ in the Extended Principle yields the original Principle of Mathematical Induction. A bit later, we will see that these two forms of induction are actually equivalent.

Before we move one, one additional comment about the format of induction proofs is in order. Generally, when we construct an induction argument, we do not set up the set S as we have done in our previous examples. Instead, if we are trying to prove a family $\{P(n)\}$ of statements, one for each integer greater than or equal to some integer n_0, we simply prove that $P(n_0)$ is true, and then prove that if $P(n)$ is true for some integer $n \geq n_0$, then $P(n+1)$ is also true. In the induction proofs throughout the rest of this appendix (and throughout the remainder of the text), we will follow this simplified format.

The Strong Form of Mathematical Induction

Preview Activity B.10. Let's return once more to our *Let's Make a Great Deal* game. We will keep the rules the same and suppose in addition that behind each door is not only a prize and a key to open the next door, but also keys to open all of the *preceding* doors. Compare and contrast this game to the previous versions of the game we have studied. How is it similar, and how is it different? Do the outcomes of the game change?

The version of the *Let's Make a Great Deal* game from Activity B.10 may seem a bit silly; after all, why do we need all of those extra keys? But let's examine how this version of the game translates to an induction proof. We still need to answer some question (prove some base case n_0) to begin. In our previous inductive steps, we then showed that $n \in S$ implies $n + 1 \in S$—that is, each door contains the key to the next door. In our new version of the game, the idea of having all of the keys to the preceding doors is analogous to assuming not only that $n \in S$, but also that n_0, $n_0 + 1$, $n_0 + 2$, ..., n are all in S. In other words, we can assume the validity of all of the previous statements, not just the nth statement. To see why this might be useful, consider the statement that every nonnegative integer n has a binary representation—that is, there exists $r \geq 0$ and integers a_r, a_{r-1}, ..., a_1, a_0, all either 0 or 1, such that

$$n = a_r 2^r + a_{r-1} 2^{r-1} + \cdots + a_2 2^2 + a_1 2 + a_0.$$

For our base case ($n = 0$) we have that $0 = 0$, and so we are done (letting $r = 0$ and $a_0 = 0$). For the inductive step, it is a bit complicated to add 1 to n (in binary) to show that $n + 1$ has a binary representation. (We would have to worry about all the possible carries.) However, if $n + 1$ is even, then $k = (n + 1)/2$ is smaller than $n + 1$. If k has a binary representation

$$k = b_s 2^s + b_{s-1} 2^{s-1} + \cdots + b_2 2^2 + b_1 2 + b_0,$$

then

$$n + 1 = 2k = b_s 2^{s+1} + b_{s-1} 2^s + \cdots + b_2 2^3 + b_1 2^2 + b_0 2,$$

and so $n + 1$ has a binary representation. If $n + 1$ is odd, then $k = n/2$ is smaller than $n + 1$. If k has a binary representation

$$k = b_s 2^s + b_{s-1} 2^{s-1} + \cdots + b_2 2^2 + b_1 2 + b_0,$$

then

$$n + 1 = 2k + 1 = b_s 2^{s+1} + b_{s-1} 2^s + \cdots + b_2 2^3 + b_1 2^2 + b_0 2 + 1,$$

and so $n+1$ has a binary representation in this case as well. By assuming that *all* of the nonnegative integers less than or equal to n have binary representations, we can fairly easily prove that $n + 1$ also has a binary representation.

Being able to assume the statement we want to prove for *all* integers less than or equal to n is at times necessary for us to carry out an induction proof. The axiom that allows us to use such a method is called the *Strong Form of Mathematical Induction.*

Axiom B.11 (Strong Form of Mathematical Induction). *Let S be a subset of $\mathbb{Z}$ containing some integer n_0. Suppose that for all $n \geq n_0$, S contains $n+1$ whenever S contains each integer m with $n_0 \leq m \leq n$. Then S contains all integers greater than or equal to n_0.*

To reiterate, the Strong Form of Mathematical Induction allows us to assume much more in our inductive hypothesis than the previous two versions do. Strong induction is useful in a variety of settings, including proving results involving certain recursively defined sequences like the Fibonacci sequence.

Recall that the Fibonacci sequence is defined by the recurrence relation

$$f_n = f_{n-1} + f_{n-2} \tag{B.7}$$

for all $n \geq 3$, with $f_1 = f_2 = 1$. The recurrence relation (B.7) is very time consuming to use to compute f_n for large values of n. However, it turns out that there is a fascinating formula that gives the nth term of the Fibonacci sequence directly, without using the relation from (B.7).

Let $\varphi = \frac{1+\sqrt{5}}{2}$ and $\overline{\varphi} = \frac{1-\sqrt{5}}{2}$. We will show that

$$f_n = \frac{\varphi^n - \overline{\varphi}^n}{\sqrt{5}}. \tag{B.8}$$

Formula (B.8) is called *Binet's Formula.** The number $\varphi = \frac{1+\sqrt{5}}{2}$ is intimately related to the Fibonacci sequence. This number also occurs often in other areas of mathematics. It was an important number to the ancient Greek mathematicians who felt that the most aesthetically pleasing rectangles had sides in the ratio of $\varphi : 1$. The Greeks called φ the *golden mean* or *golden ratio.* Formula (B.8) provides a fascinating relationship between the Fibonacci numbers and the golden ratio. It is also

*If you wonder where a formula like this comes from, the quantities φ and $\overline{\varphi}$ are eigenvalues for a certain matrix that we can use to generate the Fibonacci sequence. This formula follows in a straightforward manner.

surprising (and not at all obvious) that the expression on the right hand side of (B.8) is an integer for each positive integer n.

To prove formula (B.8), we will use mathematical induction. Note that since f_1 and f_2 are defined independent of the recursion relation, it will be necessary to verify our statement in both the $n = 1$ and $n = 2$ cases. First we will make a few observations. Note that the golden ratio φ and its conjugate $\overline{\varphi}$ are the solutions (check this!) to the quadratic equation

$$x^2 = x + 1.$$

In addition,

$$\varphi + \overline{\varphi} = 1 \qquad \text{and} \qquad \varphi - \overline{\varphi} = \sqrt{5}.$$

Therefore,

$$\varphi^{n+1} = \varphi^2 \varphi^{n-1} = (\varphi + 1)\varphi^{n-1} = \varphi^n + \varphi^{n-1}.$$

Similarly,

$$\overline{\varphi}^{n+1} = \overline{\varphi}^2 \overline{\varphi}^{n-1} = (\overline{\varphi} + 1)\overline{\varphi}^{n-1} = \overline{\varphi}^n + \overline{\varphi}^{n-1}.$$

We will use these last two identities in our proof of Binet's Formula. We will proceed by mathematical induction on n. When $n = 1$, we have

$$\frac{1}{\sqrt{5}}(\varphi + \overline{\varphi}) = \frac{1}{\sqrt{5}}\left(\frac{1+\sqrt{5}}{2} - \frac{1-\sqrt{5}}{2}\right) = \frac{\sqrt{5}}{\sqrt{5}} = 1 = f_1.$$

So equation (B.8) is true when $n = 1$. When $n = 2$, we have

$$\begin{aligned}
\frac{1}{\sqrt{5}}(\varphi^2 + \overline{\varphi}^2) &= \frac{1}{\sqrt{5}}\left[\left(\frac{1+\sqrt{5}}{2}\right)^2 - \left(\frac{1-\sqrt{5}}{2}\right)^2\right] \\
&= \frac{1}{\sqrt{5}}\left(\frac{1}{4}(1 + 2\sqrt{5} + 5) - \frac{1}{4}(1 - 2\sqrt{5} + 5)\right) \\
&= \frac{\sqrt{5}}{\sqrt{5}} \\
&= 1 \\
&= f_1.
\end{aligned}$$

So equation (B.8) is true when $n = 2$.

Since each term in the Fibonacci sequence depends on the preceding two terms, we will need to use the Strong Form of Mathematical Induction in our proof. Therefore, assume equation (B.8) is true for all positive integers m less than a given $n \geq 2$. We must show that

$$f_{n+1} = \frac{\varphi^{n+1} - \overline{\varphi}^{n+1}}{\sqrt{5}}.$$

This follows by observing that

$$\begin{aligned}
\frac{\varphi^{n+1} - \overline{\varphi}^{n+1}}{\sqrt{5}} &= \frac{1}{\sqrt{5}}\left[(\varphi^n + \varphi^{n-1}) - (\overline{\varphi}^n + \overline{\varphi}^{n-1})\right] \\
&= \frac{1}{\sqrt{5}}\left[(\varphi^n - \overline{\varphi}^n) + (\varphi^{n-1} - \overline{\varphi}^{n-1})\right] \\
&= f_n + f_{n-1} \\
&= f_{n+1}.
\end{aligned}$$

Thus, by induction, Binet's Formula is true for all $n \in \mathbb{N}$.

Note that, with Binet's Formula, we can easily compute f_n for very large values of n. For example, f_{500} is equal to

$$\begin{aligned}&1394232245616978801397243828704072839500702565876973\\&\quad 072641089629483255716228632906915576588762225212 94125.\end{aligned}$$

The Well-Ordering Principle

Preview Activity B.12.

(a) Which of the following sets contains a smallest element? Explain.

(i) $A = \{1, 2, 3, 4\}$

(ii) $B = \{n \in \mathbb{N} \mid n > 4\}$

(iii) $C = \{x \in \mathbb{Z} \mid x > 4\}$

(iv) $D = \{x \in \mathbb{Z} \mid x < 4\}$

(b) Do you believe the following statement is true or false?

Every nonempty subset of $\mathbb{N}$ has a least element.

No proof is required if you believe the statement is true, but if you believe it is false, you should be able to give a counterexample.

(c) Do you believe the following statement is true or false?

Every nonempty subset of $\mathbb{Z}$ has a least element.

No proof is required if you believe the statement is true, but if you believe it is false, you should be able to give a counterexample.

(d) Let's return to our *Let's Make a Great Deal* game. Suppose a friend of yours has won the game. Is it possible to determine which question your friend answered correctly to win? Explain.

We will begin our discussion of the Well-Ordering Principle with the following familiar proof that $\sqrt{2}$ is not a rational number:

Assume to the contrary that $\sqrt{2} = \frac{m}{n}$ for some positive integers m and $n \neq 0$ so that $\frac{m}{n}$ is in reduced form (that is, the greatest common divisor of m and n is 1). Then $n\sqrt{2} = m$, and so $2n^2 = m^2$. Thus, the prime 2 divides m^2 and so 2 also divides m. This means that $m = 2k$ for some integer k. Then $4k^2 = m^2 = 2n^2$, and so $2k^2 = n^2$. From this we see that 2 divides n, contradicting the fact that m and n have no common factors greater than 1. We can therefore conclude that $\sqrt{2}$ is not a rational number.

All of the results that we used in this proof—including those that you may not have seen before—are verified in our investigations, with one exception. This exception illustrates how easy it is to take certain mathematical results for granted. The exception in the proof is that we can always find a rational number *in reduced form* that is equal to $\frac{m}{n}$. How do we know we can do this? Recall that the rational numbers $\frac{a}{b}$ and $\frac{c}{d}$ are equal if $ad = bc$. To show that we can find a rational number in reduced form that is equal to $\frac{m}{n}$, we might consider all rational numbers $\frac{a}{b}$ that are equal to $\frac{m}{n}$ and prove that there is one so that the greatest common divisor (or gcd) of a and b is 1. In other words, let

$$S = \left\{\gcd(a, b) : \frac{a}{b} = \frac{m}{n}\right\}$$

be the set of all greatest common divisors of the numerators and denominators of fractions that are equal to $\frac{m}{n}$. Certainly S is not empty, since $\gcd(m, n)$ is in S. Also, since $\gcd(x, y) \geq 1$ for any integers x and y, not both 0, it follows that the integers in S are all greater than or equal to 1. We need to actually show that 1 is in S. If 1 is in S, then 1 will have to be the smallest element in S.

This is where our conclusion from Activity B.12 is helpful. If we assume that every nonempty subset of $\mathbb{N}$ contains a smallest integer, then S must contain a smallest integer d. Now all we have to do is show that $d = 1$. This is because if $d = 1$, then there must be a fraction $\frac{a}{b}$ equal to $\frac{m}{n}$ with $\gcd(a, b) = 1$. Since $d \in S$, there exists a rational number $\frac{a}{b}$ that is equal to $\frac{m}{n}$ with $\gcd(a, b) = d$. Now d divides both a and b, so let $da' = a$ and $db' = b$ for some positive integers a' and b'. Since $d = \gcd(a, b)$, it follows that $\gcd(a', b') = 1$. But

$$\frac{m}{n} = \frac{a}{b} = \frac{a'd}{b'd} = \frac{a'}{b'},$$

and so $\gcd(a', b')$ is in S. However, this can happen only if $d = 1$, and so 1 is the smallest element in S, and there is a fraction in reduced form that is equal to $\frac{m}{n}$.

The key to our proof that every rational number is equal to a rational number in reduced form was the assumption that the set S contained a smallest element. Based on Activity B.12, this seems like a reasonable assumption to make. The principle that allows us to make it is called the *Well-Ordering Principle*.

To thoroughly understand the Well-Ordering Principle, we first need to discuss well-ordered sets. But to talk about well-ordered sets, we need to understand ordered sets in general. This leads us to the idea of binary relations.

A *binary relation* on a set (or *relation* for short) is simply a way to compare elements in the set. For example, consider the set consisting of the citizens of the state of Michigan. We might say that one person is related to another if the two have at least one parent in common. Note that some people in this set are related and others are not. This observation illustrates the fact that a relation on a set does not need to compare *every* pair of elements in the set.

As a smaller example, let $S = \{1, 2, 3, 4\}$, and say that a and b are related in S if a divides b. In this case we have that 1 is related to 2, 3, and 4, while 2 is related only to 4. To clearly identify related pairs of elements in S, we might list all of the related elements as ordered pairs. For this relation, the resulting pairs are $(1, 2)$, $(1, 3)$, $(1, 4)$, and $(2, 4)$. The general definition of a relation on a set follows this example.

Definition B.13. A **relation** on a set S is a subset R of the Cartesian product $S \times S$. In other words, a relation on S is a set of ordered pairs, where both coordinates of each pair are elements of S.

For example, the subset $R = \{(a, a) : a \in \mathbb{Z}\}$ of $\mathbb{Z} \times \mathbb{Z}$ is the relation we call *equals*. If R is a relation on a set S, we usually suppress the set notation and write $a \sim b$, read "a is related to b," if $(a, b) \in R$. In this case, we often refer to $\sim$ as the relation instead of the set R. Sometimes we use familiar symbols for special relations. For example, we write $a = b$ if (a, b) is in the set $R = \{(a, a) : a \in \mathbb{Z}\}$.

There are several properties that relations may satisfy. For example:

- A relation $\sim$ on a set S is *reflexive* if $a \sim a$ for all $a \in S$.
- A relation $\sim$ on a set S is *symmetric* if whenever $a \sim b$ (for any a, $b \in S$), we also have $b \sim a$.
- A relation $\sim$ on a set S is *transitive* if whenever $a \sim b$ and $b \sim c$ (for any a, b, $c \in S$), we also have $a \sim c$.
- A relation $\sim$ on a set S is *antisymmetric* if whenever $a \sim b$ and $b \sim a$ (for any a, $b \in S$), then $a = b$.

Activity B.14. Determine whether each of the given relations on $\mathbb{Z}$ is reflexive, symmetric, transitive, and/or antisymmetric. Give reasons to support your answers.

(a) $R = \{(a, b) : a > b\}$

(b) $R = \{(a, b) : a^2 = b^2\}$

(c) $R = \{(a, b) : ab \geq 0\}$

(d) $R = \{(a, b) : a$ and b leave the same remainder when divided by $3\}$

Some relations, like the relation $\leq$ on $\mathbb{R}$, give us a way of organizing the elements in the sets on which they are defined in a specified way (e.g., on the number line). The next definition formalizes this idea.

Definition B.15. A set S is a **partially ordered set** (or **poset**) if there is a relation, which we will denote by $\leq$, on S such that for all $x, y, z \in S$:

(i) $x \leq x$ ($\leq$ is a reflexive relation);

(ii) if $x \leq y$ and $y \leq x$, then $x = y$ ($\leq$ is an antisymmetric relation); and

(iii) if $x \leq y$ and $y \leq z$, then $x \leq z$ ($\leq$ is a transitive relation).

A partially ordered set S is **totally ordered** if it also satisfies

(iv) either $x \leq y$, $y \leq x$, or $x = y$ ($\leq$ satisfies the trichotomy property).

What makes a partially ordered set totally ordered is that *any* two elements are related somehow, which is not necessary in a partially ordered set. An example of a partially ordered set that is not totally ordered is the set of positive integers, where a is related to b if a divides b. Examples of totally ordered sets are $\mathbb{Z}, \mathbb{Q}$, and $\mathbb{R}$, using the standard "less than or equal to" relation ($\leq$).

The Well-Ordering Principle tells us that any subset of $\mathbb{Z}$ that is bounded below contains a smallest element. To make this all precise, we need to explain what we mean by a smallest element in a set and also what bounded below means. These definitions should not be surprising.

Definition B.16. Let S be a totally ordered set, and let A be a subset of S.

- An element $m \in S$ is a **lower bound** for A if $m \leq a$ for all $a \in A$. The set A is **bounded below** if A has a lower bound in S.

- An element $a \in A$ is a **least** or **smallest** element in A if $a \leq a'$ for all $a' \in A$.

It is important to note the difference between a lower bound and a smallest element. The integer -2 is a lower bound for $\mathbb{N}$, but is not a smallest element in $\mathbb{N}$ since it is not an element of $\mathbb{N}$. Every smallest element in a set is also a lower bound for the set. However, not every set is bounded below or contains a least element. For example, the set of even integers is not bounded below. In addition, a set can be bounded below but not contain a least element. For example, the open interval $(0, 1) = \{x \in \mathbb{R} : 0 < x < 1\}$ is bounded below by 0 but does does not have a smallest element (since there is no smallest positive real number).

We have one more step before stating the Well-Ordering Principle.

Definition B.17. A totally ordered set S is **well-ordered** if every nonempty subset A of S contains a least element.

We can now formally state the Well-Ordering Principle.

Axiom B.18 (The Well-Ordering Principle). *Every nonempty subset of $\mathbb{Z}$ that is bounded below is well-ordered.*

The Well-Ordering Principle is often stated within the specific context of the natural numbers, where it implies that every nonempty subset of $\mathbb{N}$ contains a smallest element. Our version is somewhat more general and is equivalent to the following:

Axiom B.19 (The Well-Ordering Principle). *Every nonempty subset of $\mathbb{Z}$ that is bounded below contains a smallest element.*

Note that, in general, a set can have a smallest element without being well-ordered. Consider, for example, the set $\mathbb{R}^*$ of all nonnegative real numbers. Note that $\mathbb{R}^*$ has a smallest element—namely, 0—but is not well-ordered, since it contains a nonempty subset (the positive reals, for example) that does not have a smallest element. The equivalence of the two forms of the Well-Ordering Principle, as we have stated them, stems from the fact that both are universally quantified—that is, both refer to *every* nonempty subset of $\mathbb{Z}$ that is bounded below. The first is really saying that if S is a nonempty subset of $\mathbb{Z}$ that is bounded below, then every nonempty subset of S contains a smallest element. But a nonempty subset of S is still a nonempty subset of $\mathbb{Z}$ that is bounded below. For this reason, the second version of the Well-Ordering Principle is equivalent to the first.

As an example of the use of the Well-Ordering Principle, we will prove the following theorem, which we also proved in Investigation 4 as part of the Fundamental Theorem of Arithmetic. (See page 35.)

Theorem. *Every integer greater than 1 is either prime or can be factored into a product of primes.*

Proof. To use the Well-Ordering Principle, we need to define a nonempty subset of $\mathbb{Z}$ that is bounded below. To do so, we will proceed by contradiction and assume that there is an integer greater than 1 that is not prime and cannot be written as a product of primes. Let

$$S = \{n \in \mathbb{N} : n \text{ is not prime and cannot be written as a product of primes}\}.$$

Then S is nonempty by hypothesis and is bounded below (by 1). The Well-Ordering Principle tells us that S contains a smallest element m. By definition, m is not prime, so there exist integers a and b with $1 < a, b < m$ such that $m = ab$. Since m is the smallest element in S, it follows that a and b are not in S. Thus, a and b are either prime or can be written as a product of primes. Therefore, there exist positive integers r and s and primes $p_1, p_2, \ldots, p_r$ and $q_1, q_2, \ldots, q_s$ such that

$$a = p_1 p_2 \cdots p_r \quad \text{and} \quad b = q_1 q_2 \cdots q_s.$$

But then

$$m = ab = p_1 p_2 \cdots p_r q_1 q_2 \cdots q_s$$

is a product of primes, which is a contradiction, since we assumed that m could not be written as a product of primes. We can therefore conclude that every integer greater than 1 is either prime or can be factored into a product of primes. ∎

You may want to compare the above proof to the proof of the Fundamental Theorem of Arithmetic from Investigation 4, which used induction instead of the Well-Ordering Principle. It is no coincidence that both methods can be used to establish similar results. In fact, as we will see in the next section, the Well-Ordering Principle and all three different forms of the Principle of Mathematical Induction are logically equivalent.

It is also important to note that we have labeled both the principles of mathematical induction and the Well-Ordering Principle as axioms and not theorems. That is because we cannot prove any one of these principles (although, as noted above, we can prove that they are equivalent to each other), but they seem evident enough that we will assume them to be true.

The Equivalence of the Well-Ordering Principle and the Principles of Mathematical Induction

In this section, we will prove that the principles of mathematical induction and the Well-Ordering Principle are equivalent. That is, any one of these principles implies any of the others. It is important to note that Theorem B.20 does not prove any of these principles, but says that if we assume one of them to be valid, then all of the others are valid as well.

Theorem B.20. *The following are equivalent:*

(i) *The Principle of Mathematical Induction*

(ii) *The Extended Principle of Mathematical Induction*

(iii) *The Strong Form of Mathematical Induction*

(iv) *The Well-Ordering Principle.*

A word on the proof of Theorem B.20: To prove this string of equivalences, we will show that (i) implies (ii), (ii) implies (iii), (iii) implies (iv), and then (iv) implies (i). This will demonstrate that any one of the four statements implies any of the others, as can be seen by following an appropriate string of implications. (For example, to see that (iii) implies (ii), we can simply note that (iii) implies (iv), (iv) implies (i), and (i) implies (ii).) Also, the proofs of each equivalence are subtle in that both the hypotheses and conclusions are complicated statements. Because of this, we will have to be very careful about our assumptions in each case. We will begin by showing that the Principle of Mathematical Induction implies the Extended Principle of Mathematical Induction. The steps to complete this proof are outlined in the next activity.

Activity B.21. To prove that the Principle of Mathematical Induction implies the Extended Principle of Mathematical Induction, we will assume that the Principle of Mathematical Induction is true. This means that any subset S of $\mathbb{N}$ that contains 1 and has the property that $n + 1 \in S$ whenever $n \in S$ must be equal to $\mathbb{N}$.

We need to prove that the Extended Principle of Mathematical Induction is true. So we will assume that n_0 is an integer and T is a subset of $\mathbb{Z}$ such that $n_0 \in T$ and $n + 1 \in T$ whenever $n \geq n_0$ and $n \in T$. We need to prove that $\{n \in \mathbb{Z} \mid n \geq n_0\} \subseteq T$.

In order to use the Principle of Mathematical Induction, we need to construct some subset of $\mathbb{N}$ that is related to T but contains 1 as its smallest element. To do so, we can shift or re-index the elements of T so that n_0 corresponds to 1. In particular, we will define S to be the set

$$S = \{k - n_0 + 1 \in \mathbb{N} \mid k \in T\}.$$

(a) Use the assumption that $n_0 \in T$ to prove that $1 \in S$.

(b) We now need to prove that if $n \geq 1$ is in S, then $n + 1 \in S$. Let $n \geq 1$ be in S. There is a corresponding element k in T. Write down a formula for k in terms of n and n_0. Explain your reasoning.

(c) Based on our assumptions about T, what integer besides k must also be in T?

(d) Now use the result of part (c) to conclude that $n + 1 \in S$.

This proves that S contains 1 and that $n + 1 \in S$ whenever $n \in S$. Therefore, by the Principle of Mathematical Induction, $S = \mathbb{N}$. We will now use this fact to prove that $\{n \in \mathbb{Z} \mid n \geq n_0\} \subseteq T$.

By assumption, $n_0 \in T$. So we need to prove that if $x \in \mathbb{Z}$ with $x > n_0$, then $x \in T$. To this end, assume that $x \in \mathbb{Z}$ and $x > n_0$.

(e) Prove that $x - n_0 \in \mathbb{N}$ and therefore $x - n_0 \in S$.

(f) Show that part (e) implies that $x - 1 \in T$. Then explain how we can conclude that $x \in T$.

(g) Explain why we have now completed the proof that the Extended Principle of Mathematical Induction is implied by the Principle of Mathematical Induction.

We will now consider the other implications in Theorem B.20.

Proof of Theorem B.20. (ii) $\rightarrow$ (iii): We will assume that the Extended Principle of Mathematical Induction is true. We will then prove that the Strong Form of Mathematical Induction must also be true.

To prove the Strong Form, we let n_0 be an integer and assume T is a subset of $\mathbb{Z}$ such that

(1) $n_0 \in T$, and

(2) for every $n \in \mathbb{Z}$ with $n \geq n_0$, if $\{n_0, n_0 + 1, \ldots, n\} \subseteq T$, then $(n + 1) \in T$.

We then need to prove that T contains all integers greater than or equal to n_0—or, equivalently, that

$$\{x \in \mathbb{Z} \mid x \geq n_0\} \subseteq T.$$

We will use the Extended Principle of Mathematical Induction to prove the following statement:

For each natural number k with $k \geq n_0$, $\{n_0, n_0 + 1, \ldots, k\} \subseteq T$.

Since we have assumed that $n_0 \in T$, we know that $\{n_0\} \subseteq T$. Hence the base case ($k = n_0$) is true.

For the inductive step, let $k \in \mathbb{N}$ with $k \geq n_0$, and assume that

$$\{n_0, n_0 + 1, \ldots, k\} \subseteq T.$$

By what we assumed about the set T, we can conclude that $k + 1 \in T$, and therefore

$$\{n_0, n_0 + 1, \ldots, k, k + 1\} \subseteq T.$$

This proves that if $\{n_0, n_0 + 1, \ldots, k\} \subseteq T$, then $\{n_0, n_0 + 1, \ldots, k, k + 1\} \subseteq T$ is true; hence, the inductive step has been established. By the Extended Principle of Mathematical Induction, we can conclude that for each natural number k with $k \geq n_0$, $\{n_0, n_0 + 1, \ldots, k\} \subseteq T$. This proves that T contains all integers greater than or equal to n_0, which is what we needed to prove to show that the Strong Form of Mathematical Induction is true.

We have therefore shown that if the Extended Principle of Mathematical Induction is true, then the Strong Form of Mathematical Induction is also true.

(iii) $\to$ (iv): We will now show that the Strong Form of Mathematical Induction implies the Well-Ordering Principle. We will assume the Strong Form of Mathematical Induction. That is, whenever we have a subset U of $\mathbb{Z}$ such that

- $n_0 \in U$ for some integer n_0; and
- whenever $k \in U$ for all $n_0 \leq k \leq n$, then $n + 1 \in U$,

then U contains the set of all integers greater than or equal to n_0. To prove the Well-Ordering Principle, we must show that any nonempty subset of $\mathbb{Z}$ that is bounded below contains a smallest element. We will proceed by contradiction and assume there is a nonempty subset T of $\mathbb{Z}$ that is bounded below and does not contain a least element. Let m be a lower bound for T. If $m \in T$, then T contains a smallest element, namely m. So m cannot be an element of T. Let S be the set of all strict lower bounds for T—that is,

$$S = \{n \in \mathbb{Z} : n < t \text{ for all } t \in T\}.$$

Since m is a lower bound for T and $m \notin T$ it follows that $m < t$ for all $t \in T$. So $m \in S$. Suppose $n \geq m$ so that $m, m+1, m+2, \ldots, n$ are all in S. We will show $n + 1 \in S$. Since $n \in S$ we must have $n < t$ for all $t \in T$. This, however, implies that

$$n + 1 \leq t \tag{B.9}$$

for all $t \in T$. If $n + 1 = t$ for some $t \in T$, then $n + 1$ must be the smallest element in T, which cannot happen. Therefore,

$$n + 1 \neq t \tag{B.10}$$

for all $t \in T$. Combining (B.9) and (B.10) shows $n + 1 < t$ for all $t \in T$, and so $n + 1 \in S$. By the Strong Form of Mathematical Induction, we can then conclude that S contains all integers greater than or equal to m. It follows that every integer is a strict lower bound for T, and so $T = \emptyset$, a contradiction. Therefore, no such set T exists, which means that every nonempty subset of $\mathbb{Z}$ that is bounded below contains a smallest element. We have therefore shown that the Strong Form of Mathematical Induction implies the Well-Ordering Principle.

(iv) $\to$ (i): This is left to the reader in Activity B.22.

■

Concluding Activities

Activity B.22. Complete the proof of Theorem B.20 by proving that the Well-Ordering Principle implies the Principle of Mathematical Induction. (Hint: If S is a subset of $\mathbb{N}$ that contains 1 and also contains $n+1$ whenever S contains n, consider the set T of all natural numbers that are not in S.)

Exercises

(1) In calculus, we often use the fact that $\frac{d}{dx}x^n = nx^{n-1}$ for every positive integer n, but we usually don't provide a rigorous proof of this result. Use induction to verify this derivative formula. Assume the product rule if you need it.

(2) Prove that

$$1^2 + 2^2 + 3^2 + \cdots + n^2 = \frac{n(n+1)(2n+1)}{6}$$

for every positive integer n.

(3) In the mid 20th century, the mathematician George Pólya suggested the apparent paradox that all girls have eyes of the same color.† His induction argument to verify this statement is as follows:

> For $n = 1$ the statement is obviously (or "vacuously") true. It remains to pass from n to $n+1$. For the sake of concreteness, I shall pass from 3 to 4 and leave the general case for you.
>
> Let me introduce you to any four girls, Ann, Berthe, Carol, and Dorothy, or A, B, C, and D, for short. Allegedly ($n = 3$) the eyes of A, B, and C are of the same color. Consequently, the eyes of all four girls A, B, C, and D, must be of the same color; for the sake of full clarity, you may look at the diagram:
>
> $$\overbrace{A + B + C}\underbrace{ + D}.$$
>
> This proves the point for $n + 1 = 4$, and the passage from 4 to 5, for example, is, obviously, not more difficult.

A quick glance into the eyes of several girls will show that not all girls have the same eye color, so there must be a flaw in the argument. Find and explain the flaw.

(4) Consider the conjecture

$$1 + 2 + 3 + 4 + \cdots + n = \frac{1}{2}\left(n + \frac{1}{2}\right)^2. \tag{B.11}$$

†This appears in Pólya's 1954 work *Induction and Analogy in Mathematics*, volume 1 of *Mathematics and Plausible Reasoning*, Princeton University Press.

(a) Assume (B.11) is true for some positive integer n. That is, assume

$$1+2+3+4+\cdots+n=\frac{1}{2}\left(n+\frac{1}{2}\right)^2.$$

Show that (B.11) is true for the integer $n+1$.

(b) For which integers n is (B.11) a true statement? Explain. What does this exercise tell us about the importance of establishing a base case in an induction proof?

(5) (a) Experiment and conjecture a simple closed form for the sum

$$s_n=\frac{1}{1\times 3}+\frac{1}{3\times 5}+\frac{1}{5\times 7}+\cdots+\frac{1}{(2n-1)(2n+1)}$$

that is valid for every positive integer n.

(b) Use induction to prove your formula from part (a). Be explicit about which version of induction you are using.

(6) Experiment and conjecture a simple closed form for

$$\left(1-\frac{1}{4}\right)\left(1-\frac{1}{9}\right)\left(1-\frac{1}{16}\right)\cdots\left(1-\frac{1}{n^2}\right)$$

that is valid for every positive integer $n \geq 2$. Prove your conjecture.

(7) (a) Let $a_1 = 5$, $a_2 = 7$ and $a_n = 3a_{n-1} - 2a_{n-2}$ for $n \geq 3$. Experiment and conjecture a simple closed form for a_n that is valid for every positive integer n. (Hint: Compare a_n to 2^n.)

(b) Use induction to prove your formula from part (a). Be explicit about which version of induction you are using.

(8) Recall that the Fibonacci numbers f_n are defined by $f_1 = 1$, $f_2 = 1$, and $f_n = f_{n-1} + f_{n-2}$ for all $n \geq 3$. Show that every fifth Fibonacci number is divisible by 5. (In fact, something stronger is true: for any prime p, every p^{th} Fibonacci number is divisible by p.)

(9) Prove that for every positive integer n,

$$1(1!)+2(2!)+3(3!)+\cdots+n(n!)=(n+1)!-1. \tag{B.12}$$

(10) Prove that for every $n \in \mathbb{N}$, the number of subsets of a set with n elements in 2^n.

(11) Is the following statement true or false?

For all $n \in \mathbb{N}$, $(1+2+3+\cdots+n)^2 = 1^3+2^3+3^3+\cdots n^3$

If the statement is true, prove it. If it is false, find a counterexample.

(12) For which positive integers is $n!$ less than n^n? Prove your assertion.

(13) In this exercise, we will compare exponential functions to factorials. Let $a \geq 2$ be a positive integer.

(a) Show that if $a^n < n!$ for some positive integer n, then $a^{n+1} < (n+1)!$.

(b) To show that $a^n < n!$ for all n larger than some fixed integer, it remains to demonstrate that $a^n < n!$ for some positive integer n. This is a challenging problem. It is conjectured that, for $a > 3$, the sequence

$$s(a) = \text{round}\left(ae - \left(\frac{1}{2}\right)\log(2a\pi) - \frac{1}{a}\right)$$

gives the smallest positive integer n so that $a^n < n!$.[‡] (The function *round* means to round to the nearest integer.) Verify this formula for $a = 4$, $a = 5$, and $a = 6$.

(14) **Round Robin Tournaments.** Consider a tournament involving m players in which each player plays every other player just once and there are no ties. A *cycle* in the tournament is a set $\{P_1, P_2, \ldots, P_n\}$ of players so that player P_1 beats player P_2, player P_2 beats player P_3, and so on, and player P_n beats player P_1. Show that if there is a cycle in the tournament, then there is a cycle consisting of exactly three players.

(15) In this investigation, we proved that

$$1 + 2 + 3 + \cdots + n = \frac{n(n+1)}{2}$$

for every positive integer n. Then, in Exercise 2, you were asked to show that

$$1^2 + 2^2 + 3^2 + \cdots + n^2 = \frac{n(n+1)(2n+1)}{6}$$

for every positive integer n. Mathematical induction is a useful tool for verifying such formulas, but how do we actually find the formulas in the first place? In this exercise, we will consider ways to answer this question.

(a) Let's next determine a formula for the sum of cubes. Our starting place is the expansion of $(x-1)^4$. Note that

$$(x-1)^4 = x^4 - 4x^3 + 6x^2 - 4x + 1.$$

Then

$$x^4 - (x-1)^4 = 4x^3 - 6x^2 + 4x - 1.$$

Next, we will calculate each side of the previous equation as x ranges from 1 to n:

$$\begin{array}{lcccccccc}
n^4 - (n-1)^4 & = & 4n^3 & - & 6n^2 & + & 4n & - & 1 \\
(n-1)^4 - (n-2)^4 & = & 4(n-1)^3 & - & 6(n-1)^2 & + & 4(n-1) & - & 1 \\
(n-2)^4 - (n-3)^4 & = & 4(n-2)^3 & - & 6(n-2)^2 & + & 4(n-2) & - & 1 \\
\vdots & & & & \vdots & & & & \\
4^4 - 3^4 & = & 4(4)^3 & - & 6(4)^2 & + & 4(4) & - & 1 \\
3^4 - 2^4 & = & 4(3)^3 & - & 6(3)^2 & + & 4(3) & - & 1 \\
2^4 - 1^4 & = & 4(2)^3 & - & 6(2)^2 & + & 4(2) & - & 1 \\
1^4 - 0^4 & = & 4(1)^3 & - & 6(1)^2 & + & 4(1) & - & 1
\end{array}$$

Now we can add the entries on each side to find a formula for

$$1^3 + 2^3 + \cdots + n^3.$$

Complete this process, and then prove your formula by induction.

‡By Benoit Cloitre; see sequence A086824 in the On-Line Encyclopedia of Integer Sequences (`https://oeis.org/`).

(b) Repeat the process from part (a) to find a formula for

$$1^4 + 2^4 + \cdots + n^4.$$

Then prove your formula.

(16) **The Towers of Hanoi.** In an ancient city in India, so the legend goes, monks in a temple have to move a pile of 64 sacred disks from one location to another. The disks are fragile; only one can be carried at a time. A disk may not be placed on top of a smaller, less valuable disk. In addition, there is only one other location in the temple (besides the original and destination locations) sacred enough that a pile of disks can be placed there.

So the monks begin moving disks back and forth, between the original pile, the pile at the new location, and the intermediate location, always keeping the piles in order (largest on the bottom, smallest on the top). The legend is that, before the monks make the final move to complete the pile in the new location, the temple will turn to dust and the world will end.

Generalize this problem to show that if there were n disks to move, it would take a total of $2^n - 1$ moves to complete the transfer from one location to another. Should we be worried about the world coming to an end?

(17) The usual total ordering given by $\leq$ on $\mathbb{Z}$ behaves nicely with respect to addition. Show that there is *no* total ordering of $\mathbb{Z}_n$ that behaves nicely with respect to addition in $\mathbb{Z}_n$. That is, show that there is no total ordering on $\mathbb{Z}_n$ such that, for all $[a]$, $[b]$, $[c] \in \mathbb{Z}_n$, if $[a] \leq [b]$, then $([a] + [c]) \leq ([b] + [c])$. (Hint: If there is such an ordering with $[0] \leq [1]$, use transitivity to show that $[0] \leq [n-1]$, and explain why this leads to a contradiction. Then think about what similar argument needs to be made to complete the proof.)

Index